T0200548

Stiffness and Damping in Mechanical Design

Stiffness
and Damping
in Mechanical
Design

EUGENE I. RIVIN
Wayne State University
Detroit, Michigan

CRC Press is an imprint of the
Taylor & Francis Group, an **informa** business

CRC Press
Taylor & Francis Group
6000 Broken Sound Parkway NW, Suite 300
Boca Raton, FL 33487-2742

First issued in paperback 2019

© 1999 by Taylor & Francis Group, LLC
CRC Press is an imprint of Taylor & Francis Group, an Informa business

No claim to original U.S. Government works

ISBN-13: 978-0-8247-1722-3 (hbk)
ISBN-13: 978-0-367-39976-4 (pbk)
Library of Congress Card Number 99-14998

This book contains information obtained from authentic and highly regarded sources.
Reasonable efforts have been made to publish reliable data and information, but the author
and publisher cannot assume responsibility for the validity of all materials or the
consequences of their use. The authors and publishers have attempted to trace the copyright
holders of all material reproduced in this publication and apologize to copyright holders if
permission to publish in this form has not been obtained. If any copyright material has not
been acknowledged please write and let us know so we may rectify in any future reprint.

Except as permitted under U.S. Copyright Law, no part of this book may be reprinted,
reproduced, transmitted, or utilized in any form by any electronic, mechanical, or other
means, now known or hereafter invented, including photocopying, microfilming, and
recording, or in any information storage or retrieval system, without written permission
from the publishers.

For permission to photocopy or use material electronically from this work, please access
www.copyright.com (http://www.copyright.com/) or contact the Copyright Clearance Center,
Inc. (CCC), 222 Rosewood Drive, Danvers, MA 01923, 978-750-8400. CCC is a
not-for-profit organization that provides licenses and registration for a variety of users. For
organizations that have been granted a photocopy license by the CCC, a separate system of
payment has been arranged.

Trademark Notice: Product or corporate names may be trademarks or registered trademarks,
and are used only for identification and explanation without intent to infringe.

Library of Congress Cataloging-in-Publication Data

Rivin, Eugene I.
   Stiffness and damping in mechanical design / Eugene I. Rivin.
      p. cm.
   Includes bibliographical references and index.
   ISBN 0-8247-1722-8 (alk. paper)
    1. Vibration. 2. Damping (Mechanics). 3. Dynamics, Rigid.
I. Title.
TA355.R53   1999
621.8'11—dc21

                                                    99-14998

Visit the Taylor & Francis Web site at
http://www.taylorandfrancis.com

and the CRC Press Web site at
http://www.crcpress.com

Preface

Computers are becoming more and more powerful tools for assisting in the design process. Finite element analysis and other software packages constituting computer-aided design (CAD) allow quick and realistic visualization and optimization of stresses and deformations inside the component of a structure. This computer technology frees designers from tedious drafting and computational chores: it not only allows them to concentrate on general, conceptual issues of design, but also forces them to do so. Some of these issues are so-called conceptual design, reliability, energy efficiency, accuracy, and use of advanced materials. Very important conceptual issues are stiffness of mechanical structures and their components and damping in mechanical systems sensitive to and/or generating vibrations.

Stiffness and strength are the most important criteria for many mechanical designs. However, although there are hundreds of books on various aspects of strength, and strength issues are heavily represented in all textbooks on machine elements, stiffness-related issues are practically neglected, with a few exceptions. Although dynamics and vibrations, both forced and self-excited, of mechanical systems are becoming increasingly important, damping and stiffness are usually considered separately. However, frequently damping and stiffness are closely interrelated, and efforts to improve one parameter while neglecting the other are generally ineffective or even counterproductive.

This book intends to correct this situation by addressing various aspects of structural stiffness and structural damping and their roles in design. Several typical cases in which stiffness is closely associated with damping are addressed. The basic conceptual issues related to stiffness and damping are accentuated. A more detailed analytical treatment is given in cases where the results were not previously published or were only published in hard-to-obtain sources (e.g., publications in languages other than English). Many of these concepts are illustrated by practical results and/or applications (practical case studies) either in the text or as appendices and articles. The articles, mostly authored or coauthored by the author of this book, are intended both to extend coverage of some important issues and to provide practical application examples.

This book originated from course notes prepared for the "Stiffness in Design" tutorial successfully presented at four Annual Meetings of the American Society for Precision Engineering (ASPE). The contents of the book are based to a substantial degree on the author's personal professional experiences and research results.

The two parameters covered in this book are treated differently. No monographs and few if any extended chapters on stiffness have recently been published in English. However, there are several books and handbook chapters available on damping. Accordingly, although an attempt was made here to provide a comprehensive picture of the role of stiffness in mechanical design, the treatment of damping is less exhaustive. Two main groups of the many damping-related issues are addressed: (1) damping properties of contacts (joints) and power transmission systems, which are addressed only scantily in other publications, and (2) the interrelationship between stiffness and damping parameters in mechanical systems and structural materials. Thus, the damping-related sections can be considered complementary to the currently available monographs and handbooks.

Many important stiffness- and damping-related issues were studied in depth in the former Soviet Union. The results were published in Russian and are practically unavailable to the engineering community in non-Russian-speaking countries. Several of these results are covered in the book.

A general introduction to the subject matter is given in Chapter 1. General performance characteristics are described for which the stiffness and damping criteria are critical. This chapter also lists a selection of structural materials for stiffness- and damping-critical applications. Information on the influence of the mode of loading and the component design on stiffness is provided in Chapter 2.

Chapter 3 is dedicated to an important subject of nonlinear and variable stiffness (and damping) systems. Specially addressed is the issue of preloading, which is very important for understanding and controlling stiffness and damping characteristics.

Design and performance information on various aspects of normal and tangential contact stiffness, as well as of damping associated with mechanical contacts, is given in Chapter 4. Information on these subjects is very scarce in the technical literature available in English. Stiffness of mechanical components is determined not only by their own structural properties, but also by their supporting conditions and devices. Influence of the latter on both static stiffness and dynamic characteristics is frequently not well understood. These issues, as well as some issues related to machine foundations, are addressed in Chapter 5.

Chapter 6 concentrates on very specific issues of stiffness (and damping) in power transmission and drive systems, which play a significant role in various mechanical systems. Several useful techniques, both passive and active, aimed at enhancing structural stiffness and damping characteristics (i.e., reduction of structural deformations and enhancement of dynamic stability) are described in Chapter 7. Special cases in which performance of stiffness-critical systems can be improved by reduction or a proper tuning of components' stiffness are described in Chapter 8.

The issues related to stiffness and damping in mechanical design are numerous and very diverse. This book does not pretend to be a handbook covering all of them, but it is the first attempt to provide illuminating coverage of some of these issues.

In addition to the body of the book, I have included Appendices 1–3 to provide more detailed treatments and derivations for some small but important subjects. I have also provided, in their entirety, several articles from previous publications, each of which gives an in-depth treatment of an important stiffness and/or damping critical area of mechanical design.

I am very grateful to the book reviewers, who made valuable suggestions. Especially helpful have been discussions with Professor Dan DeBra (Stanford University). These discussions resulted in important changes in the book's emphasis. Suggestions by Professor Vladimir Portman (Ben Gurion University of the Negev, Israel) were also very useful. I take full responsibility for all of the shortcomings of the book and will greatly appreciate readers' feedback.

Eugene I. Rivin

Contents

  1. Rivin, E.I., "Principles and Criteria of Vibration Isolation of Machinery,"
     ASME Journal of Mechanical Design, 1979, Vol. 101, pp. 682–692.
  2. Rivin, E.I., "Design and Application Criteria for Connecting Couplings,"
     ASME Journal of Mechanical Design, 1986, Vol. 108, pp. 96–105.
  3. Rivin, E.I., "Properties and Prospective Applications of Ultra-Thin Layered
     Rubber-Metal Laminates for Limited Travel Bearings," Tribology International, 1983, Vol. 18, No. 1, pp. 17–25.
  4. Rivin, E.I., Karlic, P., and Kim, Y., "Improvement of Machining Conditions for Turning of Slender Parts by Application of Tensile Force," Fundamental Issues in Machining, ASME PED, 1990, Vol. 43, pp. 283–297.

5. Rivin, E.I., and Kang, H., "Enhancement of Dynamic Stability of Cantilever Tooling Structures," International Journal of Machine Tools and Manufacture, 1992, Vol. 32, No. 4, pp. 539–561.
6. Johnson, C.D., "Design of Passive Damping Systems," Transactions of the ASME, 50th Anniversary of the Design Engineering Division, 1995; Vol. 117(B), pp. 171–176.
7. Rivin, E.I., "Trends in Tooling for CNC Machine Tools: Machine System Stiffness," ASME Manufacturing Review, 1991, Vol. 4, No. 4, pp. 257–263.

Stiffness and Damping in Mechanical Design

1

Introduction and Definitions

1.1 BASIC NOTIONS

1.1.1 Stiffness

Stiffness is the capacity of a mechanical system to sustain loads without excessive changes of its geometry (deformations). It is one of the most important design criteria for mechanical components and systems. Although strength is considered the most important design criterion, there are many cases in which stresses in components and their connections are significantly below the allowable levels, and dimensions as well as performance characteristics of mechanical systems and their components are determined by stiffness requirements. Typical examples of such mechanical systems are aircraft wings, and frames/beds of production machinery (machine tools, presses, etc.), in which stresses frequently do not exceed 3–7 MPa (500–1,000 psi). Another stiffness-critical group of mechanical components is power transmission components, especially shafts, whose deformations may lead to failures of gears and belts while stresses in the shafts caused by the payload are relatively low.

Recently, great advances in improving strength of mechanical systems and components were achieved. The main reasons for such advances are development of high strength structural metals and other materials, better understanding of fracture/failure phenomena, and development of better techniques for stress analysis and computation, which resulted in the reduction of safety factors. These advances often result in reduction of cross sections of the structural components. Since the loads in the structures (unless they are weight-induced) do not change, structural deformations in the systems using high strength materials and/or designed with reduced safety factors are becoming more pronounced. It is important to note that while the strength of structural metals can be greatly improved by selection of alloying materials and of heat treatment procedures (as much as 5–7 times for steel and aluminum), modulus of elasticity (Young's modulus) is not very sensitive to alloying and to heat treatment. For example, the Young's modulus of stainless steels is even 5–15% lower than that of carbon steels (see Table 1.1). As a result, stiffness can be modified (enhanced) only by proper selection of the component geometry (shape and size) and its interaction with other components.

Stiffness effects on performance of mechanical systems are due to influence of deformations on static and fatigue strength, wear resistance, efficiency (friction losses), accuracy, dynamic/vibration stability, and manufacturability. The importance of the stiffness criterion is increasing due to:

1. Increasing accuracy requirements (especially due to increasing speeds and efficiency of machines and other mechanical systems)
2. Increasing use of high strength materials resulting in the reduced cross sections and, accordingly, in increasing structural deformations
3. Better analytical techniques resulting in smaller safety factors, which also result in the reduced cross sections and increasing deformations
4. Increasing importance of dynamic characteristics of machines since their increased speed and power, combined with lighter structures, may result in intense resonances and in the development of self-excited vibrations (chatter, stick-slip, etc.)

Factors 2–4 are especially pronounced for surface and flying vehicles (cars, airplanes, rockets, etc.) in which the strength resources of the materials are utilized to the maximum in order to reduce weight.

Stiffness is a complex parameter of a system. At each point, there are generally different values of stiffness k_{xx}, k_{yy}, k_{zz} in three orthogonal directions of a selected coordinate frame, three values of interaxial stiffness k_{xy}, k_{xz}, k_{yz} related to deformations along one axis (first subscript) caused by forces acting along an orthogonal axis (second subscript), and also three values of angular stiffness about the x, y, and z axes. If the interaxial stiffnesses vanish, $k_{xy} = k_{xz} = k_{yz} = 0$, then x, y, z are the *principal stiffness axes*. These definitions are important since in some cases several components of the stiffness tensor are important; in special cases, ratios of the stiffness values in the orthogonal directions determine dynamic stability of the system. Such is the case of chatter instability of some machining operations [1]. Chatter stability in these operations increases if the cutting and/or the friction force vector is oriented in a certain way relative to principal stiffness axes x and y. Another case is vibration isolation. Improper stiffness ratios in vibration isolators and machinery mounts may cause undesirable intermodal coupling in vibration isolation systems (see Article 1).

Main effects of an inadequate stiffness are absolute deformations of some

Table 1.1 Young's Modulus and Density of Structural Materials

Material	E (10^5 MPa)	γ (10^3 kg/m^3)	E/γ (10^7 m^2/s^2)
(a) Homogeneous Materials			
Graphite	7.5	2.25	33.4
Diamond	18.0	5.6	32
Boron carbide, BC	4.50	2.4	19
Silicon carbide, SiC	5.6	3.2	17.5
Carbon, C	3.6	2.25	16.0
Beryllium, Be	2.9	1.9	15.3
Boron, B	3.8	2.5	15.2
Sapphire	4.75	4.5	10.1
Alumina, Al$_2$O$_3$	3.9	4.0	9.8
Lockalloy (62% Be + 38% Al)	1.90	2.1	9.1
Kevlar 49	1.3	1.44	9.0
Titanium carbide, TiC	4.0–4.5	5.7–6.0	7.0–9.1
Silicone, Si	1.1	2.3	4.8
Tungsten carbide, WC	5.50	16.0	3.4
Aluminum/Lithium (97% Al + 3% Li)	0.82	2.75	3.0
Molybdenum, Mo	3.20	10.2	3.0
Glass	0.7	2.5	2.8
Steel, Fe	2.10	7.8	2.7
Titanium, Ti	1.16	4.4	2.6
Aluminum, Al (wrought)	0.72	2.8	2.6
Aluminum, Al (cast)	0.65	2.6	2.5
Steel, stainless (.08–0.2% C, 17% Cr, 7% Ni)	1.83	7.7	2.4
Magnesium, Mg	0.45	1.9	2.4
Wood (along fiber)	0.11–0.15	0.41–0.82	2.6–1.8
Marble	0.55	2.8	2.0
Tungsten (W + 2 to 4% Ni, Cu)	3.50	18.0	1.9
Granite	0.48	2.7	1.8
Beryllium copper	1.3	8.2	1.6
Polypropylene	0.08	0.9	0.9
Nylon	0.04	1.1	0.36
Paper	0.01–0.02	0.5	0.2–0.4
(b) Composite Materials			
HTS graphite/5208 epoxy	1.72	1.55	11.1
Boron/5505 epoxy	2.07	1.99	10.4
Boron/6601 Al	2.14	2.6	8.2
Lanxide NX − 6201 (Al + SiC)	2.0	2.95	6.8
T50 graphite/2011 Al	1.6	2.58	6.2
Kevlar 49/resin	0.76	1.38	5.5
80% Al + 20% Al$_2$O$_3$ powder	0.97	2.93	3.3
Melram (80% Mg, 6.5% Zn, 12% SiC)	0.64	2.02	3.2
E glass/1002 epoxy	0.39	1.8	2.2

components of the system and/or relative displacements between two or several components. Such deformations/displacements can cause:

Geometric distortions (inaccuracies)
Change of actual loads and friction conditions, which may lead to reduced efficiency, accelerated wear, and/or fretting corrosion
Dynamic instability (self-excited vibrations)
Increased amplitudes of forced vibrations

Inadequate stiffness of transmission shafts may cause some specific effects. The resulting linear and angular deformations determine behavior of bearings (angular deformations cause stress concentrations and increased vibrations in antifriction bearings and may distort lubrication and friction conditions in sliding bearings); gears and worm transmissions (angular and linear deformations lead to distortions of the meshing process resulting in stress concentrations and variations in the instantaneous transmission ratios causing increasing dynamic loads); and traction drives (angular deformations cause stress concentrations and changing friction conditions).

It is worthwhile to introduce some more definitions related to stiffness:

Structural stiffness due to deformations of a part or a component considered as beam, plate, shell, etc.
Contact stiffness due to deformations in a connection between two components (contact deformations may exceed structural deformations in precision systems)
Compliance $e = 1/k$, defined as a reciprocal parameter to stiffness k (ratio of deformation to force causing this deformation)
Linear stiffness vs. *nonlinear stiffness* (see Ch. 3)
Hardening vs. *softening* nonlinear stiffness (see Ch. 3)
Static stiffness k_{st} (stiffness measured during a very slow loading process, such as a periodic loading with a frequency less than 0.5 Hz) vs. *dynamic stiffness* k_{dyn}, which is measured under faster changing loads. Dynamic stiffness is characterized by a *dynamic stiffness coefficient* $K_{dyn} = k_{dyn}/k_{st}$. Usually $K_{dyn} > 1$ and depends on frequency and/or amplitude of load and/or amplitude of vibration displacement (see Ch. 3). In many cases, especially for fibrous and elastomeric materials K_{dyn} is inversely correlated with damping, e.g., see Fig. 3.2 and Table 1 in Article 1.

1.1.2 Damping

Damping is the capacity of a mechanical system to reduce intensity of a vibratory process. The damping capacity can be due to interactions with outside systems, or due to internal performance-related interactions. The damping effect for a vibratory process is achieved by transforming (dissipating) mechanical energy of the vibratory motion into other types of energy, most frequently heat, which can be evacuated from the system. If the vibratory process represents self-excited vibrations (e.g., chatter), the advent of the vibratory process can be prevented by an adequate damping capacity of the system.

In the equations of motion to vibratory systems (e.g., see Appendix 1), both intensity and character of energy dissipation are characterized by coefficients at the first derivative (by time) of vibratory displacements. These coefficients can be constant (linear or viscous damping) or dependent on amplitude and/or frequency of the vibratory motion (nonlinear damping). There are various mechanisms of vibratory energy dissipation which can be present in mechanical systems, some of which are briefly explored in Appendix 1.

Since the constant coefficient at the time-derivative of the vibratory displacement term results in a linear differential equation, which is easy to solve and to analyze, such systems are very popular in textbooks on vibration. However, the constant damping coefficient describes a so called viscous mechanism of energy dissipation that can be realized, for example, by a piston moving with a relatively slow velocity inside a conforming cylinder with a relatively large clearance between the piston and the cylinder walls, so that the resistance force due to viscous friction has a direction opposite to the velocity vector and is proportional to the relative velocity between the cylinder and the piston. In real-life applications such schematic and conditions are not often materialized. The most frequently observed energy dissipation mechanisms are hysteretic behavior or structural materials; friction conditions similar to coulomb (dry) friction whereas the friction (resistance) force is directionally opposed to the velocity vector but does not depend (or depends weakly) on the vibratory velocity magnitude; damping in joints where the vibratory force is directed perpendicularly to the joint surface and causes squeezing of the lubricating oil through the very thin clearance between the contacting surfaces (thus, with a very high velocity) during one-half of the vibratory cycle and sucking it back during the other half of the cycle; and damping due to impact interactions between the contacting surfaces. Some of these mechanisms are analytically described in Appendix 1.

Effects of damping on performance of mechanical systems are due to reduction of intensity of undesirable resonances; acceleration of decay (settling) of transient vibration excited by abrupt changes in motion parameters of mechanical components (start/stop conditions of moving tables in machine tools and of robot links, engagement/disengagement between a cutting tool and the machined part, etc.); prevention or alleviation of self-excited vibrations; prevention of impacts between vibrating parts when their amplitudes are reduced by damping; potential for reduction of heat generation, and thus for increase in efficiency due to reduced

peak vibratory velocities of components having frictional or microimpacting interactions; reduction of noise generation and of harmful vibrations transmitted
to human operators; and more.

It is important to note that while damping is associated with transforming
mechanical energy of the vibratory component into heat, increase of damping
capacity of mechanical system does not necessarily result in a greater heat generation. Damping enhancement is, first of all, changing the dynamic status of the
system and, unless the displacement amplitude is specified (for example, like
inside a compensating coupling connecting misaligned shafts; see Section 8.5.2),
most probably would cause a *reduction in the heat generation.* This somewhat
paradoxical statement is definitely true in application to mechanical systems
prone to development of self-excited vibrations, since enhancement of damping
in the system would prevent starting of the vibratory process, and thus the heat
generation, which is usually caused by vibratory displacements. This statement
is also true for a system subjected to transient vibration. Since the initial displacement of mass m in Fig. A.1.1 and the natural frequency of the system do not
significantly depend on damping in the system, a higher damping would result
in smaller second, third, etc. amplitudes of the decaying vibrations, and thus in
a lower energy dissipation.

Less obvious is the case of forced vibration when force $F = F_0 \sin \omega t$ is
applied to mass m in Fig. A.1.1. Let's consider the system in which mass m is
attached to the frame by a rubber flexible element combining both stiffness and
damping properties (hysteresis damping, $r = 1$; see Appendix 1). If amplitude
of mass m is A, then the maximum potential energy of deformation of the flexible
element is

$$V = k \frac{A^2}{2} \qquad (1.1)$$

The amount of energy dissipated (transformed into heat) in the damper c or in
the rubber flexible element is

$$\Delta V = \Psi V = \Psi k \frac{A^2}{2} \qquad (1.2)$$

At the resonance, amplitude A_{res} of mass m is, from formula (A.1.19b) at $\omega = \omega_0$ and from (A1.18)

$$A_{res} = \frac{F_0}{k\sqrt{\left(\frac{\alpha}{\pi k}\right)^2}} = \frac{F_0}{k}\frac{\pi k}{\alpha} = A_0 \frac{\pi}{\delta} \approx A_0 \frac{2\pi}{\Psi} \qquad (1.3a)$$

where $A_0 = F_0/k =$ static ($\omega = 0$) deflection of the flexible element, $\delta =$ logarithmic decrement, and for not very high damping

$$\Psi \approx 2\delta \qquad (1.3b)$$

Thus, the *energy dissipation at resonance* (or the maximum energy dissipation in the system) *is decreasing with increasing damping capacity* (increasing Y).

This result, although at the first sight paradoxical, does not depend on the character (mechanism) of damping in the system and can be easily explained. The resonance amplitude is inversely proportional to the damping parameter (Ψ, δ, etc.) because the increasing damping shifts an equilibrium inside the dynamic system between the excitation (given, constant amplitude), elastic (displacement-proportional), inertia (acceleration-proportional), and damping (velocity-proportional) forces. The amount of energy dissipation is a secondary effect of this equilibrium; the energy dissipation is directly proportional to the square of the vibration amplitude. Although this effect of decreasing energy dissipation with increasing damping is especially important at the resonance where vibratory amplitudes are the greatest and energy dissipation is most pronounced, it is not as significant in the areas outside of the resonance where the amplitudes are not strongly dependent on the damping magnitude (see Fig. A.1.3).

Effects of damping on performance of mechanical system are somewhat similar to the effects of stiffness, as presented in Section 1.1.1. Damping influences, directly or indirectly, the following parameters of mechanical systems, among others:

1. *Fatigue strength.* Increasing damping leads to reduction of strain and stress amplitudes if the loading regime is close to a resonance. It is even more important for high-frequency components of strain/stress processes, which are frequently intensified due to resonances of inevitable high frequency components of the excitation force(s) and/or nonlinear responses of the system with higher natural frequencies of the system.

2. *Wear resistance.* High (resonance) vibratory velocities, especially associated with high-frequency parasitic microvibrations, may significantly accelerate the wear process. High damping in the system alleviates these effects.

3. *Efficiency (friction losses).* Depending on vibration parameters (amplitudes, frequencies, and, especially, directivity), vibrations can increase or reduce friction. In the former case, increasing damping can improve efficiency.

4. *Accuracy and surface finish* of parts machined on machined tools. Although surface finish of the machined surface is directly affected by vibrations, accuracy (both dimensions and macrogeometry) may be directly influenced by low-frequency vibrations, e.g., transmitted from the environment (see Article 1) or may be indirectly affected by changing geometry of the cutting tool whose sharp edge(s) are fast wearing out under chatter- or microvibrations. The latter

are especially dangerous for brittle cutting materials such as ceramic and diamond tools.

    5. *Dynamic/vibration stability* of mechanical systems can be radically enhanced by introducing damping into the system.

    6. *Manufacturability*, especially of low-stiffness parts, can be limited by their dynamic instability, chatter, and resonance vibrations during processing. Damping enhancement of the part and/or of the fixtures used in its processing can significantly improve manufacturability.

Importance of the damping criterion is increasing with the increasing importance of the stiffness criterion as discussed in Section 1.1.1 due to:

    a. Increasing accuracy requirements
    b, c. Increasing use of high strength materials and decreasing safety factors, which result in lower stiffness and thus higher probability of vibration excitation
    d. Increasing importance of dynamic characteristics
    e. Increasing awareness of noise and vibration pollution

Main sources of damping in mechanical systems are:

    a. Energy dissipation in structural materials
    b. Energy dissipation in joints/contacts between components (both in moving joints, such as guideways, and in stationary joints)
    c. Energy dissipation in special damping devices (couplings, vibration isolators, dampers, dynamic vibration absorbers, etc.). These devices may employ viscous (or electromagnetic) dampers in which relative vibratory motion between component generates a viscous (velocity dependent) resistance force; special high-damping materials, such as elastomers or "shape memory metals" (see Table 1.2); specially designed ("vibroimpact") mechanisms in which coimpacting between two surfaces results in dissipation of vibratory energy (see Appendix 1); etc.

1.2 INFLUENCE OF STIFFNESS ON STRENGTH AND LENGTH OF SERVICE

This influence can materialize in several ways:

Inadequate or excessive stiffness of parts may lead to overloading of associated parts or to a nonuniform stress distribution
Inadequate stiffness may significantly influence strength if loss of stability (buckling) of some component occurs
Impact/vibratory loads are significantly dependent on stiffness
Excessive stiffness of some elements in statically indeterminate systems may lead to overloading of the associated elements

It is known that fatigue life of a component depends on a high power (5–9) of maximum (peak) stresses. Thus, uniformity of the stress distribution is very important.

Fig. 1.1 [2] shows the influence of the stiffness of rims of meshing gears on load distribution in their teeth. In Fig. 1.1a, left sides of both gear rims have higher stiffness than their right sides due to positioning of the stiffening disc/spokes on their hubs. This leads to concentration of the loading in the stiff area so that the peak contact stresses in this area are about two times higher than the average stress between the meshing profiles. In Fig. 1.1b, the gear hubs are symmetrical, but again the stiff areas of both rims work against each other. Although the stress distribution diagram is different, the peak stress is still about twice as high as the average stress. The design shown in Fig. 1.1c results in a more uniform stiffness along the tooth width and, accordingly, in much smaller peak stresses—about equal to the average stress magnitude. The diagrams in Fig. 1.1 are constructed with an assumption of absolutely stiff shafts. If shaft deformations are significant, they can sustantially modify the stress distributions and even reverse the characteristic effects shown in Fig. 1.1.

Another example of influence of stiffness on load distribution is shown in Fig. 1.2. It is a schematic model of threaded connection between bolt 1 and nut 2. Since compliances of the thread coils are commensurate with compliances of bolt and nut bodies, bending deformations of the most loaded lower coils are larger than deformations of the upper coils by the amount of bolt elongation between these coils. This leads to a very nonuniform load distribution between the coils. Theoretically, for a 10-coil thread, the first coil takes 30–35% of the total axial load on the bolt, while the eighth coil takes only 4% of the load [3]. In real threaded connections, the load distribution may be more uniform due to

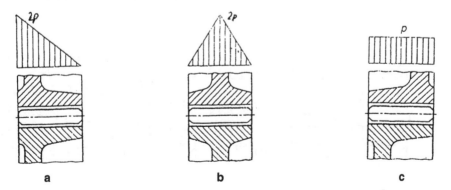

a   b   c

Figure 1.1   Contact pressure distribution in meshing gears as influenced by design of gears.

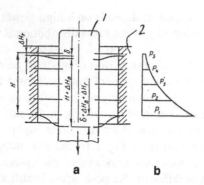

Figure 1.2 Contact pressure distribution (b) in threaded connection (a).

possible yielding of the highest loaded coils, contact deformations in the thread, and higher compliances of the contacting coils because of their inaccuracies and less than perfect contact. Thus, the first coil may take only 25–30% of the total load instead of 34%. However, it is still a very dramatic nonuniformity that can cause excessive plastic deformations of the most loaded coils and/or their fatigue failure. Such a failure may cause a chain reaction of failures in the threaded connection.

Such redistribution and concentration of loading influencing the overall deformations and the effective stiffness of the system can be observed in various mechanical systems. Fig. 1.3 [2] shows a pin connection of a rod with a tube. Since the tube is much stiffer than the rod, a large fraction of the axial load P is acting on the upper pin, which can be overloaded (Fig. 1.3a). The simplest way to equalize loading of the pins is by loosening the hole for the upper pin (Fig. 1.3b). This leads to the load being applied initially to the lower pin only.

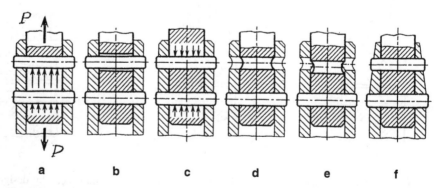

Figure 1.3 Influence of component deformations on load distribution.

The upper pin takes the load only after some stretching of the rod has occurred. Another way to achieve the same effect is by prestressing (preloading) the system by creating an initial loading (during assembly) in order to counteract the loading by force P (Fig. 1.3c). This effect can be achieved, for example, by simultaneously drilling holes in the rod and in the tube (Fig. 1.3d) and then inserting the pins while the rod is heated to the specified temperature. After the rod cools down, it shrinks (Fig. 1.3e) and the system becomes prestressed. The load equalization effect can also be achieved by local reduction of the tube stiffness (Fig. 1.3f).

1.3 INFLUENCE OF STIFFNESS AND DAMPING ON VIBRATION AND DYNAMICS

This effect of stiffness can be due to several mechanisms.

At an impact, kinetic energy of the impacting mass is transformed into potential energy of elastic deformation; accordingly, dynamic overloads are stiffness-dependent. For a simple model in Fig. 1.4, kinetic energy of mass m impacting a structure having stiffness k is

$$E = \tfrac{1}{2}mv^2 \qquad (1.1)$$

After the impact, this kinetic energy transforms into potential energy of the structural impact-induced deformation x

$$V = \tfrac{1}{2}kx^2 = E = \tfrac{1}{2}mv^2 \qquad (1.2)$$

Since the impact force $F = kx$, from (1.2) we find that

$$x = v\sqrt{\frac{m}{k}} \quad \text{and} \quad F = v\sqrt{km} \qquad (1.3)$$

Thus, in the first approximation the impact force is proportional to the square root of stiffness.

For forced vibrations, a resonance can cause significant overloads. The resonance frequency can be shifted by a proper choice of stiffness and mass values and distribution. While shifting of the resonance frequencies may help to avoid the excessive resonance displacement amplitudes and overloads, this can help only if the forcing frequencies are determined and cannot shift. In many cases this is not a realistic assumption. For example, the forcing (excitation) frequencies acting on a machine tool during milling operation are changing with the change of the number of cutting inserts in the milling cutter and with the changing spindle

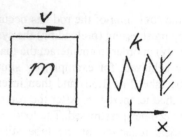

Figure 1.4 Impact interaction between moving mass and stationary spring.

speed (rpm). A much more effective way to reduce resonance amplitudes is by *enhancing damping* in the vibrating system. The best results can be achieved if the stiffness and damping changes are considered simultaneously (see Article 2 and discussion on loudspeaker cones in Section 1.6).

Variable stiffness of shafts, bearings, and mechanisms (in which stiffness may be orientation-dependent) may cause quasi-harmonic (parametric) vibrations and overloads. While variability of the stiffness can be reduced by design modifications, the best results are achieved when these modifications are combined with damping enhancement.

Chatter resistance (stability in relation to self-excited vibrations) of machine tools and other processing machines is determined by the criterion $K\delta$ (K = effective stiffness and δ = damping, e.g., logarithmic decrement). Since in many cases dynamic stiffness and damping are interrelated, such as in mechanical joints (see Ch. 4) and materials (see Ch. 3 and Article 1), the stiffness increase can be counterproductive if it is accompanied by reduction of damping. In some cases, stiffness reduction can be beneficial if it is accompanied by a greater increase in damping (see the case study on influence of mount characteristics on chatter resistance of machine tools and Ch. 8).

Deviation of the vector of cutting (or friction) forces from a principal stiffness axis may cause self-exciting vibrations (coordinate coupling) [1].

Low stiffness of the drive system may cause stick-slip vibration of the driven unit on its guideways.

1.4 INFLUENCE OF MACHINING SYSTEM STIFFNESS AND DAMPING ON ACCURACY AND PRODUCTIVITY

1.4.1 Introduction

Elastic deformations of the production (machining) system, machine tool–fixture–tool–machined part, under cutting forces are responsible for a significant fraction of the part inaccuracy. These deformations also influence productivity of the machining system, either directly by slowing the process of achieving the desired geometry or indirectly by causing self-excited chatter vibrations.

In a process of machining a precision part from a roughly shaped blank, there is the task to reduce deviation Δb of the blank surface from the desired geometry to a smaller allowable deviation Δp of the part surface (Fig. 1.5). This process can be modeled by introduction of an *accuracy enhancement factor* ζ

$$\zeta = \Delta b/\Delta p = t_1 - t_2/y_1 - y_2 \qquad (1.4)$$

where t_1 and t_2 are the maximum and minimum depth of cut; and y_1 and y_2 are the cutter displacements normal to machined surface due to structural deformations caused by the cutting forces. If the cutting force is

$$P_y = C_m t s^q \qquad (1.5)$$

then

$$\zeta = (k/C_m)/s^q \qquad (1.6)$$

where C_m = material coefficient; k = stiffness of the machining system; t = depth of cut; s = feed; and q = 0.6–0.75. For the process of turning medium-hardness steel with s = 0.1–0.75 mm/rev on a lathe with k = 20 $N/\mu m$,

$$\zeta = 150 - 30 \qquad (1.7)$$

Knowing shape deviations of the blanks and the required accuracy, the above formula for ζ allows us to estimate the required k and allowable s, or to decide on the number of passes required to achieve the desired accuracy.

Inadequate stiffness of the machining system may result in various distortions of the machining process. Some examples of such distortions are shown in

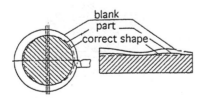

Figure 1.5 Evolution of geometry of machined parts when machining system has finite stiffness.

Fig. 1.6. The total cross sectional area of the cut is smaller during the transient phases of cutting (when the tool enters into and exits from the machined part) than during the steady cutting. As a result, deflection of the blank part is smaller during the transient phases thus resulting in deeper cuts (Fig. 1.6a, b).

Turning of a part supported between two centers requires driving of the part by a driving yoke clamped to the part (Fig. 1.6c). Asymmetry of the driving system results in an eccentricity (runout) of the part with the magnitude

$$\delta = P_z d / k_c R \tag{1.8}$$

where k_c = stiffness of the supporting center closest to the driving yoke.

Heavy traveling tables supporting parts on milling machines, surface grinders, etc., may change their angular orientation due to changing contact deformations in the guideways caused by shifting of the center of gravity during the travel (Fig. 1.6d). This also results in geometrical distortions of the part surface.

A surface deviation Δ caused by a variable stiffness of the machining system

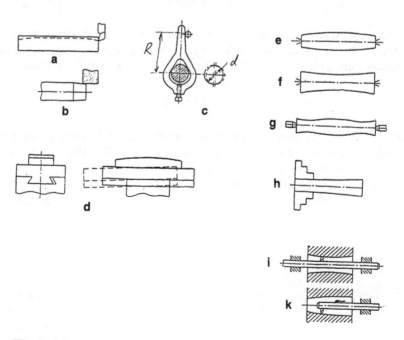

Figure 1.6 Influence of compliances in the machining system on geometry of machined parts.

can be expressed as

$$\Delta = P_y(1/k_{min} - 1/k_{max}) \qquad (1.9)$$

where k_{min} and k_{max} = low and high stiffness of the machining system and P_y = cutting force.

Figure 1.6e shows a barrel shape generated in the process of turning a slender elongated part between the rigid supporting centers; Fig. 1.6f shows a "corset" shape when a rigid part is supported by compliant centers. The part in Fig. 1.6g is slender and was supported by compliant centers. Fig. 1.6h shows the shape of a cantilever part clamped during machining in a nonrigid chuck. Fig. 1.6i shows shape of the hole bored by a slender boring bar guided by two stationary rigid supports, while Fig. 1.6k illustrates shape of the hole machined by a cantilever boring bar guided by one stationary support.

The role of stiffness enhancement is to reduce these distortions. When they are repeatable, corrections that would compensate for these errors can be commanded to a machine by its controller. However, the highest accuracy is still obtained when the error is small and it is always preferable to avoid the complications of this compensation procedure, which appropriate stiffness can accomplish.

Manufacturing requirements for stiffness of parts often determine the possibility of their fabrication with high productivity (especially for mass production). Sometimes, shaft diameters for mass-produced machines are determined not by the required strength but by a possibility of productive multicutter machining of the shafts and/or of the associated components (e.g., gears). Machining of a low-stiffness shaft leads to chatter, to a need to reduce regimes, and to copying of inaccuracies of the original blank.

Stiffness of the production equipment influences not only its accuracy and productivity. For example, stiffness characteristics of a stamping press also influence its energy efficiency (since deformation of a low-stiffness frame absorbs a significant fraction of energy contained in one stroke of the moving ram); dynamic loads and noise generation (due to the same reasons); product quality (since large deformations of the frame cause misalignments between the punch and the die and thus, distortions of the stamping); and die life (due to the same reasons). In crank presses developing the maximum force at the end of the stroke, the amount of energy spent on the elastic structural deformations can be greater than the amount of useful energy (e.g., spent on the punching operation). Abrupt unloading of the frame after the breakthrough event causes dangerous dynamic loads/noise, which increase with increasing structural deformations.

In mechanical measuring instruments/fixtures, a higher stiffness is sometimes needed to reduce deformations from the measuring (contact) force.

Deformations at the tool end caused by the cutting forces result in geometric inaccuracies and in a reduced dynamic stability of the machining process. It is important to understand that there are many factors causing deflections at the tool end. For example, in a typical boring mill, deformation of the tool itself represents only 11% of the total deflection while deformation of the spindle and its bearings is responsible for 37%, and the tapered interface between the tool-holder and the spindle hole is responsible for 52% of the total deflection [4], [5].

1.4.2 Stiffness and Damping of the Cutting Process

Background

Deformations in the machining system are not only due to the finite stiffness of the structural components, but also due to finite stiffness of the cutting process itself. The cutting process can be modeled as a spring representing effective cutting stiffness and a damper representing effective cutting damping. The stiffness and damping parameters can be derived from the expression describing the dynamic cutting force. Various expressions for dynamic cutting forces were suggested. The most convenient expression for deriving the stiffness and damping parameters of the cutting process is one given in Tobias [1]. The dynamic increment of the cutting force dP_z in the z-direction for turning operation can be written as

$$dP_z = K_1[z(t) - \mu z(t - T)] + K_2 \dot{z}(t) \qquad (1.10)$$

Here z = vibratory displacement between the tool and the workpiece, whose direction is perpendicular to the axis of the workpiece and also to the cutting speed direction in the horizontal plane; μ = overlap factor between the two subsequent tool passes in the z-direction; K_1 = cutting stiffness coefficient in the z-direction; K_2 = *penetration rate coefficient* due to the tool penetrating the workpiece in the z-direction; and $T = 2\pi/\Omega$, where Ω rev/sec is the rotating speed of the workpiece.

By assuming displacement z as

$$z(t) = A \cos \omega t \qquad (1.11)$$

where A = an indefinite amplitude constant and ω = chatter frequency, Eq. (1.10) can be rearranged as

$$dP_z = K_{cz}z + C_{cz}\frac{dz}{dt}(t) \qquad (1.12)$$

where

$$K_{cz} = K_1[1 - \mu \cos 2\pi(\omega/\Omega)] \qquad (1.13)$$

$$C_{cz} = K_1(\mu/\omega)\sin 2\pi(\omega/\Omega) + K_2 \qquad (1.14)$$

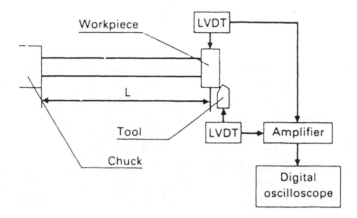

Figure 1.7 Cantilever workpiece for measuring cutting process stiffness and measurement setup.

K_{cz} and C_{cz} can be defined as effective cutting stiffness and effective cutting damping, respectively (since only the z-direction is considered, the subscript z is further omitted). The effective cutting stiffness and cutting damping are functions not only of the cutting conditions but also of the structural parameters of the machining system (stiffness and mass), which enter Eq. (1.13) and (1.14) via frequency ω. The dynamic cutting force P_z depends not only on displacement $z(t)$ but also on velocity $\dot{z}(t) = dz/dt$. The velocity-dependent term may bring the system instability when effective cutting damping $C_{cz} < 0$, and the magnitude of C_{cz} is so large that it cannot be compensated by positive structural damping.

Experimental Determination of Effective Cutting Stiffness

Experimental determination of the cutting process stiffness can be illustrated on the example of a cantilever workpiece [6]. A cantilever workpiece with a larger diameter segment at the end (Fig. 1.7) can be modeled as a single degree of freedom system with stiffness K_w without cutting and with stiffness $K_w + K_t$ during cutting, where K_w is the stiffness of the workpiece at the end and K_t is the effective cutting stiffness. Since stiffness of the cantilever workpiece is relatively small as compared with structural stiffness of the machine tool (lathe) and of the clamping chuck, chatter conditions are determined by the workpiece and the cutting process only. Thus, if the natural frequency f_w of the workpiece (without cutting) and the frequency f_c of the tool or workpiece vibration at the chatter threshold were measured, then the effective cutting stiffness can be determined using the following equation:

$$K_c = \left(\frac{f_c^2}{f_w^2} - 1 \right) K_w \qquad (1.15)$$

The frequency f_w can be measured using an accelerometer, while the chatter frequency f_c can be measured on the workpiece or on the tool using a linear variable differential transformer (LVDT) during cutting as shown in Fig. 1.7.

A cantilever bar with overhang $L = 127$ mm (5 in.) having stiffness (as measured) $K_w = 10,416$ lb/in. was used for the tests. The natural frequency $f_w \approx 200$ Hz and the equivalent mass is about 0.0065 lb-sec^2/in. The values of the effective cutting stiffness and vibration amplitude under different cutting conditions are given in Fig. 1.8a–c. It can be seen that smaller vibration amplitudes are correlated with higher effective cutting stiffness values. This validates representation of the effective cutting stiffness as a spring.

1.5 GENERAL COMMENTS ON STIFFNESS IN DESIGN

In most of the structures, their structural stiffness depends on the following factors:

Elastic moduli of structural material(s)
Geometry of the deforming segments (cross-sectional area A for tension/compression/shear, cross-sectional moment of inertia $I_{x,y}$ for bending, and polar moment of inertia J_p for torsion)
Linear dimensions (e.g., length L, width B, height H)
Character and magnitude of variation of the above parameters across the structure
Character of loading and supporting conditions of the structural components
In structures having slender, thin-walled segments, stiffness can depend on elastic stability of these segments
Joints between substructures and/or components frequently contribute the dominant structural deformations (e.g., see data on the breakdown of tool-end deflections above in Section 1.4).

While for most machine components a stiffness increase is desirable, there are many cases where stiffness values should be limited or even reduced. The following are some examples:

Perfectly rigid bodies are usually more brittle and cannot accommodate shock loads
Many structures are designed as statically indeterminate systems, but if the connections in such a system are very rigid, it would not function prop-

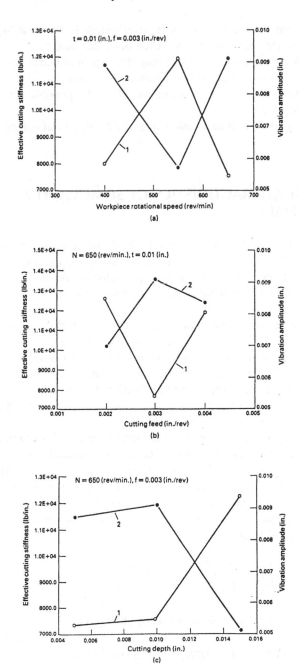

Figure 1.8 Effective cutting stiffness (line 1) and workpiece vibration amplitude (line 2) vs. (a) cutting speed, (b) feed, and (c) depth of cut.

erly since some connections might be overloaded. If the most highly
loaded connection fails, others would fail one after another

Huge peak loads (stress concentrations) may develop in contacts between
very rigid bodies due to presence of surface asperities

Stiffness adjustment/tuning by preloading would not be possible for very
rigid components

High stiffness may result in undesirable values for the structural natural frequencies

1.6 STRUCTURAL CHARACTERISTICS OF SOME WIDELY USED MATERIALS

Stiffness of a structural material is characterized by its elastic (Young's) modulus
E for tension/compression. However, there are many cases when knowledge of
just Young's modulus is not enough for a judicious selection of the structural
material. Another important material parameter is shear modulus G. For most
metals, $G = {\sim}0.4E$.

Frequently, stiffer materials (materials with higher E) are heavier. Thus, use
of such materials would result in structures having smaller cross sections but
heavier weight, which is undesirable. In cases when the structural deflections are
caused by inertia forces, like in a revolute robot arm, use of a stiffer but heavier
material can be of no benefit or even counterproductive if its weight increases
more than its stiffness and specific stiffness E/γ is the more important parameter
(see Section 7.5 for ways to overcome this problem).

Very frequently, stiffer materials are used to increase natural frequencies of
the system. This case can be illustrated on the example of two single-degree-of-
freedom dynamic systems in Fig. 1.9. In these sketches, γ, A_1, l_1, are density,
cross-sectional area, and length, respectively, of the inertia element (mass m);
A_2, l_2, h, and b are cross-sectional area, length, thickness, width, respectively, of
the elastic elements (stiffness k). For the system in Fig. 1.9a (tension/compression
elastic element) the natural frequency is

$$\sqrt{\frac{k}{m}} = \sqrt{\frac{EA_2/l_2}{\gamma A_1 l_1}} = \sqrt{\frac{E}{\gamma}} \sqrt{\frac{A_2}{A_1 l_1 l_2}} \tag{1.16}$$

For the system in Fig. 1.9b (elastic element loaded in bending)

$$m = \gamma A_1 l_1, \qquad k = 3EI/l_2^3 = (3/12)\,(Ebh^3/l_2^3) \tag{1.17}$$

thus the natural frequency is

$$\sqrt{\frac{k}{m}} = \sqrt{\frac{3}{12}\frac{Ebh^3}{l_2^3} \bigg/ \gamma A_1 l_1} = \sqrt{\frac{E}{\gamma}}\sqrt{\frac{1}{4}\frac{bh^3}{l_2^3 A_1 l_1}} \tag{1.18}$$

In both cases, the natural frequency depends on the criterion E/γ. A similar criterion can be used for selecting structural materials for many nonvibratory applications.

To provide a comprehensive information, Table 1.1 lists data on E, γ, E/γ for various structural materials (see page 3). It is interesting to note that for the most widely used structural materials (steel, titanium, aluminum, and magnesium), values of E/γ are very close.

While graphite has the second highest Young's modulus and the highest ratio E/γ in Table 1.1, it does not necessarily mean that the graphite fiber–based composites can realize such high performance characteristics. First of all, the fibers in a composite material are held together by a relatively low modulus matrix (epoxy resin or a low E metal such as magnesium or aluminum). Second of all, the fibers realize their superior elastic properties only in one direction (in tension). Since mechanical structures are frequently rated in a three-dimensional stress-strain environment, the fibers have to be placed in several directions, and this weakens the overall performance characteristics of the composite structures.

Fig. 1.10 illustrates this statement on an example of a propeller shaft for a surface vehicle [7]. Although in a steel shaft (Fig. 1.10a) steel resists loads in

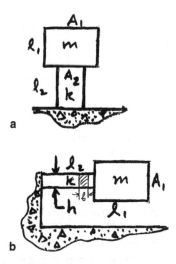

Figure 1.9   (a) Tension-compression and (b) bending single degree of freedom vibratory systems.

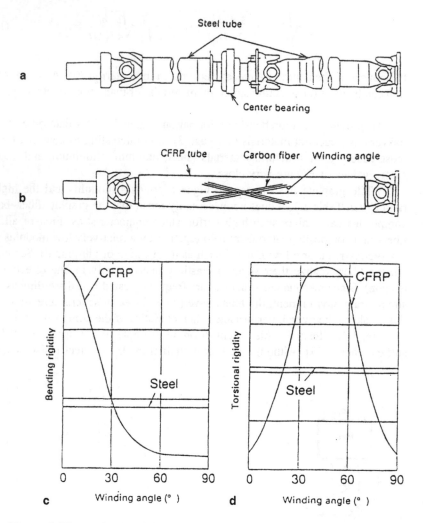

Figure 1.10 (a) Steel and (b) composite propeller shafts for automotive transmissions and comparison of their (c) bending and (d) torsional rigidity.

all directions, in a shaft made of carbon fiber reinforced plastic (CFRP) (Fig. 1.10b) there is a need to place several layers of fiber at different winding angles. Fig. 1.10c,d show how bending and torsional rigidity of the composite shaft depend on the winding angles. While it is easy to design bending or torsional stiffness of the composite shaft to be much higher than these characteristics of the steel shaft, a combination of both stiffnesses can be made superior to the steel shaft only marginally (at the winding angle, ~25 degrees).

Another example of a stiffness-critical and natural-frequency-critical components are cones and diaphragms for loudspeakers [8]. Three important material properties for loudspeaker diaphragms are:

> Large specific modulus E/γ (resulting in high natural frequencies) in order to get a wider frequency range of the speaker
> High flexural rigidity EI in order to reduce harmonic distortions
> Large internal energy dissipation (damping) characterized by the "loss factor" $\eta = \tan \beta$ (β = "loss angle" of the material; log decrement $\delta = \pi \tan \beta$) to suppress breakups of the diaphragms at resonances

Although paper (a natural fiber-reinforced composite material) and synthetic fiber-reinforced diaphragms were originally used, their stiffness values were not adequate due to the softening influence of the matrix. Yamamoto and Tsukagoshi [8] demonstrated that use of beryllium and boronized titanium (25 μm thick titanium substrate coated on both sides with 5 μm thick boron layers) resulted in significant improvement of the frequency range for high frequency and midrange speakers.

As with loudspeaker cones and diaphragms described earlier, damping of a material is an important consideration in many applications. Frequently, performance of a component or a structure is determined by combination of its stiffness and damping. Such a combination is convenient to express in the format of a criterion. For the important problems of dynamic stability of structures or processes (e.g., chatter resistance of a cutting process, settling time of a decelerating revolute link such as a robot arm, wind-induced self-excited vibrations of smoke stack, and some vibration isolation problems as in Article 1) the criterion is $K\delta$, where K is effective stiffness of the component/structure and δ is its log decrement. For such applications, Table 1.2 can be of some use. Table 1.2 lists Young's

Table 1.2 Damping (Loss Factor) and Young's Modulus of Some Materials

Material	η	E (MPa)	$E\eta$
Tinel	6.5×10^{-2}	4×10^4	2600
Polysulfide rubber (Thiokol H-5)	5.0	30	150
Tin	2×10^{-3}	6.7×10^4	134
Steel	$1 - 6 \times 10^{-4}$	21×10^4	20–120
Neoprene (type CG-1)	0.6	86.7	52
Zinc	3×10^{-4}	8×10^4	24
Aluminum	10^{-4}	6.7×10^4	6.7

modulus E, determining the effective stiffness of a component, loss factor $\eta =$ tan β, and product $E\eta$, the so-called loss modulus for some structural and energy absorbing materials. It can be seen that the best (highest) value of $E\eta$ is for a nickel titanium "shape memory" alloy Tinel ($\sim50\%$ Ni + $\sim50\%$ Ti), and the lowest value is for aluminum.

REFERENCES

1. Tobias, S.A., Machine Tool Vibration, Blackie, London, 1965.
2. Orlov, P.I., Fundamentals of Machine Design, Vol. 1, Mashinostroenie Publishing House, Moscow, 1972 [in Russian].
3. Wang, W., Marshek, K.M., "Determination of the Load Distribution in a Threaded Connector Having Dissimilar Materials and Varying Thread Stiffness," ASME J. of Engineering for Industry, 1995, Vol. 117, pp. 1–8.
4. Levina, Z.M., Zwerew, I.A., "FEA of Static and Dynamic Characteristics of Spindle Units," Stanki I instrument, 1986, No. 8, pp. 6–9 [in Russian].
5. Rivin, E.I., "Trends in Tooling for CNC Machine Tools: Tool-Spindle Interfaces," ASME Manufacturing Review, 1991, Vol. 4, No. 4, pp. 264–274.
6. Rivin, E.I., Kang, H., "Improvement of Machining Conditions for Slender Parts by Tuned Dynamics Stiffness of Tool," Intern. J. of Machine Tools and Manufacture, 1989, Vol. 29, No. 3, pp. 361–376.
7. Kawarada, K., et al. "Development of New Composite Propeller Shaft," Toyota Technical Review, 1994, Vol. 43, No. 2, pp. 85–90.
8. Yamamoto, T., Tsukagoshi, T., "New Materials for Loudspeaker Diaphragms and Cones. An Overview." Presentation at the Annual Summer Meeting of Acoustical Society of America, Ottawa, Canada, 1981, pp. 1–10.

2

Stiffness of Structural Components: Modes of Loading

2.1 INFLUENCE OF MODE OF LOADING ON STIFFNESS [1]

There are four principal types of structural loading: tension, compression, bending, and torsion. Parts experiencing tension-compression demonstrate much smaller deflections for similar loading intensities and therefore usually are not stiffness-critical. Figure 2.1a shows a rod of length L having a uniform cross-sectional area A along its length and loaded in tension by its own weight W and by force P. Fig. 2.1b shows the same rod loaded in bending by the same force P or by distributed weight $w = W/L$ as a cantilever built-in beam, and Fig. 2.1c shows the same rod as a double-supported beam.

Deflections of the rod in tension are

$$f_P^{te} = PL/EA; \quad f_W^{te} = WL/2EA \tag{2.1}$$

Bending deflections for cases b and c, respectively, are

$$f_P^{bb} = PL^3/3EI; \quad f_W^{bb} = WL^3/8EI \tag{2.2}$$

$$f_P^{bc} = PL^3/48EI \quad f_W^{bc} = 5WL^3/384EI \tag{2.3}$$

where I = cross-sectional moment of inertia. For a round cross section (diameter d, $A = \pi d^2/4$, $I = \pi d^4/64$, and $I/A = d^2/16$)

$$f^b/f^{te} = kL^2/d^2 \tag{2.4}$$

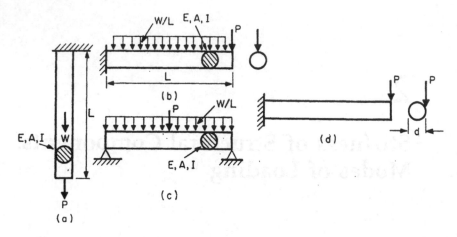

Figure 2.1 Various modes of loading of a rod-like structure: (a) tension; (b) bending in a cantilever mode; (c) bending in a double-supported mode; and (d) bending with an out-of-center load.

where coefficient k depends on loading and supporting conditions. For example, for a cantilever beam with $L/d = 20$, $(f^{bb}/f^{te})_F = 2,130$ and $(f^{bb}/f^{te})_W = 1,600$; for a double-supported beam with $L/d = 20$, $(f^{bc}/f^{te})_F \cong 133$ and $(f^{bb}/f^{te})_W \cong 167$. Thus, bending deflections are exceeding tension-compression deflections by several decimal orders of magnitude.

Figure 2.1d shows the same rod whose supporting conditions are as in Fig. 2.1b, but which is loaded in bending with an eccentricity, thus causing bending [as described by the first expression in Eq. (2.2)] and torsion, with the translational deflection on the rod periphery (which is caused by the torsional deformation) equal to

$$f^{to} = PLd^2/4GJ_p \tag{2.5}$$

where J_p = polar moment of inertia and G = shear modulus of the material. Since $J_p = \pi d^4/32$ for a circular cross section then

$$f^{to}/f^{te} = d^2/4(EA/GJ_p) = 2E/G \cong 5 \tag{2.6}$$

since for structural metals $E \cong 2.5G$. Thus, the torsion of bars with solid cross sections is also associated with deflections substantially larger than those under tension/compression.

These simple calculations help to explain why bending and/or torsional compliance is in many cases critical for the structural deformations.

Many stiffness-critical mechanical components are loaded in bending. It was shown earlier that bending is associated with much larger deformations than tension/compression of similar-size structures under the same loads. Because of this, engineers have been trying to replace bending with tension/compression. The most successful designs of this kind are trusses and arches.

Advantages of truss structures are illustrated by a simple case in Fig. 2.2 [2], where a cantilever truss having overhang l is compared with cantilever beams of the same length and loaded by the same load P. If the beam has the same cross section as links of the truss (case a) then its weight G_p is 0.35 of the truss weight G_t, but its deflection is 9,000 times larger while stresses are 550 times higher. To achieve the same deflection (case c), diameter of the beam has to be increased by the factor of 10, thus the beam becomes 35 times heavier than the truss. The stresses are equalized (case b) if the diameter of the beam is increased by 8.25 times; the weight of such beam is 25 times that of the truss. Ratio of the beam deflection f_b to the truss deflection f_t is expressed as

$$f_b/f_t \approx 10.5(l/d)^2 \sin^2 \alpha \cos \alpha \qquad (2.7)$$

Deflection ratio f_b/f_t and maximum stress ratio σ_b/σ_t are plotted in Fig. 2.3 as functions of l/d and α.

Similar effects are observed if a double-supported beam loaded in the middle

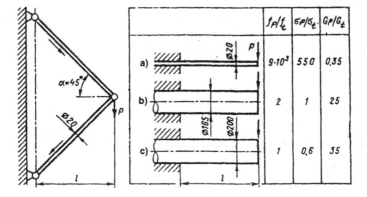

Figure 2.2 Comparison of structural characteristics of a truss bracket and cantilever beams.

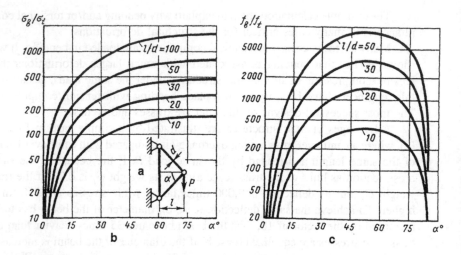

Figure 2.3 Ratios of (b) stresses and (c) deflections between a cantilever beam (diameter d, length l) and (a) a truss bracket.

of its span (as shown in Fig. 2.4a) is replaced by a truss (Fig. 2.4b). In this case

$$f_b/f_t = \ {\sim}1.3(l/d)^3 \sin^2 \alpha \cos \alpha \tag{2.8}$$

Deflection ratio f_b/f_t and maximum stress ratio σ_b/σ_t are plotted in Fig. 2.5 as functions of l/d and α. A similar effect can be achieved if the truss is transformed into an arch (Fig. 2.4c).

These principles of transforming the bending mode of loading into the tension/compression mode of loading can be utilized in a somewhat "disguised" way in designs of basic mechanical components, such as brackets (Fig. 2.6). The

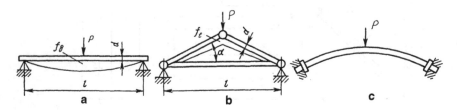

Figure 2.4 Typical load-carrying structures: (a) double-supported beam; (b) truss bridge; (c) arch.

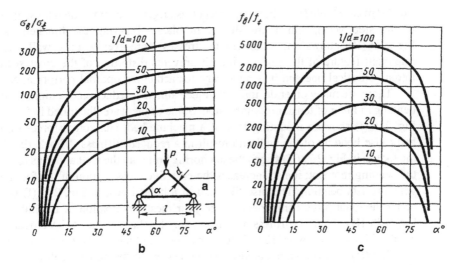

Figure 2.5 Ratios of (b) stresses and (c) deflections between (a) a double-supported beam in Fig. 2.4a and a truss bridge in Fig. 2.4b.

bracket in Fig. 2.6a(I) is loaded in bending. An inclination of the lower wall of the bracket, as in Fig. 2.6a(II), reduces deflection and stresses, but the upper wall does not contribute much to the load accommodation. Design in Fig. 2.6a(III) provides a much more uniform loading of the upper and lower walls, which allows one to significantly reduce size and weight of the bracket.

Even further modification of the "truss concept" is illustrated in Fig. 2.6b.

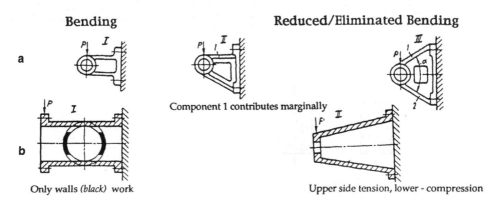

Figure 2.6 Use of tension/compression instead of bending for structural components.

Load *P* in case 2.6b(I) (cylindrical bracket) is largely accommodated by segments
of the side walls, which are shown in black. Tapering the bracket, as in
Fig. 2.6b(II), allows one to distribute stresses more evenly. Face wall f is an
important feature of the system since it prevents distortion of the cross section
into an elliptical one and it is necessary for achieving optimal performance.

There are many other design techniques aimed at reduction or elimination of
bending in favor of tension/compression. Some of them are illustrated in Fig. 2.7.
Fig. 2.7a(I) shows a mounting foot of a machine bed. Horizontal forces on the
bed cause bending of the wall and result in a reduced stiffness. "Pocketing" of
the foot as in Fig. 2.7a(II) aligns the anchoring bolt with the wall and thus reduces
the bending moment; it also increases the effective cross section of the foot area,
which resists bending. The disc-like hub of a helical gear in Fig. 2.7b(I) bends
under the axial force component of the gear mesh. Inclination of the hub as in
Fig. 2.7b(II) enhances stiffness by introducing the "arch concept." Vertical load
on the block bearing in Fig. 2.7c(I) causes bending of its frame, while in
Fig. 2.7c(II) it is accommodated by compression of the added central support.
Bending of the structural member under tension in Fig. 2.7d(I) is caused by its
asymmetry. After slight modifications as shown in Fig. 2.7d(II), its effective cross
section can be reduced due to total elimination of bending.

Some structural materials, such as cast iron, are better suited to accommodate
compressive than tensile stress. While it is more important for strength, stiffness
can also be influenced if some microcracks which can open under tension, are
present. Fig. 2.8 gives some directions for modifying components loaded in bend-

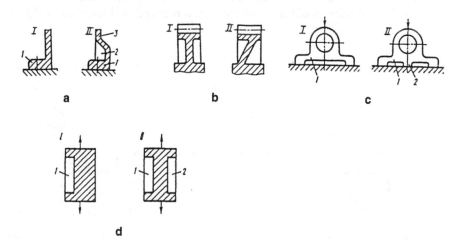

Figure 2.7 Reduction of bending deformations in structural components.

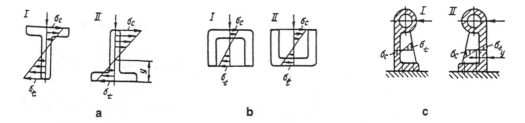

Figure 2.8 Increasing compressive stresses at the expense of tensile stresses.

ing so that maximum stresses are compressive rather than tensile. While the maximum stresses in the beam whose cross section is shown in Fig. 2.8a(I) are tensile (in the bottom section), turning this beam upside down as in Fig. 2.8a(II) brings maximum stresses to the compressed side (top). Same is true for Fig. 2.8b. A similar principle is used in transition from the bracket with the stiffening wall shown in Fig. 2.8c(I) to the identical but opposedly mounted bracket in Fig. 2.8c(II).

2.1.1 Practical Case 1: Tension/Compression Machine Tool Structure

While use of tension/compression mode of loading in structures is achieved by using trusses and arches, there are also mechanisms providing up to six degrees-of-freedom positioning and orientation of objects by using only tension/compression actuators. The most popular of such mechanisms is the so-called Stewart Platform [3]. First attempts to use the Stewart Platform for machine tools (machining centers) were made in the former Soviet Union in the mid-1980s [4].

Figure 2.9 shows the design schematic of the Russian machining center based on application of the Stewart Platform mechanism. Positioning and orientation of the platform 1 holding the spindle unit 2 which carries a tool machining part 3 is achieved by cooperative motions of six independent tension/compression actuators 4, which are pivotably engaged via spherical joints 5 and 6 with platform 1 and base plate 7, respectively.

Cooperation between the actuators is realized by using a rather complex controlling software which commands each actuator to participate in the programmed motion of the platform. One shortcoming of such a machining center is a limited range of motion along each coordinate, which results in a rather complex shape of the work zone as illustrated in Fig. 2.10.

However, there are several advantages that make such designs promising for many applications. Astanin and Sergienko [4] claim that while stiffness along

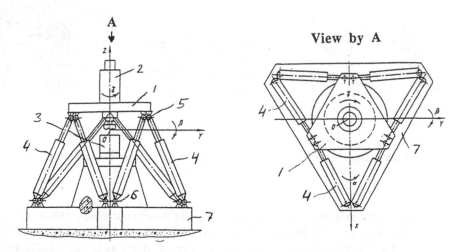

Figure 2.9 Design schematic and coordinate axes of Russian machining center based on the Stewart Platform kinematics.

the y-axis (k_y) is about the same as for conventional machining centers, stiffness k_z is about 1.7 times higher. The overall stiffness is largely determined by deformations in spherical joints 5 and 6, by platform deformations, and by spindle stiffness, and can be enhanced 50–80% by increasing platform stiffness in the x-y plane and by improving the spindle unit. The machine weighs 3–4 times less

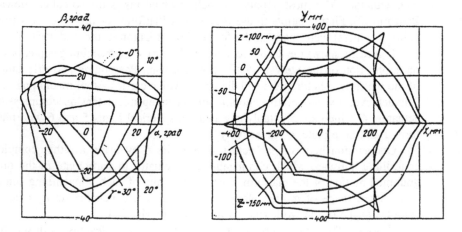

Figure 2.10 Work zone of machining center in Fig. 2.9.

than a conventional machining center and is much smaller (2–3 times smaller footprint). It costs 3–4 times less due to use of standard identical and not very complex actuating units and has 3–5 times higher feed force.

Similar machining centers were developed in the late 1980s and early 1990s by Ingersol Milling Machines Co. (Octahedral–Hexapod) and by Giddings and Lewis Co. (Variax). Popularity of this concept and its modifications for CNC machining centers and milling machines has recently been increasing [5], [6].

2.1.2 Practical Case 2: Tension/Compression Robot Manipulator

Tension/compression actuators also found application in robots. Fig. 2.11 shows schematics and work zone of a manipulating robot from NEOS Robotics Co. While conventional robots are extremely heavy in relation to their rated payload (weight-to-payload ratios 15–25 [1]), the NEOS robot has extremely high performance characteristics for its weight (about 300 kg), as listed in Table 2.1.

2.2 OPTIMIZATION OF CROSS-SECTIONAL SHAPE

2.2.1 Background

Significant gains in stiffness and/or weight of structural components loaded in bending can be achieved by a judicious selection of their cross-sectional shape. Importance of the cross-section optimization can be illustrated on the example of robotic links, which have to comply with numerous, frequently contradictory, constraints. Some of the constraints are as follows:

The links should have an internal hollow area to provide conduits for electric power and communication cables, hoses, power-transmitting components, control rods, etc.

At the same time, their external dimensions are limited in order to extend the usable workspace.

Links have to be as light as possible to reduce inertia forces and to allow for the largest payload per given size of motors and actuators.

For a given weight, links have to possess the highest possible bending (and in some cases torsional) stiffness.

One of the parameters that can be modified to comply better with these constraints is the shape of the cross section. The two basic cross sections are hollow round (Fig. 2.12a) and hollow rectangular (Fig. 2.12b). There can be various approaches to the comparison of these cross sections. Two cases are analyzed below [1]:

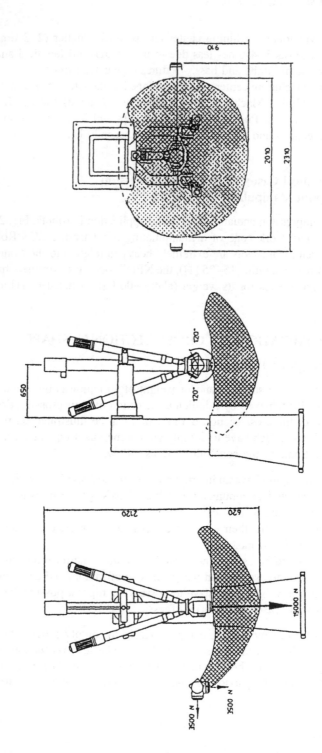

Figure 2.11 Design schematic and work zone of NEOS Robotics robot utilizing tension/compression links.

Table 2.1 Specifications of NEOS Robot

Load capacity	Handling payload	150 kg
	Turning torque	200 Nm
	Pressing, maximum	15,000 N
	Lifting, maximum	500 kg
Accuracy	Repeatability (ISO 9283)	≤ ± 0.02 mm
	Positioning	≤ ± 0.20 mm
	Path following at 0.2 m/s	≤ ± 0.10 mm
	Incremental motion	≤ 0.01 mm
Stiffness	Static bending deflection (ISO 9283.10)	
	X and Y directions	0.0003 mm/N
	Z direction	0.0001 mm/N

1. The wall thickness of both cross sections is the same.
2. The cross-sectional areas (i.e., weight) of both links are the same.

In both cases, the rectangular cross section is assumed to be a square whose external width is equal to the external diameter of the round cross section.

The bending stiffness of a beam is characterized by its cross-sectional moment of inertia I, and its weight is characterized by the cross-sectional area A. For the round cross section in Fig. 2.12a

$$I_{rd} = \pi(D_0^4 - D_i^4)/64 = \pi[D_0^4 - (D_0 - 2t)^4]/64 \cong \pi(D_0^3 t/8)(1 - 3t/D_0 + 4t^2/D_0^2) \quad (2.9)$$

$$A_{rd} = \pi(D_0^2 - D_i^2)/4 = \pi D_0 t(1 - t/D_0) \quad (2.10)$$

Hollow Round   **Hollow Rectangular**

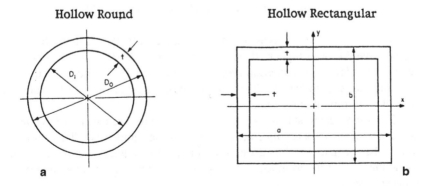

Figure 2.12 Typical cross sections of a manipulator link: (a) hollow round (ring-like); (b) hollow rectangular.

For the rectangular cross section in Fig. 2.12b, the value of I depends on the direction of the neutral axis in relation to which the moment of inertia is computed. Thus

$$I_{re,x} = ab^3/12 - (a - 2t)(b - 2t)^3/12; \quad I_{re,y} = a^3b/12 - (a - 2t)^3(b - 2t)/12 \quad (2.11a)$$

For the square cross section

$$I_{sq} = a^4/12 - (a - 2t)^4/12 \cong 2/3 \, a^3 t(1 - 3t/a + 4t^2/a^2) \quad (2.11b)$$

The cross-sectional areas for the rectangular and square cross sections, respectively, are

$$A_{re} = ab - (a - 2t)(b - 2t) = 2t(a + b) - 4t^2; \quad A_{sq} = 4at(1 - t/a) \quad (2.12)$$

For case 1, $D_0 = a$, and t is the same for both cross sections. Thus,

$$I_{sq}/I_{rd} = (2/3)/(\pi/8) = 1.7; \quad A_{sq}/A_{rd} = 4/\pi = 1.27 \quad (2.13)$$

or a square cross section provides a 70% increase in rigidity with only a 27% increase in weight; or a 34% increase in rigidity for the same weight.

For case 2 ($D_0 = a$, $A_{rd} = A_{sq}$, and $t_{rd} \neq t_{sq}$), if $t_{rd} = 0.2D_0$, then $t_{1sq} = 0.147D_0 = 0.147a$ and

$$I_{rd} = 0.0405D_0^4; \quad I_{sq} = 0.0632a^4; \quad I_{sq}/I_{rd} = 1.56 \quad (2.14a)$$

If $t_{2rd} = 0.1D_0$, then $t_{2sq} = 0.0765D_0 = 0.0765a$, and

$$I_{rd} = 0.029D_0^4; \quad I_{sq} = 0.0404a^4; \quad I_{sq}/I_{rd} = 1.40 \quad (2.14b)$$

Thus, for the same weight, a beam with the thin-walled square cross section would have 34–40% higher stiffness than a beam with the hollow round cross section. In addition, the internal cross-sectional area of the square beam is significantly larger than that for the round beam of the same weight (the thicker the wall, the more pronounced is the difference).

From the design standpoint, links of the square cross section have also an advantage of being naturally suited for using roller guideways. The round links have to be specially machined when used in prismatic joints. On the other hand, round links are easier to fit together (e.g., if telescopic links with sliding connections are used).

Both stiffness and strength of structural components loaded in bending (beams) can be significantly enhanced if a solid cross section is replaced with the cross-sectional shape in which the material is concentrated farther from the neutral line of bending. Fig. 2.13 [2] shows comparisons of both stiffness (cross-sectional moment of inertia I_0) and strength (cross-sectional modulus W) for round cross sections and for solid square vs. standard I-beam profile for the same cross-sectional area (weight).

2.2.2 Composite/Honeycomb Beams

Bending resistance of beams is largely determined by the parts of their cross sections, which are farthest removed from the neutral plane. Thus, enhancement of bending stiffness-to-weight ratio for a beam can be achieved by designing its cross section to be of such shape that the load-bearing parts are relatively thin strips on the upper and lower sides of the cross section. However, there is a need for some structural members maintaining stability of the cross section so that the

Section	Ratios $d/D , h/ho$	I/I_0	W/W_0
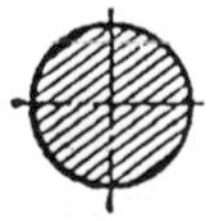	0	1	1
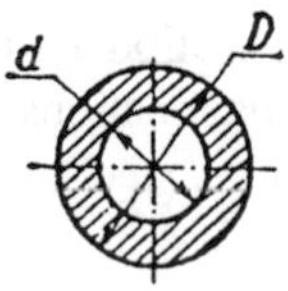	0.6	2.1	1.7
	0.8	4.5	2.7
	0.9	10	4.1
	—	1	1
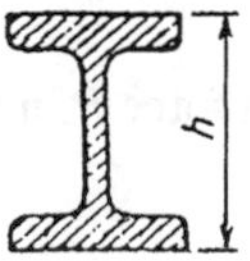	1.5	4.3	2.7
	2.5	11.5	4.5
	3.0	21.5	7.0

Figure 2.13 Relative stiffness (cross-sectional moment of inertia I) and strength (section modulus W) of various cross sections having same weight (cross-sectional area A).

positions of the load-bearing strips are not noticeably changed by loading of the beam. Rolling or casting of an integral beam (e.g., I-beams and channel beams in which an elongated wall holds the load-bearing strips) can achieve this. Another approach is by using composite beams in which the load-bearing strips are separated by an intermediate filler (core) made of a light material or by a honeycomb structure made from the same material as the load-bearing strips or from some lighter metal or synthetic material. The composite beams can be lighter than the standard profiles such as I-beams or channels, and they are frequently more convenient for the applications. For example, it is not difficult to make composite beams of any width (*composite plates*), to provide the working surfaces with smooth or threaded holes for attaching necessary components (*"breadboard" optical tables*), or to use high damping materials for the middle layer (or to use damping fillers for honeycomb structures).

It is important to realize that there are significant differences in the character of deformation between solid beams (plates) and composite beams (plates). Bending deformation of a beam comprises two components: moment-induced deformations and shear-induced deformations [7]. For beams with solid cross sections made from a uniform material, the shear deformation can be neglected for $L/h \geq 10$. For example, for a double-supported beam loaded with a uniformly distributed force with intensity q per unit length, deflection at the mid-span is [7]

$$f_{ms} = \frac{5qL^4}{384EI}\left(1 + \frac{48\alpha_{sh}EI}{5GFL^2}\right) \qquad (2.15a)$$

where E = Young's modulus, G = shear modulus, F = cross-sectional area, and α_{sh} is the so-called shear factor ($\alpha_{sh} \approx 1.2$ for rectangular cross sections, $\alpha_{sh} \approx 1.1$ for round cross sections). If the material has $E/G = 2.5$ (e.g., steel), then for a rectangular cross section ($I/F = h^2/12$)

$$f_{ms} = \frac{5qL^4}{384EI}\left(1 + 2.4\frac{h^2}{L^2}\right) \qquad (2.15b)$$

For $L/h = 10$, the second (shear) term in brackets in Eq. (2.15) is 0.024, less than 2.5%.

For a double-supported beam loaded with a concentrated force P in the middle, deformation under the force is [7]

$$f_{ms} = \frac{PL^3}{48EI}\left(1 + \frac{12\alpha_{sh}EI}{GFL^2}\right) \qquad (2.16a)$$

Again, the second term inside the brackets represents influence of shear deformation. For rectangular cross section and $E/G = 2.5$, then

$$f_{ms} = \frac{PL^3}{48EI}\left(1 + 3\frac{h^2}{L^2}\right)$$ (2.16b)

which shows slightly higher influence of shear deformation than for the uniformly loaded beam. Deformation of a cantilever beam loaded at the free end by force P can be obtained from Eq. (2.16a) if P in the formula is substituted by $2P$ and L is substituted by $2L$. For I-beams the shear effect is two-to-three times more pronounced, due to the smaller F than for the rectangular cross section beams. However, for laminated beams in which the intermediate layer is made of a material with a low G or for honeycomb beams in which F and possibly G are reduced, the deformation increase (stiffness reduction) due to the shear effect can be as much as 50%, even for long beams, and must be considered.

However, even considering the shear deformations, deformations of laminated and honeycomb beams under their own weight are significantly less than that of solid beams (for steel skin, steel core honeycomb beams about two times less). Stiffness-to-weight ratios (and natural frequencies) are significantly higher for composite and honeycomb beams than they are for solid beams.

2.3 TORSIONAL STIFFNESS

The basic strength of materials expression for torsional stiffness k_t of a round cylindrical bar or a tubular member of length l whose cross section is a circular ring with outer diameter D_0 and inner diameter D_i is

$$k_t = T/\theta = GJ_p/1 = (G/1)(\pi/32)(D_0^4 - D_i^4)$$ (2.17)

where T = torque, θ = angle of twist, G = shear modulus of the material, and J_p = polar moment of inertia. However, if the cross section is not round, has several cells, or is not solid (has a cut), the torsional behavior may change very significantly.

For a hollow solid cross section (without cuts) of an arbitrary shape (but with a constant wall thickness t) (Fig. 2.14), torsional stiffness is [8]

$$k_t = 4GA^2t/L1$$ (2.18)

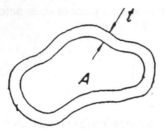

Figure 2.14   Single-cell thin-walled torsion section.

and the maximal stress is approximately

$$\tau_{max} = T/2At \qquad (2.19)$$

where A = area within the outside perimeter of the cross section, and L = peripheral length of the wall.

If this formula is applied to the round cross section (cylindrical thin-walled tube), then

$$4A^2t/L = (\pi D_0^3/8)(D_0 - D_i) \approx I_p = (\pi/32)(D_0^4 - D_i^4), \quad \text{if } (D_0 - D_i) < D_0 \quad (2.20)$$

Let this tube then be flattened out first into an elliptical tube and finally into a "double flat" plate. During this process of gradual flattening of the tube, t and L remain unchanged, but the area A is reduced from a maximum from the round cross section to zero for the double flat. Thus, the double flat cannot transmit any torque of a practical magnitude for a given maximum stress (or the stress becomes very large even for a small transmitted torque). Accordingly, for a given peripheral length of the cross section, a circular tube is the stiffest in torsion and develops the smallest stress for a given torque, since the circle of given peripheral length L encloses the maximum area A. One has to remember that formula (2.20) is an approximate one, and the stiffness of the "double flat" is not zero. It can be calculated as an open thin-walled cross section (see below).

Another interesting case is represented by two cross sections in Fig. 2.15a,b [8]. The square box-like thin-walled section in Fig. 2.15a is replaced by a similar section in Fig. 2.15b that has the same overall dimensions but also has two internal crimps (ribs). Both A and t are the same for these cross sections, but they have different peripheral lengths L ($L = 4a$ for Fig. 2.15a, $L = 16a/3$ for Fig. 2.15b). Thus, the crimped section is 33% less stiff than the square box section while being approximately 30% heavier and having greater shear stress for a given torque.

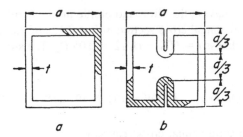

Figure 2.15 Square (box) sections (a) without and (b) with crimps. In spite of the greater weight of (b), it has the same torsional shear stress as (a) and is less stiff than (a) by a factor of 4/3.

A very important issue is torsional stiffness of elongated components whose cross sections are not closed, such as the ones shown in Fig. 2.16 [8]. Torsional stiffness of such bars with the uniform section thickness t is

$$k_t = Gbt^3/31 \tag{2.21}$$

where $b \gg t$ is the total aggregate length of wall in the section. If the sections have different wall thickness, then

$$k_t = (G/31) \sum_i b_i t_i^3 \tag{2.22}$$

where b_i = length of the section having wall thickness t_i. It is very important to note that the stiffness in this case grows only as the first power of b. It is illustra-

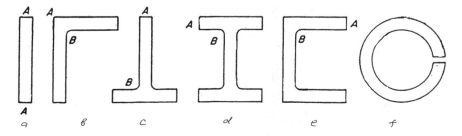

Figure 2.16 Typical cross sections to which Eq. (2.19) for torsional stiffness applies. Corners A have zero stress and do not participate in torque transmission; corners B have large stress concentrations depending on the fillet radius.

42   *Chapter 2*

tive to compare stiffness of a bar having the slit round profile in Fig. 2.16f with
stiffness of a bar having the solid annual cross section in Fig. 2.12a of the same
dimensions D_0, D_i with wall thickness $t = (D_0 - D_i)/2 = 0.05\,D_0$. The stiffness
of the former is

$$k_{t1} = (G/3\mathrm{l})[\pi(D_0 + D_i)/2][(D_0 - D_i/2)]^3 = 15.5 \times 10^{-6}GD_0^4/\mathrm{l} \qquad (2.23)$$

the stiffness of the bar with the solid annual cross section is

$$k_{t2} = GJ_p/\mathrm{l} = (G/\mathrm{l})(\pi/32)(D_0^4 - D_i^4) = 3.4 \times 10^{-2}GD_0^4/\mathrm{l} \qquad (2.24)$$

Thus, torsion stiffness of the bar with the solid (uninterrupted) annular cross
section is about 2,180 times (!) higher than torsional stiffness of the same bar
whose annular cross section is cut, so that shear stresses along this cut are not
constrained by the ends.

Another interesting comparison of popular structural profiles is made in
Fig. 2.17. The round profile is Fig. 2.17a has the same surface area as the standard
I-beam in Fig. 2.17b (all dimensions are in centimeters). Bending stiffness of the
I-beam about the x-axis is 41 times higher than bending stiffness of the round
rod with the cross section, as shown in Fig. 2.17a. Bending stiffness of the I-
beam about the y-axis is two times higher than bending stiffness of the round
rod, but its torsional stiffness is 28.5 times lower than that of the round rod.

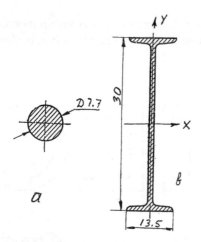

Figure 2.17   Two structural profiles having the same cross-sectional areas.

Additional information on torsional stiffness of various structural (power transmission) components is given in Chapter 6.

2.4 INFLUENCE OF STRESS CONCENTRATIONS

Stress concentrations (stress risers) caused by sharp changes in cross-sectional area along the length of a component or in shape of the component are very detrimental to its strength, especially fatigue strength. However, much less attention is given to influence of local stress concentrations on deformations (i.e., stiffness) of the component. This influence can be very significant. Fig. 2.18 [2] compares performance of three round bars loaded in bending. The initial design, case 1, is a thin bar (diameter $d = 10$ mm, length $l = 80$ mm). Case 2 represents a much larger bar (diameter $1.8d$) that has two circular grooves required by the design specifications. While the solid bar of this diameter could have bending stiffness 10 times higher than bar 1, stress concentrations in the grooves result

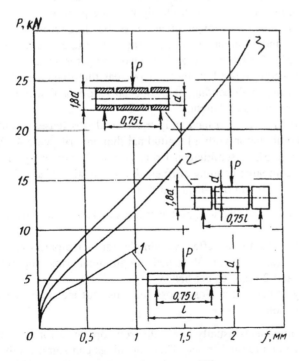

Figure 2.18   Design influence on stiffness.

in only doubling the stiffness. The stress concentrations can be substantially reduced by using the initial thin (case 1) bar with reinforcement by tightly fit bushings (case 3). This results in 50% stiffness increase relative to case 2, as well as in strength increase (the ultimate load $P_1 = 8$ KN; $P_2 = 2.1 P_1$; $P_3 = 3.6 P_1$).

2.5 STIFFNESS OF FRAME/BED COMPONENTS

2.5.1 Background

Presently, complex mechanical components such as beds, columns, and plates are analyzed for stresses and deformations by application of finite element analysis (FEA) techniques. However, the designer frequently needs some simple guidelines for initial design of these complex components.

Machine beds and columns are typically made as two walls with connecting partitions or rectangular boxes with openings (holes), ribs, and partitions. While the nominal stiffness of these parts for bending and torsion is usually high, it is greatly reduced by local deformations of walls, causing distortions of their shapes, and by openings (holes). The actual stiffness is about 0.25–0.4 of the stiffness of the same components but with ideally working partitions.

Figure 2.19 shows influence of longitudinal ribs on bending (cross-sectional moment of inertia I_{ben}) and torsional (polar moment of inertia J_{tor}) stiffness of a box-like structure [2]. The table in Fig. 2.19 also compares weight (cross-sectional area A) and weight-related stiffness. It is clear that diagonal ribs are very effective in increasing both bending and, especially, torsional stiffness for the given outside dimensions and weight.

Box-shaped beams in Fig. 2.20 have only transversal ribs (cases 2 and 3) or transverse ribs in combination with a longitudinal diagonal rib (case 4), harmonica-shaped ribs (case 5), or semidiagonal ribs supporting guideways 1 and 2 (case 6). The table compares bending stiffness k_x, torsional stiffness k_t, and weight of the structure W. It can be concluded that:

   With increasing number of ribs, weight W is increasing faster than stiffnesses k_x and k_t

   Vertical transversal ribs are not effective; simple transverse partitions with diagonal ribs (case 4) or V-shaped longitudinal ribs supporting guideways 1 and 2 (case 6) are better

   Ribs are not very effective for close cross sections, but are necessary for open cross sections

Machine frame components usually have numerous openings for accessing mechanisms and other units located inside. These openings can significantly reduce stiffness (increase structural deformations), depending on their relative dimensions and positioning. Fig. 2.21 illustrates some of these influences: δ_x and

Profile	Factors				
	I_{ben}	I_{tors}	$\dot{A}$	$\dfrac{I_{ben}}{A}$	$\dfrac{I_{tors}}{A}$
	1	1	1	1	1
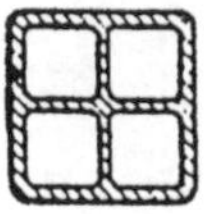	1.17	2.16	1.38	0.85	1.56
	1.55	3	1.26	1.23	2.4
	1.78	3.7	1.5	1.2	2.45

Figure 2.19   Stiffening effect of reinforcing ribs.

δ_y are deformations caused by forces F_x and F_y, respectively; δ_t is angular twist caused by torque T. Fig. 2.21 shows that:

Holes (windows) significantly reduce torsional stiffness

When the part is loaded in bending, the holes should be designed to be made close to the neutral plane (case 1)

Location of the holes in opposing walls in the same cross sections should be avoided

Holes exceeding 1/2 of the cross-sectional dimension ($D/a > 0.5$) should be avoided

The negative influence of holes on stiffness can be reduced by embossments around the holes or by well-fit covers. If a cover is attached by bolts, it would compensate for the loss of stiffness due to the presence of the hole if the preload force of each bolt is [9]

$$Q \geq [T(b_0 + l_0)]/Ffn \qquad (2.25)$$

Case #	Rib location	Kx%	Kt	W%
1	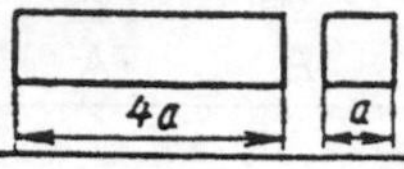 4a · a	100	100	100
2		101	103	108
3		102	109	125
4	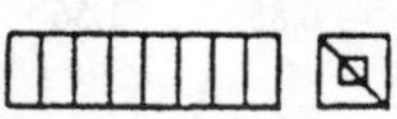	116	132	130
5	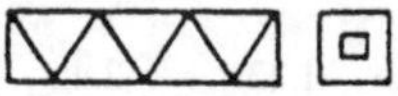	113	112	115
6	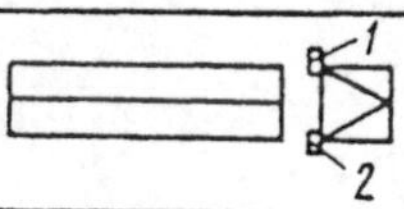 1 · 2	135	–	140

Figure 2.20 Reinforcement of frame parts by ribs.

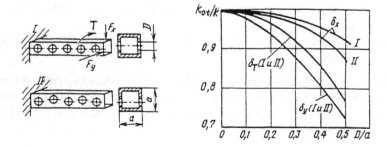

Figure 2.21 Influence of holes in frame parts of stiffness.

where F = cross-sectional area of the beam undergoing torsion; T = torque applied to the beam; b_0 and l_0 = width and length of the holes; f = friction coefficient between the cover and the beam; and n = number of bolts.

2.5.2 Local Deformations of Frame Parts

Local contour distortions due to torsional loading and/or local bending loading may increase elastic deformations up to a decimal order of magnitude in comparison with a part having a rigid partition. The most effective way of reducing local deformations is by introduction of tension/compression elements at the area of peak local deformations. Fig. 2.22a shows local distortion of a thin-walled beam in the cross section where an eccentrically applied load causes a torsional deformation. This distortion is drastically reduced by introduction of tension/compression diagonal ribs as in Fig. 2.22b.

Figure 2.23[2] shows distortion of a thin-walled beam under shear loading (a). Shear stiffness of the thin-walled structure is very low since it is determined by bending stiffness of the walls and by angular stiffness of the joints (corners). The same schematic represents the deformed state of a planar frame. The corners (joints) can be reinforced by introducing corner gussets holding the shape of the corners (Fig. 2.23b). The most effective technique is introduction of tensile (c) or compressive (d) reinforcing diagonal members (diagonal ribs in the case of a beam). Tilting of the cross section is associated with stretching/compression of the diagonal member by an increment Δ. Since tension/compression stiffness of the diagonal member(s) is much greater than bending stiffness of the wall, the

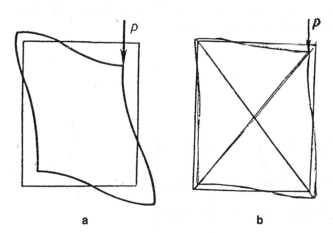

Figure 2.22 Contour distortion in a loaded thin-walled part (a) without and (b) with reinforcing ribs.

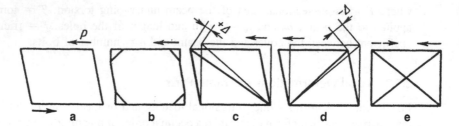

Figure 2.23   Diagonal reinforcement for shear loading.

overall shear stiffness significantly increases. Loading of the diagonal member in tension is preferable since the compressed diagonal member is prone to buckling at high force magnitudes. When the force direction is alternating, crossed diagonal members as in Fig. 2.23e can be used.

   A different type of local deformation is shown in Fig. 2.24. In this case the local deformations of the walls are caused by internal pressure. However, the solution is based on the same concept—introduction of a tensile reinforcing member (lug bolt 2) in the axial direction and a reinforcing ring 1, also loaded in tension, to prevent bulging of the side wall. These reinforcing members not only reduce local deformations, but also reduce vibration and ringing of walls as diaphragms.

2.6   GENERAL COMMENTS ON STIFFNESS ENHANCEMENT OF STRUCTURAL COMPONENTS

The most effective design techniques for stiffness enhancement of a structural component without increasing its weight are:

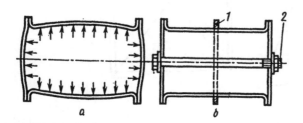

Figure 2.24   Reduction of local deformations.

Replacement of bending by tension/compression

Optimization of load distribution and support conditions if the bending mode of loading of a component is inevitable (see also Chapter 5)

Judicious distribution of the mass in order to achieve the largest cross-sectional and/or polar moments of inertia for a given mass of a component

Use of adjacent (connected) parts for reinforcement of the component: to achieve this effect, special attention has to be given to reinforcement of the areas where the component is joined with other components

Reduction of stress concentrations: in order to achieve this, sharp changes of cross-sectional shapes and/or areas have to be avoided or smoothed

Use of stiffness reinforcing ribs, preferably loaded in compression

Reduction of local deformations by introduction of ties parallel or diagonal in relation to principal sides (walls) of the component

Use of solid, noninterrupted cross sections, especially for components loaded in torsion

Geometry has a great influence on both stiffness values and stiffness models:

For short beams (e.g., gear teeth) shear deformations are commensurate with bending deformations and may even exceed them; in machine tool spindles, shear deformations may constitute up to 30% of total deformations.

For longer beams, their shear deformations can be neglected (bending deformations prevail); for example, for $L/h = 10$, where L is length and h is height of the beam, shear deformation is 2.5–3% of the bending deformation for a solid cross section, but increases to 6–9% for I-beams. Contribution from shear is even greater for multilayered and honeycomb beams.

If the cross-sectional dimensions of a beam are reduced relative to its length, the beam loses resistance to bending moments and torques, as well as to compression loads, and is ultimately becoming an elastic string.

Reduction of wall thickness of plates/shells transforms them into membranes/flexible shells that are able to accommodate only tensile loads.

Cross-sectional shape modifications can enhance some stiffness values relative to other.

Beams with open cross sections, like in Fig. 2.25a, may have high bending stiffness but very low torsional stiffness.

Slotted springs (Fig. 2.25b) may have high torsional but low bending stiffness.

Plates and shells can have anisotropic stiffness due to a judicious system of ribs or other reinforcements.

Thin-layered rubber-metal laminates [10] (see also Article 3 and Section 3.3) may have the ratio between stiffnesses in different directions (compres-

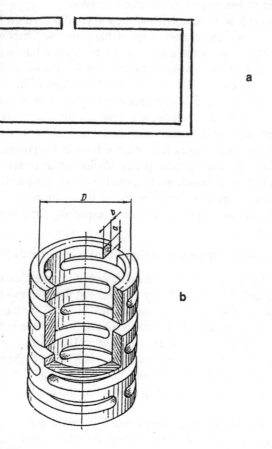

Figure 2.25 Structures with: (a) very low torsional but high bending stiffness; (b) very low bending and high torsional stiffness.

sion to shear) as high as 3000–5000. If loaded in bending, these components provide excellent damping due to a "constrained layer" effect.

REFERENCES

1. Rivin, E.I., Mechanical Design of Robots, McGraw-Hill, New York, 1988.
2. Orlov, P.I., Fundamentals of Machine Design, Vol. 1, Mashinostroenie Publishing House, Moscow, 1972 [in Russian].

3. Stewart, D., "A Platform with Six Degrees of Freedom," Proceedings of the Institute of Mechanical Engineers, 1965, Vol. 180, Part 1, No. 15, pp. 371–386.

4. Astanin, V.O., Sergienko, V.M., "Study of Machine Tool of Non-Traditional Configuration," Stanki I Instrument, 1993, No. 3, pp. 5–8 [in Russian].

5. Suzuki, M., et al., "Development of Milling Machine with Parallel Mechanism," Toyota Technical Review, 1997, Vol. 47, No. 1, pp. 125–130.

6. Pritchof, G., Wurst, K.-H., "Systematic Design of Hexapods and Other Parallel Link Systems," CIRP Annals, 1997, Vol. 46/1, pp. 291–296.

7. Timoshenko, S.P., Gere, J.M., Mechanics of Materials, Van Nostrand Reinhold, New York, 1972.

8. DenHartog, J.P., Advanced Strength of Materials, Dover Publications, Inc., Mineola, NY, 1987.

9. Kaminskaya, V.V., "Load-Carrying Structures of Machine Tools," In: Components and Mechanisms of Machine Tools, ed. by D.N. Reshetov, Mashinostroenie Publishing House, Moscow, 1973, Vol. 1, pp. 439–562 [in Russian].

10. Rivin, E.I., "Properties and Prospective Applications of Ultra Thin Layered Rubber-Metal Laminates for Limited Travel Bearings," Tribology International, Vol. 18, No. 1, 1983, pp. 17–25.

3

Nonlinear and Variable Stiffness Systems: Preloading

3.1 DEFINITIONS

Since stiffness is the ratio of the force to the displacement caused by this force, the load-deflection plot (characteristic) allows one to determine stiffness as a function of force. It is much easier to analyze both static and dynamic structural problems if the displacements are proportional to the forces that caused them, i.e., if the load-deflection characteristic is linear. However, most of the load-deflection characteristics of actual mechanical systems are nonlinear. In many cases the degree of nonlinearity is not very significant and the system is considered as linear for the sake of simplicity. A significant nonlinearity must be considered in the analysis, especially for analysis of dynamic processes in which nonlinearity may cause very specific important, and frequently undesirable, effects [1]. At the same time, there are many cases when the nonlinearity may play a useful role by allowing adjustment and controlling the stiffness parameters of mechanical systems.

There are two basic types of nonlinear load-deflection characteristics as presented in Fig. 3.1. Line 1 represents the case when the rate of increase of deflection x slows down with increasing force P. If the local (differential) stiffness is defined as ratio between increments of force (ΔP) and deflection (Δx), then

$$k = \Delta P/\Delta x \tag{3.1}$$

and the stiffness along the line 1 is increasing with the increasing load

$$k_1'' = \Delta P_1''/\Delta x_1'' > k_1' = \Delta P_1'/\Delta x_1' \tag{3.2a}$$

This is called the *hardening load-deflection characteristic.*

52

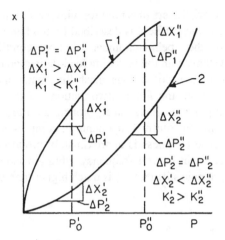

Figure 3.1 (1) Hardening and (2) softening nonlinear load-deflection characteristics.

The rate of increase of deflection x accelerates with increasing force P along line 2, thus the local stiffness along line 2 is decreasing with the increasing load

$$k_2' = \Delta P_2'/\Delta x_2' > k_2'' = \Delta P_2''/\Delta x_2'' \qquad (3.2b)$$

This is called the *softening load-deflection characteristic*.

Both types of nonlinear load-deflection characteristics allow for varying the actual stiffness by moving the *working point* along the characteristic. This can be achieved by applying a *preload force* to the system that is independent from actual process forces (payload). If the preload force is not constant but changing with changing operational conditions of the system, then there is a potential for creating a system with *controlled* or *self-adaptive stiffness*.

3.2 EMBODIMENTS OF ELEMENTS WITH NONLINEAR STIFFNESS

Nonlinear stiffness is specific for:

a. *Elastic deformations of parts whose material is not exactly described by Hooke's law.* There are many materials that exhibit nonlinear deformation characteristics. Deformations of cast iron and concrete components are characterized by softening load-deflection characteristic. Stiffness of rubber components is increasing with load (hardening) if the component is loaded in compression; stiffness is decreasing with load (softening) if the component is loaded in shear.

Very peculiar load-deflection characteristics are observed for wire mesh, felt, and other fibrous mesh elements. Their static stiffness is described by the hardening load-deflection characteristic, while their behavior under vibratory conditions (dynamic stiffness) is typical for systems with the softening load-deflection characteristic (discussed below). Deformations of fibrous mesh-like materials are caused by slippages in contacts between the fibers, deformations of the fiber material itself (natural or synthetic polymer fibers, steel, bronze, etc.) are relatively small. Static stiffness of the fibrous mesh components in compression is of a strongly hardening nonlinear type (stiffness is approximately proportional to the compression load, like in constant natural frequency vibration isolators described below). However, their dynamic stiffness k_{dyn} is much higher than static stiffness k_{st}

$$k_{dyn} = K_{dyn} k_{st} \tag{3.3}$$

where $K_{dyn} = 1-20$ is the dynamic stiffness factor. Both k_{dyn} and K_{dyn} strongly depend on amplitude of vibrations as shown in Fig. 3.2 [2]. It is known that in a

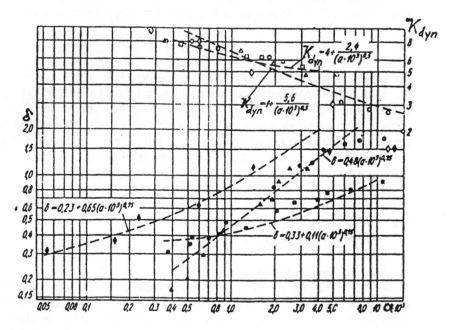

Figure 3.2 Amplitude dependence of damping (δ) and dynamic stiffness coefficient (K_{dyn}) for mesh-like materials: 0 = wire mesh, low specific load; Δ = same, high specific load; X = thin-fiber felt; x = thick-fiber felt; open symbols = K_{dyn}; solid symbols = δ.

vibratory system mass m–nonlinear spring k, its *natural frequency* (thus, effective stiffness) *is increasing with increasing amplitudes of excitation for hardening nonlinearity* of the spring and *decreasing with increasing amplitudes for softening nonlinearity* [1]. Fig. 3.2 shows that dynamic (vibratory) stiffness of mesh-like materials is decreasing with increasing amplitudes (softening nonlinearity), while for the static loading the nonlinearity is of the hardening type. Thus, the fibrous mesh-like materials have *dual nonlinearity*. Figure 3.2 also shows amplitude dependency of internal damping of the mesh-like materials.

b. *Contact deformations.* Joints loaded perpendicularly to the contact surfaces are characterized by a hardening nonlinearity due to increase of effective contact area with the increasing load. Tangential contact deformations may exhibit a softening nonlinearity. Characteristics of contact deformations are described in Chapter 4.

c. *Changing part/system geometry* due to large deformations or due to a special geometry designed in order to obtain changing stiffness. Typical examples of components having nonlinear stiffness due to changing geometry are:

Coil springs with variable pitch and/or variable coil diameter
Thin-layered rubber-metal laminates
Rubber elements with built-in constraining
Elastic elements with contact surfaces shifting with load

Figure 3.3a shows a conventional helical compression spring having constant pitch and constant diameter along its length. These design parameters result in a linear load-deflection characteristic. Variable coil diameter coil springs are shown in Fig. 3.3b–d; variable pitch/constant diameter coil spring is shown in Fig. 3.3e.

3.2.1 Nonlinear Elements with Hardening Characteristics

Deformation pattern of a variable coil diameter spring is shown in Fig. 3.4. Since axial compression of a coil is proportional to the third power of its diameter, the larger coils would deform more than the smaller coils and thus would eventually flatten and reduce the effective number of coils. As a result, stiffness of the spring increases with load (hardening characteristic). In some designs, each coil can fit inside the preceding larger coil, so that the spring would become totally flat at its ultimate load.

In a variable pitch spring in Fig. 3.3e, coils with the smaller pitch gradually touch each other and thus reduce the effective number of coils, while the coils with the larger pitch are still operational. The result is also a hardening load-deflection characteristic.

Figure 3.3f shows a nonlinear coil spring 1 loaded in torsion. The applied torque causes reduction of its diameter, and variable diameter core 2 allows one

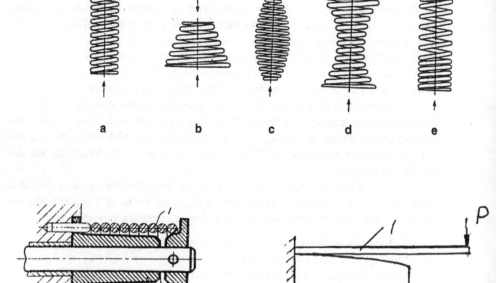

Figure 3.3 Coil springs with (a) linear and (b–g) nonlinear characteristics: a = cylindrical spring with constant pitch and constant diameter; b, c, d = variable diameter springs (b = conical; c = barrel; d = hourglass); e = constant diameter/variable pitch spring; f = nonlinear torsion spring; g = nonlinear flat spring having a shaped support surface.

Figure 3.4 Deformation of conical spring.

to change the number of active coils while the torque is increasing, thus creating a hardening characteristic "torque-twist angle." This is a typical example of an elastic element whose contact surfaces with other structural components are changing with increasing load. A similar concept is used for a nonlinear flat spring shown in Fig. 3.3g, whose effective length is decreasing (and thus stiffness is increasing) with increasing deformation. Another example of achieving nonlinear load-deflection characteristic by changing contact surfaces is presented by deformable bodies with curvilinear external surfaces. Frequently, rubber elastic elements are designed in such shapes that result in changing their "footprint" with changing load (e.g., spherical or cylindrical elements [3–5]).

Rubber-like (or elastomeric) materials have unique deformation characteristics because their Poisson's ratio $v = 0.49 - 0.4995 \approx 0.5$. Since the modulus of volumetric compressibility is

$$K = G/(1 - 2v) \tag{3.4}$$

then it is approaching infinity when v is approaching 0.5. Materials with $v = 0.5$ are not changing their volume under compression, thus rubber is practically, a "volumetric incompressible material." Change of volume of a compressed rubber component can occur only due to minor deviations of the Poisson's ratio from $v = 0.5$. Accordingly, compressive deformation of a rubber component can occur only if it has free surfaces so that bulging on these free surfaces would compensate the deformation in compression. Thus, compression deformation under a compression force P_z of cylindrical rubber element 1 in Fig. 3.5a, which is bonded to metal end plates 2 and 3, can develop only at the expense of the bulging of element 1 on its free surfaces. If an intermediate metal layer 4 is placed in the middle of and bonded to rubber element 1 as in Fig. 3.5b, thus dividing it into two layers 1′ and 1″, the bulging becomes restricted and compression deformation under the same force P_z is significantly reduced. For a not very thin layer, $d \leq 5\text{–}10h$, the apparent compression (Young's) modulus of a cylindrical rubber element bonded between the parallel metal plates can be calculated as [3]

$$E = 3mG(1 + kS^2) \tag{3.5}$$

where S = so-called shape factor defined as the ratio of loaded surface area A_l to free surface areas A_f of the element; G = shear modulus; k and m = coefficients depending on the hardness (durometer) H of rubber, with $k = 0.93$ and $m = 1$ at $H = 30$, $k = 0.73$ and $m = 1.15$ at $H = 50$, and $k = 0.53$ and $m = 1.42$ at $H = 70$. For an axially loaded rubber cylinder having diameter d and height h

$$S = A_l/A_f = (\pi d^2/4)/\pi dh = d/4h \tag{3.6}$$

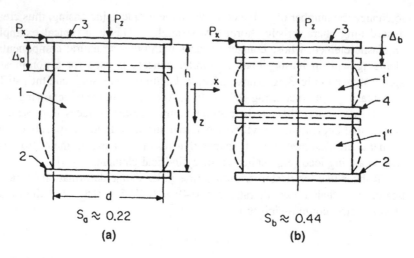

$$S_a \approx 0.22 \qquad\qquad S_b \approx 0.44$$

(a)   (b)

Figure 3.5 Shape factor influence on compression deformation of bonded rubber element.

The stiffening nonlinearity of the compressed rubber elements can be enhanced if elements with cross sections varying along the line of force application are used. The most useful elements are cylindrical elements loaded in the radial direction as well as spherical and ellipsoidal elements. Such streamlined shapes result in the lowest stresses for given loads/deformations, and in reduced creep, as it was shown in [4], [5]. These features also contribute to enhancement of fatigue life of the rubber components. Stiffness of a cylindrical or spherical rubber element under radial compression (Fig. 3.6) increases with increasing load due to three effects contributing to a gradual increase of the shape factor S: increasing "footprint" or the loaded surface area; increasing cross-sectional area; and decreasing free surface on the sides due to reduced height. Figure 3.7a [4] shows the load-deflection characteristic of a rubber cylinder $L = D = 1.25$ in. (38 mm) under radial compression (line 2). This load-deflection characteristic can be compared with line 1, which is the load-deflection characteristic of the same rubber cylinder (whose faces are bonded to the loading surfaces) under axial compression, in which case the contact areas are not changing.

It can be shown (see Article 1) that performance of vibration isolators significantly improves if they have the so-called constant natural frequency (CNF) characteristic when the natural frequency of an object mounted on the isolators does not depend on its weight W. The natural frequency f_n of a mass (m) − spring (k) system is

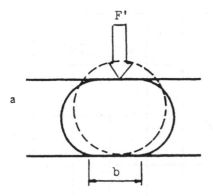

Figure 3.6 Compression of spherical rubber element.

$$f_n = \frac{1}{2\pi} \sqrt{\frac{k}{m}} = \frac{1}{2\pi} \sqrt{\frac{kg}{W}} \qquad (3.7)$$

where g = acceleration of gravity. Accordingly, in order to assure that f_n = a constant for any weight, stiffness k must increase proportionally to the weight load W on the isolator. It represents a nonlinear elastic element with a special hardening characteristic for which

$$\Delta P/\Delta x = k = AW \qquad (3.8)$$

where A = a constant. Such a characteristic can be achieved by radial compression of cylindrical or spherical rubber elements as illustrated in Fig. 3.7b. The load range within which this characteristic occurs is described in ratio of the maximum P_{max} and minimum P_{min} loads of this range. For cylindrical/spherical rubber elements P_{max}/P_{min} = 3 to 5. Although this is adequate in many applications, in some cases (such as for isolating mounts for industrial machinery), a broader range is desirable.

A much broader range P_{max}/P_{min} in which the CNF characteristic is realized can be achieved by designing a system in which bulging of the rubber element during compression is judiciously restrained. Bulging of the rubber specimen on the side surfaces can be restrained by designing interference of two bulging surfaces within the rubber element and/or by providing rigid walls. Both approaches were used in the design of a popular nonlinear vibration isolator for industrial machinery shown in Fig. 3.8 [6] (see also Article 1). Its elastic element 3 is comprised of two rubber rings 3′ (external) and 3″ (internal), separated by an

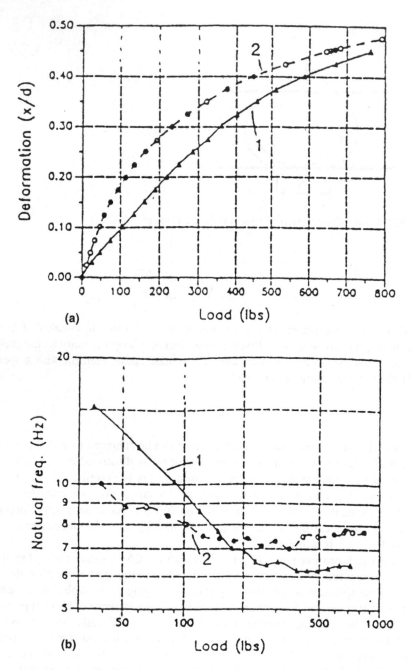

Figure 3.7 (a) Deformation and (b) load-natural frequency characteristics of rubber cylinder ($D = L = 1.25$ in.) when loaded in (1) axial and (2) radial directions.

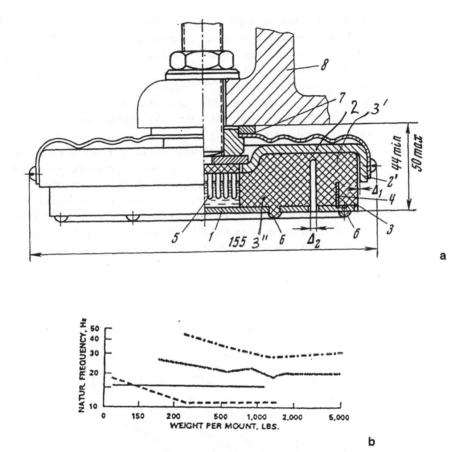

Figure 3.8 Constant natural frequency vibration isolator: 1 = bottom cover; 2 = top cover; 2′ = lid; 3 = rubber elastic element; 4 = transversal reinforcing ring (rib); 5 = viscous damper; 6 = rubber friction rings; 7 = level adjustment unit; 8 = foot of the installed machine. (All dimensions in millimeters.)

annular clearance Δ_2. Rings 3′ and 3″ are bonded to lower 1 and upper 2 metal covers. When the axial load P_z (weight of the installed machine 8) is small, each ring is compressing independently and rubber can freely bulge on the inner and outer side surfaces of both rings. At a certain magnitude of P_z, the bulges on the inner surface of ring 3′ and on the outer surface of ring 3″ are touching each other, and further bulging on these surfaces is restrained. At another magnitude of P_z, the bulge on the outer surface of ring 3′ touches lid 2′ of top cover 2, thus also restraining bulging. Initially, there is an annular clearance Δ_1 between lid

2' and the external surface of ring 3'. Both restraints result in a hardening nonlinear load-deflection characteristic whose behavior can be tailored by designing the clearances Δ_2 between rings 3' and 3", and Δ_1 between the outer surface of ring 3' and lid 2'. Plots in Fig. 3.8b illustrate load-natural frequency characteristics of several commercially realized CNF isolators of such design. It can be seen that the ratios P_{max}/P_{min} as great as 100:1 have been realized.

While the load range within which the CNF characteristic occurs for the streamlined rubber elements, as in Fig. 3.7b, is not as wide as in Fig. 3.8b, the design is much simpler and easier to realize than for a mount in Fig. 3.8a.

Thin-layered rubber-metal laminates have nonlinear properties which are very interesting and important for practical applications [7] (see Article 3). Further splitting and laminating of the block in Fig. 3.5 leads to even higher stiffness. When the layers become very thin, on the order of 0.05–2 mm (0.002–0.08 in.), compression stiffness becomes extremely high and highly nonlinear, as shown in Fig. 3.9. Both stiffness and strength (the ultimate compressive load) are greatly

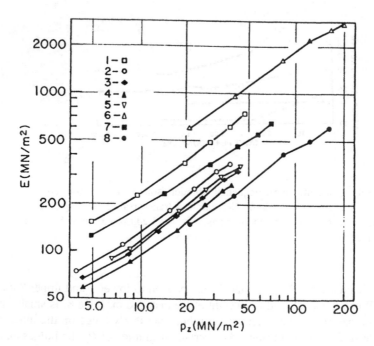

Figure 3.9 Compression modulus of ultrathin-layered rubber-metal laminates. Rubber layer thickness: (1) 0.16 mm; (2) 0.33 mm; (3) 0.39 mm; (4) 0.53 mm; (5) 0.58 mm; (6) 0.106 mm; (7) 0.28 mm; (8) 0.44 mm. Test samples (1)–(5) have brass intermediate layers; (6)–(8) have steel interlayers.

influenced by deformations and strength of the rigid laminating layers (usually metal). The thin-layered rubber-metal laminates fail under high compression forces when the yield strength of the metal layers is exceeded and the metal disintegrates. When the rigid laminating layers made of a high strength steel were used, static strength values as high as 500–600 MPa (75,000–90,000 psi) have been realized for rubber layer thickness of 0.5–1.0 mm (0.006–0.04 in.). This unique material changes its stiffness by a factor of 10–50 during compressive deformation of only 10–20 μm.

Another special feature of the rubber-metal laminates is anisotropy of their stiffness characteristics. Since shear deformation (under force P_x in Fig. 3.5) is not associated with a volume change, shear stiffness does not depend on the design of the rubber block, only on its height and cross-sectional area. As a result of this fact, shear stiffness of the laminates stays very low while the compression stiffness increases with the thinning of the rubber layers. Ratios of stiffness in compression and shear exceeding 3000–5000 are not difficult to achieve.

3.2.2 Nonlinear Elements with Softening Characteristic

All the examples described above represent mechanical systems with hardening load-deflection characteristics. It is more difficult to obtain a softening load-deflection characteristic in a mechanical system. Usually, some ingenious design tricks are required. Figure 3.10a shows such specially designed device; its load-deflection characteristic is shown in Fig. 3.10b. The system consists of linear spring 1 and nonlinear spring 2 having a hardening load-deflection characteristic. These springs are precompressed by rod (drawbar) 3. Stiffness k of the set of springs acted upon by force P is the sum of stiffnesses of spring 1 (k_1) and 2

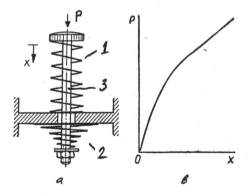

Figure 3.10   (a) Spring device with (b) softening load-deflection characteristic.

(k_2), $k = k_1 + k_2$. With increasing load P, deformation of nonlinear spring 2 is diminishing and, accordingly, its stiffness is decreasing. Since stiffness k_1 is constant, the total stiffness is also decreasing as in Fig. 3.10b. This process continues until spring 2 is completely unloaded, after which event the system becomes linear.

Another design direction for realizing nonlinear systems with the softening load-deflection characteristic is by using thin-walled elastic systems that are usually capable of having two or several elastically stable configurations, i.e., capable of collapsing the fundamental stable configuration. Figures 3.11b and c [8] shows stages of compression of a thin-walled rubber cylinder in Fig. 3.11a. The first

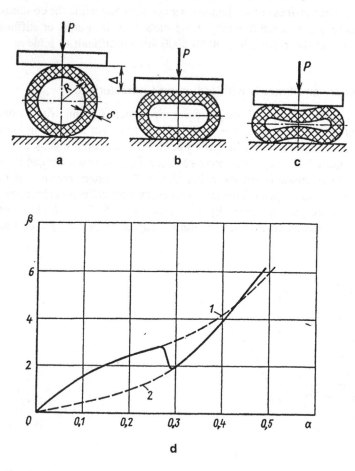

Figure 3.11 Compression of thin-walled rubber cylinder.

stable configuration (Fig. 3.11b) is characterized by the softening load-deflection characteristic 1 in Fig. 3.11d; the second stable configuration (Fig. 3.11c) has the hardening load-deflection characteristic 2 and there is a sizable segment of "negative stiffness" describing the collapsing process of the cylinder. The realizable load-deflection characteristic is shown by the solid line; the dimensionless coordinates in Fig. 3.11d are $\alpha = \Delta/R$ and $\beta = 12PR^2/E\delta^3$.

Some additional examples of elements with softening and/or neutral (zero stiffness) load-deflection characteristics and their practical applications are described in Section 8.5.

3.2.3 Practical Case 3 [9]

Figure 3.12 shows a somewhat different embodiment of a spring system with the softening load-deflection characteristic, which is used as a punch force simulator for evaluation dynamic and noise-radiation characteristics of stamping presses. In this design, Belleville springs 2 are preloaded by calibrated (or instrumented) bolts 1 between cover plates 3 and 4 to the specified load P_s. The simulator is installed on the press bolster instead of the die. During the downward travel of the press ram, it contacts the head 6 of the simulator and on its further travel down it unloads the bolts from the spring-generated force. As the bolts are being unloaded, the ram is gradually loaded. This process of the press ram being exposed to the full spring load is accomplished while the ram travels a distance equal to the initial deformation of the bolts caused by their preloading of the force P_s. This initial deformation is very small (high stiffness segment) and can be adjusted by changing length/cross section of the bolts. After this process is complete, the ram is further compressing the springs (low stiffness segment). Thus, this device having the softening load-deflection characteristic provides sim-

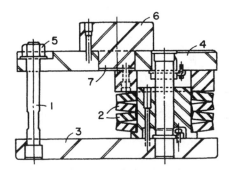

Figure 3.12 Punch force simulator.

ulation of the punching process consisting of a very intense loading during the "breakthrough" process and much less intense loading afterwards.

3.3 STIFFNESS MANAGEMENT BY PRELOADING (STRENGTH-TO-STIFFNESS TRANSFORMATION)

The contribution of stiffness to performance characteristics of various mechanical systems is diverse. In some cases, increase of stiffness is beneficial; in other cases, a certain optimal value of stiffness has to be attained. To achieve these goals, means for adjustment of stiffness are needed. In many instances, the stiffness adjustment can be achieved by *preloading* of the components responsible for the stiffness parameters of the system. Preloading involves intentional application of internal forces to the responsible components, over and above the payloads. The proper preloading increases the stiffness but may reduce strength or useful service life of the system due to application of the additional forces. Thus, some fraction of the overall strength is transformed into stiffness.

Preloading can regulate stiffness both in linear and nonlinear systems. Preloading of linear systems can be illustrated on the example of Fig. 3.13.

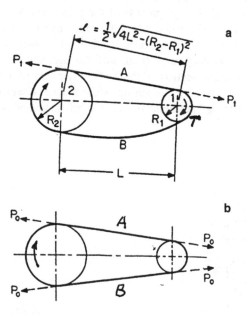

Figure 3.13 Belt drive (a) without and (b) with preload. A = active branch, B = slack branch.

In a belt drive in Fig. 3.13a, torque T applied to driving pulley 1 results in application of a tangential force P_t to driven pulley 2 by tensioning the leading (active) branch A of the belt. This tensioning leads to stretching of the branch A by an increment Δ. The idle branch B accommodates this elongation Δ by getting more slack. Stiffness k_{bd} of the belt drive is a ratio of torque T on the driving pulley divided by the angular deflection α of the pulley if the driven pulley is kept stationary (locked). In this case, it is more convenient to derive an expression for compliance $e_{bd} = 1/k_{bd}$. It is then obvious that

$$e_{bd} = \alpha/T = (\Delta/R_1)/(P_t R_1) = \Delta/P_t R_1^2 = (P_t L_{ef}/EA)/P_t R_1^2 = L_{ef}/R_1^2 EA \qquad (3.8)$$

In this drive, only the branch A is transmitting the payload (tangential force); the other branch B is loose. The drive can be preloaded by pulling the pulleys apart during assembly so that each branch of the belt (both the previously active branch A and the previously loose branch B) is subjected to the *preload force* P_0 even if no payload is transmitted. In this case, application of a torque to the driving pulley will increase tension of the active branch A of the belt by the amount P_A and will reduce the tension of the previously passive (loose) branch B by the amount P_B. These forces are

$$P_A = P_B = (1/2\ T)/R_1 \qquad (3.9)$$

Total elongations of branches A and B are, respectively

$$\Delta_a = (P_0 + P_A)L_{ef}/EA; \quad \Delta_b = (P_0 - P_B)L_{ef}/EA \qquad (3.10)$$

Incremental elongations of branches A and B caused by the payload are, respectively

$$\Delta_1 = P_A L_{ef}/EA; \quad \Delta_2 = -P_B L_{ef}/EA \qquad (3.11)$$

and the angular compliance of the preloaded drive is

$$e'_{bd} = L_{ef}/2R_1^2 EA \qquad (3.12)$$

or one-half of the compliance for the belt drive without preload. In other words, preload has doubled stiffness of the belt drive, which is a system with a linear load-deflection characteristic.

It is important to note that the active branch in a preloaded drive is loaded with a larger force ($P_0 + P_t/2$) than in a nonpreloaded drive ($P_t < P_0 + P_t/2$). Thus, only belts with upgraded strength limits can be preloaded, and, accordingly, the described effect can be viewed as an example of *strength-to-stiffness transformation.*

While belts have approximately linear load-deflection characteristics, chains are highly nonlinear, both roller chains (Fig. 3.14a) and silent chains (Fig. 3.14b). Experimental data show that with increasing the load from $0.1P_r$ to P_r, the tensile stiffness of a silent chain increases about threefold (P_r, rated load) (Fig. 3.14c).

The statics of a chain drive are very similar to those of a belt drive. If the chain is installed without initial preload, like the belt in Fig. 3.13a, only the active branch will be stretched by the full magnitude of the tangential force $P_t = T/R_1$. The tensile deformation is described by Eq. (3.8), in which $A = l_c d$ is the effective

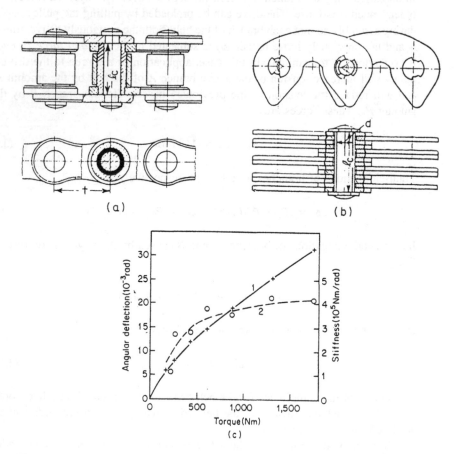

Figure 3.14 Power transmission chains: (a) roller chain; (b) silent chain; (c) torque-angular (1) deflection characteristic and (2) stiffness of a silent chain drive.

cross-sectional area (l_c = total load-carrying width of the chain, d = diameter of the roller-supporting axles for the roller chains or the laminae-supporting axles for silent chains) (Figs. 3.14a and b), and $E = t/k_{ch}$ is the effective elastic modulus (t = chain pitch and k_{ch} = compliance factor of the chain, which is a function of the tensile load P_t). Thus

$$e_{ch} = k_{ch}(P_t)L/R_1^2At \qquad (3.13)$$

If a preload is applied, the chain drive stiffness would immediately double, which is described [10], by an expression similar to Eq. (3.12)

$$e_{ch} = k_{ch}(P_0)L/2R_1^2At \qquad (3.14)$$

However, in addition to this direct effect of the preload, with increasing preload P_0, stiffness of each branch increases because of the hardening nonlinear load-deflection curve (Figs. 3.1 and 3.14c) as reflected in the expression $k_{ch}(P_0)$. Thus, an increase in the preload would additionally enhance the stiffness, up to three times for a typical roller chain. The total stiffness increase, considering both effects, can be up to 5–6 times.

Another advantage of the preloaded chain drive is the stability of its stiffness. Since only one branch transmits payload in a nonpreloaded chain and this branch is characterized by a nonlinear hardening characteristic, stiffness at low payloads is low and stiffness at high payloads is high. Thus, the accuracy of a device driven by a chain drive would be rather poor at relatively low torques. In a preloaded drive, stiffness for any payload $P_t < 2P_0$ will have the same (high) value, since the "working point" on the load-deflection plot in Fig. 3.1 (or torque-angular deflection plot in Fig. 3.14c) would always correspond to load P_0.

This statement is reinforced by the data in Section 4.3, Eqs. (4.9) and (4.10), on the preloading of *flat joints* in order to increase their stiffness in case of moment loading. Although contact deformations are characterized by nonlinear load-deflection curves, preloaded joints are essentially linear when a moment is applied.

Of course, preloading of flat joints leads to increased stiffness not only for the moment loading but for the force loading as well. Indeed, loading with an incremental force $\pm P_1$ of a joint preloaded with a compressive force P_0, $P_1 < P_0$, would result in stiffness roughly associated with the working point P_0, which is higher than the stiffness for the working point $P_1 < P_0$ (Fig. 3.1). In addition, preloaded joints can work even in tension if the magnitude of the tensile force does not exceed P_0.

Similar, but numerically much more pronounced effects occur in *rubber-metal laminates* described above and in Rivin [7] (Article 3).

The concept of internal preload is very important for *antifriction bearings*, especially in cases when their accuracy of rotation and/or their stiffness have to be enhanced. The accuracy of rotation under load depends on the elimination of backlash as well as on the number of the rolling bodies participating in the shaft support. The deflections of the shaft (spindle), which also affect accuracy, are also influenced by deformations of bearings.

Elastic displacements of shafts in antifriction bearings consist of elastic deformations of rolling bodies and races calculated by using Hertz formulae (see Table 4.1), as well as of deformations of joints between the outer race and the housing and the inner race and the shaft. Since contact pressures between the external surfaces of both the races and their counterpart surfaces (shaft, housing bore) are not very high, contact deformations have linear load-deflection characteristics (see Section 4.4.1, cylindrical joints). The total compliance in these joints that are external to the bearing is [10]

$$\delta'' = (4/\pi)(Pk_2/db)(1 + d/D) \quad \text{(m)} \tag{3.15}$$

where P = radial force, N; d and D = inner and outer diameters of the bearing, m; b = width of the bearing, m; and $k_2 = 5$ to 25×10^{-11} m^3/N. Lower values of k_2 are representative for high-precision light interference cylindrical or preloaded tapered fits and/or high loads, and higher values are representative for regular precision tight or stressed fits, with reamed holes and finish ground shaft journals and/or light loads.

Hertzian deformations of rolling bodies in ball bearings are nonlinear and can be expressed as

$$\delta_b' = (0.15 - 0.44d) \times 10^{-6} P^{2/3} \quad \text{(m)} \tag{3.16}$$

For roller bearings

$$\delta_r' = \sim k_1 P \quad \text{(m)} \tag{3.17}$$

where $k_1 = 0.66 \times 10^{-10}/d$ for narrow roller bearings; $0.44 \times 10^{-10}/d$ for wide roller bearings; $0.41 \times 10^{-10}/d$ for normal-width tapered roller bearings; and $0.34 \times 10^{-10}/d$ for wide tapered roller bearings [10].

As a general rule, δ'' is responsible for 20–40% of the overall deformation at low loads (precision systems) and for 10–20% at high loads.

Accordingly, the internal preload of roller bearings (which can be achieved by expanding the inner race, e.g., by using a tapered fit between the bearing and the shaft) (Fig. 3.15e) brings results similar to those achieved in belt drives, namely, doubling their radial stiffness (reducing δ_r' in half) because of the elimination of backlash and "bringing to work" idle rollers on the side opposite to the area compressed by the radial force. In ball bearings, the effect of the internal preload on δ_b' is rather similar to that in chain drives and consists of (1) doubling the number of active balls and (2) shifting the working point along the nonlinear load-deflection curve.

The positive effects of internal preload in antifriction bearings are (in addition to the stiffness increase) elimination of backlash and enhancement of rotational accuracy because of more uniform loading of the rolling bodies and better averaging of the inevitable inaccuracies of the races and the rolling bodies. The negative effects are higher loading of the bearing components and its associated higher working temperature, as well as faster wear and higher energy losses (it can be expressed as "part of the strength had been used to enhance stiffness").

Since (1) Hertzian deformations are responsible for only 60–90% of the total bearing compliance [the balance being caused by external contact deformations

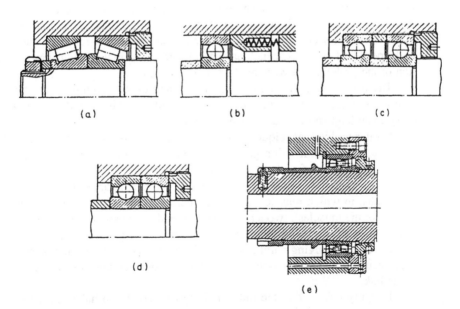

Figure 3.15 Some practical techniques for preloading rolling friction bearings: (a) preload by tightening a nut; (b) preload maintained by a spring; (c) preload caused by unequal spacers for inner and outer races; (d) preload by reducing width of outer races; (e) preload by squeezing inner race on a shallow cone.

described by Eq. (3.15)] and (2) the latter are not affected by the internal preload, then in stiffness-critical cases bearing preload has to be accompanied by more stringent requirements to the assembly of the bearings with the shafts and housings. The resulting reduction of k_2 in Eq. (3.15) would assure the maximum stiffness-enhancement effect.

Figure 3.15 shows some techniques used to create the preload in the bearings. Figure 3.15a shows preload application by tightening a threaded connection; Fig. 3.15b shows preload by springs; and Figs. 3.15c and d show preload by a prescribed shift between the rolling bodies and the races achieved through the use of sleeves of nonequal lengths between the outer and inner races (c) or by machining (grinding) the ends of one set of races (d). In the case shown, the width of each of the outer races is reduced by grinding. A technique similar to the one shown in Fig. 3.15d is used in "four-point contact" bearings. For preloading cylindrical roller bearings in stiffness-critical applications, their inner races are fabricated with slightly tapered holes (Fig. 3.15e). Threaded preload means are frequently used for tapered roller bearings, which are more sturdy and less sensitive to overloading than ball bearings. Since some misalignment is always induced by the threaded connections, so the technique is generally used for nonprecision units. In cases where a high precision is required, the load from a threaded load application means is transmitted to the bearing through a tight-fitted sleeve with squared ends, as shown in Fig. 3.15e (or very accurate ground threads are used). Springs (or an adjustable hydraulic pressure) are used for high- and ultra-high-speed ball bearings, which have accelerated wear that would cause a reduction of the preload if it were applied by other means.

The necessary preload force P_0 is specified on the premise that after the bearing is loaded with a maximum payload, all the rolling bodies would still carry some load (no clearance is developing).

Similar preload techniques, and with similar results, are used in many other cases, such as in antifriction guideways, ball screws, traction drives [11].

A special case of preload, especially important for manipulators, is used in so-called antagonist actuators (Fig. 3.16a). In such actuators (modeled from human limbs, in which muscles apply only pulling forces developing during contraction of the muscles), two tensile forces F_1 and F_2 are applied at opposite sides of a joint. A general schematic of such an actuator is given in Fig. 3.16b. Link L is driven by applying parallel tension forces F_1 and F_2 to driving arms 1 and 2, respectively, which are symmetrical and positioned at angles $90° + \alpha$ relative to the link.

To apply torque T to the link, the following condition has to be satisfied:

$$T = -F_1 a \cos(\theta + \alpha) + F_2 a \cos(\theta - \alpha) \tag{3.18a}$$

or

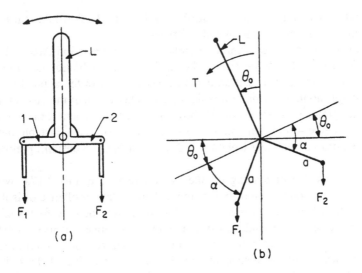

Figure 3.16 (a) Antagonist actuator and (b) its schematic.

$$T/F_1 - F_2/F_1 a \cos(\theta - \alpha) - a \cos(\theta + \alpha) \qquad (3.18b)$$

It follows from Eq. (3.18b) that one tension (e.g., F_1) can be chosen arbitrarily and then the required second tension (F_2) can be determined from the ratio

$$F_2/F_1 = [(T/F_1 + a \cos(\theta + \alpha)]/a \cos(\theta - \alpha) \qquad (3.18c)$$

The stiffness of the actuator is

$$dT/d\theta = k = F_1 a \sin(\theta + \alpha) - F_2 a \sin(\theta - \alpha)$$
$$= F_1 a[\sin(\theta + \alpha) - F_2/F_1 \sin(\theta - \alpha)] \qquad (3.19)$$

It follows from Eq. (3.19) that for a given F_2/F_1, stiffness is directly proportional to the absolute value of the tensile force F_1. Since a higher stiffness value is usually preferable, the highest tensile forces allowable by the strength of arms 1 and 2, by the joint bearings, by the available hydraulic pressures (for hydraulically driven actuators), etc., have to be assigned. Of course, the overall stiffness of the actuator system would be determined, in addition to Eq. (3.19), by the stiffness (compliance) of the other components, such as hydraulic cylinders and piping (see Chapter 6). As a result, an increase of tensions F_1 and F_2 is justified only to the point at which an increase of stiffness k by Eq. (3.19) would still result in a meaningful increase in the overall stiffness.

Advantages of nonlinear vibration isolators are addressed in Article 1 and in Rivin [2], and are briefly described above. If nonlinear vibration isolators are used directly, their stiffness self-adjusts in accordance with the applied weight load. However, nonlinear vibration isolators and other resilient mounts can also be used with preload. Figure 3.17a [2] shows an adjustable mount for an internal combustion engine 4 (only one mounting foot is shown). The adjustable mount is composed of two nonlinear isolators 1, 2, (e.g., of Fig. 3.8 design), preloaded by bolt 5.

It is obvious that preloading of one isolator (e.g., 2 in Fig. 3.17a) would result in a parallel connection of bolt 5 and isolator 2, thus resilience of isolator 2 would be lost. When two isolators 1 and 2 are used, mounting foot 4 of the engine is connected with foundation structure 3 via resilient elements of isolators 1 and 2, and the stiffness of bolt 5 plays an insignificant role in the overall stiffness breakdown for the combined mount. However, tightening of the bolt increases loading of both isolators, and thus is moving the "working points" along their load-deflection characteristics. If the load-deflection characteristic is of a hardening type (the most frequent case), then tightening of bolt 5 would result in the increasing stiffness. A typical case is shown in Fig. 3.17b. It is important to note that a preloaded combination of two identical nonlinear resilient elements, as in Fig. 3.17a, becomes a linear compound element of adjustable stiffness. When foot 4 in Fig. 3.17a is displaced downwards by an increment Δx, the load on the lower isolator 2 is increasing by ΔP_2 while load on the upper isolator 1 is decreasing by ΔP_1. If isolators 1 and 2 have identical hardening load-deflection

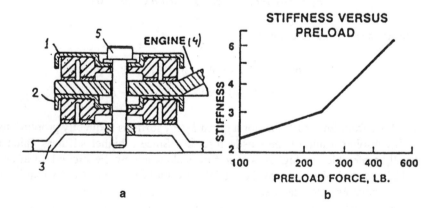

Figure 3.17 Variable stiffness vibration isolator composed of two nonlinear (constant natural frequency) isolators.

characteristics, then stiffness of the upper isolator is decreasing by Δk_1. Since the initial loading P_0 and the initial stiffness k_0 of both isolators are the same, $\Delta P_1 = \sim\Delta P_2$ and $\Delta k_1 = \sim\Delta k_2$, thus the total stiffness

$$k_1 - \Delta k_1 + k_2 + \Delta k_2 = 2k_0 - \Delta k_1 + \Delta k_2 = \sim 2k_0 = \text{constant} \qquad (3.20)$$

Thus, preloading of the properly packaged pair of nonlinear isolators allows one to construct an isolator whose stiffness can be varied in a broad range. Such an isolator can serve as an output element of an active stiffness control system or a vibration control system. Figure 3.18 shows test results of a similar system in which radially loaded rubber cylinders were used as the nonlinear elements. Just two turns of the preloading bolt changed the natural frequency from 20 to 60 Hz, which corresponds to the ninefold increase in stiffness. The similar effect is employed for tuning the dynamic vibration absorber for the chatter-resistant boring bar in Article 4.

It is important to note that preloading of structural joints (such as in bearings and interfaces; see Chapter 4), as well as preloading of elastomeric elements, may result in reduction of damping.

Use of internal preload (i.e., the strength-to-stiffness transformation concept) for enhancement of bending stiffness is addressed in Chapter 7.

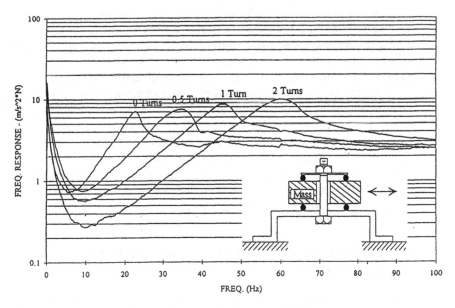

Figure 3.18 Use of preloaded rubber cylinders for tuning vibration control devices.

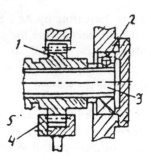

Figure 3.19 System with variable stiffness (bearing 2 is fit on splined shaft 3).

3.4 SOME EFFECTS CAUSED BY VARIABLE STIFFNESS

Although nonlinear systems are characterized by stiffness varying with the loading conditions, in some mechanical systems stiffness changes with the changing geometry. Although this effect is typical for linkage mechanisms in which the links are not very stiff, it can also be present and can create undesirable effects in many other basic mechanical systems.

Figure 3.19 shows a segment of a gearbox in which ball bearing 2 is mounted on a splined shaft 3. Since races (rings) of antifriction bearings are usually very thin, they easily conform with the profile of the supporting surface. Thus, the inner race of bearing 2 has different local effective stiffness in the areas supported by the splines and in the areas corresponding to spaces (valleys) between the splines. Even minute stiffness variations of the inner race under balls during shaft rotation may cause undesirable, even severe, parametric vibrations.

Similar effects may be generated by a gear or a pulley having a nonuniform stiffness around its circumference and tightly fit on the shaft. Figure 3.20 shows an oscillogram of vibration of the housing of a gear reducer caused by a five-spoke gear which was interference fit on its shaft [12]. Effective stiffness of the gear rim is very different in the areas supported by the spokes and in the areas between the spokes. This resulted in a beat-like pattern of vibration amplitudes varying in the range of 3:1. Such amplitude variation can cause intensive and annoying noise and other damaging effects.

3.5 SYSTEMS WITH MULTIPLE LOAD-CARRYING COMPONENTS

There are numerous cases in which the loads acting on a mechanical system are unequally distributed between multiple load-carrying components. In these

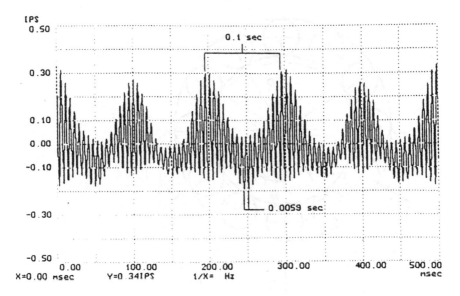

Figure 3.20 Vibration caused by a five-spoke gear that is interference fit on its shaft.

statically indeterminate systems, evaluation of the overall stiffness becomes more involved. Three systems described below are typical examples of the systems with multiple load-carrying components. Understanding of these systems may help in analyzing other systems with multiple load-carrying components. The system analyzed in Section 3.5.1 (antifriction bearing with multiple rolling bodies) is characterized by high rigidity of the supporting surfaces (inner and outer races), which are reinforced by the shaft and housing, respectively. The spoked bicycle wheel analyzed in Section 3.5.2 has a relatively low stiffness of the rim. These differences lead to very different patterns of load/deformation distribution.

3.5.1 Load Distribution Between Rolling Bodies and Stiffness of Antifriction Bearings [1]

Radial load R applied to an antifriction bearing is distributed nonuniformly between the rolling bodies (Fig. 3.21). In the following, all rolling bodies are called "balls," even though all the results would also apply to roller bearings. If the bearing is not preloaded, the load is applied only to the balls located within the arc, not exceeding 180 deg. The most loaded ball is one located on the vector of the radial force R (the central ball). The problem of evaluating loads acting on each rolling body is a statically indeterminate one. The balls symmetrically located relative to the plane of action of R are equally loaded. The force acting

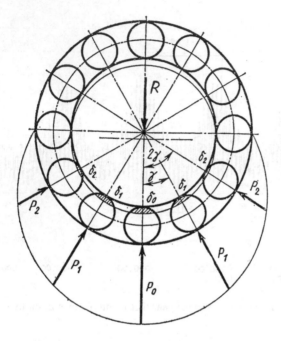

Figure 3.21 Model of load distribution between rolling bodies in antifriction bearing.

on the most highly loaded central ball is P_0. On the balls located at a pitch angle γ to the load vector is P_1, on balls located at 2γ the vector is P_2, at angle $n\gamma$ the vector is P_n, and so on. Here n is one-half of the balls located in the loaded zone. It is assumed for simplicity of the derivation that the balls are symmetrical relative to the plane containing the force R. The equilibrium condition requires that the force R is balanced by reactions of the loaded balls, or

$$R = P_0 + 2P_1 \cos \gamma + 2P_2 \cos 2\gamma + \cdots + 2P_n \cos n\gamma \qquad (3.21)$$

In addition to Eq. (3.21) describing static equilibrium, equations for deformations should be used in order to solve the statically indeterminate problem. Since the bearing races are supported by the housing (the outer race) and by the shaft (the inner race), their bending deformations can be neglected. It is assumed that there is no radial clearance in the bearing. In such case, it can be further assumed that the radial displacements between the races at each ball, caused by contact deformations of the balls and the races, are equal to projections of the total displacement δ_0 of the inner race along the direction of R on the respective radii

$$\delta_1 = \delta_0 \cos \gamma; \quad \delta_2 = \delta_0 \cos 2\gamma; \quad \delta_i = \delta_0 \cos i\gamma \tag{3.22}$$

where i = number of a ball.

For ball bearings, total contact deformation δ of the balls and the races under force P is (see Chapter 4)

$$\delta = cP^{2/3} \tag{3.23}$$

where c = a constant.

Substituting expressions (3.23) into (3.22), we arrive at

$$P_1 = P_0 \cos^{3/2} \gamma; \quad P_2 = P_0 \cos^{3/2} 2\gamma; \cdots P_i = P_0 \cos^{3/2} i\gamma \tag{3.24}$$

Substitution of Eq. (3.24) into equilibrium Eq. (3.21) results in

$$R = P_0 \left(1 + 2 \sum_{i=1}^{i=n} \cos^{3/2} i\gamma \right) \tag{3.25}$$

From this equation P_0 can be determined. It is convenient to introduce a coefficient k

$$k = \frac{z}{1 + 2 \sum_{i=1}^{i=n} \cos^{5/2} i\gamma} \tag{3.26}$$

then

$$P_0 = kR/z \tag{3.27}$$

where z = total number of balls. Knowledge of P_0 allows one to determine the radial deformation and stiffness of the bearing by using Eq. (3.23).

For bearings with the number of balls $z = 10 - 20$, $k = 4.37 \pm 0.01$. If the bearing has a clearance, then the radial load is accommodated by balls located along an arc lesser than 180 deg., which results in the load P_0 on the most loaded ball being about 10% higher than given by Eq. (3.27). Because of this, for single row radial ball bearings it is assumed that $k = 5$ and $P_0 = 5R/z$.

In spherical double-row ball bearings there is always some nonuniformity of radial load distribution between the rows of balls. To take it into consideration, it is usually assumed that $P_0 = 6R/z \cos \alpha$, where α = angle of tilt of contact normals between the balls and the races.

Radial force on each ball for the radial loading in angular contact ball bearings is greater than in radial ball bearings by a factor $1/\cos \beta$, where β = contact angle between the balls and the races.

For roller bearings the solution is similar, but contact deformations of the rollers and the races are approximately linear

$$\delta = c_1 P \tag{3.28}$$

where c_1 = a constant (see Chapter 4). Similarly to ball bearings

$$P_0 = kR/z$$

And similarly

$$k = \frac{z}{1 + 2 \sum\limits_{i=1}^{i=n} \cos^2 i\gamma}$$

For roller bearings having $z = 10$ to 20, average value of $k = 4$, considering radial clearance, should be increased to $k = 4.6$. For double row bearings, $k = 5.2$, to take into consideration nonuniformity of load distribution between the rows of rollers.

Load distribution between the rolling bodies can be made more uniform (thus reducing P_0 and δ_0, i.e., enhancing stiffness) by modification of the bore in the housing into which the bearing is fit. The bore should be shaped as an elliptical cylinder elongated in the prevailing direction of the radial load.

If the bearing is preloaded, each rolling body is loaded by a radial force P_{pr} caused by the preload, even before the radial load R is applied. As a result, all rolling bodies (along the 360 degree arc) are participating in the loading process, with the bodies in the lower half of Fig. 3.21 experiencing increase of their radial loading, and the bodies in the upper half experiencing reduction of their radial loading. The process is very similar to the preloading process of a belt drive (see Section 3.3). Such pattern of the load sharing will continue until at least one of the rolling bodies in the upper 180 degree arc becomes unloaded. The deformation of a preloaded bearing under a radial force R can be analyzed using the same approach as for bearings not preloaded, but derivations become more complex since the expression for deformation of the ball caused by forces acting on each bearing becomes, instead of Eq. (3.23)

$$\delta_i = c(P_{pr} \pm P_i)^{2/3} \tag{3.29}$$

3.5.2 Loading of Bicycle Wheels [14]

The bicycle wheel is subjected to high static (weight of the rider) and dynamic (inertia on road bumps, torques for acceleration, braking, and traction, etc.)

forces. Its predecessor—a wagon wheel (Fig. 3.22)—has relatively strong wooden spokes and rim. The spokes in the lower part of the wheel accommodate (by compression) the loads transmitted to them by the rim and the hub, just like the bearings discussed earlier. The bike's wheel needed to be much lighter, thus wood was replaced by high strength metals that allowed one to dramatically reduce cross sections of both the spokes and the rim. In fact, the spokes possess necessary strength while having very small cross sections equivalent to thin wire (Fig. 3.23). However, such thin spokes cannot withstand compression due to buckling. The solution was found in prestressing the spokes in tension, so that the tensile preload force on each spoke is higher than the highest compressive force to be applied to the spoke during ride conditions. With such a design, the spoke would never loose its bending stiffness if the specified loads are not exceeded. However, the lateral stiffness of the spoke is decreasing when the tension is reduced by high radial compressive forces (see Section 7.4.1). In such a condition, a relatively small lateral force caused by turning, for example, may lead to collapse of the wheel.

Contrary to the antifriction bearings which were discussed earlier, the rim is a relatively compliant member. Since the total force applied to the rim by the prestressed spokes can be as great as 5000 N (1100 lb), the rim is noticeably compressing, thus reducing the effective preload forces on the spokes. Pressurized tires also apply compressive pressures to the rim equivalent to as much as

Figure 3.22 Spoked wagon wheel.

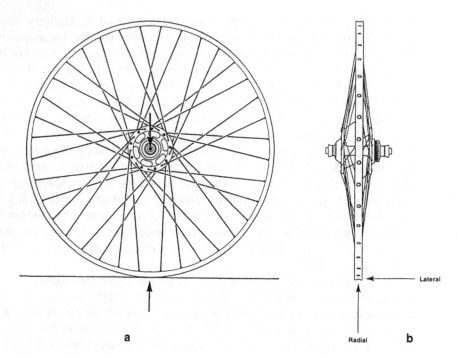

Figure 3.23   Spoked bicycle wheel: (a) front view; (b) side view.

7–15% of the spoke load. The tension of the spokes changes due to driving and braking torques, which cause significant pulling and pushing forces on the oppositely located spokes. Combination of the vertical load with torque-induced pushing and pulling loads results in local changes in spoke tension, which appear as waves on the rim circumference. These effects are amplified by the spoke design as shown in Figs. 3.23: to enhance the torsional stiffness, the spokes are installed not radially but somewhat tangentially to the hub (Fig. 3.23a); to enhance lateral stiffness and stability of the wheel, the spokes are installed in a frustoconical manner, not in a single plane (Fig. 3.23b).

When a vertical load is applied to the wheel, the spokes in the lower part of the wheel are compressed (i.e., their tension is reduced).The spokes in the upper half of the wheel are additionally stretched, but even the spokes in the midsection of the wheel are increasing their tension since the rim becomes somewhat oval. This ovality is not very significant—the increase of the horizontal diameter is about 4% of the deformation at the contact with the road—but it has to be considered. Since the rim is not rigid, its flattening at the contact with the road leads to reduction of effective stiffness of the wheel. It is in agreement with a suggestion in Section 3.5.1 to enhance stiffness of a bearing by increasing the vertical diameter of the bore.

It is important to note that performance loads (radial, torque, braking, and turning loads) cause significant distortion of the rim (Fig. 3.24), which, in turn, results in very substantial deviation from load distribution for an idealized model (like the bearing model in Fig. 3.21).

While the problem of load distribution in and deformations of the real-life spoked bicycle wheel is extremely complex, there is a closed-form analytical solution for load and bending moment distribution along the wheel rim and for deformations of the rim with some simplifying assumptions [15]. These assumptions are as follows: the road surface is flat and rigid; the spokes are radial and are coplanar with the rim; and the number of spokes n is so large that the spokes can be considered as a continuous uniform disc of the equivalent radial stiffness. With these assumptions, the problem becomes a problem of a radially loaded ring with elastic internal disc. Every point of the rim would experience radial

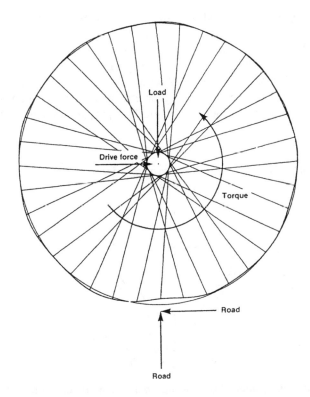

Figure 3.24 Distortion of bicycle wheel rim under radial and torque loads.

reaction force from the spokes proportional to deflection w at this point. There are $n/2\pi R$ spokes per unit length of the rim circumference (R = radius of the wheel). Deformation w and bending moment M along the rim circumference due to vertical force P acting on the wheel from the road are

$$w = \frac{PR^3}{4\alpha\beta E_1 I}\left(\frac{2\alpha\beta}{\pi a^2} - A \cosh \alpha\phi \cos \beta\phi + B \sinh \alpha\phi \sin \beta\phi\right) \qquad (3.30)$$

$$M = -\frac{PR}{2}\left(\frac{1}{\pi a^2} + A \sinh \alpha\phi \sin \beta\phi + B \cosh \alpha\phi \cos \beta\phi\right) \qquad (3.31)$$

The force acting on a spoke is obviously

$$P_s = \frac{E_2 F}{R} w \qquad (3.32)$$

Here E_1 and I = Young's modulus and cross-sectional moment of inertia of the rim, respectively; E_2, F, and $l \approx R$ = Young's modulus, cross-sectional area, and length of spoke; $a = (R^2 n/2\pi)(E_2 F/E_1 I)$; $\alpha = \sqrt{(a-1)/2}$; $\beta = \sqrt{(a+1)/2}$; and angular coordinate ϕ of a point on the rim is counted from the top point of the wheel. Formulas (3.30)–(3.32) allow one to evaluate, at least in a first approximation, the influence of various geometric and material parameters on force and bending moment distributions. Figure 3.25 [15] shows these distri-

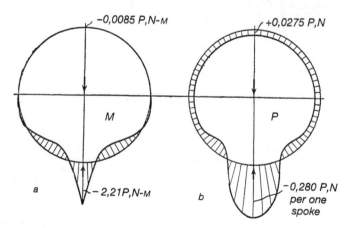

Figure 3.25 Calculated distribution of (a) bending moment and (b) radial forces along the circumference of a bicycle wheel.

butions for $R = 310$ mm; $I = 3000$ mm^4; $n = 36$; spoke diameter $d = 2$ mm ($F = 3.14$ mm); $E_1 = E_2 = 2 \times 10^5$ N/mm^2 (steel). After the maximum static and dynamic forces on the wheel are estimated/measured, the data from Fig. 3.25b can be used to specify the necessary initial tension of each spoke P_t. This tension should be safely greater than the maximum possible compressive force acting on the spoke, which for the parameter listed above is $0.280\ P_{max}$.

3.5.3 Torsional Systems with Multiple Load-Carrying Connections

Such connections are typical for various types of power transmission couplings. Their performance is analyzed in Article 2, where expressions for both torsional and lateral compensating stiffness are derived. An important feature of such connections with less than six radial protrusions (''spider legs'') is the fact that they generate dynamic loading on the connected shafts even when the misalignment vector is constant.

REFERENCES

1. DenHartog, J.P., Mechanical Vibrations, McGraw-Hill, New York, 1952.
2. Rivin, E.I., "Passive Engine Mounts—Directions for Future Development," SAE Transactions, 1985, pp. 3.582–3.592.
3. Freakley, P.K., Payne, A.R., Theory and Practice of Engineering with Rubber, Applied Science Publishers, London, 1978.
4. Rivin, E.I., Lee, B.-S., "Experimental Study of Load-Deflection and Creep Characteristics of Compressed Rubber Components for Vibration Control Devices," ASME J. Mechanical Design, 1994, Vol. 116, No. 2, pp. 539–549.
5. Lee, B.-S., Rivin, E.I., "Finite Element Analysis of Load-Deflection and Creep Characteristics of Compressed Rubber Components for Vibration Control Devices," ASME J. Mechanical Design, 1996, Vol. 118, No. 3, pp. 328–336.
6. Rivin, E.I., "Resilient Anti-Vibration Support," U.S. Patent 3,442,475.
7. Rivin, E.I., "Properties and Prospective Applications of Ultra-Thin-Layered Rubber-Metal Laminates for Limited Travel Bearings," Tribology International, 1983, Vol. 16, No. 1, pp. 17–25.
8. Lavendel, E.E. (ed.), Vibration in Engineering, Vol. 4, 1981, Mashinostroenie Publishing House, Moscow [in Russian].
9. Rivin, E.I., "Cost-Effective Noise Abatement in Manufacturing Plants," Noise Control Engineering Journal, 1983, Nov.-Dec., pp. 103–117.
10. Rivin, E.I., Dynamics of Machine Tool Drives, Mashinostroenie Publishing House, Moscow, 1996 [in Russian].
11. Rivin, E.I., Mechanical Design of Robots, McGraw-Hill, New York, 1988.

12. Taylor, J., "Improvement in Reliability and Maintainability Requires Accurate Diagnosis of Problems," P/PM Technology, 1995, No. 12, pp. 38–41.
13. Reshetov, D.N., Machine Elements, Mashinostroenie Publishing House, Moscow, 1974 [in Russian].
14. Brandt, J., The Bicycle Wheel, Avocet Inc., Menlo Park, NJ, 1996.
15. Feodosiev, V.I., Selected Problems and Questions on Strength of Materials, Nauka Publishing House, Moscow, 1973 [in Russian].

4

Contact (Joint) Stiffness and Damping

4.1 INTRODUCTION

When two components are in contact, their actual area of contact is generally very small. When components with substantially different curvature radii are compressed together (such as spheres, cylinders, and toruses with flat surfaces), the contact area is small (ideally a point or a line) because of their geometry (Figs. 4.1a and b). Analysis of contact deformations in the contact between streamlined parts with initial point or line contact area can be performed using the Hertz formulas. Formulas for some important cases are given in Table 4.1 [1,2]. Formulas in Table 4.1 were derived with the following assumptions: stresses in the contact zone are below the yield strength; contact areas are small relative to dimensions of the contacting bodies; and contact pressures are perpendicular to the contact areas. These formulas also consider structural deformations of the contacting bodies.

When the contact surfaces are flat or shaped with identical or slightly different curvature radii, the actual contact area is small because of the roughness and waviness of the contacting surfaces (Fig. 4.1c). Because only the outstanding areas of surface asperities or waves are in contact, under small loads only a small fraction of one percent of the nominal contact surface is in actual contact (this fact causes also loss of thermal conductivity between the contacting bodies). Accordingly, contact deformations could be very substantial—commensurate with or even exceeding the structural deformations. It is especially true for high-precision mechanical systems, as well as for general-purpose machines that have numerous joints. The largest contact deformations are associated with joints providing for the relative motion of the components, especially while they are

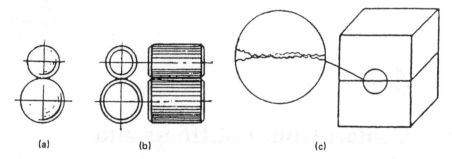

(a)   (b)   (c)

Figure 4.1 Important cases of contact deformations: (a, b) compression of surfaces with substantially different curvature radii; (c) compression of surfaces with nominally conforming curvature radii.

experiencing low loads (e.g., at the final positioning stages of moving components in machine tools and robots when the highest accuracy and the highest stiffness are required). The role of contact deformations in machine tools can be illustrated by an example: contact deformation in typical sliding guideways of a machine tool is about 10^{-11} m^3/N (10 μm per 1 MPa or 400 μinch per 150 psi contact pressure). It is equivalent to a compression deformation of a 1 m (40 in.) long cast-iron bar under the same pressure. Contact deformations between conforming surfaces are described in Section 4.2.

An intermediate case is contacting between quasi-conforming surfaces (surfaces with slightly different curvatures). In such cases, the total deformations are determined by a combination of structural deformations of the contacting bodies and of deformations of the surface asperities. This case is addressed for the cylindrical connections with clearances in Section 4.4.1a.

Due to the nonlinearity of contact deformations, fast Fourier transform (FFT)-based dynamic experimental techniques for mechanical systems, in which contact deformations play a substantial role, should be used with caution. In many cases, more reliable results can be obtained by static testing (see Section 7.1).

4.2 CONTACT DEFORMATIONS BETWEEN CONFORMING SURFACES

Contact deformations between conforming surfaces influence to a significant degree:

1.   Vibration and dynamic loads
2.   Load concentration and pressure distribution in contact areas

Table 4.1 Displacements in Point and Line Contacts

Sketch of contact	Auxiliary coefficients		Approach of contacting bodies
	A	B	
Sphere and plane 	$(1/2R)$	$(1/2R)$	$178\sqrt[3]{\dfrac{P^2}{R}\left(\dfrac{1-\nu_1^2}{E_1}+\dfrac{1-\nu_2^2}{E_2}\right)^2}$
Sphere and cylinder 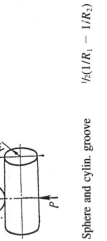	$1/2\,R_1$	$\tfrac{1}{2}(1/2R_1+1/2R_2)$	$1.41n_\Delta\sqrt[3]{P^2\dfrac{2R_2+R_1}{R_1R_2}\left(\dfrac{1-\nu_1^2}{E_1}+\dfrac{1-\nu_2^2}{E_2}\right)^2}$
Sphere and cylin. groove 	$\tfrac{1}{2}(1/R_1-1/R_2)$	$1/2R_1$	$1.41n_\Delta\sqrt[3]{P^2\dfrac{2R_2-R_1}{R_1R_2}\left(\dfrac{1-\nu_1^2}{E_1}+\dfrac{1-\nu_2^2}{E_2}\right)^2}$

Table 4.1 Continued

Sketch of contact	Auxiliary coefficients		Approach of contacting bodies
	A	B	
Sphere and thor. groove	$\tfrac{1}{2}(1/R_1 - 1/R_2)$	$\tfrac{1}{2}(1/R_1 + R_3)$	$1.41 n_\Delta \sqrt[3]{P^2 \left(\dfrac{2}{R_1} - \dfrac{1}{R_2} + \dfrac{1}{R_3} \right) \left(\dfrac{1 - \nu_1^2}{E_1} + \dfrac{1 - \nu_2^2}{E_2} \right)^2}$
Roller bearing	$\tfrac{1}{2}(1/R_2 - 1/R_4)$	$\tfrac{1}{2}(1/R_1 + 1)$	$1.41 n_\Delta \sqrt[3]{P^2 \left(\dfrac{2}{R_1} + \dfrac{1}{R_2} + \dfrac{1}{R_3} - \dfrac{1}{R_4} \right) \left(\dfrac{1 - \nu_1^2}{E_1} + \dfrac{1 - \nu_2^2}{E_2} \right)^2}$

Perpendicular cylinders

$1/2R_2$ $1/2R_1$

$$1.41 n_\Delta \sqrt[3]{P^2 \frac{2R_2 + R_1}{R_1 R_2}\left(\frac{1-\nu_1^2}{E_1} + \frac{1-\nu_2^2}{E_2}\right)^2}$$

Parallel cylinders

$\frac{1}{2}(1/R_1 + 1/R_{23})$ —

$$\frac{2P}{\pi l}\left[\frac{1-\nu_1^2}{E_1}\left(\log_e \frac{2R_1}{b} + 0.41\right) + \frac{1-\nu_2^2}{E_2}\left(\log_e \frac{2R_2}{b} + 0.41\right)\right]$$

Roller between plates

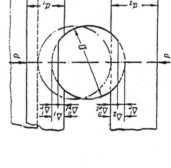

$$\Delta_r = 4\frac{P}{l}\frac{1-\nu_1^2}{\pi E}\left(\log_e \frac{2D}{b} + 0.41\right)$$

$$\Delta_{pl} = \frac{P}{l}\frac{1-\nu_2^2}{\pi E_2}\left(\log_e \frac{16d^2}{b^2} - 1\right)$$

$$b = 1.6\sqrt{\frac{P}{l}D\left(\frac{1-\nu_1^2}{E_1} + \frac{1-\nu_2^2}{E_2}\right)}$$

Δ_r = shrinking of roller

Δ_{pl} = denting of each plate

A/B	1.00	0.404	0.250	0.160	0.085	0.067	0.044	0.032	0.020	0.015	0.003
n_Δ	1.00	0.957	0.905	0.845	0.751	0.716	0.655	0.607	0.546	0.510	0.358

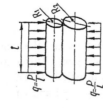

3.   Relative and absolute positioning accuracy of the units

The negative effect of contact deformations is obvious and is due to increased compliance in structures that is caused by the structural joints. Role of contact deformations in the overall breakdown of deformations in mechanical systems can be illustrated by the following examples. Effective torsional compliance of power transmission systems (gearboxes) is comprised of three major sources: torsion of the shafts (~30%); bending of the shafts (~30%); and contact deformations in the connections, such as keys and splines, and bearings (~40%) [3] (see also Chapter 6). Contact deformations in spindle units of machine tools are responsible for 30–40% of the total deformations at the spindle end. Contact deformations in carriages, cantilever tables, etc., constitute up to 80–90% of the total, and those in moving rams about 40–70% [4].

However, the positive effects of contact deformations also have to be considered. They include more uniform pressure distribution in joints and dissipation of vibratory energy (damping). Damping in joints between conforming surfaces is an important design feature in mechanical systems subjected to vibrations (see Section 4.7).

If the contact surfaces are perfectly flat (e.g., like contact surfaces of gage blocks) or if the curvature of one surface perfectly matches (conforms with) the curvature of its counterpart surface, and both surfaces have a high degree of surface finish, the magnitudes of contact deformations are insignificant. In a contact between two surfaces that are neither perfectly flat nor perfectly matching in curvature, and in which the average height of micro-asperities is comparable with amplitudes of waviness, contact deformations are caused by Hertzian deformations between the apexes of the contacting waves, by flattening of the contacting waves and by similar deformations of microasperities representing roughness, and by squeezing lubrication oil from between the asperities.

There has recently been some progress in the computational analysis of interactions between nonideal contacting surfaces. However, because of the extremely complex nature of the surface geometry, the reliable design information is still based on empirical (experimental) data. For accurately machined and carefully matched flat contact surfaces of a relatively small nominal joint area (100–150 cm^2 or 15–23 sq. in.) contact deformations between cast iron and/or steel parts are nonlinear (Fig. 4.2) and can be expressed as

$$\delta = c\sigma^m \qquad (4.1)$$

where σ = average contact pressure in MPa, and δ = contact deformation in micrometers (10^{-6}m; 1 μm = 40 μin). This expression describes a nonlinear hardening load-deflection characteristic with a degree of nonlinearity represented by the exponent m. For contacting steel and cast-iron surfaces, m = ~0.5. The

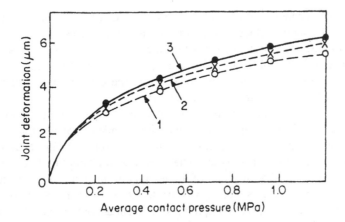

Figure 4.2 Load-deflection characteristics for flat deeply scraped surfaces (overall contact area 80 cm^2). (1) No lubrication, (2) lightly lubricated (oil content 0.8×10^{-3} g/cm^2), and (3) richly lubricated (oil content 1.8×10^{-3} g/cm^2).

contact deformation coefficient c is 5.0–6.0 for deep scraping, 3.0–4.0 for medium-deep scraping, 0.6–1.0 for fine scraping, 0.5–1.0 for fine turning and grinding, and 0.2 for lapping [4]. As can be seen from Fig. 4.2, lubrication results in larger deformations. Since formula (4.1) is empirical, it is approximate and parameters m and c may fluctuate.

Contacts between cast-iron or steel and plastic materials are characterized by m = ~0.33 in Eq. (4.1) and by larger deformations, and c = 10–18 for average scraped surfaces.

For very small joints (contact area 2–3 cm^2 or ~0.3–0.5 sq. in.) the waviness is hardly detectable and, accordingly, contact stiffness is about five times higher (or deformations δ for the same pressure σ are approximately five times smaller) [4]. Such small joints are typical for key connections [3,5], index pins, dead stops, etc.

For larger contacting surfaces, deformations are larger because of difficulties in matching the surfaces. For large contacting surfaces, the exponent in (4.1) is closer to 1.0 (an almost linear load-deflection curve) because of higher local pressures and because of the more pronounced role of structural deformations of the contacting components. Out-of-flatness of the order of 10–15 μm leads to 2–2.5 times higher values of the coefficient c in (4.1).

Contact stiffness is very sensitive to the method of surface finishing. There is some data showing that ball burnishing can increase contact stiffness by a factor of 1.5–2, and additional diamond smoothing can additionally increase it by about the same factor. Orthogonal positioning of machining traces (lay lines) on contacting surfaces increases contact stiffness by about 40% as compared with a joint between the same surfaces with parallel machining traces.

Contact stiffness for vibratory loads (dynamic stiffness) is the same as static stiffness if the joint is not lubricated. However, for lubricated joints, because of viscous resistance to the fast squeezing of oil from the contact area, dynamic stiffness is about 50% higher than the static stiffness.

There were several studies on influence of thin film gaskets in the joints on their stiffness [6]. Study of gaskets 0.1 mm thick made of various materials (polyvinylchloride, polyethylene, aluminum and lead foils, rubber, etc.) has shown that use of the gaskets results in higher stiffness and damping. For example, use of a neoprene gasket in joints of a knee-type milling machine resulted in increase of dynamic stiffness from 1.5×10^7 to 2.6×10^7 N/m (83×10^3 to 140×10^3 lbs/in.), and increase of damping (as a percentage of the critical damping) from 2.5% to 3.2%. It has been found also that in the studied range of joint pressures 2.5–6.0 MPa (420–900 psi), stiffness does not significantly depend on the load, i.e., the joints with gaskets exhibit an approximately linear load-deflection characteristic. There are indications that a similar stiffening effect is observed when the contacting surfaces are coated with a soft metal coating such as tin, lead, indium, silver, and gold. This effect of the thin, relatively soft gaskets and coatings can be explained by their equalizing action, since they compensate effects of microasperities and also, to some degree, of waviness.

Because the contact deformations are usually characterized by nonlinear hardening load-deflection characteristics, the contact stiffness can also be enhanced by increasing contact pressures. This can be achieved by *increasing* the load on the contact and/or by *reducing* the surface area of the joint. The incremental contact deformations of reasonably well machined flat surfaces ($R_a \leq 4$ μm) can be neglected for contact pressures above 100 MPa.

4.3   USE OF CONTACT STIFFNESS DATA IN STRUCTURAL ANALYSIS

For design computations, small increments of contact deflections can be considered as proportional to the contact pressure

$$\delta = k(\sigma)\sigma \qquad (4.2)$$

with k being dependent on the pressure magnitude. Displacements of rigid parts such as brackets and massive carriages can be computed by considering their rotation and linear displacement relative to the supporting components due to contact deformations in their connections.

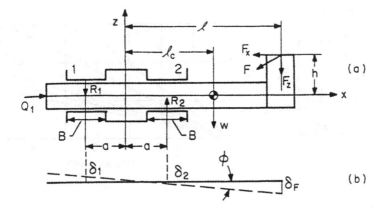

Figure 4.3 (a) Rigid ram in a sliding joint and (b) deflected position of its axis.

An example in Fig. 4.3(a) represents a rigid ram that is sliding in a prismatic joint consisting of two support areas 1 and 2. External force F is applied to the end with an offset h, and actuating force is $Q_1 = F_x + (R_1 + R_2)f$, where $f = $ friction coefficient. Reaction forces $R_{1,2}$ acting on support areas can be found from equations of static equilibrium as

$$R_1 = \frac{W(l_c - a) + F_z(l - a) - F_x h}{2a} \tag{4.3}$$

$$R_2 = \frac{Wl_c + F_z(l + a) - F_x h}{2a} \tag{4.4}$$

The contact pressures and deformations in the support areas are

$$\sigma_1 = \frac{R_1}{aB}; \qquad \sigma_2 = \frac{R_2}{aB} \tag{4.5}$$

$$\delta_1 = k(\sigma_1)\sigma_1; \qquad \delta_2 = k(\sigma_2)\sigma_2 \tag{4.6}$$

$$\delta_F = \delta_1 - \phi(l + 2a) = \delta_1 - \left(\frac{\delta_1 + \delta_2}{2a}\right)(l + 2a)\delta \tag{4.7}$$

Here B = width of the ram, δ_F = deflection at the point of application of force F, σ_1 and σ_2 = contact pressures in support areas 1 and 2, δ_1 and δ_2 = contact deformations in the areas 1 and 2, and ϕ = angular deflection of the ram caused by contact deformations (Fig. 4.3b).

A general case of a joint between two rigid parts involves loading with a

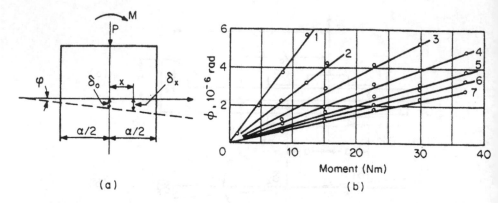

Figure 4.4 (a) Flat joint loaded with compressive force P and moment M and (b) its torque versus angular deflection characteristics at various P. Contact area 75 cm^2, surface scraped. Average pressure: (1) $\sigma = 0.055$ MPa, (2) 0.11 MPa, (3) 0.22 MPa, (4) 0.42 MPa, (5) 0.68 MPa, (6) 1.0 MPa, and (7) 1.35 MPa.

central force P and a moment M (Fig. 4.4a) [3]. Equilibrium conditions can be written as

$$B\int_{-a/2}^{a/2} \sigma_x dx = P \tag{4.8a}$$

$$B\int_{-a/2}^{a/2} \sigma_x x dx = M \tag{4.8b}$$

where B = width of the joint and σ_x = contact pressure at a distance x from the line of action of force P. From Eq. (4.1), $\sigma_x = (\delta_x/C)^{1/m}$, where $\delta_x = \delta_0 + \phi x$ is the elastic displacement between the contacting surfaces in the cross section x, δ_0 = elastic displacement under the force P, and ϕ = angular displacement in the joint.

It was shown by Levina and Reshetov [4] that resolving Eqs. (4.8a) and (4.8b) gives for the following for $m = 0.5$:

$$\frac{\phi a}{\delta_0} \cong 6\frac{M}{Pa} \qquad \delta_0 \cong c\sigma^{0.5} \tag{4.9}$$

and for $m = 0.33$

$$\frac{\phi a}{\delta_0} \cong 6\frac{M}{Pa} \qquad \delta_0 \cong c\sigma^{0.33} \tag{4.10}$$

where $\sigma = P/aB$.

The maximum and minimum values of the contact pressure for $m = 0.5$ are

$$\sigma_{max} = \frac{\delta_0^2}{c^2}\left(1 + \frac{\phi a}{2\delta_0}\right)^2 \cong \sigma\left(1 + \frac{3M}{Pa}\right)^2 \qquad (4.11a)$$

$$\sigma_{min} = \frac{\delta_0^2}{c^2}\left(1 - \frac{\phi a}{2\delta_0}\right)^2 \cong \sigma\left(1 - \frac{3M}{Pa}\right)^2 \qquad (4.11b)$$

For $m = 0.33$

$$\sigma_{max} = {\sim}\sigma[1 + 2(M/Pa)]; \qquad \sigma_{min} = {\sim}\sigma[1 - 2(M/Pa)] \qquad (4.12)$$

From the expressions for σ_{min} in Eqs. (4.11) and (4.12), it can be concluded that the joint "opens up" if $M_{max} > Pa/3$ for $m = 0.5$ or if $M_{max} > Pa/2$ for $m = 0.33$.

Equations (4.9) and (4.10) show that for a constant average pressure σ, angular deflection ϕ is proportional to moment M, or a *nonlinear joint preloaded with a force P has linear load-deflection characteristics for the moment loading up to* M_{max}.

An expression for the overall angular deflection for a rectangular joint can be easily derived from Eqs. (4.9) and (4.10) as

$$\phi \cong 12m\frac{\delta_0 M}{Pa^2} \qquad (4.13)$$

Substituting expressions for the moment of inertia and for the surface area of the joint, $I = Ba^3/12$ and $A = aB$ into Eq. (4.14)

$$\phi \cong m\frac{A\delta_0}{P}\frac{M}{I} \qquad (4.14)$$

Using the expression for δ_0 from Eqs. (4.10) and (4.11) in which $\sigma = P/A$, the following expression for ϕ can be written

$$\phi = k\frac{M}{I} \qquad (4.15)$$

where

$$k = cm\sigma^{m-1} \qquad (4.16)$$

It was proven and confirmed by experiments in Levina [4] that Eqs. (4.15) and (4.16) hold for $m = 1.0$, 0.5, and 0.33, not only for rectangular joints, but also for hollow rectangular, ring-shaped, and round joints. For $m < 1$, k is diminishing with increasing σ (or P). Thus, preloaded joints behave as *linear angular springs stiffening with increasing preload force P*. Figure 4.4b shows the correlation between the computations using Eq. (4.15) and the experiments [4].

Effect of the preload on angular (moment) stiffness of a flat joint can be illustrated on a simple model in Fig. 4.5. The system in Fig. 4.5a is composed of bar 1 supported by two identical nonlinear springs 2 and 3, $k_2 = k_3 = k$. Fig. 4.5b shows the (hardening) load-deflection characteristic of springs 2 and 3. When bar 1 is loaded by a vertical force $2P_a$ at the midpoint, stiffness of each spring is $k_2\,(P_a) = k_3(P_a) = k(P_a)$, and vertical stiffness of the system is $k_2 + k_3 = 2k(P_a)$. Thus, the system has a nonlinear stiffness in the vertical direction.

If the bar is loaded also with a moment M in addition to the force $2P_a$, this moment will be counterbalanced by two additional reaction forces R_{2M} and R_{3M} having the same magnitudes but opposite directions

$$R_{2M} = -\,R_{3M} = R_M = M/1$$

Due to these additional forces, the stiffness of spring 2 increases from $k(P_a)$ to $k(P_a + R_M) = k(P_a) + \Delta'k$, and stiffness of spring 3 decreases from $k(P_a)$ to

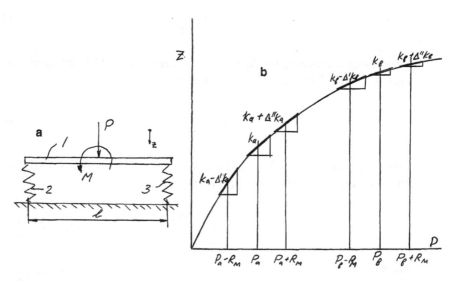

Figure 4.5 Model of a nonlinear joint loaded with a moment: (a) bar supported by nonlinear springs; (b) load-deflection characteristic of the springs.

$k(P_a - R_M) = k(P_a) - \Delta''k$ (see Fig. 4.5b). The moment-induced deformation of spring 2 is $\Delta z_2 = R_M/k_2 = R_M/[k(P_a) + \Delta'k]$, and of spring 3 it is $\Delta z_3 = -R_M/k_3 = -R_M/[k(P_a) - R_M\Delta''k]$.

The angular stiffness of the system is $k_\alpha = M/\alpha$, where α is the angular deflection (tilt) of bar 1 caused by moment M

$$\alpha = \frac{\Delta z_2 - \Delta z_3}{l} = \frac{1}{l}\left[\frac{R_M}{k(P_a) + \Delta'k} + \frac{R_M}{k(P_a) - \Delta''k}\right]$$

$$= \frac{R_M}{l}\frac{k(P_a) + \Delta'k + k(P_a) - \Delta''k}{[k(P_a) + \Delta'k][k(P_a) - \Delta''k]} \cong \frac{2R_M}{lk(P_a)} \tag{4.17}$$

since $\Delta'k \approx \Delta''k$ if $R_M \ll P_a$ (e.g., in the case of angular vibrations). Consequently,

$$k_{\alpha a} = M/\alpha = (R_M l)/[2R_M/lk(P_a)] = \sim \tfrac{1}{2}k(P_a)l^2 \tag{4.17a}$$

If the normal force is increasing to $P_b > P_a$ (Fig. 4.5b), then, analogously,

$$k_{\alpha b} = \sim\frac{1}{2}k(P_b)l^2 \tag{4.17b}$$

Equations (4.17a) and (4.17b) show that the angular stiffness is increasing with increasing P but does not depend on M, or is *linear relative to M*.

If both structural and contact stiffness are commensurate, the computational procedure is more complex. Since one of the joined components is usually much more rigid than another, such computations can be performed with the assumption that the compliant component is a beam or a plate on a continuous elastic foundation (bed). Even for discrete contact points (like in a case of a sliding ram supported by rollers), substitution of continuous elastic foundation for discrete resilient mounts is fully justified. Computations were reported [2] of the deflection diagrams for a beam with $EI = 3.6 \times 10^3$ Nm^2 and a concentrated load in the middle for two support conditions: (1) eight discrete resilient supports, each having stiffness $k = 1.37 \times 10^8$ N/m (corresponds to the stiffness of a steel roller with a 0.01 m diameter) and installed $a = 0.02$ m apart and (2) a distributed elastic foundation with stiffness $K = k/a = 0.68 \times 10^{10}$ N/m. The difference between the computed deflections of the two beams did not exceed 0.9%.

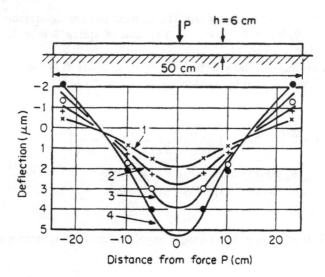

Figure 4.6 Deflection profile of a beam caused by joint compliance (lines are computation and points are experiment): (1) average pressure $\sigma = 0.04$ MPa, $k(\sigma) = 1.8 \times 10^{-11}$ m³/N; (2) $\sigma = 0.1$ MPa, $k = 1.0 \times 10^{-11}$ m³/N; (3) $\sigma = 0.2$ MPa, $k = 0.65 \times 10^{-11}$ m³/N; (4) $\sigma = 0.4$ MPa, $k = 0.45 \times 10^{-11}$ m³/N.

Figure 4.6 [3] shows a substantial effect of the joint compliance on a beam deflection as well as a good correlation between the experimental data and calculations based on considering the contact as an elastic foundation (bed). Such computations are important for sliding slender beams, such as robotic links in prismatic joints, sliding bars, and sleeves in machine tools. Elastic deflections of such components computed without considering joint deformations can be up to three times smaller than actual deflections, which can be determined using the elastic foundation technique.

Contact deformations in guideways are closely associated with local structural deflections of parts. Contact deformations can rise substantially (up to several times) because of distortions in guideways caused by local structural deformations.

From the information on contact stiffness given above, some recommendations on its enhancement can be made:

Better accuracy and surface finish to reduce values of c

Proper orientation of machining traces on the contacting surfaces

Adding thin plastic or soft metal gaskets or soft metal plating of the contacting surfaces

Using smaller contact areas and/or higher specific loads (pressures) to take advantage of hardening nonlinearity

Using preload to take advantage of hardening nonlinearity

Optimization of the shape of contacting surfaces to increase their moments of interia *I* (for cases of moment loading)

4.3.1 Factors Influencing Stiffness of Bolted Joints

Stiffness of structural connections in mechanical systems, such as in machine tool frames, is determined by contact deformations between the connected surfaces and, thus, by contact pressures. Frequently, the contact pressures are generated by tightening (preloading) bolts that pass through smooth holes in one part and engage with the threaded holes in the other part (base), thus creating the joint-tightening force between the bolt heads and the thread. Although some generic features of behavior of the bolted joints were described above in Eqs. (4.8)–(4.16), there are many factors to be considered in designing high-stiffness bolted connections. Some of these factors were studied experimentally [15].

The test rig is shown in Fig. 4.7. The studied joint between steel cantilever beam 4 and base 2 is tightened by two bolts 3. Base 2 is bolted to foundation 1 by several large bolts. External load *P* at the beam end creates moment loading of the joint and deflection *y* at the beam end is measured. Force *P* can be applied downward (stiffness $K_d = P/y$) or upward (stiffness K_u).

Three configurations of the beam were tested as shown in Figs. 4.8a–c: a beam with uniform thickness $h = 30$ mm along its length (Fig.4.8a); a beam with a thicker ($h = 37$ mm) cross section at the joint area (Fig. 4.8b); and a beam modified from Fig. 4.8b by recessing areas at the bolt holes in order to reduce the contact area (Fig, 4.8c). The base is shown in Fig. 4.8d. Both surface

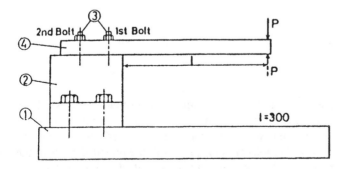

Figure 4.7   Test setup for study of bolted joints.

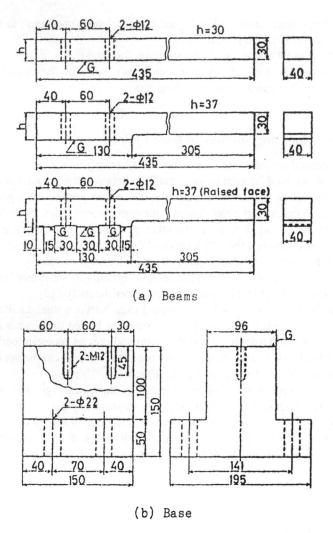

(a) Beams

(b) Base

Figure 4.8 Tested components of the bolted joint: (a) beams; (b) base.

roughness ($R_{\max}$) and waviness were within 2 μm (ground surfaces). Three sets of bolts were tested: short (length of shank $l = 30$ mm), medium ($l = 37$ mm), and long ($l = 44$ mm), but with the same length ($l = 14$ mm) of the threaded segment. Accordingly, the bolts are engaging with the threaded holes of the base in their upper, middle, and lower segments (Fig. 4.9). Loading of the bolts was

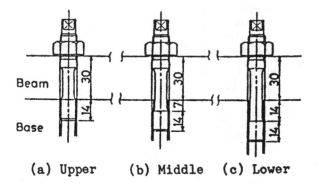

Figure 4.9 Tested variants of positioning the bolt thread relative to the base: (a) upper; (b) middle; (c) lower.

measured by strain gages, and pressure distribution along the joint area was measured by an ultrasonic probe.

Figure 4.10 shows stiffness values K_d and K_u as functions of the bolt preload force for three lengths of bolts. The much smaller values of K_u as compared with K_d are to be expected since the downward loading is associated with increasing contact pressures. However, the stiffness increase for longer bolts (both K_u and K_d) is not an obvious effect. It can be explained by a more uniform pressure distribution in the joint that is fastened by more resilient longer bolts, which also apply load to the base further from its surface (Fig. 4.11).

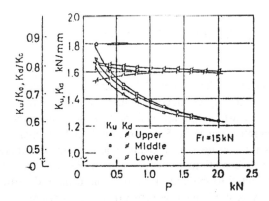

Figure 4.10 Effect of thread positioning on joint stiffness.

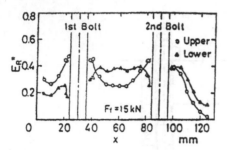

Figure 4.11   Effect of thread positioning on contact stress distribution in the joint.

Figure 4.12 illustrates incremental changes of preload force ΔF_t in the front ("first") and rear ("second") bolts as a result of applying upward (Figs. 4.12a and b) and downward (Figs. 4.12c and d) forces to the beam (short bolts). The data in Fig. 4.12, especially in Fig. 4.12a, emphasizes importance of high initial tightening force F_t in the bolts in order to prevent opening of the joint. Due to nonlinearly of the joint deformation, increments ΔF_t are much smaller (for the same P) when the initial F_t is larger.

Thickness of the beam in the joint area is a very important factor in determining the joint stiffness. Figure 4.13 shows significant improvements, especially for the most critical parameter K_u achieved by changing thickness of the beam in the joint area from $h = 30$ mm to $h = 37$ mm. Reduction of the joint area as in Fig. 4.8c results in higher contact pressures for the same preload force F_t.

This results in higher values of the joint stiffness, especially for low external forces, P, as shown in Fig. 4.14.

4.4   DISPLACEMENTS IN CYLINDRICAL/CONICAL CONNECTIONS

Cylindrical and conical connections have special importance for design since they are widely used both for sliding connections when the male and female parts are fit with a clearance (bearings, guideways), and for holding (clamped) connections when there is an interference fit. Since frequently only a small fraction of an elongated part is interacting with a cylindrical or conical (tapered) hole, both radial and angular deformations in the connection are important. The angular deformations are projected to the end of the part and usually determine the effective stiffness of the system. Contact deformations in cylindrical and tapered connections cannot be analyzed by Hertz formulas since those were derived for cases when the contact area is small in relation to curvature radii of the con-

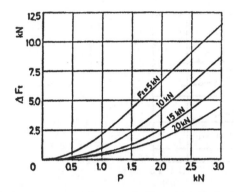

(a) Upward bending (the first bolt)

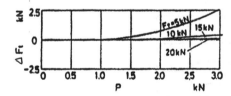

(b) Upward bending (the second bolt)

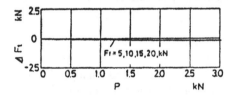

(c) Downward bending (the first bolt)

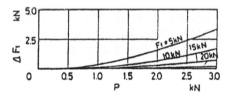

(d) Downward bending (the second bolt)

Figure 4.12 Effect of moment load applied to the joint on bolt preload.

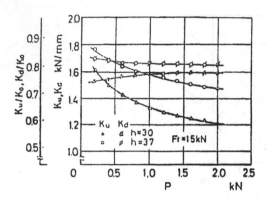

Figure 4.13 Effect of beam thickness on joint stiffness.

tacting bodies. These deformations cannot be directly analyzed using the basic notions of contact stiffness approach as described in Section 4.2 since they are influenced by the varying inclination between the contacting surfaces along the circumference of contact surfaces.

Contact deflections in cylindrical joints depend on the allowance (clearance or interference) magnitude. Since an increased clearance means a greater difference in curvature radii and, as a result, a reduction in the contact arc, it also leads to a steep increase in contact deformations.

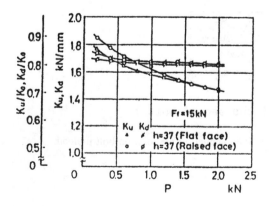

Figure 4.14 Effect of contact surface on stiffness of a bolted joint.

4.4.1 Cylindrical Connections

Connections with Clearance Fits [4]

Cylindrical connections with clearance fits are illustrated in Fig. 4.15a. Radii r_1 and r_2 of the hole and pin surfaces, respectively, are very close to each other; the clearance per diameter $\Delta = 2(r_1 - r_2)$ is several orders of magnitude smaller than r_1, r_2. Thus, the length of the contact arc is commensurate with the r_1, r_2. Although Hertz formulas are not applicable in such cases, they can be used as an approximation when loads are small and clearances are relatively large, so that

$$q/\Delta \le 500\text{--}1{,}000 \text{ N/m} \cdot \mu m = 0.5\text{--}1.0 \cdot 10^3 \text{ MN/m}^2 (70\text{--}140 \text{ lb/in.} \cdot \text{mil}) \quad (4.18$$

where q, in N/m (lb/in.), is load per unit length of the connection; and Δ, in μm (mil), is clearance (1 mil = 10^{-3} in.).

For a more general case of higher loads and/or smaller clearances, deformations of the housing in which the hole is made are important. Analysis of this problem was performed by D. N. Reshetov by cutting a cylinder from the housing and assuming that interaction of this cylinder with the housing is as shown in Fig. 4.15b as B (cosinusoidal load distribution along 180° arc). Fig. 4.16a shows ratio of maximum p_{max} to mean p_{av} contact pressures and contact angle φ_0 as functions of q/Δ and d_0/d (d_0 = external diameter of the cylinder cut from the housing; d = nominal diameter of the connection), and Fig. 4.16b shows shaft displacements. The housing is made from cast iron and the shaft (pin) is made

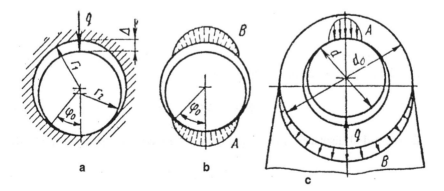

Figure 4.15 (a) Cylindrical connection with a clearance fit loaded with a distributed radial force q; (b) assumed load distribution (A, acting load, B, reacting load).

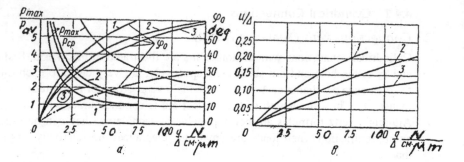

Figure 4.16 Deformation parameters of cylindrical connection due to deformations of housing and shaft: (a) length of contact arc φ_0 and contact pressure concentration P_{max}/P_{av}; (b) displacements u/Δ; $1 - d_0/d = 2$; $2 - d_0/d = 3$; $3 - d_0/d = \infty$.

of steel. Comparison of lines 1 and 2 with line 3 shows that the influence of the thickness of the housing is significant.

Analysis illustrated by Fig. 4.16 assumed that both the pin and the hole are perfectly smooth, but real parts have microasperities and waviness whose deformations may add significantly to the overall deformations. Contact compliance of the surfaces with microasperities/waviness can be considered separately assuming that the associated deformations are much larger than deformations of the shaft/housing system, which, for this stage of the analysis, is considered rigid.

Two correlations between the contact pressures p and deformations δ are considered:

$$\delta = kp \tag{4.19a}$$

$$\delta = cp^{0.5} \tag{4.19b}$$

In contact of two cylindrical surfaces with slightly differing r_1 and r the radial clearance at angle φ (Fig. 4.17a) is

$$\Delta\phi = 0.5\Delta (1 - \cos\phi) \tag{4.20}$$

Elastic radial displacement at angle φ to the direction of load q (Fig. 4.17b)

$$\delta_\varphi = \delta \cos \varphi - \Delta_\varphi., \tag{4.21}$$

where δ = displacement in the direction of the acting load q.

For linear contact compliance [Eq. (4.19a)], $\delta_\varphi = kp$, where k is the contact

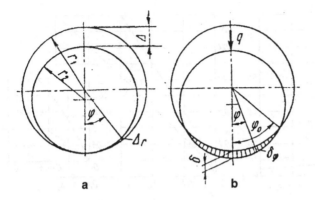

Figure 4.17 Model for analysis of contact deformations of cylindrical connections with clearance: (a) initial contact; (b) contact under load.

compliance coefficient, and the vertical component of contact pressure at angle φ is

$$p_\varphi = p \cos \varphi = \delta_\varphi \cos \varphi / k \qquad (4.22)$$

The total vertical load can be obtained by integrating p_φ along the contact arc $2\varphi_0$

$$q = (d/4k)(2\delta + \Delta)(\varphi_0 - \sin \varphi_0 \cos \varphi_0) \qquad (4.23)$$

Half-angle of contact arc φ_0 is determined from the condition $\delta_{\varphi 0} = 0$, and from Eqs. (4.20) and (4.21)

$$\cos \varphi_0 = \Delta/(2\delta + \Delta) \qquad (4.24)$$

Thus, two unknowns δ and φ_0 can be determined from Eqs. (4.23) and (4.24). Maximum pressure in the direction of force is

$$p_{max} = p_{av}[2(1 - \cos \varphi_0)]/(\varphi_0 - \sin \varphi_0 \cos \varphi_0) \qquad (4.25)$$

where $p_{av} = q/d$. If there is no clearance, then $\Delta = 0$ and

$$\varphi_0 = \pi/2; \quad \delta = 4kq/\pi d; \quad p_{max} = (4/\pi)p_{av} \qquad (4.26)$$

Thus, $\pi d/4$ can be called an "effective width" of the connection.

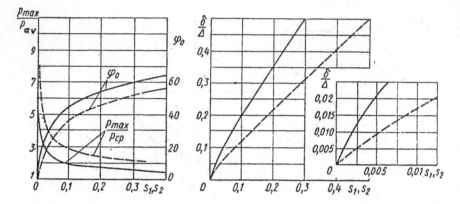

Figure 4.18   Parameters of contact deformation of cylindrical connection.

Figure 4.18 gives φ_0, p_{max}/p_{av}, and δ/Δ as functions of dimensionless parameter $s_1 = qk/d\Delta$, where q is in 10^3 N/m, d in 10^{-2}m, Δ in 10^{-6}m (μm), and k in 10^{-2} m/MPa.

For $\varphi_0 \leq 10^0$, $\sin \varphi_0 \approx \varphi_0$, $1 - \cos \varphi_0 \approx \varphi_0^2/2$, and after integration

$$\delta/\Delta = 0.83(qk/d\Delta)^{2/3} \tag{4.27}$$

If the contact deformation is nonlinear, $\delta = cp^{0.5}$, then analogously,

$$p_{av} = [(\delta \cos \phi - \Delta_\phi)^2 \cos \phi]/c^2 \tag{4.28}$$

$$q = \{[d(2\delta + \Delta)^2]/4c^2\} (\sin \varphi_0 - (1/3) \sin^3 \varphi_0 - \varphi_0 \cos\varphi_0) \tag{4.29}$$

Also, $\cos \varphi_0 = \Delta/(2 \delta + \Delta)$ [Eq. (4.24)] and

$$p_{max} = \delta^2/c^2 = (q/d) [(1 - \cos \varphi_0)^2/ \sin \varphi_0 - (1/3) \sin^3 \varphi_0 - \varphi_0 \cos \varphi_0 \tag{4.30}$$

For $\Delta = 0$

$$\delta = C \sqrt{\frac{3q}{2d}}; \qquad p_{max} = 1.5 \, p_{av} \tag{4.31}$$

For a small contact arc

$$p_{max} \approx \frac{15q}{8d} \sqrt{\frac{\Delta}{\delta}}; \qquad \frac{\delta}{\Delta} \approx \sqrt[5]{\frac{q^2c^4}{d^2\Delta^4}} \qquad \varphi_0 = 2\sqrt{\frac{\delta}{\Delta}} \tag{4.32}$$

Values δ/Δ, φ_0 and p_{max}/p_{av} are shown in Fig. 4.18 by broken lines as functions of dimensionless $s_2 = (c/d)\sqrt{2q/d}$

Coefficients k and c were measured for shaft/ring connections with various clearances. Special measures were taken to reduce solid body deformations of the rings (such as making the outer diameter equal to three-times shaft diameter or applying load by a massive elastic semiring to distribute the load along a wide area). Both cast iron (HB 180) and hardened steel (HRC 42) rings had been tested with hardened steel (HRC 45) shafts. Contact surfaces were machined to $R_a = 0.2$ μm, nonroundness was much less than the clearance magnitudes, and tests were performed with $q = 60$–400×10^3 N/m (340–2,260 lb/in.). The test results are shown in Fig. 4.19a in comparison with analytical results. Plots in Fig. 4.18 were used to determine dimensionless parameters s_1, s_2 for each δ/Δ, and then k and c were determined.

Tests for high intensity loading (Fig. 4.19b) were performed with shafts having $R_a = 0.4$ μm and holes having $R_a = 0.4$–1.6 μm.

At high loading intensity, $q > 200 \times 10^3$ N/m (1,130 lb/in.) and contact arc $2\varphi_0 > 80°$–$90°$, the best correlation between computational and test results is for the nonlinear dependence [Eq. (4.19b)] with values of coefficient c close to its values for the flat joints having similar surface finish. For cases in Fig. 4.19a, $c = 0.18$, and for cases in Fig. 4.19b, $c = 0.35$–0.45.

At low loads, $q < 200 \times 10^3$ N/m (1,130 lb/in.), $2\varphi_0 < 90°$, and values of

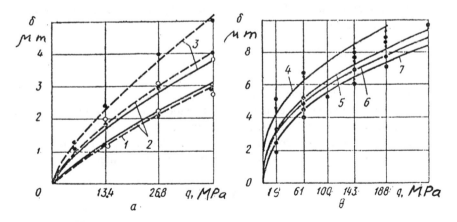

Figure 4.19 Elastic contact displacements in cylindrical connections shaft/ring with clearance. (a) Low specific loads; (b) High specific loads: dots, test data; lines, computation; solid lines, steel-to-steel; broken lines, steel-to-cast iron. Clearance Δ per diameter: 1, 3 μm; 2, 13 μm; 3, 29 μm; 4–7, 20–30 μm. Dimensions $d \times b$ mm: 1–3, 55 $\times$ 25; 4, 75 $\times$ 8; 5, 110 $\times$ 8; 6, 80 $\times$ 8; 7, 90 $\times$ 8.

c are increasing with increasing load, from $c = 0.08$–0.12 to 0.18 due to reduced influence of surface waviness for small contact surfaces, less than 5–6 cm^2 (~1 sq. in.) for the tested specimens. At low loads, better correlation was observed when the linear dependence [Eq. (4.19a)] was used. For the steel/cast iron pair coefficient *k* was 0.25–0.29 μm/MPa for small clearance 3–10 μm and 0.3–0.32 for larger clearances (15–30 μm), while for the steel/steel pair $k = 0.17$–0.2 for 10–15 μm clearance and $k = 0.21$–0.24 for 29 μm clearance. These values also correlate well with values of *k* for flat ground joints hardened steel/cast iron (0.3 μm/MPa) and hardened steel/hardened steel (0.2 μm/MPa). Computed values of δ in Fig. 4.11a were determined using the linear dependence for low loads ($k = 0.3$ μm/MPa for steel/cast iron, $k = 0.2$ μm/MPa for steel/steel), and using non-linear dependence for high loads ($c = 0.7$–0.9 μm/MPa$^{0.5}$).

Combined influence of contact and solid body deformations can be obtained assuming their independence. Relative importance of the component deformations is determined by criterion

$$\gamma = k\pi^2 E/4(1 - \nu^2)\ 10^4 d \tag{4.33}$$

where k = contact compliance coefficient (μm/MPa); E = Young's modulus of the housing (MPa); ν = Poisson's ratio; and d = connection diameter (cm).

When $\gamma \leq 0.5$, displacements in the connection are determined largely by deformations of the shaft and the housing and plots in Fig. 4.16 can be used. When $\gamma \geq 10$, the determining factor is contact deformations and plots in Fig. 4.18 can be used. At $0.5 < \gamma < 10$, both components are commensurate and the total deformation

$$u = \xi\delta \tag{4.34}$$

where ξ = a correction factor given in Fig. 4.20.

Real shafts and holes have some deviations from the round shape that may be commensurate with the clearance magnitude (Fig. 4.21). The influence of these deviations on contact deformations in the connection can be relatively easy to analyze if a sinusoidal pattern of nonroundness is assumed

$$\Delta_{c\varphi} = \Delta_c \cos n\varphi \tag{4.35}$$

where Δ_c = maximum deviation from the cylindrical shape. This assumption is in many cases very close to actual shape of the surface. The surface radius at angle φ is

$$\rho = r\Delta_c \cos n\varphi \tag{4.36}$$

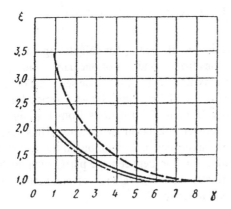

Figure 4.20 Correction factor for computing total deformations in cylindrical connections with clearances.

and by changing n, typical basic patterns in Fig. 4.21 can be represented. For connections in which one surface is assumed to be perfectly round and the other having deviations as per Eq. (4.36), the initial clearance at angle φ is (Figs. 4.22a and b)

$$\Delta\varphi = 0.5\Delta(1 - \cos\varphi) - \Delta_c(\cos n\varphi - \cos\varphi) \qquad (4.37)$$

and the correlations between δ, q, and φ_0 are as follows:

$$qk/d\Delta = (\varphi_0 - \sin\varphi_0\cos\varphi_0)/4\cos\varphi_0$$
$$+ (\Delta_c/2\Delta)\{[\sin(n - 1)\varphi_0]/(n - 1) + [\sin(n + 1)\varphi_0]/(n + 1)$$
$$- \cos n\varphi_0/\cos\varphi_0(\varphi_0 + \sin\varphi_0\cos\varphi_0)\} \qquad (4.38)$$

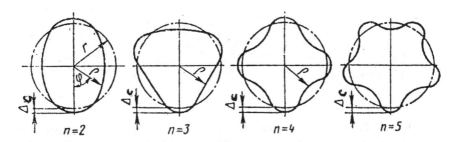

Figure 4.21 Basic patterns of nonroundness.

$$\delta/\Delta = 1/2 \cos \varphi_0 - 1/2 + (\Delta_c/2\Delta)(1 - \cos n\varphi_0/\cos \varphi_0) \qquad (4.39)$$

Plots in Fig. 4.22c give correction coefficient c_c for calculating an increase of elastic deformation due to deviations from roundness of one part of the connection for two basic patterns of deviations along axis z_1. It can be seen that deformations may increase 1.4–1.7 times for small loads and 1.1–1.5 times for large loads. Loading in z_2 direction gives different deformations from loading along z_1 since in most cases the initial contact will be in two points. Thus, usually displacements in the z_1 direction are larger than at $\Delta_c = 0$, but in the z_2 direction they are smaller. If there is a relative rotation between the parts, such stiffness variation can create undesirable effects, such as escalation of parametric vibrations or scatter of machining/measuring errors.

Interference-Fit Cylindrical Connections

Usually, displacements and pressures caused by the interference are much larger than displacements and pressures caused by external forces. The initial interference-fit pressures create a preloaded system similar to ones described in Chapter 3. Under external loading of the connection, total pressures on one side are increasing, while on the opposite side they are decreasing, and magnitudes of incremental elastic deformations in two diametrically opposed points are equal. While the maximum displacements caused by the external force are along the direction of force, distribution of the displacements along the circumference can be assumed to be cosinusoidal.

For large magnitudes of the interference fit pressures p_i, approximately

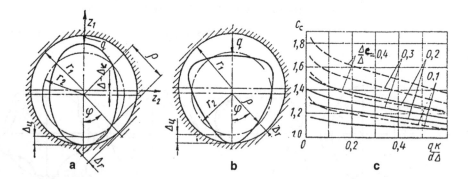

Figure 4.22 Contacting between out-of-round surfaces: (a, b) typical shapes; (c) correction coefficient for deformations of cylinders with deviations from roundness. solid lines, oval shape (case a); broken line, triangular shape (case b).

$$\delta \approx (d\delta/dp)_p = p_i p \approx kp \qquad (4.40)$$

where $p = 2q/\pi d$ is maximum pressure from the external load, having approximately the same magnitude on the loaded and unloaded sides, and $k = 0.5 \, cp_i^{-0.5}$ is the contact compliance coefficient.

Elastic displacement in the connection thus can be determined as

$$\delta = 2kq/\pi d \qquad (4.41)$$

Dependence between the magnitude of interference (diametral interference Δ) and pressure in the interference fit is, considering contact displacements

$$\Delta = 10^4 p_i d\{[(1 + \zeta_1^2)/(1 - \zeta_1^2) - \nu_1]/E_1 + [(1 + \zeta_2^2)/(1 - \zeta_2^2) + \nu_2]/E_2\} + 2p_i^{0.5} \qquad (4.42)$$

where subscripts 1 and 2 relate to the male and female parts, respectively; E_1 and E_2 = Young's moduli (MPa); ν_1 and ν_2 = Poisson's ratios; $\zeta_1 = d_h/d$; $\zeta_2 = d/d_0$; d = connection diameter (cm); d_h = diameter of a hole in the male part (cm); and d_0 = external diameter of the female part (cm).

Coefficients k were experimentally determined for shaft/ring connections similar to ones used for study of connections with clearances, with various interferences. Figure 4.23a shows load-radial deflection characteristics while Figs. 4.23b,c can be used to determine value k as a function of the relative interference Δ/d (Fig. 4.23b) and interference pressure (Fig. 4.23c). The dashed line in Fig. 4.23c represents correlation $k = 0.5 \, cp_i^{0.5}$, computed for $c = 0.47$. This value of c is also typical for flat joints and for cylindrical connections with a clearance for the same surface finish.

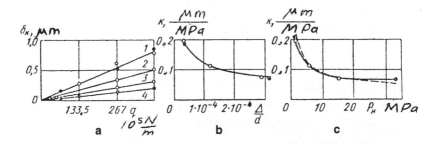

Figure 4.23 (a) Elastic displacements and (b, c) contact compliance coefficients in cylindrical interference fit connections shaft/ring. 1, Diametral interference $\Delta = 2$ μm; 2, $\Delta = 7$ μm; 3, $\Delta = 13$ μm; 4, $\Delta = 16$ μm; surface finish $R_a = 0.2$ μm. ● = Hardened steel shaft, cast iron ring; 0 = hardened steel shaft and ring.

4.4.2   Elastic Displacements in Conical (Tapered) Connections

Conical connections of mechanical components are very popular since they provide self-centering of the connection and also allow one to realize easily disassembleable interference-fit connections. They are frequently used in machine tools and measuring instruments (see, e.g., Fig. 4.24) as well as in other precision apparatuses. Deformations in such connections are due to large magnitudes of *bending moments* caused by long overhangs of cantilever tools, etc., connected by the taper, and also due to *bending deformations* of the cantilever tools that cause very nonuniform pressure distributions inside the connection. Angular deformation θ_0 at the end of the connection due to contact deformations may cause large "projected" deformations at the tool end (Fig. 4.24c). Frequently, these projected displacements may even exceed bending deformations of the connected parts (tools, mandrels, etc.) on the overhang. Elastic deflection under the load is

$$\delta = \delta_p + \delta_0 + \theta_0 L \qquad (4.43)$$

where δ_p = deformation of the tool/mandrel of length L itself if it is considered "built-in" at the cross section 0, δ_0 = radial deformation in the connection, and $\theta_0 L$ = projected angular deformation.

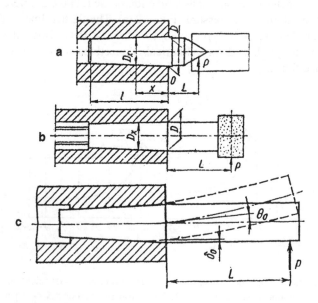

Figure 4.24   Schematics of conical connections: (a) dead center in lathe spindle; (b) grinding mandrel in grinding machine spindle; (c) deformations in conical connection.

Test Data [4]

An experimental study was performed on spindles of internal grinders [4]. The spindle was sturdily supported, especially under the front end, and mandrels with Morse #3 taper were inserted and axially preloaded by a drawbar. A load P was applied at the free end of the mandrel and angular deflection at the mouth of the spindle hole was measured as shown in Fig. 4.25. Sometimes, the total displacement at the end of the mandrel was also measured. The testing was performed at small loads, typical for precision machines. Surface finish of the tapers was for the mandrel $R_a = 0.4$ µm and for the spindle hole $R_a = 0.2$ µm. In all tested connections (with one exception) a "fitting dye" test has shown full conformity between the male and female tapers. The same mandrel was tested in several spindles to assess influence of manufacturing errors. The test results are shown in Fig. 4.26; similar tests were performed with gage-quality tapers Morse #3 and #4, which were lapped together with the respective holes.

Tests were performed in [7] on machining center spindles, both for 7/24 taper #50 toolholders and on hollow HSK toolholders (taper 1/10) (Fig. 4.27), providing simultaneous taper and face contact due to radial deformation of the hollow taper when the axial (drawbar) force is applied.

All these tests have shown the following.

1. Load-angular deflection characteristic is linear since the connection is preloaded.

2. A stronger preload results in increased stiffness (smaller angular deflection). This effect is more pronounced on imperfect connections (low stiffness connections 1 and 4 in Fig. 4.26), wherein increase of the drawbar force improves the contact conformity. For connections with a good initial fit (and high initial stiffness) elastic displacements do not depend significantly on the preload (con-

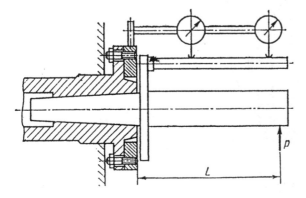

Figure 4.25 Setup for measuring angular deflections in tapered connection.

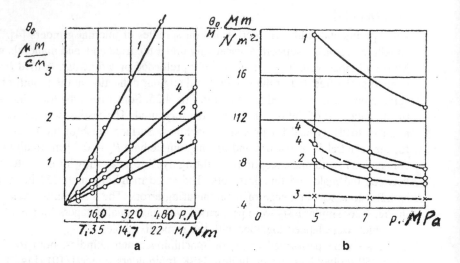

Figure 4.26 Test results for tapered connection mandrel/spindle (Morse #3): 1–4, numbers of the tested connections. (a) Inclination angle at the mouth, interference pressure 7MPa (1,000 psi); P = external force, M = bending moment at the mouth; (b) angular compliance θ_o/M as functions of interference pressure p; dashed lines represent test results for rotated (90°) mandrel relative to spindle.

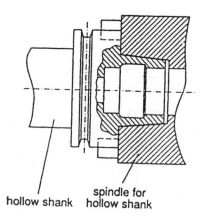

hollow shank   spindle for
              hollow shank

Figure 4.27 Toolholder/spindle interface with a shallow hollow taper (HSK).

nection 3). Such "saturation" of the stiffness values is confirmed by testing of steel 7/24 tapers #50 performed elsewhere [7] (Fig. 4.28). It can be seen that deformations of the tool end do not diminish noticeably at preloading (drawbar, entry) force exceeding 20,000 N for taper #50.

3. Manufacturing deviations are important (see following section). Deflections for several connections for the same test conditions and a good dye-tested conformity differ by 50–100% (lesser differences for high drawbar forces). Stiffness of the same connections after rotating of the mandrel is also changing noticeably (25–30% for connection 4). It is interesting to note that other tests [7] have shown, in some cases, a deterioration of stiffness for very precise connections (Fig. 4.29) (tolerances for taper grades AT3, AT4, and AT5 are equivalent to angular deviations 8, 13, and 21 angular seconds, respectively, for both male and female tapers). These results can be explained by a possibility of some uncertainty of the contact area location for a very precise connection, since structural deformations may exceed the very small dimensional variations. The deformations caused by the radial force are due, to a large extent, to the angular mismatch between the male and female tapers. Accordingly, the radial stiffness can be

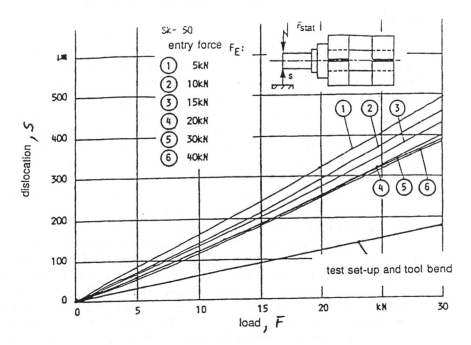

Figure 4.28 Influence of axial preload (drawbar force) on stiffness of 7/24 taper connection.

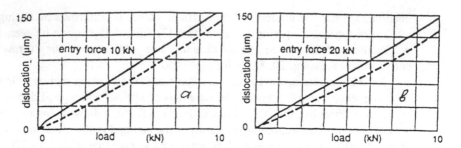

Figure 4.29 Influence of fabrication accuracy of 7/24 taper connection on its stiffness. Solid lines, combination toolholder/spindle AT3/AT3; dashed line, combination AT4/AT5

significantly enhanced by bridging the clearance at one end of the connection resulting from the angular mismatch by elastic elements (see Section 8.2).

4. Radial stiffness of the lapped connections is higher than that of the ground connections, and the scatter between the 12 tested connections did not exceed 25–30%.

5. A typical load-deflection characteristic of a 7/24 taper #50 in axial direction is shown in Fig. 4.30 [7]. Increase in the axial force causes a very significant axial deformation, which is equivalent to axial stiffness 0.6–0.8 μm/kN (the larger value at the linear region). The axial stiffness is not strongly dependent on accuracy of the connection, but is dependent on lubrication condition (lower stiffness for lubricated connections). Such axial behavior, detrimental for accuracy, is due to the initial clearance between the face of the spindle and the flange of the tapered toolholder. Both axial stiffness and axial accuracy problems are solved by using the elastic taper providing a simultaneous taper/face contact (see line 2 in Fig. 4.30 and Section 8.3).

6. The radial position accuracy (runout) of a component (such as a toolholder) connected to another component (e.g., spindle) by a tapered connection is uncertain due to tolerances on the taper angles and due to wear of the connection caused by repeated connections/disconnections. For a typical case in which the taper angle of the male part is steeper than the taper angle of the female part (discussed below), the wear pattern is the so-called bell mouthing of the spindle hole. This uncertainty can be significant as shown in Fig. 4.31 [7]. The uncertainty decreases with the increasing axial (drawbar) force (Fig. 4.31b), and is more pronounced for used (worn out) tapers (Fig. 4.31c). The uncertainty is increasing with increasing angular mismatch. It can be significantly reduced by bridging the angular mismatch or by achieving a simultaneous taper/face contact (see Section 8.3).

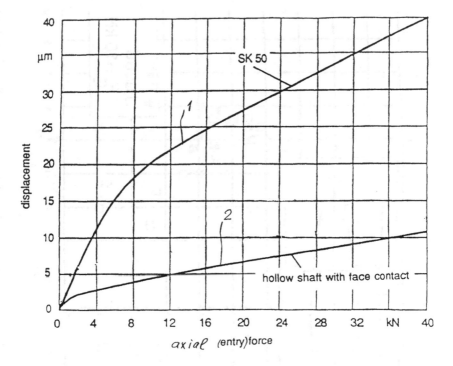

Figure 4.30 Axial displacement of (1) #50 taper and (2) HSK toolholder with face contact with the spindle vs. axial force.

7. A preloaded taper connection results in shrinking of the internal (male) part and expansion of the external (female) part of the connection. Such an expansion is undesirable since it can change clearances in the spindle bearings and degrade their performance. One study [7] demonstrated that expansion of the spindle by an axially preloaded solid 7/24 taper is very small and can be neglected. However, deformation of the spindle during insertion of a hollow taper with a small taper angle, as in Fig. 4.27, can be significant due to a greater mechanical advantages creating large radial forces necessary for radial contraction of the hollow taper in order to obtain the simultaneous taper and face contacts. The spindle expansion is shown in Fig. 4.32 for different modifications of the hollow taper/face interface [7]. Especially important is expansion under bearings, measuring point 4 in Fig. 4.32. Reduction of the radial stiffness of the locating taper surface of the toolholder, as in the design in Section 8.3, completely eliminates this effect.

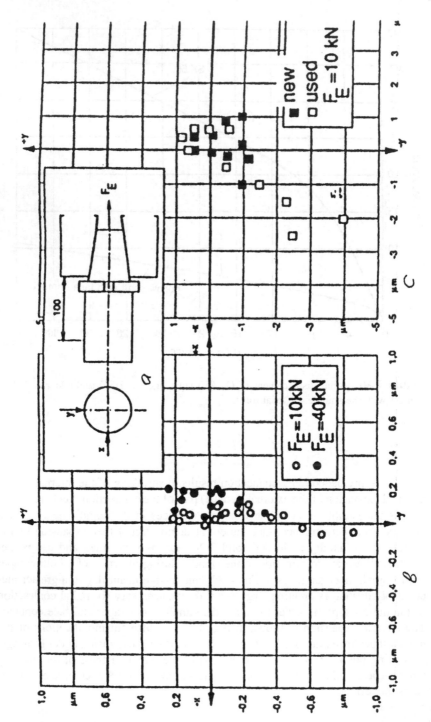

Figure 4.31 (a) Radial repeatability of #50 tapers measured at 100 mm from spindle face; (b) repeatability as a function of axial force; (c) repeatability as a function of condition. F_E = axial (drawbar) force.

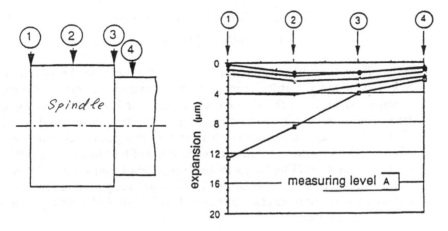

Figure 4.32 Bulging of spindle caused by insertion of HSK-type toolholders at 15 kN axial force: (a) positions of measuring points; (b) test results.

Computational Evaluation of Contact Deformations in Tapered Connections [4]

A mandrel or a cantilever tool can be considered as a cantilever beam on an elastic bed, which is provided by the surface contact between the male and female tapers. The connection diameter (i.e., width of the elastic bed B_x) varies along its length, so

$$D_x = D(1 - 2\alpha x/D); \quad EI_x = E\pi D_x^4/64; \quad B_x = \pi D_x/2 \qquad (4.44)$$

where 2α = taper angle; D = larger (gage) diameter of the connection; and D_x = diameter at distance x from the mouth. Since there is always an initial axial preload, it can be assumed that the contact arc is the whole circumference πD_x, and that the pressure-deflection characteristic is linear. The deflected shape of such connection (deflection y in cross section x) is described by differential equation

$$\frac{d^2}{dx^2}\left(EI_x \frac{d^2 y}{dx^2}\right) + \frac{B_x}{k}y = 0 \qquad (4.45)$$

Its boundary conditions are as follows: at $x = 0$, $d^2 y/dx^2 = M/EI$, $d^3 y/dx^3 = P/EI$; at $x = 1$, $d^2 y/dx^2 = 0$, $d^3 y/dx^3 = 0$. Here $M = PL$ and P are moment and force at the mouth of the connection, respectively.

Radial displacement δ_o and angular displacement θ_o at the mouth are

$$\delta_0 = (0.2P\beta k/B)(\beta LC_1 + C_2), \mu m \qquad (4.46a)$$

$$\theta_0 = (0.2P\beta^2 k/B)(2\beta LC_3 + C_u), \mu m/cm, \qquad (4.46b)$$

where P = force (N), L = distance from the force to the mouth; β = $\sqrt[4]{(B \times 10^2)/(4EIk)}$ = stiffness index of the connection (1/cm); E = Young's modulus (MPa); $I = \pi D^4/64$ = cross-sectional moment of inertia of the male taper at the mouth (cm^4); $B = 0.5 \pi D$ = effective width of the connection; C_1, C_2, C_3, C_4 (Table 4.2) = correction coefficients considering influence of the variable diameter. For a cylindrical shrink-fit connection and for Morse tapers at $\beta l > 6$, C_1, C_2, C_3, C_4 = ~1. Contact compliance coefficient k ($\mu m \times cm^2/N$) was measured for connections of parts having $R_a = 0.4 \mu m$ and plotted in Fig. 4.33. For a good fitting of the connected tapers, $k = 0.1–0.15 \mu m/MPa$; for a poor fitting, $k = 0.8–1.0 \mu m/MPa$.

Influence of Manufacturing Errors

Inevitable manufacturing errors may cause a nonuniform contact or even a partial loss of contact in the connection. Typical errors are (Fig. 4.34) as follows.

Deviation from roundness of the cross sections
Deviations from straightness of the side surface
Deviations from the nominal taper angle

The most important parameter is angular difference between the male and female tapers, when the contact exists only on a partial length l_e. Two cases may exist:

The hole has a steeper taper angle than the mandrel
The hole has a more shallow angle than the mandrel

In the former case, the contact area is at the smaller diameter, and the effective

Table 4.2 Coefficients C_1, C_2, C_3, and C_4

Morse taper					Steep (7/24) taper				
$\gamma = \beta l$	C_1	C_2	C_3	C_4	$\gamma = \beta l$	C_1	C_2	C_3	C_4
2	1.23	1.2	1.2	1.23	1.5	2.34	2.06	1.70	2.34
3	1.10	1.08	1.015	1.10	2	2.16	1.94	1.35	2.16
4	1.06	1.04	1.01	1.06	2.5	1.65	1.64	1.17	1.65
5	1.05	1.03	1.0	1.05	3	1.45	1.48	1.07	1.45
6	1.04	1.03	1.0	1.04	3.5	1.34	1.36	1.05	1.34
					4	1.30	1.34	1.04	1.30

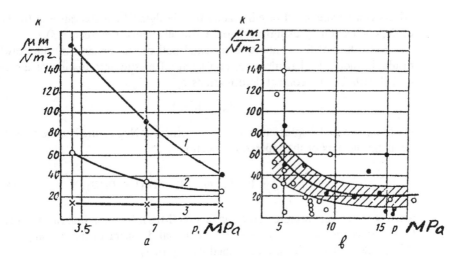

Figure 4.33 Contact compliance coefficients for preloaded conical connections: (a) mandrel/spindle (connections 1, 2, 3); (b) gage tapers (hatched area represents 90% probability). ● = taper Morse #3; ○ = taper Morse #4.

length of the overhang is increasing, thus negatively influencing the bending stiffness. In the latter case, the contact area is at the larger diameter and stiffness of the connection is affected not very significantly when the loading is not too intense. However, at high loading (or dynamic loading, as during milling operations), the short frontal contact area behaves like a pivot. Small angular motions of the mandrel cause a fast wear of the contact area ("bell mouthing" of the spindle), fretting corrosion, and increased runout at the free end of the mandrel. These ill

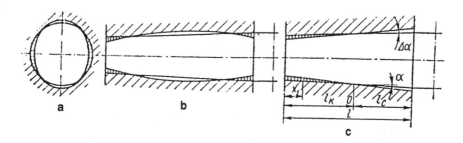

Figure 4.34 Shape and fit errors of tapered connections: (a) nonroundness; (b) nonstraightness; (c) angle differential.

effects can be alleviated or eliminated by "bridging" the clearance at the back of the connection by using precision elastic elements (see Sect. 8.3).

Preloading of the connection by axial pull of the mandrel by the drawbar results in radial displacements δ_r consisting of elastic deformation of the connected parts δ_1, and contact deformations δ_2

$$\delta_{rx} = \delta_1 + \delta_2 \tag{4.47}$$

where δ_{rx} = radial displacement at the coordinate x. Neglecting change of diameter along the axis of the tapered connection

$$\delta_1 = pD_{ef} \times 10^3/E, \ \mu\text{m}; \quad D_{ef} = D/(1 - D_2/D_0^2) \tag{4.48}$$

where p = contact pressure in the connection (MPa); D = nominal (mean) diameter of the connection (mm); and D_0 = external diameter of the housing (mm). Contact deformations δ_2 are described by Eq. (4.1).

Computational analysis has shown that for realistic contact pressures and dimensions

$$p = A\delta_{rx}^n \tag{4.49}$$

For ground and lapped parts c = 0.3; A and n are given in Table 4.3.

If the radial interference in the initial cross section $x = 0$ is δ_{ro}, then in the cross section with axial coordinate x it is

$$\delta_{rx} = \delta_{ro} - x\Delta\alpha \tag{4.50}$$

where $\Delta\alpha$ = angle differential in μm/cm. Accordingly, contact pressure p_x and total axial preloading (drawbar) force P_{db} are

$$p_x = A\delta_x^n \tag{4.51}$$

$$P_{db} = \int_0^{l_k} p_x \pi D_x \tan(\alpha + \varphi)dx = \frac{\pi A \delta_{ro}^{n+1} D \tan(\alpha + \varphi)}{(n + 1)\Delta\alpha} \tag{4.52}$$

Table 4.3 Values of A and n

D_{ef}(cm)	2	3	4	5	7	10
A	3.9	3.1	2.6	2.1	1.7	12.5
n	1.5	1.35	1.33	1.25	1.2	1.18

where α = one-half of the taper angle and φ = friction angle ($f = \tan \varphi \approx 0.25$ for nonlubricated connections). Equation (4.52) is correct for $l_k \le l$, where the length of contact zone l_k is determined from the condition of disappearing contact pressure at some cross section o.

$$\delta' = \delta_{ro} - l_k\Delta\alpha = 0 \tag{4.53}$$

$$l_k = \delta_{ro}/\Delta\alpha, \text{ cm} \tag{4.54}$$

or

$$\frac{l_k}{l} \approx \sqrt[n+1]{\frac{(n + 1)p}{A(l\Delta\alpha)^n}} \tag{4.55}$$

Fig. 4.35a shows l_k/l for steel Morse tapers as a function of parameter $\Delta = l\Delta\alpha$ at various D_{ef} and P, and a function of Δ/δ_{av}, where $\delta_{av} = \sqrt[n]{P/A}$ (Fig. 4.35b). Product $\Delta = l\Delta\alpha$ represents linear dimension deviation accumulated along the length of the connection, δ_{av} is average contact deformation during preloading, and Δ/δ_{av} is the ratio of the linear error to the average contact deformation.

Thus, elastic deflection under load P (Fig. 4.24c) for an ideal Morse taper connection (or a cylindrical ''shrink fit'' connection) can be calculated as

$$\delta = 10^2 PL^3/3EI + 0.02P\beta k(1 + 2\beta L + 2\beta^2 L^2)/B, \text{ } \mu m \tag{4.56a}$$

where I = cross-sectional moment of inertia of the cantilever mandrel along its length L and coefficients $C_1, C_2, C_3, C_4 = 1$.

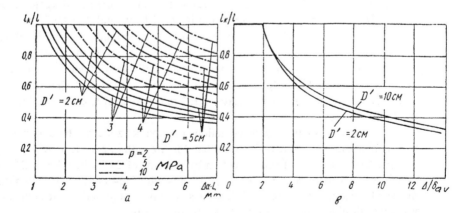

Figure 4.35 Relative length of contact zone l_k/l as function of (a) $\Delta\alpha$ and (b) $\Delta\delta_{av}$.

Influence of angular assumed mismatch on stiffness can be approximately evaluated by substituting l_k as the length of the elastic bed and $(L + l_c)$ as the length of the overhang, where $l_c = L - l_k$. Then formula (4.56a) becomes

$$\delta = 10^2 P(L + l_c)^3/3EI + 0.02Pk\beta'[T + 2\beta'(L + l_c) + 2(\beta')^2(L + L_c)^2]/B', \mu m \qquad (4.56b)$$

where

$$\beta' = \sqrt[4]{\frac{10^2 B'}{4EI'k}} \qquad (4.57)$$

It is assumed in Eq. (4.56b) that diameter of the cantilever part on the interval l_c is approximately equal to diameter of the cylindrical part. Effective parameters B' and I' should be calculated for the initial cross section at the distance l_c from the mouth of the connection.

EXAMPLE Morse #3 taper connection ($l = 7.3$ cm, $D_{ef} = 3$ cm). Let's find out at what value of $\Delta\alpha$ and p the contact zone is along the whole length, $l_k/l = 1$. From Fig. 4.35b, this condition is satisfied if $\Delta/\delta_{av} < 1.9$. Then, at $\Delta\alpha = 5''$ (25 μrad or 0.24 μm/cm) it is required to create contact pressure $p = 3$ MPa, at $\Delta\alpha = 10''$ (50 μrad or 0.48 μm/cm) $p = 7.5$ MPa, and at $\Delta\alpha = 15''$ (75 μrad or 0.72 μm/cm) $p = 12$ MPa.

Thus, for realistic values of $p = 3$–8 MPa only very small deviations of the angle are allowable. Actually, the allowable $\Delta\alpha$ are even less since there is need to create a finite contact pressure at the mouth to prevent opening of the connection under load.

If at $p = 5$ MPa $\Delta\alpha = 10''$ (50 μrad), then $\Delta = l\Delta\alpha = 3.5$ μm. From Fig. 4.35a, $l_k/l = 0.82$ and $l_c = l(1 - l_k/l) = 1.3$ cm. At the mouth (gage) cross section, $D = 2.38$ cm, $I = 1.58$ cm^4, and $B = 3.74$ cm, and $\beta = 0.62$ 1/cm. At the cross section at the distance $l_c = 1.3$ cm from the mouth, $D' = 2,38–0.05l_c = 2.31$ cm, $I' = 1.4$ cm^4, $B' = 3.62$ cm, $\beta = 0.64$ 1/cm. If the external force is applied at 7 cm from the mouth, then elastic deflection under the force P at $\Delta\alpha = 0$ is $\delta = 0.067 P$ (μm), but for $\Delta\alpha = 10''$ and $p = 5$MPa then $\delta = 0,11 P$ or is 1.6 times greater. At $\Delta\alpha = 15''$ and $p = 5$MPa, elastic deflections are 2.5 times greater.

Even for very small angular deviations the contact zone can be only partial, causing much larger elastic deflections than for $\Delta\alpha = 0$. Experiments with Morse #4 taper have shown stiffness reduction by a factor of 2–3 when $\Delta\alpha = 2''$ and by a factor of 4–6 when $\Delta\alpha = 8''$. Experiments with 7/24 taper have shown that change of the angle α by $\Delta\alpha = 27''$ (for the same hole) increased elastic deflection at the mouth nine times, and by $\Delta\alpha = 40''$ by 15 times.

For the angular difference in the opposite direction (the hole is more shallow than the taper on the mandrel), stiffness is not sensitive to the mismatch. The tests have demonstrated that deviations $\Delta\alpha$ up to 60" do not influence significantly the connection stiffness. However, as it was mentioned above, such deviations can cause a faster wear and increased radial runout at the end.

Finite Element Modeling of 7/24 Taper Connection [8]

Results similar to those described above were obtained by Tsutsumi et al. in an extensive computational (finite element) and experimental analysis of the 7/24 taper connection [8]. This study addressed both axial displacement of the taper inside the hole under the axial (drawbar) force P_t and bending deformations at the spindle end under the radial force F. The design and manufacturing parameters whose influence was explored included ratio of the outside diameter of the spindle D to the gage (maximum) diameter of the connection d_0; accuracy of the gage diameter; surface finish of the connected male and female tapers R_a; friction coefficient f in the connection; and magnitude of drawbar force P_t. The study was performed for taper #40. It was found that the axial displacement for D/d_0 = 2 for a very high friction (f = 0.6) and surface finish R_a = 0.5 μm (20 μin.) was 17 μm (0.0007 in.) for P_t = 5 kN (1,100 lbs), 21 μm (0.00085 in.) for P_t = 10 kN (2,200 lbs.), and 23 μm (0.0009 in.) for P_t = 15 kN (3,300lbs.). These deformations are mostly due to radial expansion of the spindle walls and, to a much smaller extent, contraction of the male taper. Deviation of d_0 by $\pm$ 25 μ (0.00 1 in.), which represents a mismatch of the taper angles of the hole and the male taper, led to a further increase of the axial displacement by up to 20–30%. Increase of surface roughness to R_a = 1.2 μm (48 μin.) led to another 25% increase in the axial displacement. Reduction of f from 0.6 to 0.1 resulted in a 50% increase in the axial displacement.

The data of [8] on axial displacements in 7/24 tapered connections compares well with experimental data on axial displacements of various interfaces from Hasem et al. [9] (Fig. 4.36). Figure 4.36 indicates that not only flat joints and "curvic coupling" connections have much smaller axial deformation that the 7/24 tapered connection, but they also return to their original nondeformed configurations when the axial force is removed. On the contrary, the 7/24 taper connection is held in its deformed condition after the axial force is removed due to friction forces. The curvic coupling connection is an engagement of two identical flat spiral gears that are held together by the axial force. One gear is fastened to the spindle flange and another is fastened to the tool. Other characteristics of this connection are given in Section 4.8.3.

It is interesting to note that a change in the outside diameter of spindle D has only a marginal effect on the axial displacement: at D/d_0 = 4 it is only 10% less than at D/d_0 = 2. However, bending deformation of the tapered insert (toolholder) as characterized by shear displacement between the male and female

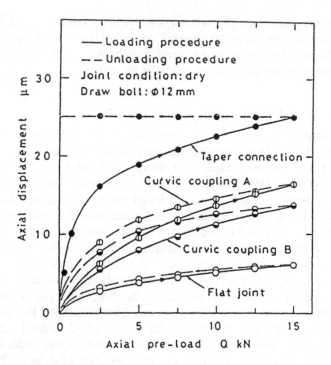

Figure 4.36 Axial deformation of various toolholder/spindle interfaces.

tapers (Fig. 4.37), is noticeably influenced by changes in D. The most dramatic increase of bending deformation is observed at $D/d_0 < 2$, due to increasing local deformations of spindle walls.

Some General Comments on Tapered Connections

The major advantage of the tapered connections is their self-centering while providing relatively easy disassemblable interference-fit joint. There are two major types of the taper connections. The connections with shallow tapers (eg., Morse tapers) are self-locking and do not require axial tightening. However the male Morse taper has to be axially pushed out of the hole with a significant force for disassembly. Another disadvantage of shallow taper connections is a very strong dependence of the relative axial position of the male and female parts on their diameter. For example, only a 10 μm diametral difference between two male tapers (e.g., at the gage diameter) results in about 100 μm difference in their

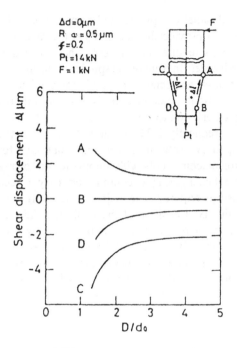

Figure 4.37 Influence of spindle diameter on bending deformation of toolholder.

axial positions. As a result, the Morse connections are not as popular presently as they were in the past when majority of machine tools were manually operated, extra seconds required for disassembling the connection were not critical, and the scatter of axial positioning between several toolholders could be compensated, if necessary, by manual adjustments.

Now one can see a resurgence of shallow taper connections for toolholders/ spindle interfaces of high-speed and/or high-accuracy machine tools (Kennametal "KM," HSK, Sandvik "Capto," etc.). One of the major requirements for the advanced interfaces is to provide an ultimate axial accuracy by assuring a simultaneous taper/face contact with the spindle. It is interesting to note that a drawback of the shallow taper connection—sensitivity of the relative axial position of the male and female tapers to minor variation of the gage diameter— has been turned to its advantage by making the male taper compliant. The high sensitivity of axial position to the diameter change allows for a significant adjustment of the axial position with relatively small deformations of the male taper. Still, designing even a shallow taper for a simultaneous taper/face connection requires extremely tight tolerances and creates many undesirable effects [10].

The taper should have an enhanced radial compliance to be able to provide for the face contact with the spindle. To provide the compliance, the tapered part must be hollow, as in Fig. 4.27. This requires one to move the tool clamping device to the outside part of the toolholder (in front of the spindle), thus increasing its overhang and reducing its stiffness; a special "kick-out" mechanism is required for disassembly of the connection; the connection is 2–4 times more expensive than the standard "steep taper" connection.

Connections with steep (most frequently 7/24, so-called Caterpillar) tapers are not self-locking and are clamped by application of a significant axial force. This feature is very convenient for machine tools with automatic tool changers since the assembly/disassembly (tool changing) procedure is very easy to accomplish and no "kick-out" mechanism is required. The major shortcomings of this connection are indeterminacy of the axial position of the toolholder (discussed above); practical impossibility to achieve the taper/face contact, even by making the taper body compliant due to a relatively low sensitivity of the axial position to diameter changes; and micromotions (more pronounced than in shallow taper connections) leading to fretting corrosion and fast wear at heavy lateral loading. These shortcomings have been overcome by introduction of external elastic elements (see Section 8.3).

4.5 CONTACT DEFORMATIONS CAUSED BY MISALIGNMENT OF CONNECTED PARTS

Besides deformations in point or line contacts (Hertzian deformations) and deformations in contacts between extended nominally conforming surfaces (addressed in Sections 4.2–4.4), stiffness of contact interfaces can also be influenced by misalignments and resulting interactions between the contacting bodies. This is especially, but not only, important for tapered fits when it is practically impossible to achieve perfect identity of the taper angles in the male and female components. The important embodiments of tapered contacts are interfaces between dead and live centers and center holes in the part ends for machining of oblong parts on turning and grinding machines.

Axial contact stiffness between centers and center holes was studied by Kato et al. [11]; the system studied is shown in Fig. 4.38. Axial deformations of the connections have been analyzed using the finite element analysis approach with a variable mesh size model (Fig. 4.39) for various misalignments, connection diameters, and friction coefficients. While the center angle in all cases was 60°, the centerhole was assumed to be each of three sizes: 59.5°, 60° (perfect fit), and 60.5°. For the connection having diameter 10 mm and length 4.5 mm, 0.5°

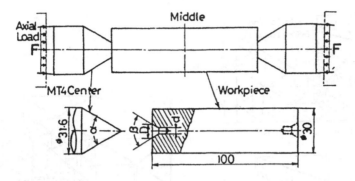

Figure 4.38   Workpiece supported by two centers.

misalignment is equivalent to 20 μm clearance between the parts. Figure 4.40 shows deformation patterns for these three cases and Fig. 4.41 gives load-deflection characteristics for the axial loading.

It is obvious from Fig. 4.41 that the connection with the perfect angular match is the stiffest. However, the connection in which the female taper has a more shallow angle (59.5°) has the closet stiffness to the perfect match case, only 1.35 times lower for the friction coefficient $f = 0.7$ and about 2.0 times lower for a more realistic $f = 0.3$ (for the connection diameter 10 mm). The connection in which the female taper has a steeper angle (60.5°) has a much lower stiffness, 2.6 times lower for $f = 0.7$ and 3.2 times lower for $f = 0.3$.

Qualitatively similar conclusions were derived in the study of larger tapers (Morse and "steep" 7/24 taper) used for toolholder/spindle interfaces [4].

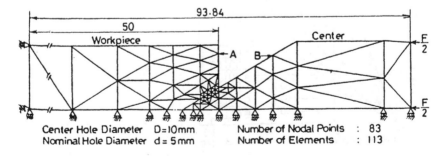

Figure 4.39   Finite element mesh for center–center hole interface.

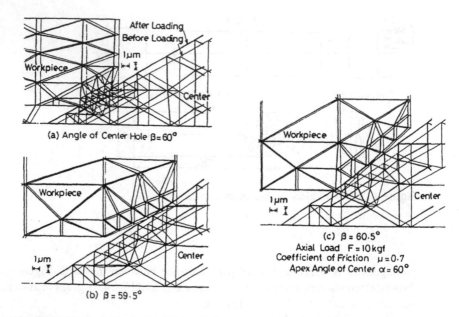

(a) Angle of Center Hole β=60°

(b) β = 59.5°

(c) β = 60.5°
Axial Load F = 10 kgf
Coefficient of Friction μ = 0.7
Apex Angle of Center α = 60°

Figure 4.40 Deformation patterns of center–center hole interface with different misalignments.

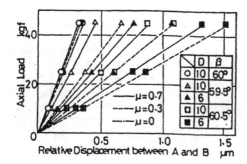

Figure 4.41 Load-deflection characteristics of axial deformation of center–center hole interface.

4.6 TANGENTIAL CONTACT COMPLIANCE

If a joint is loaded with a force acting tangentially to the contacting surfaces, the connected parts will initially be elastically displaced so that after removal of the force their initial position is restored. After the force increases above the elastic limit of the joint, some plastic (nonrestoring) displacements develop, and then the breakthrough occurs (a continuous motion commences). Tangential compliance of joints, although much lower than the normal compliance addressed above, is becoming more and more important with increasing stiffness requirements to precision machines and measuring apparatuses. It is also important for understanding mechanisms of functioning of the joints, their strength, fretting corrosion, and damping. Section 4.6.1 describes the results of an experimental study of tangential compliance based on the work of Levina and Reshetov [4], while Section 4.6.2 addresses analytical studies of mechanisms of dynamic behavior (stiffness and damping) of joints loaded in the normal direction and subjected to axial dynamic loads.

4.6.1 Experimental Study of Tangential Compliance of Flat Joints [4]

An extensive experimental study of tangential compliance was performed by Levina [4] on flat annular joints loaded by a moment (torque) applied within the joint plane while the joint was preloaded with a normal force. The connected cast-iron parts were very rigid so that their deformations could be considered negligible. The contact surfaces were machined by fine turning, grinding, scraping with various depth of the dips, and lapping. Both carefully deoiled surfaces and surfaces lubricated by light industrial mineral oil were tested.

There is a big difference between the first and repeated tangential loading. At the first loading of the joint, the load-displacement characteristic is highly nonlinear (Fig. 4.42a). When the tangential force is removed, the unloading branch of the characteristic is parallel to the loading branch (the same stiffness) but there is a substantial hysteresis. At the following force applications (Fig. 4.42b), the process is linear for loads not exceeding the loads in the first loading process, thus

$$\delta_\tau = k_\tau \tau \tag{4.58}$$

where δ_τ = tangential displacement (μm); τ = specific tangential load (MPa); and k_τ = tangential compliance coefficient (mμ/MPa). For the plastic region, the tangential compliance coefficient is

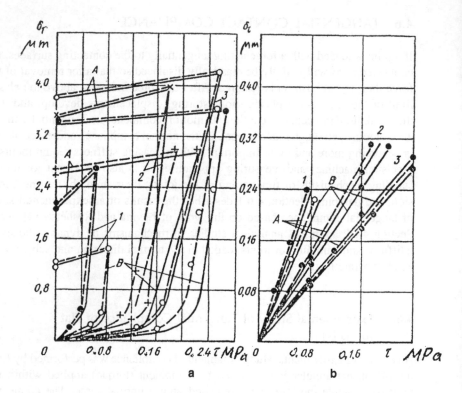

Figure 4.42 Displacement δ_τ vs. tangential stress τ in a flat annular joint with surface area 51 cm^2 (7.9 sq. in.). (a) Scraping, depth of dips 4–6 μm, number of spots 15–6 per sq in.; (b) grinding, surface finish R_a = 0.4 μm; solid lines, nonlubricated joint; dashed lines, lubricated joint. Normal pressure: 1, σ = 0.5 MPa (70 psi); 2, σ = 1.0 MPa (140 psi); 3, σ = 1.5 MPa (210 psi).

$$k_{\tau p} = (20\text{--}25)k_\tau \qquad\qquad (4.59)$$

The magnitude of the tangential displacement before commencement of the continuous motion condition is increasing with increasing normal pressure and can reach 20–30 μm at the first loading. The important parameters for applications are maximum tangential stress τ_e of the elastic region and τ_{max} at the breakthrough event. The elastic range of the joint can be characterized by coefficient

$$f_e = \tau_e/\sigma \qquad\qquad (4.60)$$

Table 4.4 Coefficients f_e and f

Surface condition	No lubrication		With lubrication	
	f_e	f	f_e	f
Fine turning, $R_a = 2\ \mu m$	0.13	0.25	0.13	0.25
Rough grinding	0.12	0.18	0.12	0.18
Grinding and lapping	0.17	0.35	0.14	0.30
Scraping, 8–10 μm dips	0.12	0.22	0.12	0.22
Fine scraping, 1–2 μm dips	0.14	0.28	0.1	0.24

where σ = normal pressure. Table 4.4 gives f_e for annular specimen with the joint area 51 cm², as well as friction coefficient

$$f = \tau_{max}/\sigma \tag{4.61}$$

obtained from the same tests. It can be concluded that elastic displacements occur for the loads about one-half of the static friction forces. However, the maximum elastic displacement is only a small fraction of the total tangential displacement before the breakthrough. For lubricated surfaces f_e and f are slightly smaller (for the light oil used in the tests), especially for high normal pressures (the lubricant is squeezed out). While tangential characteristics of the lubricated joints are not changing in time, f_e for dry joints increases 25–30% with increasing time of preloading from 5 min to 2.5 h.

The magnitude of τ_{max} depends on the rate of application of the tangential loading. For a joint having an area of 225 cm² (35 sq. in.), increase in the rate of loading from 0.003 mm/s to 0.016 mm/s reduced τ_{max} from 0.07 to 0.055 MPa, but the magnitude of the total displacement before the breakthrough did not change much.

If the contacting surfaces are also connected by pins or keys, tangential stiffness in the elastic region does not change a great deal since the tangential stiffness of the joint is usually much higher than stiffness of the pins/keys. However, in the plastic region, the joint stiffness is decreasing and influence of the pins becomes noticeable. Figure 4.43 shows results of testing of a 51 cm² joint with and without holding pins. At small tangential forces, the forces are fully accommodated by the joint itself, and load-deflection characteristics with and without pins coincide. At the forces exceeding the elastic limit of the joint, its role in accommodating the external forces is gradually decreasing while role of the pins is increasing. When the force exceeds the static friction force in the joint, it is fully accommodated by the pins (another linear section of the load-deflection curve).

The joint-pins system is statically indeterminate. Load sharing between the

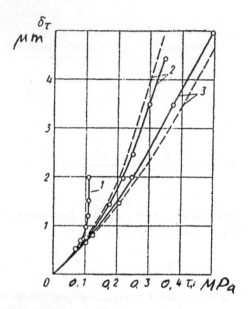

Figure 4.43 Influence of holding pins on tangential compliance: 1, joint without pins; 2, two pins of 12 mm diameter; 3, two pins of 16 mm diameter. Solid lines, experiments; dashed lines, computation.

joint and the pins can be found by equating their displacements. Each pin can be considered as a beam on an elastic bed.

The total elastic displacements are

$$\delta = \delta_0/(1 + \eta) \tag{4.62}$$

where δ_0 = elastic displacement in the joint without pins and η = stiffness enhancement coefficient

$$\eta = \frac{k_\tau}{k} \frac{z B_{ef}}{2\beta F_j} \tag{4.63}$$

where z = number of pins; $B_{ef} = \pi d/2$ = effective width of the "elastic bed" of a pin; d = average pin diameter (cm); and F_j = surface area of the joint (cm²). Also,

$$\beta = \sqrt[4]{\frac{B_{ef} \times 10^4}{4EIk}} = \sqrt[4]{\frac{0.04}{d^3 k}}, \frac{1}{cm} \tag{4.64}$$

where I = cross-sectional moment of inertia of a pin (cm⁴); E = Young's modulus of pin material (MPa); and k = coefficient of normal contact compliance of the joint between pin and hole (μm/MPa).

This analysis is validated for the case of Fig. 4.43 (broken lines) when the assumed values of compliance coefficients were $k = 0.5$ μm/MPa; $k_\tau = 2.0$ μm/MPa; and $k_{\tau p} = 40$ μm/MPa.

In precision systems, it is important to realize such conditions when each joint is loaded in its elastic region. Usually these joints are designed from the condition that the external forces do not exceed static friction forces. In many cases such an approach is inadequate since loading above f_e can lead to significant plastic deformations. It is especially important for dynamically loaded joints. To use this approach in design, it is important to know k_τ as a function of the surface finish (machining quality) and normal preload, which is given in Fig. 4.44.

Use of the above data for actual machine tool units resulted in good correlation between calculated and tested data for τ_e and k_τ. The value of $k_\tau = 3$ μm/MPa as measured for the joint guideways–spindle head of a jig borer (Fig. 4.45) compares well with $k_\tau = 3.5$ μm/MPa measured on the scraped specimen in the lab.

For calculations of the total tangential displacement in a joint, the following formula considering both elastic and plastic deformations can be used

$$\delta = \delta_e + k_{\tau p}(\tau - \tau_e) \qquad (4.65)$$

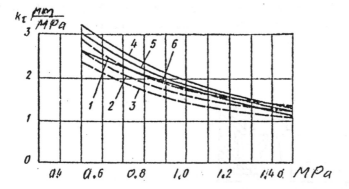

Figure 4.44 Tangential contact compliance coefficients for specimens with joint area 57 cm 1, Fine turning, $R_a = 3.2$ μm; 2, grinding, $R_a = 0.4–0.8$ μm; 3, grinding with lapping, $R_a = 0.1$ μm; 4,5,6, scraping, depth of dip 8–10, 4–6, 1–2 μm, respectively.

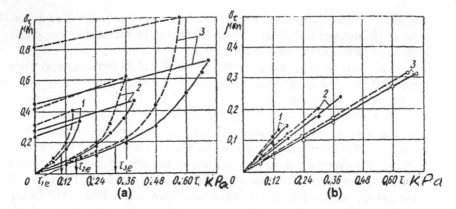

Figure 4.45 Tangential displacements in sliding guideways at (a) first and (b) repeated loading. 1, σ = 0.09 MPa; 2, σ = 0.18 MPa; 3, σ = 0.35 MPa. Solid lines, no lubrication; dashed lines, lubricated.

EXAMPLE.   Find the allowable force on a dead stop (Fig. 4.46). Normal pressure in joints 1 (surface area F_1 = 3.2 cm^2) and 2 (F_2 = 1.6 cm^2) is applied by two bolts, size M8 (8 mm diameter), each preloaded to p_0 = 5000–6000 N (1100–1300 lbs), thus creating joint pressures σ_1 = 30 MPa (4200 psi) and σ_2 = 60 MPa (8400 psi). Tangential stresses in the joints from the external force P applied to the stop are

$$\tau_1 = \tau_2 = P/(F_1 + F_2) \qquad (4.66)$$

The allowable load (within the elastic region of the joints) at f_e = 0.1 is P_{max} = 1450 N. If P_{max} is exceeded, there is a danger of displacement of the stop due to nonrestoring plastic deformations.

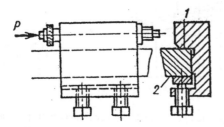

Figure 4.46   Dead stop for a machine tool.

Tangential loading of joints preloaded by normal forces may generate a complex distribution of tangential stresses within the joint, a combination of elastic, plastic, and slippage zones (see Section 4.6.2). This can apply to both flat and cylindrical (interference fit between a shaft and a bushing) joints.

The slippage zone can cause fretting corrosion. These zones can be identified if the tangential contact phenomena are considered. Depending on the load magnitude, the following cases are possible with the increasing load:

1. Elastic along the whole length ($\tau < \tau_e$)
2. Elasto-plastic with both elastic and plastic zones ($\tau > \tau_e$)
3. Plastic along the whole length ($\tau < \tau_{max}$)
4. Plastic and slippage zones ($\tau = \tau_{max}$)

Figure 4.47a shows computed distribution of contact tangential stresses in a cylindrical joint (shown in Fig. 4.47b) ($d = 25$ mm, $l = 20$ mm, ground surfaces with finish $R_a = 0.2$ μm; press fit with contact pressure $\sigma = 50$ MPa). The shaft is loaded by torque T at the end. Analysis without considering tangential contact compliance would conclude that the slippage starts from very low torque magnitudes, but in reality it starts only at high torque magnitudes.

Energy dissipation (damping) can also be determined by computations [12,13] (see also Section 4.6.2). Energy dissipation is mostly concentrated in the plastic and slippage zones.

If joints are loaded with dynamic forces outside their elastic regions, then

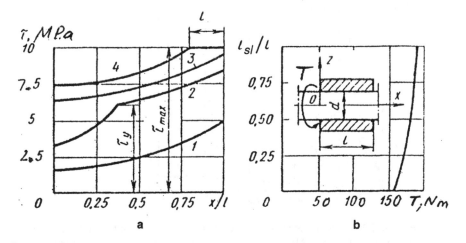

Figure 4.47 (a) Tangential stress distribution in cylindrical press-fit joint (1, $T = 50$ N-m; 2, $T = 120$ N-m; 3, $T = 150$ N-m; 4, $T = 170$ N-m). (b) Length l_{sl} of slippage zone.

residual displacements can develop. Depending on type and intensity of loading, these displacements can stabilize at a certain level or grow continuously, up to thousands of μm.

Study of the behavior of joints under tangential impacts was performed for cylindrical interference-fit joints d = 25 mm loaded by an axial force or a torque, and on flat annular joints (d_0 = 62 mm, d_i = 42 mm) loaded by torque [12]. Initial normal pressures in the flat joints were σ = 13–16 MPa; in the cylindrical joints they were σ = 10–100 MPa. Some results are shown in Figs. 4.48–4.50.

When the energy of each impact cycle is low, displacements in the joints are fully elastic and do not accumulate (grow). Beyond the elastic region, displacements always accumulate. If the energy of one impact E only slightly exceeds the energy of elastic deformation E_{min}, then the process of accumulation of the residual displacement is slowing down. After a certain number of impact cycles the process is stabilized, not reaching the magnitude of displacement δ_0 at the breakthrough (lines 1–3 in Fig. 4.48). If the single impact energy is significantly greater than the elastic deformation energy E_{min}, the displacement gradually grows up to δ_0, and after that the surfaces abruptly commence a relative motion (lines 4–5 in Fig. 4.48). The motion can result in displacement measured in hundreds of μm. Displacement magnitude δ is proportional to the number of impacts before occurrence of the breakthrough event (Fig. 4.49a and b). If the impact loading continues, displacements again accumulate until a new slip occurs (Fig. 4.50). At a certain level of impact energy E_{max}, the initial slippage and resulting

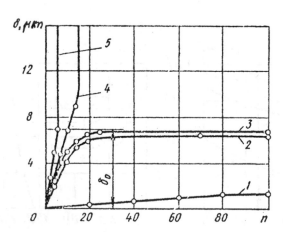

Figure 4.48 Circumferential displacement δ in inference-fit cylindrical joint between steel parts with σ = 18 MPa, R_a = 0.2 μm loaded by impact torque vs. number of impacts n; l/d = 3; E_{max} = 0.61 N-m; E_{min} = 0.21 N-m. 1, E = 0.24 N-m; 2, E = 0.27 N-m; 3, E = 0.34 N-m; 4, E = 0.42 N-m; 5, E = 0.49 N-m.

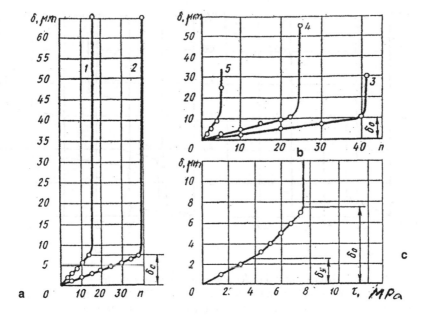

Figure 4.49 Axial displacement δ in cylindrical interference-fit joints: (a, b) impactloading; (c) static loading. (a, c) Pressure due to interference $\sigma = 30$ MPa, $E_{min} = 0.25$ N-m, $E_{max} = 1.05$ N-m; (b) $\sigma = 50$ MPa, $E_{min} = 0.6$ N-m, $E_{max} = 2.4$ N-m; 1, $E = 0.61$ N-m; 2, $E = 0.415$ N-m; 3, $E = 1.05$ N-m; 4, $E = 1.35$ N-m; 5, $E = 1.65$ N-m.

motion start after the first impact. For all tested joints, $E_{max}/E_{min} \approx 4$. These microslippage events are responsible for the fretting corrosion.

    The results discussed above are recorded for a large time interval between the impacts, more than one second. For higher impact frequency, displacements always accumulate without a visible slippage. The total displacement may reach thousands of μm, much larger than displacement before the breakthrough event under static loading.

    High-frequency (10–60 Hz) pulsating ($0-T_{max}$) torque was applied to a cylindrical interference-fit connection. Interference pressure for the test specimens was 2.5–50 MPa. In these tests, accumulation of residual displacement is slowing down and the displacement magnitude during each cycle is continuously diminishing (Fig. 4.51). The accumulated displacement converges to a certain limiting magnitude and the slippage does not develop. The residual displacement δ_1 at the first torque pulse is 5–7 to 0.5 μm, depending on magnitudes of the normal pressure σ and amplitude of the loading torque T. At $\sigma = 30$ MPa and $T = 32$ N-m, $\delta_1 = 2.5$ μm; if $T = 26.5$ N-m then $\delta = 1$ μm, and if $T = 24$ N-m then $\delta_1 = 0.5$ μm. As for impact loading, if the torque amplitude is increasing, at some

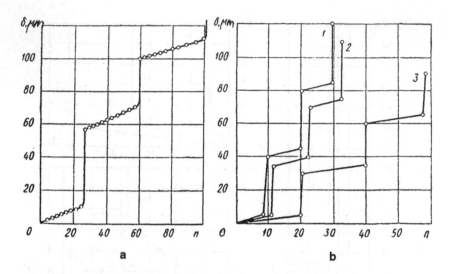

Figure 4.50 Displacement δ under impact loading: (a) cylindrical interference-fit joint of steel parts, $R_a = 0.2$ μm under axial force, σ = 30 MPa; (b) flat annular cast iron preloaded joint, surfaces scraped, σ = 16 MPa, $E_{max} = 1.05$ N-m, $E_{min} = 0.26$ N-m. 1, $E = 0.82$ N-m; 2, $E = 0.61$ N-m; 3, $E = 0.35$ N-m.

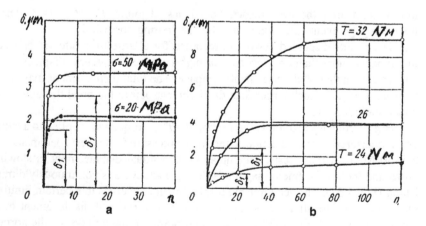

Figure 4.51 Circumferential displacement δ in interference-fit cylindrical connection between steel parts with $R_a = 0.2$ μm loaded by pulsating torque (0–T) vs. number of cycles n: (a) $T = 200$ N-m, varying σ; (b) σ = 30 MPa, varying T.

amplitudes the slippage develops at the first pulse; at small torque amplitudes the displacements do not leave the elastic zone.

If the joint is again subjected to the pulsating torque after the first exposure to the pulsating torque (during which the slippage had not developed), then the residual displacement is reduced and the torque causing continuous slippage becomes 1.25–1.5 times greater ("work hardening" of the joint). However, if during the first exposure there had been slippage, the work hardening does not develop.

Stabilization of the joint displacements under pulsating and impact loading with energy of one cycle not significantly exceeding E_{min} can be explained by increasing of the effective contact area (due to cyclical elastoplastic deformation, squeezing out the lubricant, enhanced adhesion, etc.). During high intensity impact loading, repeated plastic deformation results, finally, in a breakdown of frictional connections. The process may repeat itself since new frictional connections are developing. Another important factor is presence of high-frequency harmonics in the impact spectrum. They can result in reduction of resistance forces.

4.6.2 Dynamic (Stiffness and Damping) Model of Tangential Compliance

A basic model of a flat joint is shown in Fig. 4.52 [12]. Thin elastic strip 1 is pressed to rigid base 2 with pressure p. Strip 1 is axially loaded by a force αP cyclically varying from P_{min} ($\alpha = r$) to P ($\alpha = 1$). The maximum magnitude P of the force is assumed to be smaller than the friction force between strip 1 and base 2

$$P < fpbl \qquad (4.67)$$

where f = friction coefficient in the joint and b and l = width and length of strip 1, respectively. The friction is assumed to be dry (Coulomb) friction; the strip is assumed to have a linear load-deflection characteristic (to comply with the Hooke's law). Condition (4.67) means that the strip cannot slide as a whole along

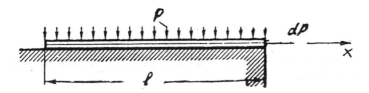

Figure 4.52 Axial loading of elastic strip 1 pressed to rigid base 2.

the base. To analyze displacements between strip 1 and base 2, it is important to note that:

(a)   The *friction force is equal to its ultimate magnitude per unit length*

$$q = fpb \qquad (4.68)$$

in all areas within which strip 1 is slipping along base 2. These are the only areas where the strip is being deformed due to assumed dry friction.

(b)   The *friction force is zero* in the areas within which the strip is not being deformed. It follows from the fact that in these areas the strip is not loaded, while the friction force, if present, would load the strip.

Thus, friction force in the joint is either q or 0. The adopted Hooke's law for deformation and Coulomb's law for friction absolutely exclude a possibility that in some area there is a friction force which is nonzero and less than q.

Let's consider three basic phases of the load change:

1.   The force αP is increasing from zero to maximum magnitude P, or $0 \leq \alpha \leq 1$.
2.   The force is decreasing from maximum magnitude P to minimum magnitude P_{min}, or $r \leq \alpha \leq 1$.
3.   The force is increasing from minimum magnitude P_{min} to maximum magnitude P, or $r \leq \alpha \leq 1$.

The last two phases are continuously repeating for cyclical variation of the force in $[P_{min}, P]$ interval.

First Phase of Loading

The length of the strip deformation area (slippage zone) is determined by the strip equilibrium condition and is (see Fig. 4.53a)

$$a_1 = \alpha P/q \qquad (4.69)$$

With a gradual increase of force αP, the length of this zone is increasing; according to Eq. (4.67)

$$a_{1_{max}} = P/q < l \qquad (4.70)$$

In accordance with Hooke's law, the relative elongation (strain) of an element of the deformed zone is

$$du_1/dx = N/EF \qquad (4.71)$$

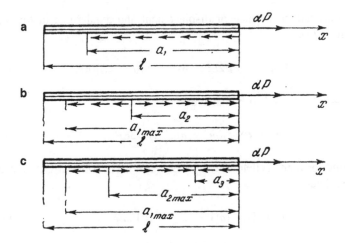

Figure 4.53 Forces acting on elastic strip: (a) loading (phase 1); (b) unloading (phase 2); (c) repeated loading (phase 3).

where $u_1(x, \alpha)$ = displacement of the cross section with coordinate x along the x-axis; $N = N(x, \alpha)$ = tensile force in cross section x of the strip; EF = tensile rigidity of the strip; and F = cross-sectional area of the strip. From equilibrium of an element dx it follows that

$$dN/dx = q \tag{4.72}$$

Substituting (4.71) into (4.72), we arrive at

$$d^2u_1/dx^2 = q/EF \tag{4.73}$$

The integral of Eq. (4.73) is

$$u_1 = A_1 + B_1 x + qx^2/2EF \tag{4.74}$$

The boundary conditions for the cross section dividing the deformed and undeformed zones are

$$u_1(l - a_1, \alpha) = 0; \quad du_1(l - a_1, \alpha)/dx = 0 \tag{4.75}$$

These conditions reflect absence of displacement and axial force at this cross section. From (4.75),

$$A_1 = [q(l - a_1)^2]/2EF; \quad B_1 = -[q(l - a_1)]/EF \tag{4.76}$$

Substituting (4.76) into (4.74), we arrive at

$$u_1(x, \alpha) = [q(l - a_1 - x)^2]/2EF \tag{4.77}$$

where a_1 depends on a and is determined by Eq. (4.69). The following expressions will be needed later:

$$u_1(x, 1) = [q(1 - a_{1_{max}} - x)^2]/2EF \tag{4.78}$$

$$u_1(1, \alpha) = \alpha^2 P^2/2qEF \tag{4.79}$$

Equation (4.78) describes distribution of displacements u_1 along the deformed zone in the end of the first phase of loading. Equation (4.79) defines displacement of the end cross section during the first phase of loading.

Second Phase of Loading

When the force αP starts to decrease, end elements of the strip start to shift in the negative x-direction; accordingly, friction forces in the positive x-direction appear. The equilibrium condition for the strip (Fig. 4.53b) is

$$\alpha P + qa_2 - q(a_{1_{max}} - a_2) = 0 \tag{4.80}$$

This expression allows one to find the length of the "back-shift zone" as

$$a_2 = P(1 - \alpha)/2q \tag{4.81}$$

and

$$a_{2_{max}} = P(1 - r)/2q \tag{4.82}$$

To the left of the "back-shift zone," the displacements and friction forces existing at the end of the first phase of loading remain frozen.

The equilibrium condition of some element within the "back-shift zone" is different from Eq. (4.72) and is

$$dN/dx = -q \tag{4.83}$$

Substituting (4.71) into (4.83), we arrive at

$$d^2u_2/dx^2 = -q/EF \qquad (4.84)$$

whose integral is

$$u_2 = A_2 + B_2x - qx^2/2EF \qquad (4.85)$$

This solution should comply with the following boundary conditions:

$$u_2(l - a_2, \alpha) = u_1(l - a_2, 1) \qquad (4.86a)$$

$$du_2(l - a_2, \alpha) = du_1(l - a_2, 1)/dx \qquad (4.86b)$$

These conditions reflect the identity of displacements and axial forces at the cross section $x = l - a_2$, which divides zones of "back-shift" and "direct" displacements. To construct the right-hand parts of expressions (4.86a) and (4.86b), Eq. (4.78) should be used. Thus,

$$A_2 = (q/EF)[(l - a_{1_{max}})^2 - 2(l - a_2)]; \quad B_2 = (q/EF)(l - 2a_2 + a_{1_{max}}) \qquad (4.87)$$

Consequently,

$$u_2(x, \alpha) = (q/2EF)[(a_2 - a_{1_{max}} - x)^2$$
$$- (l - a_2 - x)^2 - x^2 + 2(l - a_2)(a_2 - a_{1_{max}})] \qquad (4.88)$$

where a_2 is function of a and is determined by Eq. (4.81). The following expressions will be useful at the next phase:

$$u_2(x, r) = \frac{1 + 2r - r^2}{4qEF} P^2 - \frac{(l - x)rP}{EF} - \frac{q(l - x)^2}{2EF} \qquad (4.89)$$

$$u_2(l, \alpha) = \frac{1 + 2\alpha - \alpha^2}{4qEF} P^2 \qquad (4.90)$$

Equation (4.89) describes distribution of displacements u_2 within the "back-shift zone" at the end of the second phase of loading, while Eq. (4.90) represents shifting of the end cross section during the whole second phase.

Third Phase of Loading

In the beginning of the third phase the positive displacements are again developing in the end zone. Distribution of the friction forces is illustrated in Fig. 4.53c. The equilibrium condition for the strip is

$$\alpha P - (a_{1_{max}} - a_{2_{max}})q - a_3 q + (a_{2_{max}} - a_3)q = 0 \qquad (4.91)$$

From Eq. (4.91) the length of the end zone is determined as

$$a_3 = \frac{\alpha - r}{2q} P \qquad (4.92)$$

The differential equation again becomes as in Eq. (4.73), and its integral is

$$u_3 = A_3 + B_3 x + qx^2/2EF \qquad (4.93)$$

Boundary conditions in the cross section with the coordinate $x = l - a_3$ are

$$u_3(l - a_3, \alpha) = u_2(l - a_3, r); \qquad du_3(l - a_3, \alpha)/dx = du_2(l - a_3, r)/dx \quad (4.94)$$

and these boundary conditions lead to

$$A_3 = \frac{(1 + 2r - r^2)P^2}{4qEF} + \frac{q}{2EF}[(l - 2a_3)^2 - 2a_3^2]; B_3 = \frac{q}{EF}\left[\frac{rP}{q} + 2a_3 - l\right] \quad (4.95)$$

Thus,

$$u_3(x, \alpha) = \frac{(1 + 2r - r^2)P^2}{4qEF} - \frac{(l - x)rP}{EF} - \frac{q(l - x)^2}{2EF} + \frac{q(l - a_3 - x)^2}{EF} \quad (4.96)$$

and displacement of the end cross section $x = l$ is expressed as

$$u_3(l, \alpha) = \frac{1 - 2\alpha r + 2r + \alpha^2)P^2}{4qEF} \qquad (4.97)$$

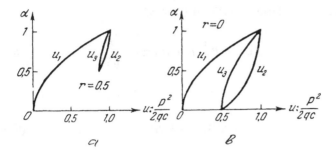

Figure 4.54 Hysteresis loops: (a) $r = 0.5$; (b) $r = 0$.

Figure 4.54 illustrates the functions $u_1(l, \alpha)$, $u_2(l, \alpha)$, and $u_3(l, \alpha)$ for cases when $r = 0.5$ and $r = 0$. In both cases, there are close hysteresis loops. The area inside the loop represents part Ψ of the work performed by the force αP that is lost (dissipated). This area can be computed as

$$\Psi = \int_{P_{min}}^{P} [u_2(l, \alpha) - u_3(l, \alpha)]d(\alpha P) = P \int_{r}^{1} (u_2 - u_3)d\alpha \qquad (4.98)$$

Substituting (4.90) and (4.97) into (4.98), we arrive at

$$\Psi = \frac{P^3(1 - r)^3}{12qEF} \qquad (4.99)$$

This result is easier to understand if the amplitude of force αP is designated as P_v, with $P(1 - r) = 2P_v$. Then Eq. (4.99) becomes

$$\Psi = \frac{2P_v^3}{3qEF} \qquad (4.100)$$

This expression shows that energy dissipation depends only on the cyclical component P_v and does not depend on the median (d.c.) component of the force $P_m = (P_{min} + P)/2$. For practical calculations, it is important to express the energy dissipation in relative terms, as a relative energy dissipation ψ that is expressed as the ratio of Ψ to the maximum potential energy V_{max} of the system. For small ψ,

$$\psi \cong 2\delta \qquad (4.101)$$

where δ = logarithmic decrement. From Fig 4.52,

$$V_{max} = Pu_{max} - P \int_0^1 u_1(l, \alpha)d\alpha = \frac{2P^2}{3qEF} \qquad (4.102)$$

and

$$\psi = \Psi/V_{max} = P_v^2/P^2 \qquad (4.103a)$$

In order to determine stiffness of the system in Fig. 4.52 and to analyze dynamic characteristics of the system as functions of the displacement amplitude, displacement amplitude of the end cross section of the strip can be expressed, using Eq. (4.97), as

$$u_v = u_3(l, 1) - u_3(l, r) = \frac{(1 - 2r + 2r + 1)P^2}{4qEF} - \frac{(1 - 2r^2 + 2r + r^2)P^2}{4qEF} \qquad (4.104)$$

$$= \frac{(1 - r)^2 P^2}{4qEF} = \frac{4P_v^2}{4qEF} = \frac{P_v^2}{qEF}$$

which is obvious from Fig. 4.54. Combining (4.103a) and (4.104), we get

$$\psi = \frac{4qEF}{P^2}u_v \sim 2\delta \qquad (4.103b)$$

This amplitude dependence of log decrement δ is very similar with test data for mesh-like and elastomeric materials in Fig. 3.2. For a symmetric cycle ($r = -1$, $P_v = P$), from Eq. (4.103a) $\psi = 1$. Such large values of damping parameters are typical for the systems with hysteresis. However, they have to be treated carefully since the model in Fig. 4.52 is rather simplistic. For example, if the strip in Fig 4.52 continues further to the right, this extension would contribute to the value of V_{max}, thus resulting in reduction of ψ in Eq. (4.103a). Equation (4.103b) indicates that a higher damping effect can be achieved (for the same displacement amplitude u_v) by increasing friction coefficient f and normal pressure p [see Eq. (4.68)].

Figure 4.54 indicates that the *static stiffness* k_{st} of the system in Fig. 4.52 (stiffness associated with the first phase of loading from $\alpha P = 0$ to $\alpha P = P$) is significantly smaller than *dynamic stiffness* k_{dyn} associated with the cyclical loading (the second and third phases of loading). This is also typical for mesh-like and elastomeric materials, for example, those represented in Fig. 3.2. The average value of static stiffness for the first phase of loading is the ratio of the maximum force P to maximum displacement of the end $u_1(l, 1)$. Using Eq. (4.79),

$$k_{st} = u_1(l, 1)/P = 2qEF/P \qquad (4.105a)$$

or the system is softening nonlinear (see Chapter 3) with the static stiffness decreasing inversely proportional to the applied force P. The average value of dynamic stiffness is the ratio of the amplitude of cyclical force P_v to displacement amplitude u_v or

$$k_{dyn} = P_v/u_v = P_v/(P_v^2/qEF) = qEF/P_v \qquad (4.105b)$$

From Eqs. (4.105a) and (4.105b), the dynamic-to-static stiffness ratio K_{dyn} (see Chapter 3) can be derived as

$$K_{dyn} = k_{dyn}/k_{st} = 2P/P_v = 4/1 - r \qquad (4.106)$$

It follows from Eq. (4.106) that the dynamic-to-static stiffness ratio for the model in Fig. 4.52 is increasing with decreasing amplitude of the cyclical force (or, consequently, amplitude of vibratory displacement). This also is in a qualitative agreement with test data for the mesh-like and elastomeric materials in Fig. 3.2. It can be noted that the mesh-like materials consist of wires or fibers pressed to each other by design as well as by the performance-related forces.

Similar analyses were performed by Panovko [12] for axial and torsional loading of cylindrical press-fit joints, for axial and torsional loading of riveted connections, for elasto-frictional connections (a model as in Fig. 4.52 in which the interaction between the strip and the base is not only frictional, but also elastic).

4.7 PRACTICAL CASE STUDY OF A MODULAR TOOLING SYSTEM

A modular tooling system Varilock shown in Fig. 4.55 was found to have excessive runouts at the end of tool (up to 50–100 μm) if several spacers are used [14]. This prevented its use for long overhang tools without lengthy adjustments. The system uses round couplings with annular contact surfaces ($D_0 = 63$ mm, $D_i = 46.5$ mm, and surface area $A = 2.9 \times 10^3$ mm^2) interrupted by a deep key slot. The joints between the components are tightened by M19 ($\sim$3/4 in.) bolts with the manufacturer-recommended tightening torque 150 lb-ft (200 N-m). Axial force F_a generated by this torque was calculated using an approximate formula

$$T = 0.2 \, F_a D \qquad (4.67)$$

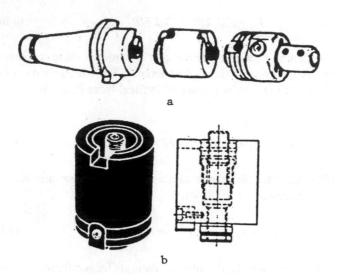

a

b

Figure 4.55 Modular tooling system Varilock with axial tightening.

where T = tightening torque and D = bolt diameter. For T = 200 N-m and D = 0.019 m, F_a = 52,600 N (~12,000 lbs), and the specific pressure in the joint is F_a/A = 18 MPa (2700 psi).

Inspection of the components has shown that due to the interrupted surface caused by the presence of the key slot, the radial edges of the key slot are protruding by 2.5–5.0 μm.

Finite element analysis of compression of the spacer having a key slot has shown that due to presence of the key, compression at the side of the key slot (0.012 mm) was greater than at the diametrically opposing side (0.007 mm). Thus, the difference in compression deformations on the opposing sides of the contact surface of the spacer is 0.005 mm. Since both connected components have this deformation difference, the total asymmetry is 0.005 × 2 = 0.01 mm. This asymmetry results in the angular displacements between the two connected parts

$$0.01/63 = 0.16 \times 10^{-3}$$

This angular deformation is due to the solid body compression of the contacting parts. In addition to this, there are also contact deformations of the surfaces. Since the surfaces are finely ground, c = ~0.5 (see Section 4.2 above). However, due to the presence of protrusions whose role is similar to other asperities on the surface, the value of c would be different for different sides of the joint. For the side with the key slot/protrusions, it can be taken as for the fine-to-medium deep scraping (depth of dips 3–10 μm), $c \approx 2$. From Eq. (4.1), for this side

$$\delta = 2 \times 18^{0.5} = 8.5 \text{ μm}$$

while for the other side

$$\delta = 0.5 \times 18^{0.5} = 2.1 \text{ μm}$$

Thus, the additional angular displacement due to the contact deformation is

$$(0.0085 - 0.0021)/63 = 0.1 \times 10^{-3} \text{ rad}$$

The overall angular displacement in one joint per 200 N-m tightening torque is $(0.16 + 0.1) 10^{-3} = 0.26 \times 10^{-3}$ rad, or 0.13×10^{-5} rad/N-m. This angular deformation is approximately proportional to the specific pressure (tightening torque).

The initial position of the joint (with no tightening torque applied) is also inclined from the theoretical axis of the modular toolholder due to the presence of the protrusions. For the protrusions 5 μm high, the initial inclination is

$$0.005/63 = 7.9 \times 10^{-5} \text{ rad}$$

This inclination is in the opposite direction relative to the torque-induced inclination. Thus, the total inclination at torque T is

$$\alpha = (0.13 \times 10^{-5} T - 7.9 \times 10^{-5}) \text{rad}$$

The minimum (zero) inclination would develop when

$$T = 7.9/0.13 = 61 \text{ Nm}$$

Inclinations α in the joints are projected to the end-of-tool thus creating eccentricity and runout. With one extension spacer 120 mm long, displacement at the end will be $\Delta = 120\alpha$, with two assembled spacers $\Delta = 240\alpha$, etc. The toolholder 80 mm long adds to this displacement, thus for one spacer $\Delta = 200\alpha$, for two spacers $\Delta = 320\alpha$, and so on. The end runout is 2Δ.

Fig. 4.56a shows the predicted runout of several assemblies. The runout is fast increasing with increasing length of the assembly. At $T = 60$ N-m, the runout changes its direction (zero runout). Zero inclination does not necessarily mean zero runout due to other causes of runout, but it is associated with a much smaller runout.

These conclusions were tested by first measuring the assembly with one, two, three, and four joints at $T = 20$ N-m, then all combinations were tested at 40 N-m, 80 N-m, 120 N-m, and 190 N-m. The results are shown in Fig. 4.56b. The plots are similar to the prediction but the measured runouts are much lower than calculated. Still, runouts at the recommended $T = 200$ N-m are hardly acceptable. Also, it was known that runouts on the shop floor vary in a broad range and can become much larger.

The main conclusion of the analysis is that *there is an optimal magnitude of the tightening torque* (about 95 N-m from the test results, one-half of the manufacturer-recommended value).

4.8 DAMPING OF MECHANICAL CONTACTS

4.8.1 Damping in Flat Joints

There are three major processes contributing to loss of energy during vibratory processes (damping) in mechanical systems:

1. Loss of energy in material of the component (material damping)
2. Loss of energy in contact areas between the fitting parts (external friction during micro motions, e.g., see Sec. 4.6, and external and internal friction during deformation of micro asperities on the contacting surfaces)
3. Loss of energy in layers of the lubricating oil, due to both viscous friction in the oil layer and viscous friction between microasperities

Usually, energy dissipation in contacts is much more intense than inside structural materials. The exceptions are materials with extremely high internal hysteresis, such as nickel-titanium alloys (see Sec. 8.3) and elastomers and other specially designed polymers whose Young's modulus is too low for structural applications.

Results of a major study on energy dissipation in mechanical joints are presented in Levina and Reshetov [4]. After numerous experiments there were established damping parameter values and empirical correlations for evaluation of energy dissipation in joints considering materials of the components being joined, surface finish, amount and viscosity of oil in the joint, frequency and amplitude of vibrations, and shape and dimensions of the joint.

It is important to realize that character of damping in joints, as well as of material damping, is different from the classic "viscous damping" model (see

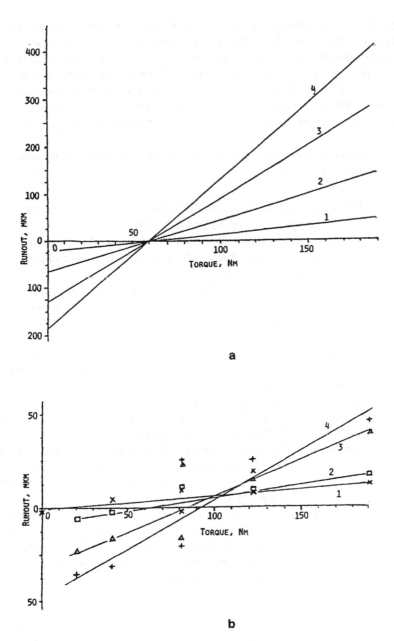

Figure 4.56 Runout of Varilock modular tooling system: (a) predicted; (b) measured. 1, Taper and one spacer; 2, same, but two spacers; 3, same, three spacers; 4, taper, three spacers, and tool head.

Appendix 1). The damping is due, at least, in a first approximation, to velocity-independent hysteresis of loading/unloading process, e.g., as discussed in Sec. 4.6, rather than due to velocity-dependent viscous friction in a viscous damper. Accordingly, such important dimensionless damping parameters as "damping ratio" c/c_{cr} and "log decrement" δ do not depend on the mass of the vibratory system and can characterize damping of a joint or of a material. Another convenient damping parameter for hysteresis-based damping is "loss angle" β, or $\tan \beta$.

Experimental findings described by Levina and Reshetov [4] and others can be summarized as follows:

1. Where better surface finish results in higher contact stiffness, the damping decreases with improved surface finish.
2. Damping (log decrement δ) in joints between steel and cast iron parts has essentially the same magnitude since the main source of energy dissipation is external friction.
3. In nonlubricated joints, damping does not depend on the pressure in the joint in the range of pressure 0.1–2 MPa (15–300 psi). For steel or cast iron joints with scraped or ground surfaces, $\delta = {\sim}0.075$; for cast iron–polymer joints, $\delta = {\sim}0.175$.
4. In lubricated joints, energy dissipation is growing with the increasing amount and viscosity of oil and decreasing with increasing mean pressure σ_m (Fig. 4.57). The amount of oil in joints tested for Fig. 4.57 is typical for the boundary friction condition.
5. Joints with scraped and ground surfaces *with the same surface finish* (same R_a) exhibit practically same damping.
6. Damping (log decrement δ) does not depend strongly on frequency and amplitude of vibration (frequency and amplitude were varied in the range 1–10). This is characteristic for the hysteresis-induced damping with $n = 1$ (see Appendix 1).
7. Damping does not depend significantly on the joint dimensions (in the experiments, the joint area was changing in the range of 2.2:1, and its cross-sectional moment of inertia in the range of 5:1). Damping is slightly increasing with increasing width of the contact area.
8. For joints with narrow contact surfaces,

$$\delta = \frac{A}{\sqrt[3]{\sigma_m}}$$ (4.107)

where $A = 0.31$ for joints lubricated by industrial mineral oil of average viscosity and $A = 0.21$ for very thin oil; σ_m is in MPa.

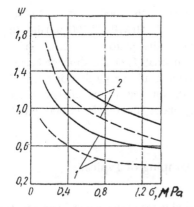

Figure 4.57 Damping in flat joints: cast iron, cast iron and steel, and steel. 1, Rectangular joints with narrow contact surfaces; 2, wide annular joints; solid lines, lubrication with medium viscosity industrial mineral oil; dashed lines, low viscosity industrial mineral oil. The amount of oil in the joints 10 gram/m².

9. Damping for tangential vibration for joints preloaded with $\sigma_m = 50\text{--}80$ MPa (750–1,200 psi) is $\delta = 0.04\text{--}0.05$ without lubrication and $\delta = 0.3\text{--}0.37$ for lubricated joints.

4.8.2 Damping in Cylindrical and Tapered Connections

Damping in cylindrical and tapered connections is mostly due to normal and tangential displacements and due to local slippages between the shaft and the sleeve.

Damping in nonlubricated preloaded tapered connections as well as in interference-fit cylindrical connections is $\delta = 0.01\text{--}0.05$ for vibrations perpendicular to the axis of the connection. It does not significantly depend on normal pressure. Damping in lubricated connections is about $\delta = 0.12\text{--}0.35$, and depends on the normal pressure as

$$\delta = 0.4\,\sigma^{-0.2} \tag{4.108}$$

where σ = normal pressure in the connection (MPa). As for the flat joints, δ is amplitude-independent.

When a long sleeve (length L) is fit on a relatively low stiffness shaft (diameter D), energy dissipation may increase due to microslippages between the sleeve and the shaft caused by bending vibrations of the shaft. For large ratios L/D and low normal pressures, $\sigma < 20$ MPa (3,000 psi), the energy loss caused by contact deformations is relatively small, up to 20% of the total; at σ = 20–30 MPa (3,000–4,500 psi) it raises to 40–50%, and at σ > 50 MPa (7,500 psi) the slippage zone has practically disappeared and all energy dissipation is due to contact deformations. In more common connections with small L/D, the slippage has already disappeared at σ = ~30 MPa (4,500 psi) and all energy dissipation is due to contact deformations. Optimal contact pressure, at which energy dissipation (damping) is the most intensive, is usually about 10–40 MPa (1,500–6,000 psi).

4.8.3   Comparison of Stiffness and Damping of Typical Mechanical Contacts

One experimental study [9] compares several types of mechanical contacts (interfaces) by their stiffness (K), damping (log decrement δ), and the chatter-resistance criterion $K\delta$ as functions of the axial preload force Q.

Fig. 4.58 shows plots of static bending stiffness for annular flat joint (outer diameter 63 mm, inner diameter 36 mm); three sizes of 7/24 taper connection (#30—gage diameter 32 mm; #40—gage diameter 44.5 mm; #45—gage diameter 57.2 mm); and two types of curvic coupling connections (outer diameter 63 mm, inner diameter 36 mm), type A having 20 teeth and type B having 24 teeth. The bending stiffness was measured at the end of a 200 mm long, 45 mm diameter bar; the theoretical bending stiffness of the built-in bar is given as a reference.

Figure 4.59 gives the damping data for the same mechanical interfaces. Comparison of Figs. 4.58 and 4.59 shows that while the bending stiffness is increasing with the increasing axial preload, the damping is decreasing with the increasing axial preload. These trends indicate that some optimal value of the axial preload may be determined to maximize the product $K\delta$ of stiffness and damping values.

Table 4.5 was calculated using data from Figs. 4.58 and 4.59. It shows that while the flat joint is the stiffest, the curvic coupling A demonstrates the best dynamic quality (value at the chatter resistance criterion $K\delta$) while being not far behind in stiffness at high magnitudes of axial preload force. The general trend is for reduction of damping with the increasing axial preload. For flat and tapered connections, the value of $K\delta$ is also decreasing with the increasing axial preload. For the curvic coupling connections, the bending stiffness increase with the increasing axial preload may be more pronounced than the reduction of damping, and some value for the axial preload (10 kN for the data in Table 4.5) corresponds to the highest value of $K\delta$.

It should be remembered, however, that there are situations in which the value of stiffness (or the value of damping) is the most important parameter. Thus, selection of the type and size of the connection and of the axial preload force magnitude has to be made with consideration for these factors.

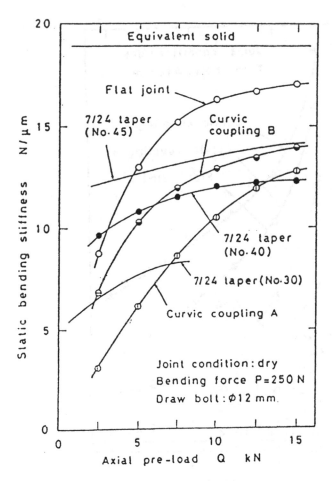

Figure 4.58 Effect of axial preload on static bending stiffness of various interface systems.

4.8.4 Energy Dissipation in Power Transmission Components

Energy dissipation in key and spline connections is due to normal and tangential deformations and to slippages on the contact surfaces. In a key connection, angular deformations are associated with tangential displacements on the cylindrical contact surface between the shaft and the sleeve and with normal deformations between the key and the key slot. Radial deformations in the connection are associated with normal displacements in the cylindrical contact and tangential

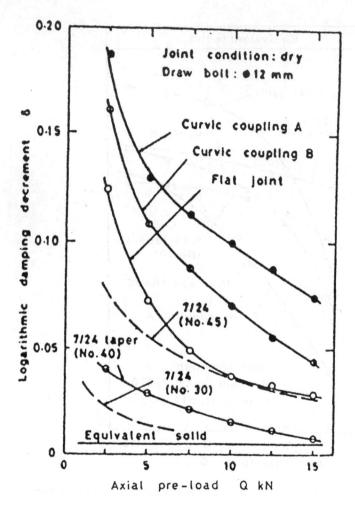

Figure 4.59 Damping of various toolholder/spindle interfaces.

deformations along the key. Energy dissipation in a tight key or spline connection is relatively low and is characterized by $\delta = 0.05-0.1$. In a sliding spline connection, $\delta = 0.3-0.4$ due to larger slippage.

Energy dissipation in antifriction bearings during bending vibrations of the shafts is due to elastic deformations at contacts of the rolling bodies with the races, at contacts between the races and the shaft and the housing caused by radial and angular displacements of the shaft, and partly due to friction between

Table 4.5 Stiffness, Damping, and Chatter-Resistance Criterion for Various Mechanical Interfaces

Q, kN	K, N/μm			δ			Kδ		
	5	10	15	5	10	15	5	10	15
Solid		18.5			0.006			0.11	
Flat joint	13	16.5	17.3	.075	.04	.03	.98	.66	.52
#45	12.5	13	14	.06	.04	.03	.75	.52	.42
#40	10.5	12	12.5	.03	.02	.01	.32	.24	.125
#30	7.5	7.9	—	.02	.01	—	.15	.08	—
Curvic coupling A	5.5	10.2	13	.13	.1	.075	.72	1.0	.98
Curvic coupling B	7.6	12.5	14	.11	.072	.045	.84	.9	.63

the rolling bodies and the cage. In tapered roller bearings an additional contributor is friction at the ends of the rollers.

It was observed during experiments that damping (δ) in antifriction bearings is amplitude- and frequency-independent. For bearings installed in single-bearing units with clearances not exceeding 10 μm, an average $\delta = 0.1–0.125$ for ball bearings; $\delta = 0.15–0.2$ for single- and double-row roller bearings; and $\delta = 0.15–0.2$ for tapered roller bearings at small angular vibrations. At low shaft stiffness resulting in larger angular vibrations, damping at tapered roller bearings is significantly higher. At high levels of preload in tapered roller bearings, $\delta = 0.3–0.35$. For double ball bearing units, $\delta = 0.1–0.15$; for a bearing unit consisting of one radial and one thrust ball bearings, $\delta = 0.25–0.3$. Energy dissipation at larger clearances in bearings (20–30 μm) and without constant preload is much more intense due to impact interactions (see Appendix 3), and can reach $\delta = 0.25–0.35$.

Use of this data in evaluating damping characteristic of a whole transmission system or other structures is described in Section 6.5.

REFERENCES

1. Pisarenko, G.S., Yakovlev, A.P., and Matveev, V.V., Strength of Materials Handbook, Naukova Dumka, Kiev, 1975.
2. Rivin, E.I., "Some Problems of Guideways Analysis," Research on Machine Tools, Mashgiz Publishing House, Moscow, 1955, pp. 101–113 [in Russian].

3. Rivin, E.I., Mechanical Design of Robots, McGraw-Hill, New York, 1988.
4. Levina, Z.M., and Reshetov, D.N., Contact Stiffness of Machines, Mashinostroenie, Moscow, 1971, [in Russian].
5. Rivin, E.I., "Compilation and Compression of Mathematical Model for Machine Transmission," ASME Paper 80-DET-104, ASME, New York, 1980.
6. Eibelshauser, P., and Kirchknopf, P., "Dynamic Stiffness of Joints," Industrie-Anzeiger, 1985, Vol. 107, No. 63, pp. 40–41 [in German].
7. "Valenite STS Quick Change System: Technical Information," 1993 [Translation of a Technical Report from RWTH Aahen].
8. Tsutsumi, M., Nakai, R., and Anno, Y., "Study of Stiffness of Tapered Spindle Connections," Nihon Kikai Gakkai Rombunsu [Trans. of Japan Society of Mechanical Engineers], 1985, C51(467), pp. 1629–1637 [in Japanese].
9. Hasem, S., Mori, J., Tsutsumi, M., and Ito, Y., "A New Modular Tooling System of Curvic Coupling Type," Proceedings of the 26th International Machine Tool Design and Research Conference, MacMillan Publishing, New York, 1987, pp. 261–267.
10. Agapiou, J., Rivin, E., and Xie, C., "Toolholder/Spindle Interfaces for CNC Machine Tools," Annals of the CIRP, 1995, Vol. 44/1, pp. 383–387.
11. Kato, M., et al, "Axial Contact Stiffness between Center and Workpiece," Bulletin of the Japan Society of Precision Engineering, 1980, Vol. 14, No. 1, pp. 13–18.
12. Panovko, Ya. G., "Internal Friction in Vibrating Elastic Systems," Fizmatgiz, Moscow, 1960 [in Russian].
13. Reshetov, D.N., and Kirsanova, V.A., "Tangential Contact Compliance of Machine Elements," Mashinovedenie, 1970, No. 3 [in Russian].
14. Rivin, E.I., and Xu, L., "Toolholder Structures—A Weak Link in CNC Machine Tools," Proceedings of the 2nd International Conference on Automation Technology, Taipei, Taiwan, July 4–6, 1992, Vol. 2, pp. 363–369.
15. Kobayashi, T., Matsubayashi, T., "Considerations on the Improvement of Stiffness of Bolted Joints in Machine Tools," Bulletin of JSME, 1986, Vol. 29, No. 257, pp. 3934–3937.

5

Supporting Systems and Foundations

The effective stiffness of mechanical systems is determined not only by stiffness characteristics of their components but also by designed interactions between the components. Influence of interactions is both due to elastic deformations in the connections between the interacting components (*contact deformations*; see Chapter 4) and due to the location, size, and shape of the connection areas (*supporting conditions*). Relatively slight changes in the supporting conditions can result in up to 1–2 decimal orders of magnitude changes of deformations under the specified forces, as well as in very significant changes in stress conditions of the components.

The influence of parameters of the supporting elements on effective damping of the supported component or structure is more difficult to assess. It depends not only on parameters (stiffness and damping) of the supporting elements, but also on rather complex correlations between dynamic characteristics of the supported system/component and the supporting system. In some cases this influence can be analyzed, at least approximately, in a close form. Such cases are represented by an analysis of influence of the mounting (vibration isolation) system on dynamic stability (chatter resistance) of a machine tool given in Article 1. However, in most cases influence of the supporting structure on the effective damping has to be analyzed by computational dynamic modeling/analysis.

This chapter addresses effects of the supporting systems/foundations on static stiffness/deformations of the supporting systems. Some important effects of the supporting conditions on damping are discussed in Section 5.3.

5.1 INFLUENCE OF SUPPORT CHARACTERISTICS

Figure 5.1 [1] compares stiffness of a uniform beam of length l loaded with a concentrated force P or distributed forces of the same overall magnitude and with a uniform intensity $q = P/l$. Maximum deformation f and stiffness k are

$$f = Pl^3/aEI; \qquad k = P/f = aEI/l^3 \qquad (5.1)$$

where E = Young's modulus of the beam material, I = cross-sectional moment of inertia, and a = coefficient determined by the supporting conditions. The cantilever beam (cases 5 and 6) is the least stiff and the double built-in beam (cases 3 and 4) is the stiffest, 64 times stiffer than the cantilever beam for the case of concentrated force loading and 48 times stiffer for the case of distributed loading.

Such simple cases are not always easy to identify in real life systems because of many complicating factors.

Figure 5.2 [1] compares 14 designs of joint areas between piston, piston pin, and connecting rod of a large diesel engine. The numbers indicate relative values of deformation f of the piston pin under force P and of maximum stress σ in the pin as fractions of deformation $f_i = Pl^3/48EI$ and stress $\sigma_1 = Pl/4W$ in the pin considered as a double-supported beam (the first case). It can be seen that seemingly minor changes of the piston and the connecting rod in the areas of interaction (essentially, design and positioning of stiffening ribs and/or bosses) may result in changes of deformation in the range of $\sim125:1$ and of stresses in the range of $\sim12:1$

However, it has to be noted that models in Fig. 5.2 are oversimplified since they do not consider clearances in the connections. Figure 5.3 illustrates the in-

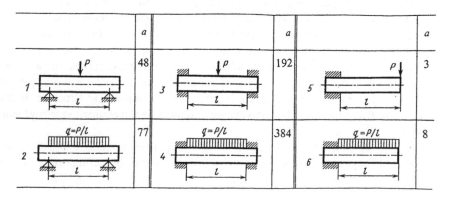

Figure 5.1 Influence of loading/supporting schematic on stiffness.

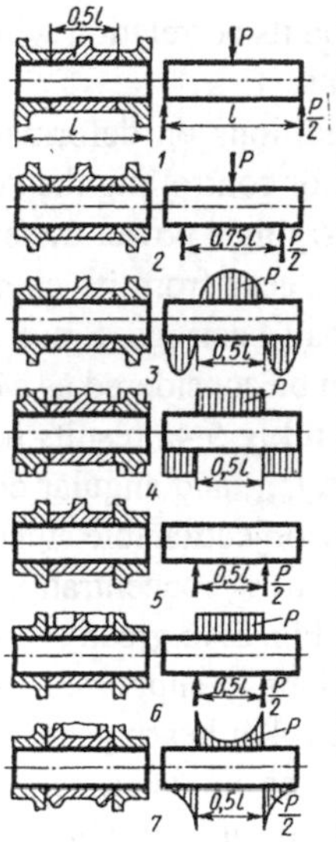

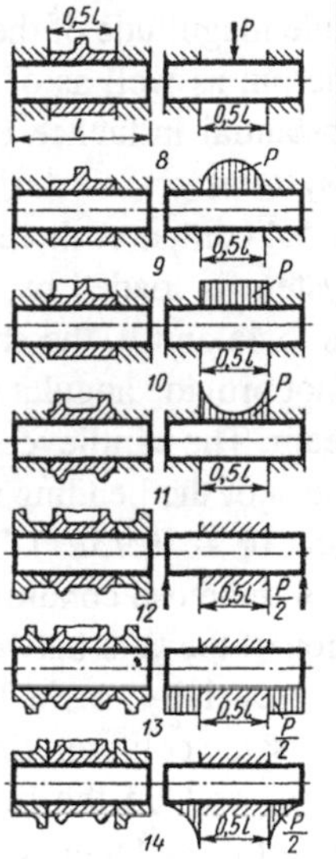

	σ	f		σ	f
1	1	1	8	0,25	0,031
2	0,75	0,42	9	0,21	0,023
3	0,56	0,36	10	0,083	0,0156
4	0,5	0,25	11	0,04	0,008
5	0,5	0,125	12	0,5	0,125
6	0,25	0,078	13	0,25	0,047
7	0,12	0,04	14	0,12	0,021

Figure 5.2 Influence of assembly design on stiffness.

fluence of the clearances. When the connection is assembled with tight fits (Fig. 5.3a) the loading of the pin model can be described as a double-built-in beam. A small clearance (Fig. 5.3b) results in a very significant change of the loading conditions caused by deformations of the pin and by the subsequent nonuniformity of contacts between the pin and the piston walls and between the pin and the bore of the connecting rod. For a larger clearance (Fig. 5.3c), the contact

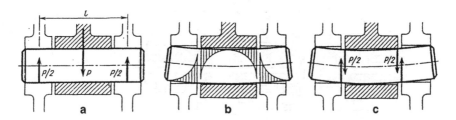

Figure 5.3 Influence of fit between interacting parts on loading schematic.

areas become edge contacts and the effective loading pattern (thus, deformations and stresses in the system) is very different from the patterns in Figs. 5.3a and b. The meaning of the terms "small" and "large" clearance depends not only on the absolute magnitude of the clearance, but also on its correlation with deformation of the pin as well as of two other components.

The substantial influence of the supporting conditions on deformations in mechanical systems gives a designer a powerful tool for controlling the deformations. Figure 5.4 compares three designs of a shaft carrying a power transmission gear. The system is loaded by force P acting on the gear from its counterpart gear. In Figs. 5.4a and b, the shaft is supported by ball bearings. Since the ball bearings do not provide angular restraint, the shaft can be considered as a double-supported beam. The cantilever location of the gear in Fig 5.4a results in excessive magnitudes of the bending moment M, deflection f_{max}, and angular deformation of the gear ($\phi = Pt^2/3EI$). The latter is especially objectionable since it leds to distortion of the mesh conditions and to significant stress concentrations in the mesh. Location of the gear between the supports as in Fig. 5.4b greatly reduces M and f_{max} and, especially, ϕ. In Fig. 5.4c, the same shaft is supported by roller bearings in the same configuration as in Fig. 5.4b. The roller bearings resist angular deformations and, in the first approximation, can be modeled as a built-in support, thus further reducing deflection f_{max} by the factor of four while reducing the maximum bending moment M_{max} (and thus maximum stresses) by the factor of two. In both cases of Figs. 5.4b and c, due to symmetrical positioning of the gear, angular deflection or the gear is zero and thus the meshing process is not distorted. There would be some, relatively minor, angular deflection and distortion of the meshing process if the gear were not symmetrical.

Large angular deflections of gears mounted on cantilever shafts result from

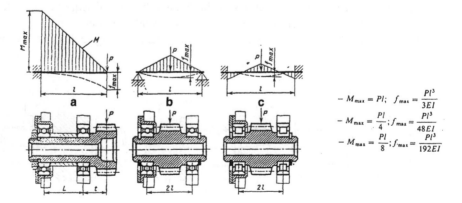

$$-M_{max} = Pl; \quad f_{max} = \frac{Pl^3}{3EI}$$

$$-M_{max} = \frac{Pl}{4}; f_{max} = \frac{Pl^3}{48EI}$$

$$-M_{max} = \frac{Pl}{8}; f_{max} = \frac{Pl^3}{192EI}$$

Figure 5.4   Influence of selection and configuration of supporting elements.

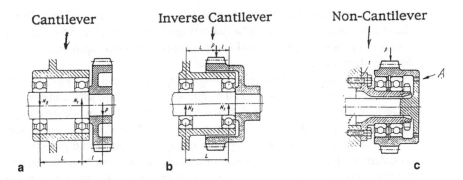

Figure 5.5 Reduction of reaction forces in supporting elements by design.

deformations of the double-supported shaft, but also due to contact deformations of the bearings (see Chaps. 4, 6) caused by reaction forces. A very important feature of cantilever designs, as in Fig. 5.5a, is the fact that the reaction forces, especially N_1, can be of a substantially higher magnitude than the acting force P as illustrated by Fig. 5.6a. In Fig. 5.5a, the greatest deformation is in the front bearing 1 accommodating a large reaction force N_1. As can be seen from Fig. 5.6a, the reaction forces are especially high when the span L between the supports (bearings) is less than $2l$ (l = length of the overhang). In addition to high magni-

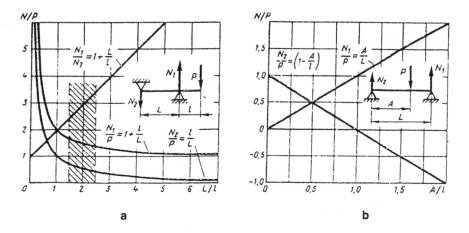

Figure 5.6 Reaction forces at supports of a double-supported beam with (a) out-of-span (cantilever) loading and (b) in-span loading.

tudes, the reaction forces for the model in Fig. 5.6a have opposite directions, thus further increasing the angular deflection of the shaft in Fig. 5.5a. The reaction forces in cases when the external force is acting within the span are substantially less, as shown in Fig 5.6b.

In some cases the force-generating component, such as a gear, can not be mounted within the span between the supports. Sometimes, such components can be reshaped so that the force is shifted to act within the span, while the component is attached outside the span, like the gear in Fig. 5.5b (*"inverse cantilever"*).

Even better performance of power transmission gears and pulleys, as well as precision shafts (e.g., the spindle of a machine tool) can be achieved by total separation of forces acting in the mesh, belt preloading forces, etc., and torques transmitted by the power transmission components. Such design is shown in Fig. 5.5c where the gear is supported by its own bearings mounted on the special embossment of the shaft housing (gear-box, headstock). The shaft also has its own bearings, but the connection between the gear and the shaft is purely torsional via coupling A, and the forces acting in the gear mesh are not acting on the shaft. In this case, the gear is maintained in its optimal nondeformed condition, and the shaft is not subjected to bending forces.

The same basic rules related to influence of supporting conditions on deformations/deflections apply to piston–pin–connecting rod assembly in Fig. 5.2 and to power transmission gears on shafts in Figs. 5.4 and 5.5. They are true also for machine frames and beds mounted on a floor or a foundation, a massive table mounted on a bed, etc. Although these rules have been discussed for beams, they are (at least qualitatively) similar to plates. For example, for round plates, *"built-in"* support (Fig. 5.7b) reduces deformation 7.7 times as compared with *"simple edge"* support (Fig. 5.7a).

It is very important to correctly model the support conditions of mechanical components in order to improve their performance. Incorrect modeling may significantly distort the actual force schematics and lead to very wrong estimations of the actual stiffness. Figure 5.8 [1] presents a case of a shaft supported by two sliding hydrodynamic bearings A and B, and loaded in the middle by a radial

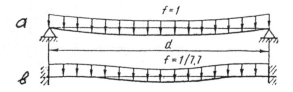

Figure 5.7 Influence of design of the support contour on deformation of round plates under distributed (e.g., weight) loading: (a) simple support contour; (b) built-in contour.

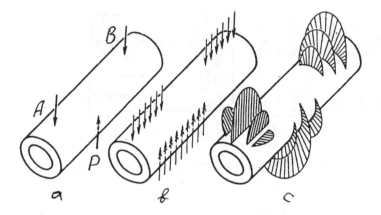

Figure 5.8 Loading schematics of a double-supported shaft.

force P also transmitted through a hydrodynamic bearing. This model describes, for example, a ''floating'' piston pin in an internal combustion engine. The actual pressure distribution along the length of the load-carrying oil film in the bearings is parabolic (left diagrams in Fig. 5.8c). The peak pressures are 2.5–3.0 times higher than the nominal (average) pressures. In transverse cross sections the pressure is distributed along a 90–120° arch (right and center diagrams in Fig. 5.8c).

Comparison of actual loading diagrams in Fig. 5.8c with simplified mathematical models in Figs. 5.8a and b shows that the schematic in Fig. 5.8a overstates the deformations and stresses, while the schematic in Fig. 5.8b understates them. None of the models a and b consider transverse components of the loads and associated deformations and stresses. It is important to remember that a more realistic picture of the loading pattern in Fig. 5.8c may change substantially in the real circumstances due to elastic deformations of both the shaft and the bearings, excessive wedge pressures, etc. Design of the front end of a diesel engine crankshaft in Fig. 5.9a [1] experienced such distortions. While the nominal loading on the front journal was relatively low, the bearing was frequently failing. It was discovered that the hollow journal was deforming and becoming elliptical under load. The elliptical shape of the journal resulted in reduction of the hydrodynamic wedge in the bearing and deterioration of its load-carrying capacity. The design was adequately improved by enhancing stiffness of the journal by using a reinforcing plug (Fig. 5.9b).

Figure 5.10 illustrates typical support conditions for power transmission shafts supported by sliding bearings and antifriction ball bearings [2]. Although the bearings are usually considered as simple supports, it is reasonably correct only for the case of a single bearing (Fig. 5.10a). For the case of tandem ball

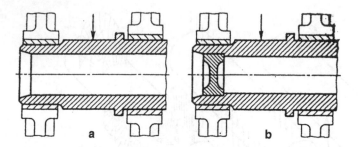

Figure 5.9 Stiffness enhancement of front-end journal of crankshaft.

bearings (Fig. 5.10b), the bulk of the reaction force is accommodated by the bearing located on the side of the loaded span (the inside bearing). The outside bearing is loaded much less, and might even be loaded by an oppositely directed reaction force if there is a distance between two bearings. Thus, it is advisable to place the support in the computational model at the center of the inside bearing or at one-third of the distance between the bearings in one bearing support, to-

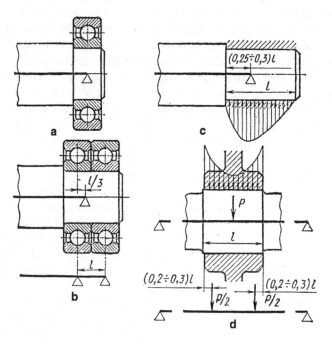

Figure 5.10 Computational models of shafts.

wards the inside bearing (Fig. 5.10b). Due to shaft deformations, pressure from sliding bearings onto the shaft is nonsymmetrical (Fig. 5.10c), unless the bearing is self-aligning. Accordingly, the simple support in the computational model should be shifted off-center and located (0.25–0.3) l from the inside end of the bearing (Fig. 5.10c).

Radial loads transmitted to the shafts by gears, pulleys, sprockets, etc., are usually modeled by a single force in the middle of the component's hub (Fig. 5.10d). However, the actual loading is distributed along the length of the hub, and the hub is, essentially, integrated with the shaft. Thus, it is more appropriate to model the hub-shaft interaction by two forces as shown in Fig. 5.10d. Smaller shifts of the forces $P/2$ from the ends of the hub are taken for interference fits and/or rigid hubs; larger shifts are taken for loose fits and/or nonrigid hubs.

5.2 RATIONAL LOCATION OF SUPPORTING AND MOUNTING ELEMENTS

The number and locations of supporting elements have a great influence on deformations/stiffness of the supported system. Frequently a ''flimsy''system may have a decent effective stiffness if it is optimally supported.

One of the optimization principles for locating the supports is *balancing of deformations* within the system so that the maximum deformations in various ''peak points'' are of about the same magnitudes. For simple supported beams the balancing effect is achieved by placing the supports at so-called Bessel's points. This approach is illustrated in Fig. 5.11 showing a beam loaded by a uniformly distributed load (e.g., by the weight load). Replacement of simple supports in Fig. 5.11a with built-in end fixtures (Fig. 5.11b) reduces the maximum

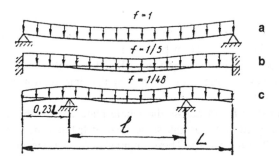

Figure 5.11 Influence of type and location of supports on maximum deflection of a double-supported beam.

deformation by a factor of 5. However, placing just simple supports at the Bessel's points located about 0.23L from the ends of the beam, or $L/l = 1.86$ ($L =$ length of the beam, $l =$ distance between the supports) reduces the maximum deformation 48 times (Fig. 5.11c). It must be noted, however, that instead of one point of the system where the maximum deformation is observed as in the cases of Figs. 5.11a and b (at the midspan), in the case of Fig. 5.11c there are three points where the deformation is maximum—midspan and two ends.

The same principle can be applied in more complex cases of plates. Stiffness of plate-like round tables of vertical boring mills was studied by Rivin [3]. Since the round table is rotating in circular guideways, the critical performance factor is relative angular displacements in the radial direction between the guiding surfaces of the table and of its supporting structure (base). If the base is assumed to be rigid (since it is, in its turn, attached to the machine foundation), then the radial angular deformation of the table along the guideways is the critical factor. There are several typical designs of the round table system, such as:

The round table is supported only by the circular guideways

There is a central support (thrust bearing) that keeps the center of the table at the same level as the guideways regardless of loading

There is a central unidirectional support that prevents the center of the table from deflection downwards but not upwards

For the first design embodiment, three loading configurations were considered:
1. Load uniformly distributed along the outer perimeter (diameter D_o) of the table (total load P) creates the same angular deformation at the guideways (diameter D_g) as the same load P uniformly distributed along the circle having diameter $D_g/2$. Angular compliances A for both loading cases are plotted in Fig. 5.12a as functions of $\beta = D_o/D_g$. The angular compliance parameter A is

$$A = PD_g/16\pi N$$

where $N =$ so-called cylindrical stiffness of the plate. The angular compliances for these two loading cases become equal at $\beta_a = 1.4$.
2. The plate is loaded with its uniformly distributed weight (Fig. 5.12b) (this case is similar to the case of Fig 5.11c). The angular deformation at the guideways vanishes at $\beta_b = 1.44$.
3. Load P uniformly distributed across the whole plate creates the same angular deformation at the guideways as the same load distributed across the circle having diameter $D_g/2$ (Fig. 5.12c). These angular deformations become equal at $\beta_c = 1.46$.

Although $\beta_a = \sim\beta_b = \sim\beta_c = \sim1.43$, there might be cases when different loading conditions are of interest and/or application of loads inside and outside

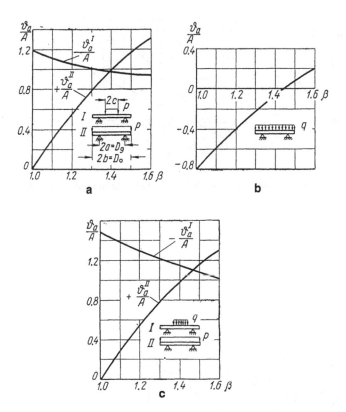

Figure 5.12 Angular compliance at the support contour of a loaded plate without a central support.

the supporting circle have different frequency and/or importance. In such cases, the selected value of β can be modified.

The similar plots for the design embodiment with a bidirectional central support are given in Figs. 5.10a and b for two cases of loading:

1. Weight load, where there is no angular deformation at the guideways at $\beta_a = 1.33$ (Fig. 5.13a)
2. Uniformly distributed load P across the whole plate and uniformly distributed load across the circle of diameter $D_g/2$ (Fig. 5.13b) with $\beta_b = 1.15$

For the design with a unidirectional central support (Fig. 5.14), the optimal location of the guideways for the distributed loading is even closer to the periphery, $\beta = 1.06$.

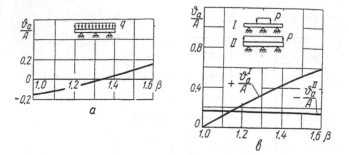

Figure 5.13 Angular compliance of a loaded plate with a bidirectional central support at the contour.

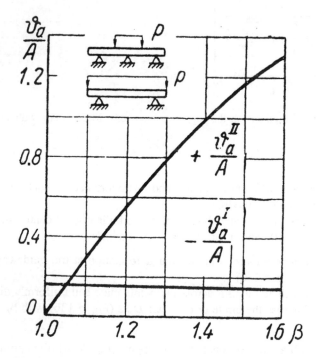

Figure 5.14 Angular compliance of a loaded plate with a unidirectional central support at the support contour.

Beds/frames of machines and other mechanical systems may have ribs and other reinforcements and may have box-like structures. However, the same basic principles can be applied. Mounting of complex beds can reduce deformations due to moving loads (e.g., tables) even with a smaller number of mounts, if their locations are judiciously selected.

Test Case 1

Test case 1 can be illustrated by a real-life case of installing a relatively large cylindrical (OD) grinder whose footprint is shown in Fig. 5.15 [4]. The 3.8 m (approximately 13 ft.) long bed had excessive deformations causing errors in positioning the machined part when a very heavy table carrying the spindle head and the part was traveling along the bed. Also, excessive positioning errors developed when the heavy grinding head was traveling in the transverse direction. To reduce the deformations, the machine manufacturer recommended that the machine be leveled on 15 rigid leveling wedges as shown by o in Fig. 5.15. The leveling protocol required lifting of the table and monitoring of the leveling condition by measuring the straightness of the guideways. However, due to substantial friction in the wedges after the leveling procedure was completed, the wedges were in a strained condition. Vibrations caused by overhead cranes, by passing transport,

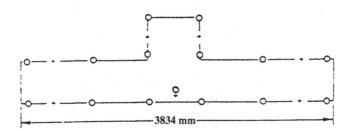

Type of Installation	Longitudinal Transverse Axis Rotation			Longitudinal Axis Rotation		
	Wheelhead position			Table position		
	Front	Rear	Diff.	Left	Right	Diff.
On 15 rigid wedge-type mounts.....	-3.0	-0.5	2.5	0	-1.0	1.0
On 7 isolators ($f_{v_z} - 20$ Hz)	+3.3	+0.8	2.5	+0.8	+0.3	0.5

Figure 5.15 Influence of mount locations on effective stiffness of cylindrical (OD) grinder.

by accidental impacts, etc., relieve the strains. As a result, the wedges change their heights. Since the releveling procedures are very expensive, the machine had to be isolated from vibrations. There are two major isolation techniques: (1) to install a massive and rigid foundation block on elastic elements, and then to mount the machine, using rigid wedges, on the foundation block; and (2) to mount the machine on the floor using vibration isolating mounts. Although the second strategy is much less expensive and allows one to quickly change the layout, it is suitable only for machines with rigid beds not requiring reinforcement from the foundation. The OD grinder in Fig. 5.15 definitely does not have an adequately rigid bed if 15 vibration isolators were placed instead of 15 rigid wedges. However, there was an attempt made to apply the load balancing or the Bessel's points concept to this machine. This was realized by replacing 15 mounting points with 7 points as shown by **x** in Fig. 5.15. The effective stiffness of the machine increased dramatically, so much so that it could be installed on 7 resilient mounts (vibration isolators), which eliminated effects of outside excitations on leveling, as well as on the machining accuracy. The table in Fig. 5.15 shows deformations (angular displacements) between the grinding wheel and the machined part during longitudinal travel of the table and during transverse travel of the grinding head. It is remarkable that the deformations are *smaller* when this large machine was installed on 7 resilient mounts than when it was installed on 15 rigid wedges.

Influence of the number and locations of the mounting elements on relative deformations between the structural components (e.g., between the grinding wheel and the part in the grinding machine in Fig. 5.15) is due to changing reaction forces acting from the mounts on the structure. These reaction forces cause changes of the structural deformation pattern. It is important to distinguish between two cases of the deformations caused by the mount reactions. In the first case (A), the supported structure is a single part or unit in which the mass distribution does not change in time. In the other case (B), there are changes of the mass distribution after the structure is installed on its supports/mounts, such as parts having significantly different weight being machined on machine tools or moving massive parts within the structure (such as tables, columns, links, etc.).

(A) A bed or a frame part is finish-machined while resting on supports (mounts) which are, in their turn, placed on a foundation. If the number of the supports is three (*"kinematic support"*), then their reaction forces do not depend on variations of the foundation surface profile, on variations of height and stiffness of the individual mounts, and on surface irregularities of the supporting surface or of the part itself. The reaction forces are fully determined by position of the center of gravity (C.G.) of the part and by coordinates of the supports relative to the C.G. If the part/machine is reinstalled on a different foundation using different supports but the same locations of the supports, the reaction forces would not change. As a result, relative positions of all components within the structure would be stable to a very high degree of accuracy.

However, in most cases there are more than three supports. In such cases the system is a *"statically indeterminate"* one, and relocation, even without changes in mass distribution within the system, would cause changes in the force distribution between the supports/mounts. The force distribution changes result in changes of the deformation map of the structure due to inevitable differences in relative height of the mounts and their stiffnesses, minor variations of stiffness and flatness of the foundation plates, etc.

If the part is finish-machined while mounted on more than three supports, and then relocated (e.g., to the assembly station), the inevitable change in the mounting conditions would result in distortions of the machined surfaces, which are undesirable. This distortions can be corrected by leveling of the part (adjusting the height of each support until the distortions of the reference surface are within an acceptable tolerance). The same procedure must be repeated at each relocation of the part or of the assembly.

If a precision maintenance of the original/reference condition of the system is required, but the reference (precisely machined) surface is not accessible, e.g., covered with a table, then leveling is recommended while the obstructing component is removed, like in the above example with the cylindrical grinder. This is not very desirable (but often unavoidable) since not only is the procedure very labor intensive and time consuming, but also the leveling procedure is performed for a system loaded differently than the actual system.

Although increasing the number of supports for a frame part of a given size reduces the distances l between the supports and reduces deformations (which are inversely proportional to l^3), the system becomes more sensitive to small deviations of the foundation surface and condition and to variations of height of the supports. These variations will significantly affect reaction forces and, since the system becomes more rigid with the increasing number of supports, any small deviation may result in large force (and thus deformations) or conversely may result in "switching off" of some supports.

Test Case 2

The table in Fig. 5.16 [5] shows reactions in the mounts (vibration isolators equipped with a reaction-measuring device) under a vertical milling machine (Fig. 5.16a) and under a horizontal milling machine (Fig. 5.16b). Two cases for each installation were compared: first the machine was installed without leveling, and then leveled by adjusting the heights of the mounts until the table of the machine (the reference surface) becomes horizontal. One has to remember that the differences in the reaction forces on various mounts create bending moments acting on the structure and deforming it. It should also be noted that if these machines were installed on more rigid (e.g., wedge-type) mounts, the ranges of

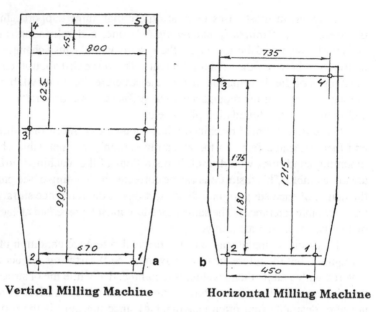

Vertical Milling Machine   **Horizontal Milling Machine**

Mounting Points

Mount Reactions, kN

Machine	Condition	1	2	3	4	5	6
a	Not levelled	4.0	20.9	5.0	3.7	17.7	5.9
	Levelled	9.25	9.1	11.5	10.1	8.9	9.9
b	Not levelled	9.1	12.2	9.8	6.6		
	Levelled	11.3	11.5	7.3	7.5		

Figure 5.16 Influence of leveling on force reactions on mounts for (a) vertical and (b) horizontal milling machines. Dimensions in millimeters.

load variation would be much wider, and distortions in the structure, correspondingly, would be much more pronounced.

Another way of maintaining the same conditions of the part and/or of the assembled unit is by measuring actual forces on the supports at the machining station for the part and/or at the assembly station for the unit. At each relocation the supports can be adjusted until the same reaction forces are reached. This approach is more accurate since stiffness values of the mounts would have a lesser influence. It was successfully used for precision installations of both the gear hobbing machine (weight 400 tons) for machining high precision large (5.0 m diameter) gears for quiet submarine transmissions [6] and for the large gear blanks placed on the table of the machine. The machine was installed on 70 load-sensing mounts while the gear blank was installed on 12 load-sensing mounts. The loads on each mount were continuously monitored.

(B) The situation is somewhat different if mass/weight distribution within the structure is changing. Figure 5.17 shows variations of load distribution between rigid (wedge) mounts under a large turret lathe (weight 24,000 lbs.) while a heavy turret carriage (weight 2000 lbs.) is traveling along the bed. If the bed were absolutely rigid, the whole machine would tilt without any relative displacements of its constitutive components. Since the bed is not absolutely rigid, travel of the heavy unit would also cause structural deformations within and between the components. These structural deformations would be reduced if the mounts

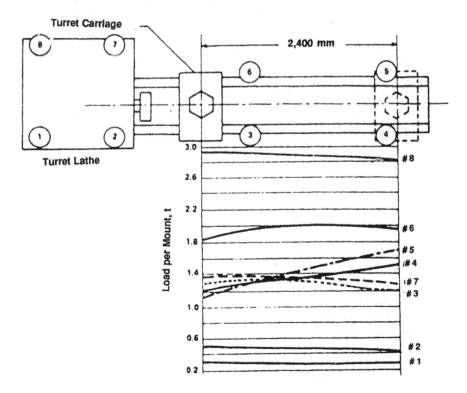

Figure 5.17   Influence of travelling heavy turret carriage on mount reactions of a turret lathe.

are rigid and provide reinforcement of the structure by reliably connecting it to the foundation. However, the rigid mounts (wedge mounts, screw mounts, shims, etc.) are prone to become strained in the course of the adjustment procedure. When the strain is relieved by vibration and shocks generated inside or outside the structure, their dimensions may change as in the above example of the OD grinder in Fig. 5.15.

As for group (A), leveling of these structures is easier and more reliable when the weight load distribution between the mounts is recorded for a certain reference configuration. In this configuration, all traveling components should be placed at the specified positions and specified additional components should be attached. For production systems the part or die weights should also be specified.

It is important to understand that the *"kinematic support"* condition in these circumstances *does not guarantee the perfect alignment* of the structural components since relocation of a heavy component inside the structure is causing changes in mount reaction forces and thus generates changing bending moments that may cause structural deformations. Accordingly, the structures experiencing significant mass/weight distribution changes must be designed with an adequate stiffness and the mounting points locations must be judiciously selected.

This is illustrated by Fig. 5.18[7], which shows the change of angular deflection between the table and the spindle head of a precision jig borer between the extreme left and right positions of the moving table for different numbers and locations of mounts. While the minimal deflection corresponds to one kinematic support configuration, the maximum deflection corresponds to another kinematic support configuration.

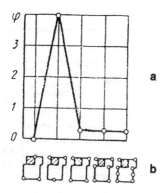

Figure 5.18 (a) Relative angular deformations (in arc sec) between table and spindle head of a precision jig borer for (b) different number and location of mounts

5.3   OVERCONSTRAINED (STATICALLY INDETERMINATE) SYSTEMS

The mounting systems discussed in the previous section are typical statically indeterminate systems if more than three mounting points are used. Statically indeterminate (overconstrained) systems are frequently used, both intentionally and unintentionally, in mechanical design. Static indeterminacy has a very strong influence on stiffness. Excessive connections, if properly applied, may serve as powerful stiffness enhancers, and may significantly improve both accuracy and load carrying capacity. However, they also may play a very detrimental role and lead to a fast deterioration of a structure or a mechanism.

Effects of overconstraining depend on the design architecture, geometric dimensions of the structures, and performance regimes. For example, overconstraining of guideways 1 and 2 of a heavy machine tool (Fig. 5.19a)[8] plays a very positive role in increasing stiffness of the guideways. This is due to the fact that the structural stiffness of the guided part 3 is relatively low because of its large dimensions and its local deformations accommodate uneven load distribution between the multiple guiding areas. On the other hand, guideways 1 and 2 for a lighter and relatively rigid carriage 3 in Fig. 5.19b are characterized by some uncertainty of load distribution that cannot be compensated by local deformations of carriage 3 due to its high rigidity.

Excessive connections, if judiciously designed, may significantly improve

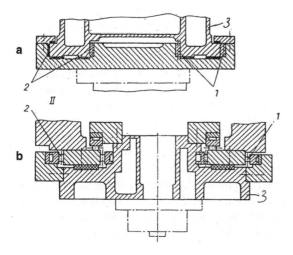

Figure 5.19   Overconstrained guideways for (a) a heavy carriage and (b) a medium size carriage.

dynamic behavior of the structure/mechanisms, such as to enhance chatter resistance of a machine tool. In the spindle unit with three bearings (Fig. 5.20a), presence of the intermediate bearing (the rear bearing is not shown) may increase chatter resistance as much as 50%. However, this beneficial effect would develop only if this bearing has a looser fit with the spindle and/or housing than the front and rear bearings. In such a case, the "third bearing" would not generate high extra loads in static conditions due to uncertainty caused by static indeterminacy, but would enhance stiffness and, especially, damping for vibratory motions due to presence of lubricating oil in the clearance. While the spindle is able to self-align in the clearance both under static loads and at relatively slow rpm-related variations, viscosity of the oil being squeezed from the clearance under high frequency chatter vibrations would effectively stiffen the connection and generate substantial friction losses (damping action).

A similar effect can be achieved by using an intermediate plain bushing that is fit on the spindle with a significant clearance (0.2–0.4 mm per diameter) and filled with oil (Fig. 5.20b) [8]. High viscous damping provided by the bushing dramatically reduced vibration amplitudes of the part being machined (Fig. 5.20c), which represents an enhancement of dynamic stiffness of the spindle. The price for the higher stiffness and damping in both cases is higher frictional losses and more heat generation in the spindle.

A similar effect can be achieved also in linear guideways. Carriage 1 in Fig. 5.21 is supported by antifriction guideways 2. The (intermediate) supports 3 and 4 are flat plates having clearance $\delta = 0.20$–0.03 mm with the base plate 5. While the intermediate supports do not contribute to the static stiffness of the system, they enhance its dynamic stiffness and damping due to resistance of oil filling the clearance to "squeezing out" under relative vibrations between carriage 1 and base 3.

Overconstraining (and underconstraining) of mechanical structures and mechanisms can be caused by thermal deformations, by inadequate precision of parts and assembly, and by design errors. Figure 5.22 shows round table 1 of vertical boring mill supported by hydrostatic circular guideway 2 and by central antifriction thrust bearing 3. At low rpm of the table the system works adequately, but at high rpm (linear speed in the guideway 8–10 m/s) heat generation in the hydrostatic guideway 2 causes thermal distortion of the round table as shown by broken line in Fig. 5.22. This effect distorts the shape of the gap in the hydrostatic support and the system fails to provide the required stiffness. The performance (overall stiffness) would *improve if thrust bearing 3 is eliminated* or if the system were initially distorted in order to create an inclined clearance in the guideway with the inclination directed opposite to the distortion caused by the thermal deformation of the round table.

Although thermal deformations in the system shown in Fig. 5.22 caused loss of linkage and resulted in an underconstrained system, frequently thermal

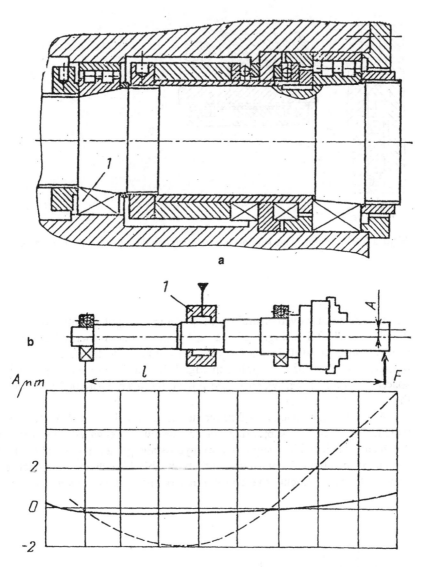

Figure 5.20 (a) Use of intermediate bearing 1 to enhance damping (to improve chatter resistance); (b) Influence of plain bushing having 0.2–0.4 mm clearance on the fundamental mode of vibration of spindle (dashed line, without damping bushing; solid line, with damping bushing).

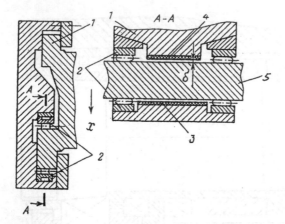

Figure 5.21   Intermediate support in antifriction guideways for enhancing dynamic stability.

deformations cause overconstraining. For example, angular contact ball bearings 2 and 3 for shaft 1 in Fig. 5.23 may get jammed due to thermal expansion of shaft 1, especially if its length $l \geq (8-12) \, d$. Similar conditions can develop for tapered roller bearings. Jamming in Fig. 5.23 can be prevented by replacing spacer 4 with a spring, which would relieve the overconstrained condition and the associated overloads.

A similar technique is used in the design of a ball screw drive for a heavy table in Fig. 5.24. The supporting bracket for nut 3 is engaged with lead screw 2, which propells table 1 supported by hydrostatic guideways 5 and 6. Bracket 4 contains membrane 7 connecting nut 3 with table 1. Such an intentional reduction of stiffness prevents overconstraining of the system, which may develop

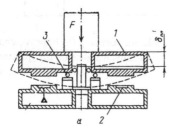

Figure 5.22   Loss of linkage (and stiffness) due to thermal deformations.

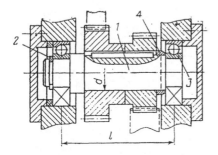

Figure 5.23 Shaft supported by two oppositely directed angular contact bearings.

due to uneven clearances in different pockets of hydrostatic guideways or to misalignment caused by imprecise assembly of the screw mechanism, among other things.

Frequently, the overconstrained condition develops due to design mistakes. Figure 5.25 shows a bearing unit in which roller bearing 1 accommodates the radial load on shaft 2 while angular contact ball bearing 3 accommodates the thrust load. In the original design, both bearings are subjected to the radial loading, thus creating overconstraining. It would be better to remove the excessive

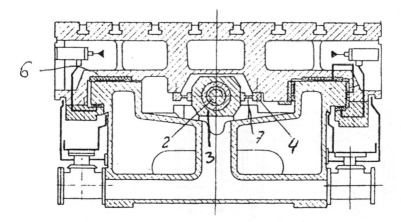

Figure 5.24 Intentional reduction of stiffness (elastic membrane 7) to enhance performance of screw drive.

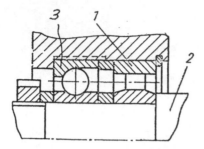

Figure 5.25 Overconstrained bearing unit.

constraints by dividing functions between the bearings as shown in Fig. 5.25 by a broken line. In the latter design, radial bearing 1 does not prevent small axial movements of shaft 2 thus making accommodation of the thrust load less uncertain.

Figure 5.26 shows a spindle unit in which the bearing system is underconstrained. Position of the inner race of the rear bearing 1 is not determined since the position of the inner race is not restrained in the axial direction. Due to the tapered fit of the inner race it can move, thus detrimentally affecting stiffness of the spindle. The design can be improved by placing an adapter ring between inner race 2 and pulley 3. The adapter ring can be precisely machined or, even better, be deformable in order to accommodate thermal expansion of the spindle.

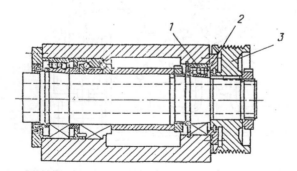

Figure 5.26 Spindle unit with underconstrained rear bearing 3.

5.4 INFLUENCE OF FOUNDATION ON STRUCTURAL DEFORMATIONS

5.4.1 General Considerations

Deformations of a structure that is kinematically supported (on three mounting points) do not depend on stiffness of mounts and/or of the foundation block, but only on location of the mounts. In many cases there are more than three mounts or the structure (e.g., frame of a machine) is attached to the foundation block by a supporting area rather than by discrete mounts. In such cases, stiffness of the foundation as well as number and location of the supporting mounts may have a profound influence on the effective stiffness of the structure.

For precision systems, location of the mounts must be assigned in such a way that influence of structural deformations on accuracy of measurements/processing is minimal. The supporting mounts and design of the foundation are especially critical for machine frames having large horizontal dimensions.

If a short bed is installed on discrete feet or mounts, the foundation block resists both linear and angular displacements of the feet. Influence of connection stiffness between the feet and the foundation on deformations of the bed depends on loads acting on the bed, and it is more pronounced for beds fastened to the foundation. In most cases this influence does not exceed 10–20% and can be neglected.

However, stiffness of a long bed supported by an adequate number of mounts on the shop floor plate [monolithic concrete plate 0.15–0.3 m (6–12 in.) thick] is increasing by 30–40% even without anchoring and/or grouting. Anchoring or grouting may result in 2–4 times stiffness increase. Fastening of the bed to an individual foundation block [0.5–1.5 m (20–60 in.) thick] may result in stiffness increase up to 10 times [9].

Computational evaluation of deformations in the system: machine bed–foundation is performed in order to analyze their influence on accuracy of the system and to appreciate its stability in time. Some important modes of deformation of this system are torsion (e.g., twisting of the bed under the moving weight force of a heavy table moving in the transverse direction) and bending (e.g., in a long broaching machine due to reaction to the cutting forces).

The computational model of a machine bed fastened to an individual foundation is a beam on an elastic bed. To analyze deformations of machine bases mounted on the floor plate, the plate should be considered as a variable stiffness plate laying on an elastic bed. The higher stiffness of the plate would be in the area of the machine base installation due to the stiffening effect of the installed base on the foundation plate. A machine base installed on mounts that are not anchored/grouted to the plate should be considered as a beam on elastic supports whose deformations depend, in turn, on deformations of the plate on an elastic bed. Such models are extremely complex, but in many cases a qualitative analysis is adequate for selecting parameters of the installation system. The approximate method [9] replaces the plate on a elastic bed with a beam having effective width B_{ef} and effective length L_{ef}, which are determined so that the displacements under each mounting point are approximately the same as for the bed installed on the plate.

The effective width B_{ef} of an infinitely long beam can be calculated for one force acting on the plate and on the equivalent beam, and for a more realistic case when two forces are acting, one on each side of the bed. For this case [9],

$$B_{ef} \approx B + 13h \qquad (5.3)$$

where B = width of the bed and h = thickness of the foundation plate. This simple expression results in the difference not exceeding 20% between deformation of the plate and the beam in the area around the force application.

The effective length L_{ef} of the beam is determined from the condition that the relative deflections of the finite and infinite length beams are close to each other within a certain span. As a first approximation, L_{ef} is determined in such a way that for a beam loaded with a concentrated force in the middle, the beam with length L_{ef} would have a deflection under the force relative to deflections at the beam's ends equal to deflection of the infinitely long beam under the force relative to deflections at the points at distances $L_{ef}/2$ from the force. Under these conditions,

$$L_{ef} \approx 3.46 \sqrt[4]{\frac{E_f I_{fy}(1 - v_b^2)}{E_b}} \qquad (5.4)$$

where E_f = Young's modulus of the plate material; $I_{fy} = B_{ef} h^3/12$ = cross-sectional moment of inertia of the beam; and E_b and v_b = Young's modulus and Poisson's ratio, respectively, of the elastic bed (soil) on which the plate/beam is supported. Although expressions for B_{ef} and L_{ef} given above are derived for the case of the bending, they are also applicable for analyzing torsional deformations [9].

Analysis of the foundation's influence on stiffness of a machine base using the "effective beam" approach is performed with an assumption that the beam is supported by the elastic bed of the Winkler type (intensity of the distributed reaction force from the elastic bed at each point is proportional to the local deformation at this point). The foundation beams are classified in three groups depending on the stiffness index

$$\lambda = \frac{L}{2} \sqrt[4]{\frac{kB}{4EI}} \qquad (5.5)$$

where L and B = length and width of the beam, EI = its bending stiffness, and the elastic bed coefficient is

$$k \approx \frac{0.65 E_B}{B_{ef}(1 - v_B^2)} \sqrt[12]{\frac{E_B B_{ef}^4}{E_f I_{fy}}} \tag{5.6}$$

If $\lambda < 0.4$, then the beam is a *rigid beam*; for $0.6 < \lambda < 2$ it is a *short beam*; and for $\lambda > 3$ — a *long beam* [9]. An overwhelming majority of machine installations can be considered short beams. The long beam classification is applicable to cases when a long machine base is installed on an individual foundation block without anchoring or grouting.

It has been suggested [9] that one must consider influence on structural stiffness of the installed machine of a foundation block or plate reduced to the equivalent beam by using *stiffness enhancement coefficients R*. Such coefficients can be determined for both bending and torsion. They are valid for analyzing deformations caused by so-called *balanced forces*, e.g., by cutting forces in a machine tool which are contained within the structure. Weight-induced forces are not balanced within the structure. They are balanced by reaction forces on the interfaces between the machine base and the foundation and between the foundation and the soil or a substructure. Distribution of these reaction forces depends on stiffness of the bed and the foundation. If the supported beam (the machine base) has an infinite stiffness, $EI \to \infty$, or the elastic bed is very soft, $k \to 0$, then the reaction forces have a quasi-linear distribution.

Structural stiffening due to elastic bed coefficient $k > 0$ can be estimated by first assuming uniform distribution of the reaction forces under the foundation block. Relative deformations of the bed are computed as for a beam acted upon by the main loading system (weight loads) and by the secondary loading system (uniformly distributed reaction), with the beam (bed) connected to the foundation block by rigid supports. Influence of deviation of actual distribution of the reaction forces from the assumed uniform distribution is considered by introduction of the stiffness enhancement coefficient.

EXAMPLE It is required to find deformation of the base of a boring mill or a large machining center from the weight of a heavy part attached to the table located in the midspan of the bed. The weight of the part is balanced by distributed reaction forces uniformly distributed across the base of the foundation block (Fig. 5.27). Considering weight of the part W_p as a concentrated force, deformation under the force relative to ends of the base (where rigid mounts are located) is [9]

$$f = \frac{W_p L}{128 EI R_B}\left[1 - \frac{4}{3}\left(\frac{L_f - L}{L_f}\right) + \frac{1}{3}\left(\frac{L_f - L}{L_f}\right)^4\right] \tag{5.7}$$

Here R_B is bending stiffness enhancement coefficient (discussed below).

Similarly, torsional deformation of the bed caused by weight $W_t + W_p$ of the table with the part that travels transversely for a distance l_1 can be determined. The torque $T = (W_t + W_p) l_1$ is counterbalanced by the reaction torque uniformly distributed along the bottom of the foundation block (Fig. 5.27). The angle of twist of the loaded cross section relative to ends of the base is then

$$\phi = \frac{TL}{8GJR_t}\left(2 - \frac{L}{L_f}\right) \tag{5.8}$$

Here J = polar moment of inertia of the cross section of the base and R_t = *torsional stiffness enhancement coefficient* (discussed below).

It is possible to use these simplistic expressions for calculating (in the first approximation) *deformations* of the structures. However, use of the similar expressions for calculating *stresses* is absolutely unacceptable.

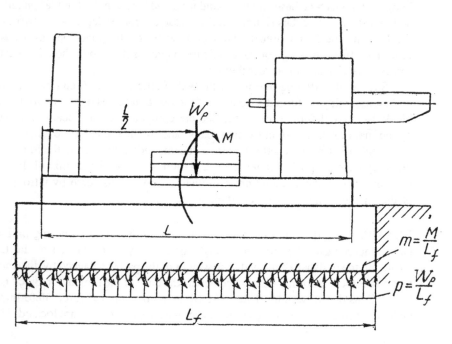

Figure 5.27 Computational model for determining deformations of heavy boring mills/ machining centers under weight loads.

5.4.2 Machines Installed on Individual Foundations or on Floor Plate [9]

If the machine base is anchored or grouted to the floor/foundation, deformations between the base and the floor can be neglected. It is assumed that the base and the foundation are deforming as one body relative to the axis passing through the effective center of gravity (C.G.) of the overall cross section of the system "base-foundation." Position of the effective C.G. is determined considering different values of Young's moduli for the machine base (E) and for the foundation (E_f). For machines installed on the monolithic floor plate, the analysis replaces it with an "effective"/equivalent foundation beam whose width and length are given by Eqs. (5.3) and (5.4). Distance from the upper surface of the base (e.g., guideways for machine tools) to the axis of the system base-foundation is

$$Z_{c.g.} = \frac{(F_b Z_b + F_f Z_f \xi)}{(F_b + F_f \xi)} \tag{5.9}$$

Here $\xi = E_f/E$; F_b and F_f = cross-sectional areas of the base and the foundation, respectively; and Z_b and Z_f = vertical distances from the upper surface of the bases to cross-sectional centers of gravity of the base and the foundation, respectively.

Bending stiffness of the system in the vertical plane X–Z is dependent on the *effective rigidity*

$$(EI_y)_{ef,v} = EI + E_f I_f + E[F_b(Z_b - Z_{c.g.})^2 + F_f \xi (Z_f - Z_{c.g.})^2] \tag{5.10}$$

Bending stiffness in the horizontal plane X–Y is

$$(EI_z)_{ef,v} \approx EI_{bZ} + E_f I_{fz} \tag{5.11}$$

Bending deformations in the horizontal plane for this group of machines are very small and are usually neglected.

The torsional stiffness of the system machine base foundation is

$$(GJ)_{ef} \approx GJ_b + G_f J_f + 0.83 \frac{G(z_f - z_b)^2}{\dfrac{\beta_b}{F_b} + \dfrac{\beta_f}{F_f \xi}} \tag{5.12}$$

where

$$\beta_b = 1 + \frac{l^2 G F_b}{14.4 \, EI_b}; \qquad \beta_f = 1 + \frac{l^2 G_f F_f}{14.4 \, E_f I_{fz}} \tag{5.13}$$

Here EI_b and EI = bending stiffness of the machine base in the horizontal and vertical planes, respectively; EJ = torsional stiffness of the base; I_{fy} and I_{fz} = cross-sectional moments of inertia of the foundation about the horizontal y and vertical z axes passing through the C.G. of the cross section of the foundation block; J_f = polar moment of inertia of the cross section of the foundation block; l = (average) distance between the anchor bolts fastening the base to the foundation; E_f and G_f = elastic moduli of the foundation material; F_b = cross-sectional area of walls of the base; and F_f = cross-sectional area of the foundation block.

Stiffness enhancement coefficients for the system machine bed-foundation due to supporting action of the soil for bending in the vertical place R_b and for torsion R_t are

$$R_b = 1 + 0.123\lambda^4; \qquad R_t = 1 + s_t \lambda_t^2 \tag{5.14}$$

Here

$$\lambda = \frac{L_f}{2} \sqrt[4]{\frac{K_f}{4(EI_x)_{ef,v}}}; \qquad \lambda_t = L_f \sqrt[4]{\frac{K_{ft}}{(GJ)_{ef}}} \tag{5.15}$$

For machines installed on a monolithic floor plate

$$L_f \approx 0.035 \sqrt[4]{\frac{(EI_x)_{ef,v} \cdot (1 - v_s^2)}{E_s}} \tag{5.16}$$

Here v_s = coefficient of transverse deformation of soil (equivalent to Poisson's coefficient). It varies from $v_s = 0.28$ for sand to $v_s = 0.41$ for clay soil. Values of elastic modulus of soil E_s are given in Table 5.1.

Coefficient $s_t = 0.02$–0.08; smaller values are to be used when the base is loaded with one concentrated torque in the midsection and with the counterbal-

Table 5.1 Elastic Modulus of Soils

Grade of soil	E_s(MPa)
Loose sand	150–300
Medium density sand	200–500
Gravel (not containing sand)	300–800
Clay	
hard	100–500
medium hard	40–150
plastic	30–80

ancing torque that is uniformly distributed along the length; larger values are taken when the bed is loaded by two mutually balancing torques or by one torque at the end that is counterbalanced by the uniformly distributed torque along the length. K_f and K_{ft} are stiffness coefficients of the interface between the foundation block and soil for bending and for torsion:

$$K_f \cong \frac{\pi}{2 \log(4\alpha)} \frac{E_s}{B_f \, (1 - v_s^2)}, \qquad \alpha = L_f/\beta_f \qquad (5.17)$$

$$K_{ft} = \frac{\pi}{16} \frac{E_s B_f^2}{(1 - v_s^2)} \qquad (5.18)$$

For bending in the horizontal plane, $R \approx 1$.

EXAMPLE [9]   Table 5.2 contains computed data for deformations of the base of a horizontal boring mill (spindle diameter 90 mm) caused by weight of its table with part [$W = 39,000$ N (18,700 lbs.)]. The system parameters are as follows: $l_1 = 1$ m; $EI = 7 \times 10^8$ N-m^2; $GJ = 5 \times 10^8$ N-m^2; $(EI_x)_{ef.v} = 25 \times 10^8$ N-m^2; $F_b = 0.085$ m^2; stiffness of each mount (when the machine is not tied to the foundation block by anchor bolts) $k_m = 5 \times 10^8$ N/m; number of mounts $n = 14$; surface area of the supporting (foot) surface of the machine $= 1.5$ m^2; distance from the supporting surface to C.G. of the bed $c_b = 0.35$ m; $L = 5$ m; elastic modulus of soil $E_s = 120$ MPa; $v_s = 0.35$; and $s_t = 0.02$.

Table 5.2 shows the critical influence of thickness of the foundation block, especially when the stiffness of the connection between the machine and the foundation is high (anchored). If the base is not anchored on a 1.4 m thick foundation its deformation is about the same as deformation of the bed anchored to a

Table 5.2   Computed Deformation of Base of Horizontal Boring Mill ($D_{sp} = 90$ mm)

Deformation of bed	Attachment to foundation	Thickness (height) of foundation, H_f (m)			
		1.4[a]	1[a]	0.4[a]	0.4[b]
Deflection	anchored	0.0022	0.0043	0.015	0.017
f (mm)	not anchored	0.016	0.021	0.035	0.037
Twist	anchored	0.0023	0.0046	0.02	0.014
ϕ (mm/m)	not anchored	0.029	0.029	0.041	0.032

[a] Individual foundation: $L_f = 5.7$ m; $B_f = 2.1$ m.
[b] Floor plate foundation: $B_{ef} = 4.3$ m.
Source: Ref. 8

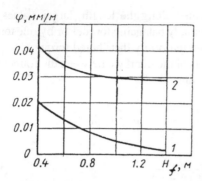

Figure 5.28 Relative angular deformations between table and spindle of horizontal boring mill (1) anchored and (2) not anchored to foundation block as function of thickness of foundation block.

0.4 m thick foundation block. These results are illustrated in Fig. 5.28. This influence is not as great when the foundation block is not very rigid (small thickness) since deformations of the nonrigid foundation are comparable with deformations of the mounts.

While the bending deformations are smaller when the machine is attached to an individual foundation block, torsional deformations are substantially smaller on the floor plate. When the machine is anchored to an individual foundation block, soil stiffness does not influence significantly the bending stiffness of the system. However, soil stiffness becomes a noticeable factor when the machine is installed on the floor plate. At $E_s = 20$ MPa, bending deformations are about two times larger than for $E_s = 120$ MPa. This effect can be explained as follows: on soils having low elastic modulus, a relatively large area of the floor plate responds to the external forces since the soil reaction is weak. Thus, curvature of the plate under the machine is steeper. It can be concluded that individual foundations are especially effective with low stiffness soils.

5.5 DEFORMATIONS OF LONG MACHINE BASES

Beds having large longitudinal dimensions are relatively easily deformable since height of the base is about the same for both short and long bases. Two factors are the most critical for effective stiffness of the long beds: deformations of the bed caused by weight of moving heavy components (e.g., a table carrying a part being machined or a gantry of a planomilling machine, e.g., Fig. 5.29). These deformations can be somewhat reduced by increasing size (thickness) of the foundation block, but this direction for improvement is limited. Very thick and long foundation blocks are becoming counterproductive since their production cycle is very long, they are very expensive, they may have high magnitudes of residual stresses causing distortions of both the block and the base, and they are extremely sensitive to even very small temperature gradients. Monitoring of loads on the mounts under the base is not practical since the loads depend on the weight of the part and are continuously changing in the process of shifting of heavy units. The most effective way of enhancing effective stiffness of the long bases is by using an active mounting system maintaining a constant distance between the supporting surface of the base and the reference frame that is not subjected to loading and is maintained at a constant temperature (see Section 7.6).

Another important factor affecting performance of the long beds anchored to the foundation blocks, is nonuniform *sagging* of the foundation along its length in time, causing distortion of the mounting surface of the foundation. The sagging is due to densification of the soil under the foundation. The amount of sagging or *set* depends on the specific load (pressure) on the soil, on the overall size of the foundation, and on compressibility of the soil. The rate of sagging/set depends mostly on water permeability of the soil and on creep resistance of the core structure of the soil.

Foundations for long machines are loaded nonuniformly, e.g., load under the gantry in a planomilling machine in Fig. 5.29 is much greater that at other locations. Accordingly, the sagging process is also nonuniform thus resulting in changing curvature of the foundation and of the base in time. The maximum differentials between the level changes along the foundation depend on properties of the soils, the rigidity of the base-foundation system, and the loading parameters. It is important to assure that releveling of the machine is scheduled so that during the time interval between the leveling procedures the change in curvature of the long bed does not exceed the tolerance. The rate of sagging is declining in time exponentially (initially, the rate of sagging is fast and then gradually slows down). The fastest stabilization is observed for soils with high water permeability, such as sands; the slowest stabilization is for the water-impermeable soils, such as clays. To utilize this exponential decrease in the sagging rate, heavy foundations are sometimes artificially stabilized before installing the machine, by loading them with "dead weights" whose total weight exceeds by 50–100% the weight of the machine to be installed, and "aging" the loaded foundation for up to 4–5 months.

The fully stabilized deformation due to the soil sagging of planomiling machine tool [$L = 7.7$ m, $B = 1.4$ m; $H_f = 1.0$ m; weight of the gantry $W = 120,000$ N (26,500 lbs.)] installed on a mixed sand + clay soil is 0.17 mm. Depending on the soil properties, this machine had to be releveled every 6–12 months in

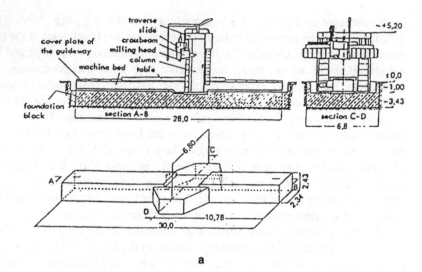

a

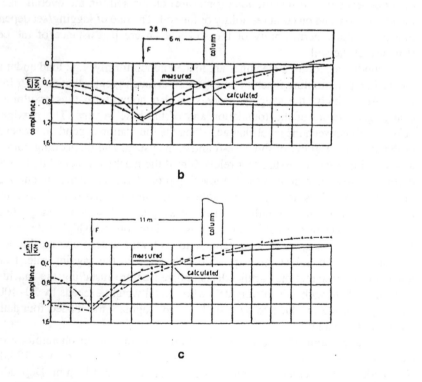

b

c

Figure 5.29 Deformations of base of a plano-milling machine under weight of moving gantry.

order to achieve a relatively weak tolerance for straightness of the guideways, 0.04 mm (0.00115 in.) [9]. The most effective technique to reduce deformations of long heavy structures caused by sagging of soil is by using an active mounting system with a reference frame isolated from the soil sagging effects. Such system is used, for example, for maintaining a stable level of foundation at the Kansai International Airport in Osaka, Japan [10]. The airport building is about 1.7 km long and is supported by a box-like foundation built on a relatively weak soil of an artificial island. The measuring system is continuously monitoring levels of the building columns (with accuracy 1 mm). When the nonuniform sagging is detected, powerful hydraulic jacks (capacity of 300 tons each) are lifting the columns to allow for correcting shims to be placed under them.

REFERENCES

1. Orlov, P.I., Fundamentals of Machine Design, Vol. 1, Mashinostroenie Publishing House, Moscow, 1972 [in Russian].
2. Reshetov, D.N., Machine Elements, Mashinostroenie Publishing House, Moscow, 1974 [in Russian].
3. Rivin, E.I., "Stiffness Analysis for Round Table-Bed System of Vertical Boring Mills," Stanki i Instrument, 1955, No. 6, pp. 16–20 [in Russian].
4. Rivin, E.I., "Vibration Isolation of Precision Equipment," Precision Engineering, 1995, Vol. 17, No. 1, pp. 41–56.
5. Rivin E.I., and Skvortzov, E.V.
6. "Photo Briefs," Mechanical Engineering, 1966, Vol. 88, No. 2.
7. Poláček, M., "Determination of Optimal Installation of Machine Tool Bed with Help of Modeling," Maschinenmarkt, 1965, No. 7, pp. 37–43 [in German].
8. Bushuev, V.V., "Load Application Schematics in Design," Stanki i Instrument, 1991, No. 1, pp. 36–41 [in Russian].
9. Kaminskaya, V.V., "Machine Tool Frames," Components and Mechanisms of Machine Tools, D.N. Reshetov, ed., Mashinostroenie, Moscow, 1972, Vol. 1, pp. 459–562 [in Russian].
10. Kanai, F., Saito, K., Kondo, S., Ishikawa, F., "Correction System for Nonuniform Sagging of a Floating Airport Foundation," Yuatsu to Kukiatsu [Journal of Japanese Hydraulic and Pneumatic Society], 1994, Vol. 25, No. 4, pp. 486–490 [in Japanese].

6

Stiffness and Damping of Power Transmission Systems and Drives

6.1. BASIC NOTIONS

Power transmission systems and drives are extremely important units for many machines and other mechanical devices. The stiffness of these systems might be a critical parameter due to several factors and depending on specifics of the device. Some of these factors are as follows:

1. Natural frequencies of power transmission systems and drives may play a significant role in vibratory behavior of the system, especially if some of them are close to other structural natural frequencies. Correct calculations of torsional natural frequencies depend on accurate information on inertia and stiffness of the components. Although calculation of inertias is usually a straightforward procedure, calculation of stiffness values is more involved.

2. In precision devices, correct angular positioning between the driving and driven elements is necessary. Inadequate stiffness of the connecting drive mechanism may disrupt proper functioning of the device.

3. Self-excited vibrations of positioning and production systems frequently develop due to inadequate stiffness of the drive mechanisms. Some examples include stick-slip vibrations of carriages and tables supported by guideways, chatter in metal-cutting machine tools, and intense torsional vibrations in drives of mining machines.

The word combination *effective stiffness* or *effective compliance* is frequently used in analyses of power transmission systems and drives. This definition means a numerical expression of the response (deflection) of the structure at a certain important point (e.g., the arm end of a manipulator) to performance-induced forces (forces caused by interactions with environment, inertia forces, etc.). Such a response, i.e., the effective stiffness and compliance, is a result of five basic factors:

1. Direct structural deformations of load-transmitting components (e.g., torsional deformations of shafts, couplings, etc.). These can be idealized for computational purposes such as beams and rods and also with idealized loading and support conditions.
2. Contact deformations in connections between the components or within the components (e.g., splines, bearings).
3. Bending deformations of shafts, gear teeth, etc., caused by forces in power transmissions (gear meshes, belts, etc.) and reduced to torsional deformations.
4. Deformations in the energy-transforming devices (motors and actuators) caused by compressibility of a working (energy delivering) medium in hydraulic and pneumatic systems, deformations of an electromagnetic field in electric motors, etc.
5. Modifications of numerical stiffness values caused by kinematic transformations between the area in which the deformations originate and the point for which the effective stiffness is analyzed.

This chapter addresses basic issues that have to be considered in order to perform the stiffness analysis of a power transmission or a drive system, and describes some cases of analytical determination of breakdown of deformations in complex real-life machines.

Although there are substantial resources for enhancement of effective stiffness of power transmissions and drives, utilization of these resources requires a clear understanding of the sources of structural compliance, computation of a breakdown of compliance, identification of weak links, and direct design efforts toward their improvement. Any efforts wrongly targeted to improve components and design features that are not the major contributors to the overall compliance would result in a waste. This can be substantiated by a simple example. If a component deformation is responsible for 50% of the overall deflection, making it two times stiffer would reduce the overall compliance (increase stiffness) about 30%. However, if a component deformation is responsible for only 5% of the overall deflection of the system, doubling its stiffness would reduce overall compliance only by an insignificant 2.5%. Costs of both treatments usually do not significantly depend on the component roles and are comparable in both cases.

Transmission and drive systems usually have chain-like structures. Since the most important goal is reduction of deflections in structural chains caused by a force or torque applied at a certain point, the use of *compliance* values instead of *stiffness* values is natural in many cases since the *compliance breakdown* is equivalent to *deflection breakdown*.

Stiffness of *bearings*, while important for analysis of power transmission systems and drives, has also other implications in designing precision mechanical devices. It is addressed in more detail in Section 6.5.

Another universally important component of power transmission systems and drives is *couplings*. Basic characteristics of couplings and principles of designing with couplings are described in Article 2.

6.2 COMPLIANCE OF MECHANICAL POWER TRANSMISSION AND DRIVE COMPONENTS

6.2.1 Basic Power Transmission Components

Compliance of mechanical elements employing joints and/or nonmetallic (e.g., elastomeric) parts can be nonlinear. It is assumed that their nonlinearity can be approximated by the following expression:

$$x = x_o(P/P_r)^n \qquad (6.1)$$

where P = acting load (torque); P_r = rated load (torque); x_o = deflection (linear or angular) of the element under load P_r; x = deflection under load P; and n = nonlinearity exponent. Usually, empirical values in expressions for compliance are given for a load magnitude $P = 0.5P_r$.

Data on *torsional compliance of shafts* e_s of various shapes and cross sections are compiled in Table 6.1 [1].

The *equivalent torsional compliance* of *key* and *spline connections* e_k is caused by contact deformations in the connection and is described by the following expression [2]:

$$e_k = k_k/D^2Lhz \quad \text{(rad/N-m)} \qquad (6.2)$$

where D = nominal diameter (for *spline* connections or *toothed clutches* it is the mean diameter); L = active length of the connection; h = active height of a key, a spline, or a clutch tooth; and z = number of keys, splines, or teeth. The joint compliance factor k_k is 6.4×10^{-12} m^3/N for a square key, 13×10^{-12} for a Woodruff key, and 4×10^{-12} for toothed clutches (even lower values for smaller z and better machining). Nonlinearity exponent $n = \sim 2/3$ for key connections and $n = \sim 1/2$ for spline connections.

Conventional key and spline connections are characterized by relatively large clearances (''play'') that are designed into the connection in order to facilitate the assembly procedure. This feature creates a highly nonlinear load-deflection characteristic of the connection as illustrated by line 1 in Fig. 6.1 [3]. The amount

Table 6.1   Torsional Compliance of Shafts

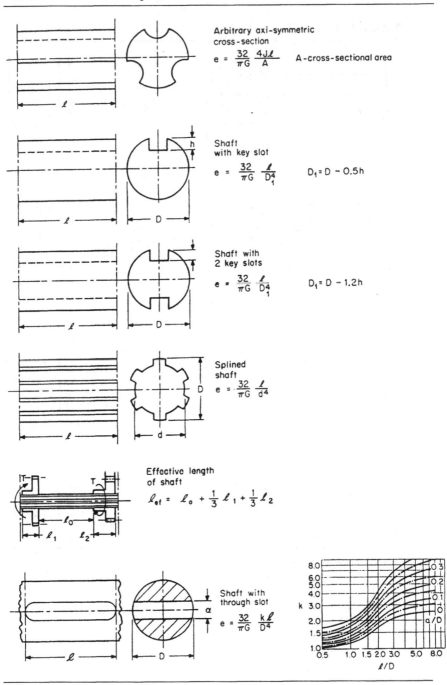

Arbitrary axi-symmetric
cross-section

$$e = \frac{32}{\pi G} \frac{4J\ell}{A}$$ A-cross-sectional area

Shaft
with key slot

$$e = \frac{32}{\pi G} \frac{\ell}{D_1^4}$$ $D_1 = D - 0.5h$

Shaft with
2 key slots

$$e = \frac{32}{\pi G} \frac{\ell}{D_1^4}$$ $D_1 = D - 1.2h$

Splined
shaft

$$e = \frac{32}{\pi G} \frac{\ell}{d^4}$$

Effective length
of shaft

$$\ell_{ef} = \ell_0 + \frac{1}{3}\ell_1 + \frac{1}{3}\ell_2$$

Shaft with
through slot

$$e = \frac{32}{\pi G} \frac{k\ell}{D^4}$$

Table 6.1   Continued

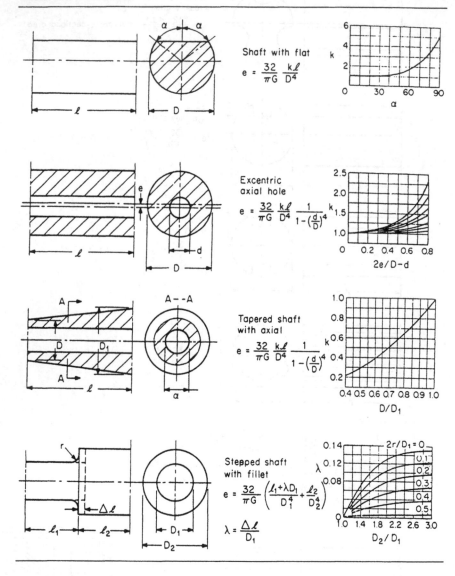

Shaft with flat

$$e = \frac{32}{\pi G} \frac{k\ell}{D^4}$$

Excentric axial hole

$$e = \frac{32}{\pi G} \frac{k\ell}{D^4} \frac{1}{1 - \left(\frac{d}{D}\right)^4}$$

Tapered shaft with axial

$$e = \frac{32}{\pi G} \frac{k\ell}{D^4} \frac{1}{1 - \left(\frac{d}{D}\right)^4}$$

Stepped shaft with fillet

$$e = \frac{32}{\pi G} \left(\frac{\ell_1 + \lambda D_1}{D_1^4} + \frac{\ell_2}{D_2^4} \right)$$

$$\lambda = \frac{\Delta \ell}{D_1}$$

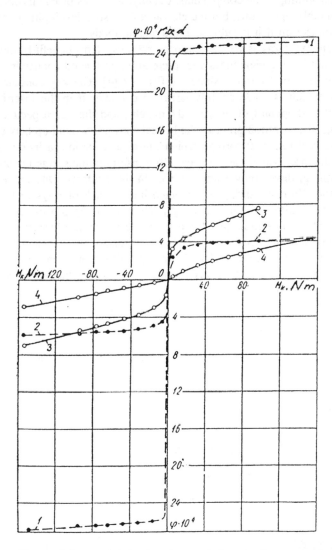

Figure 6.1 Load-deflection characteristics of spline connections: 1, sliding spline connection with play; 2, same as 1 with reduced play; 3, ball spline connection with play; 4, same as 3 but with 11 μm preload.

of play can be reduced, if critically required for functioning of the system, by more accurate machining of the components. Usually, it creates problems during assembly/disassembly operations, but the clearances are still developing during the life of the connection if it is subjected to dynamic loads.

Many precision systems as well as servocontrolled systems benefit from using connections not having clearances and, preferably, preloaded connections.

One no-play key connection is shown in Fig. 6.2 [4]. The key consists of helical spring 5, which is made slightly larger in diameter than the inscribed circle of the combined opening of slots 3 and 4 in hub 1 and shaft 2, respectively. Before (or during) the insertion process, the helical spring is wound up between its ends (like a helical torsional spring), which causes a reduction of its outside diameter below the inscribed circle diameter. After insertion, the spring unwinds and fills the opening, developing friction forces in the contacts. Due to these friction forces, the "flimsy" spring is cemented into a stack of rings that can accommodate very significant radial forces [5]. The unwinding process creates a preload thus completely eliminating the play. Since the spring key transmits the load by compression in contact with a concave surface of the slot, which has only a slightly larger curvature radius than the spring radius, stress concentrations in the connection are significantly reduced (about six to seven times reduction according to the test data)

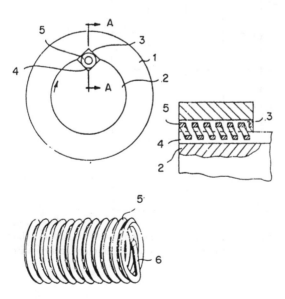

Figure 6.2 Helical spring key connection: 1, sleeve; 2, shaft; 3, key slot in sleeve; 4, key slot in shaft; 5, helical spring; 6, optional tongue for twisting spring 5.

Another type of no-play connection is a ball-spline connection, e.g., as shown in Fig. 6.3. Ball splines are used for reducing friction and eliminating play in axially mobile connections. If the ball diameter is larger than the inscribed circle of the cross section of combined grooves (splines) in the shaft and the bushing, the connection becomes preloaded and its play is eliminated (line 4 in Fig. 6.1) while the connection still retains its mobility. It can be said, however, that torsional stiffness of a ball-spline connection is lower than of a similar size conventional (sliding) spline connection. Torsional compliance of this connection is [3]

$$e_{bs} = \frac{4k \sqrt[3]{Q^2/d}}{D \sin \alpha} \qquad (6.3a)$$

$$Q = 2T/z_s zD \sin \alpha \qquad (6.3b)$$

where T = torque applied to connection; Q = force acting on one ball; z_s = number of splines; z = number of balls in one groove; D = diameter of connection; α = angle of contact; $K = 0.021$ (m/N^2)$^{1/3}$ for $r_1/r = 1.03$ and $K = 0.027$ for $r_1/r = 1.10$ (see Fig. 6.3c).

An important contributor to the equivalent torsional compliance of multishaft geared power-transmission systems is the compliance e_b caused by the *bending of the shafts*, the *elastic displacements in the bearings*, and the *bending and contact deflections of the gear teeth* [1].

Deformations of these elements lead to relative angular displacements between the meshing gears 1 and 2 and between 3 and 4 (Figs. 6.4a and b), thus the elastic member representing this compliance e_b in the mathematical model should be inserted between the inertia members representing the meshing gears. More detailed modeling of the dynamic processes in the gear trains, in which both masses and moments of inertia of the gears in an intercoupled bending-

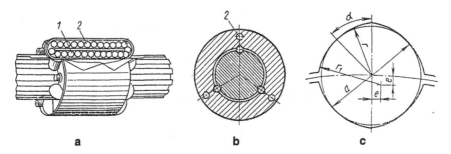

Figure 6.3 Ball-spline connection: (a) general view; (b) cross section; (c) ball in "gothic arch" spline grooves. 1, Working balls; 2, return channel.

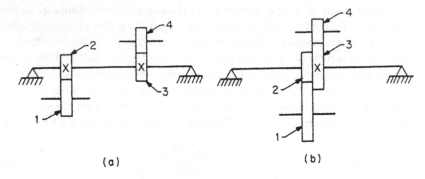

Figure 6.4 Various configurations of a gear train.

torsional system are considered (e.g., [6]), gives a better description of high-frequency modes of vibration. However, it is not usually required for the compliance breakdown analysis (and for analysis of the lower modes of vibration, which usually are the most important ones).

For calculation of e_b four steps can be followed:

1. Total vector deflection $\mathbf{y}_i$ of a shaft under the i th gear caused by all forces acting on the shaft is calculated. Sleeves on the shaft reduce its bending deflection by factor k_b (Fig. 6.5).

2. Vector displacement δ_i of the ith gear caused by compliance of bearings is calculated (Fig 6.6) as

$$\delta_i = (\delta_B - \delta_A)[a/(a + b)] + \delta_A = \delta_B[a/(a + b)] + \delta_A[a/(a + b)] \qquad (6.4)$$

where $\delta_A = e_A\mathbf{P}_A$ and $\delta_B = e_B\mathbf{P}_B \cdot \mathbf{P}_A$ and $\mathbf{P}_B$ = vector reactions at bearings A and B from all the forces acting on the shaft; and e_A and e_B = compliances of the bearings A and B (discussed below).

3. Total (vector) linear displacement of the i the gear is

$$\Delta_i = \mathbf{y}_i + \delta_i$$

Relative displacement between the meshing ith and $(i + 1)$th gears is

$$\Delta_{i,i+1} = \Delta_i - \Delta_{i+1}$$

As can be seen from Fig. 6.7, the resulting relative angular displacement of the two gears referred to the i th gear is

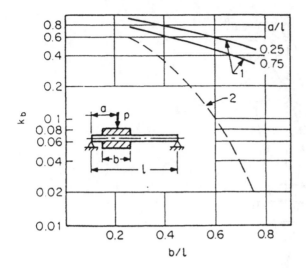

Figure 6.5 Enhancement of shaft-bending stiffness caused by bushings: (1) clearance fit and (2) interference fit.

$$\alpha_i = [\Delta^T_{i,i+1} + \Delta^R_{i,i+1} \tan(\alpha + \rho)]/R_i \qquad (6.5)$$

where $\Delta^T_{i,i+1}$ and $\Delta^R_{i,i+1}$ = tangential and radial components, respectively, of the vector $\Delta_{i,i+1}$; R_i = pitch radius of the ith gear; α = pressure angle; and ρ = friction angle (tan ρ = ~0.1).

    4. Equivalent torsional compliance is

$$e_b = \alpha/T_i = [\Delta^T_{i,i+1} + \Delta^R_{i,i+1} \tan(\alpha + \rho)]/R_i^2 P_i^T + e_m \qquad (6.6)$$

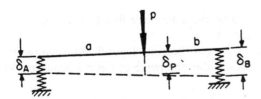

Figure 6.6 Gear displacement caused by bearing deflections.

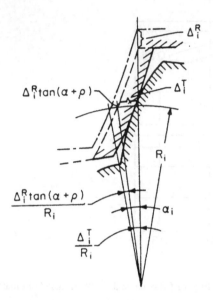

Figure 6.7 Relative angular displacement of gears caused by their translational displacements.

where T_i = torque transmitted by the ith gear; P_i^T = tangential force acting on the ith gear; and

$$e_m = k_m/bR_i^2 \cos^2\alpha \qquad (6.7)$$

is the compliance of the *gear mesh* caused by the bending and contact deformations of the engaging teeth [2]. In Eq. (6.7), b = tooth width and factor k_m = deflection of the engaged pair of teeth under the normal pressure that is acting on the unit of tooth width (for steel spur gears $k_m = \sim 6 \times 10^{-11}$ m²/N, for steel helical and chevron gears $k_m = \sim 4.4 \times 10^{-11}$, and for polyamide spur gears $k_m = \sim 5$ to 10×10^{-10}).

The compliance of antifriction bearings and their fitting areas can be calculated from Eqs. (3.15), (3.16), and (3.17).

Compliance of multipad hydrodynamic bearings (specifically, multipad bearings with screw adjustment of pads within the range of journal diameters $30 < d < 160$ mm) is

$$e_{hd} = \sim 5 \times 10^{-2}/d^{2.1} \qquad \text{(mm/N)} \qquad (6.8)$$

The compliance of hydrostatic bearings without special stiffness–enhancement servocontrol system is

$$e_{hs} = \sim\Delta/3p_p F \qquad (6.9)$$

where Δ = diametrical clearance in the bearing; p_p = pressure in the hydrostatic pocket; and F = surface area of the pocket [7].

The compliance of a *belt* or a *steel band drive* e_{bd} is described by Eqs. (3.8) and (3.12) where the effective length is [2]

$$L_{ef} = L + 0.03v(R_1 + R_2) = \sqrt{L_c - (R_1 - R_2)^2} + 0.03v(R_1 + R_2) \qquad (6.10)$$

Here L = actual length of a belt branch (the distance between its contact points with pulleys); L_c = distance between the centers of pulleys 1 and 2; v = linear speed of the belt (m/s); and $R_{1,2}$ = radii of the pulleys (m). The second term in the expression for L_{ef} reflects the influence of centrifugal forces.

The tensile modulus E of V-belts with cotton cord increases gradually when the belt is installed in the drive and tensioned and could be up to 100% higher than if it was measured by tensioning a free belt. This effect is of lesser importance for synthetic cords. With this effect considered, static modulus E = 6 to 8 $\times$ 10^2 MPa for V-belts with cotton cord; 2 $\times$ 10^2 for V-belts with nylon cord; 1.4 $\times$ 10^2 for flat leather belts; 1.4 $\times$ 10^2 for knitted belts made of cotton; 2 $\times$ 10^2 for knitted belts made of wool; 2 $\times$ 10^2 for rubber-impregnated belts; and 23 to 38 $\times$ 10^2 for high-speed thin polymer belts. For laminated belts (e.g., polymeric load-carrying layer with modulus E_1, thickness h_1, and leather friction layer with E_2 and h_2), the effective modulus is

$$E = (E_1 h_1 + E_2 h_2)/(h_1 + h_2) \qquad (6.11)$$

The compliance of *synchronous* (*timing*) *belts* is a combination of two parts: the compliance of the belt branches and the teeth compliance

$$e_{tb} = L/aR_1^2 EF + k_{tb}/bR_1^2 \qquad (6.12)$$

where b = belt width; F = cross-sectional area (between the teeth); E = 6 to 40 $\times$ 10^3 MPa = effective tensile modulus depending on cord material and structure; k_{tb} = factor of tooth deformation; k_{tb} = $\sim4.5 \times 10^{-10}$ m^3/N for a $1/2$ to $3/4$ in. pitch belt; and a = 1 for a belt drive without preload and a = 2 for a preloaded drive. The compliance of timing belts can be reduced by using a high-modulus cord material (e.g., steel wire) and/or by making the belt thinner, thus increasing the role of the cord in the effective modulus of elasticity. Also, k_{tb} can be reduced by design means (modification of tooth thickness and profile).

The compliance of a *chain drive* e_{cd} is expressed by Eqs. (3.13) and (3.14). When a chain is used without preload, $k_{cd} = 0.8$ to 1.0×10^{-12} m³/N for roller chains and 20 to 25×10^{-12} m³/N for silent chains.

Figure 6.8 gives formulas for torsional and radial stiffness of some basic designs of torsionally flexible elastomeric couplings; additional data on couplings can be found in Article 2. In Fig. 6.8, G = shear modulus of the elastomer; $K_{dyn} = k_{dyn}/k_{st}$ = ratio between dynamic (k_{dyn}) and static (K_{st}) stiffness of the elastomer.

Stiffness of Ball Screws

Ball screws (Fig. 6.9) are widely used for actuating devices in mechanical systems. They have high efficiency, in the range of 0.8–0.95, as compared with 0.2–0.4 for power screws with sliding friction. Both efficiency and stiffness of ball

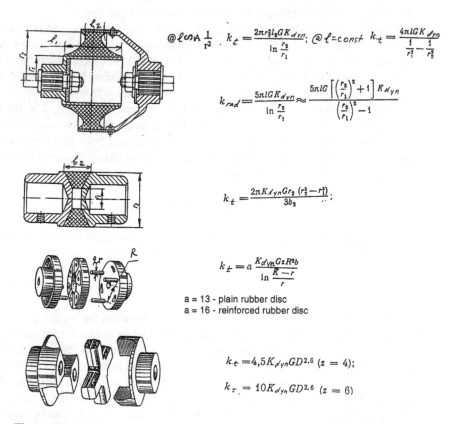

$$@ \ell \bowtie \frac{1}{r^2} \cdot \quad k_t = \frac{2\pi r_2^3 l_2 G K_{dyn}}{\ln \frac{r_2}{r_1}}; \quad @ \ell = const \quad k_t = \frac{4\pi l G K_{dyn}}{\frac{1}{r_1^2} - \frac{1}{r_2^2}}$$

$$k_{rad} = \frac{5\pi l G K_{dyn}}{\ln \frac{r_2}{r_1}} \approx \frac{5\pi l G \left[\left(\frac{r_2}{r_1} \right)^2 + 1 \right] K_{dyn}}{\left(\frac{r_2}{r_1} \right)^2 - 1}$$

$$k_t = \frac{2\pi K_{dyn} G r_2 (r_2^3 - r_1^3)}{3b_2}; \cdots$$

$$k_t = a \frac{K_{dyn} G z R^2 b}{\ln \frac{R-r}{r}}$$

a = 13 - plain rubber disc
a = 16 - reinforced rubber disc

$$k_t = 4{,}5 K_{dyn} G D^{2.6} \ (z = 4);$$

$$k_t = 10 K_{dyn} G D^{2.6} \ (z = 6)$$

Figure 6.8   Stiffness characteristics of some widely used torsionally flexible couplings.

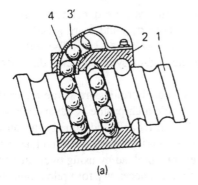

Figure 6.9 Antifriction ball screw design.

screws can be adjusted by selection of the amount of preload. A judicious preloading allows to use ball screws both as a non-self-locking transmission (low preload) and as a self-locking transmission (high preload).

There are many thread profiles used for ball screws. Although the profiles that are generated by straight lines in the axial cross section (trapezoidal, square, etc.) are the easiest to manufacture, they have inferior strength and stiffness characteristics as compared with the curvilinear cross sections shown in Figs. 6.10a and b. For a profile generated by straight lines, curvature radius of the thread surfaces is $R_2 = \infty$ and $1/R_2 = 0$, thus both maximum contact (Hertzian) stress

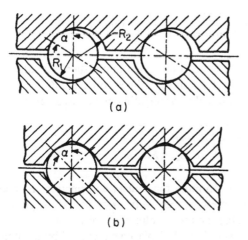

Figure 6.10 Thread profiles for ball screws: (a) semicircular and (b) gothic arch.

σ_{max} (strength) and contact deformation δ (stiffness) are dependent only on the ball radius R_1. For the curvilinear profiles, both σ_{max} and δ are greatly reduced since usually $R_2 = {\sim}1.03{-}1.05 R_1$ and $(1/R_1 - 1/R_2) \ll 1/R_1$. The state-of-the-art ball screws use larger balls to enhance the load-carrying capacity, the stiffness, and the efficiency.

The two most frequently used cross-sectional profiles are semicircular (Fig. 6.10a) and gothic arch (Fig. 6.10b). The gothic arch profile is advantageous because of the small differential between R_1 and R_2 and, additionally, because of the doubling of the number of contact points (thus increasing the rated load capacity and stiffness). It also allows one to generate preload by using oversized balls. If a semicircular thread is used, the second nut is necessary for preloading. However, the gothic arch profile is used rather infrequently because of higher manufacturing costs.

The *axial compliance* of ball screws can be a significant factor in the compliance breakdown of a drive. The determining factors for the ball-screw compliance are contact deflections, tension-compression deformations of (long) screws, and deformations in thrust bearings. Bending and shear of the thread element can be neglected.

Contact deflections are very different for nonpreloaded and preloaded mechanisms. For a *nonpreloaded* screw the axial displacement caused by an external axial force Q (daN) is [8]

$$\delta = 2CQ^{2/3}/Z_{ef}^{2/3} \sin^{5/3}\alpha \cos^{5/3}\gamma \quad \text{(cm)} \tag{6.13}$$

and the compliance is

$$e = \frac{d\delta}{dQ} = \frac{1.33C}{Z_{ef}^{2/3} \sin\alpha \cos^{5/3}\lambda} \frac{1}{\sqrt[3]{Q}} \quad \text{(mm/N)} \tag{6.14}$$

It can be seen from Eq. (6.14) that the compliance diminishes (or stiffness is increasing) with the increasing load (hardening nonlinearity). Here Z_{ef} = effective number of balls (the actual number of balls participating in the force transmission)

$$C = 0.775 \sqrt[3]{\frac{E_1 + E_2}{E_1 E_2}\left(\frac{1}{R_1} - \frac{1}{R_2}\right)}$$

is elastic coefficient from Hertzian formula for contact deformation; E_1 and E_2 = Young's moduli for ball and screw, respectively; α = contact angle of the ball in the assembly (see Fig. 6.10); and γ = helix angle of the screw.

For a *preloaded* screw, the overall axial displacement caused by the contact deflection δ is

$$\delta = \frac{CQ}{1.5 Z_{ef} \sin^2 \alpha \, \cos^2 \lambda \, \sqrt[3]{P_p}} \quad \text{(cm)} \tag{6.15}$$

and

$$e = \frac{\delta}{Q} = \frac{C}{1.5 Z_{ef} \sin^2 \alpha \, \cos^2 \lambda \, \sqrt[3]{P_p}} \quad \text{(mm/N)} \tag{6.16}$$

where P_p = preload force. Thus, with the initial preload, screw compliance does not depend on the applied load (in accordance with the conclusions in Chapter 3), and is much lower than it is for a nonpreloaded ball screw. Since, with larger ball diameters, the initial efficiency is higher and higher preload forces can be used, a higher stiffness can also be achieved.

EXAMPLE $\alpha = 45$ deg, $\lambda = 4.5$ deg, $Z_{ef} = 50$, $P_p = 0$ or $300\ N$, and $Q = 5,000\ N$.

1. For no preload, $P_p = 0$, compliance by Eq. (6.14) is

$$e_a = 0.022C \text{ mm/N, stiffness } k_a = 45.4/C \text{ N/mm}$$

2. If $P_p = 300$ N, compliance by Eq. (6.16) is

$$e_b = 0.0086C \text{ mm/N, } k_b = 116.3/C \text{ N/mm}$$

or only under 40% of the compliance without preload.

The compliance depends significantly on the screw accuracy. If, because of variations in the ball diameters or inaccuracies of the threads, there is a difference Δ between the minimum and maximum clearance between the individual balls and thread surface and an axial force Q causes a displacement δ, not all the actual balls would take equal part in transmitting force Q but only the *effective number* of balls

$$Z_{ef} = K_z Z \tag{6.17}$$

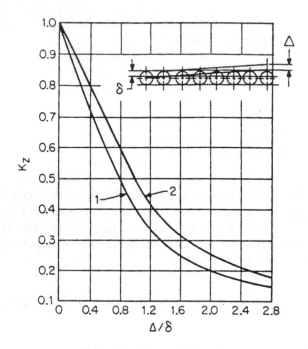

Figure 6.11 Influence of ball-screw accuracy on effective number of active balls: 1, without preload; 2, with preload.

where Z = actual number of balls in the system and K_z = coefficient depending on δ/Δ, which can be taken from Fig. 6.11. For a reasonably accurate ball screw, at high loads (close to the rated loads)

$$Z_{ef} = 0.7Z \qquad (6.17')$$

6.2.2 Compliance of Pneumatic System Components

Pneumatic actuators are very popular because of their simplicity, their very beneficial economics, and the possibility of their being used in circumstances in which electrical systems can create a safety hazard. One of the important disadvantages of pneumatic actuators is their reduced stiffness and natural frequencies caused by compressibility of air.

A generic pneumatic actuator—a linear cylinder-piston system—is sketched

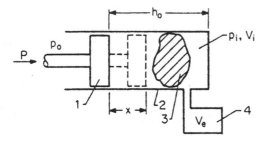

Figure 6.12 Basic pneumatic actuator

in Fig. 6.12. Piston 1 is moving inside cylinder 2 of any cross-sectional shape 3 (usually round or square) and having cross-sectional area A, instantaneous internal active volume V_i and attached external volume V_e (e.g., associated with plumbing and with ancillary units 4); instantaneous pressure p_i (excess over environmental pressure p_e) is acting in both volumes V_i and V_e; and external force P is applied to the piston rod. The increment of work dW of force P on an incremental piston displacement dx is equal to an incremental change in internal energy, thus

$$dW + d(p_i V) = P\,dx + p_i dV + V\,dp_i = 0 \qquad (6.18a)$$

where $p_i V$ = internal energy of the system. Usually pressure variations are quite small because of the system connection to the compressed air line and can be neglected, thus

$$P = -p_i(dV/dx) \qquad (6.18b)$$

For the most general polytropic process of the gas state variation

$$(p_{i,0} + p_e)/(p_i + p_e) = p_{a,0}/p_a = (V/V_0)^\gamma \qquad (6.19)$$

where p_a = absolute pressure; γ = politrope exponent; and subscript 0 is assigned to the initial state.

Substituting p_i from Eq. (6.19) into Eq. (6.18b) gives

$$P = (dV/dx)\,[p_e - p_{a,0}(V_0/V)^\gamma] \qquad (6.20)$$

Substituting

$$V = V_0 - Ax \tag{6.21}$$

into Eq. (6.20) gives

$$P = A[p_{a,0}V_0^\gamma/(V_0 - Ax)^\gamma - p_e] \tag{6.22a}$$

or

$$P = A\{p_{a,0}/[V_0 - (Ax/V_0)]^\gamma - p_e\} \tag{6.22b}$$

The value of γ depends on thermal exchange conditions between the working gas and the environment and is a function of the volume change rate in the system. Two limiting cases are:

1. Very low rates (static loading conditions, loading frequency below ~ 0.5 Hz) when there is an equilibrium between internal and external gas temperatures ("isothermal conditions," $\gamma = 1$), and
2. Very high rates (vibratory conditions at frequencies higher than ~ 3 Hz) when no energy exchange between internal and external gas can be assumed ("adiabatic conditions," $\gamma = 1.41$ for air).

The stiffness of the system can be easily derived by differentiating Eq. (6.22b) as

$$k = dP/dx = (\gamma A^2 p_{a,0}/V_0)/(1 - Ax/V_0)^{\gamma+1} \tag{6.23a}$$

or, if the vibratory variation of volume is small, $Ax \ll V_0$, and

$$k = \gamma A^2 p_{a,0}/V_0 = \gamma A(P_r + Ap_e)/V_0 \tag{6.23b}$$

where $P_r = Ap_{i0}$ is rated load of the actuator.

Since P_r is specified for a given actuator, stiffness can be increased by increasing the cross-sectional area of the piston with a simultaneous reduction of the working pressure p_{i0}. Of course, a subsequent increase of the size and mass of the actuator should be considered. Another, and in many cases, more effective way of increasing stiffness is by reducing the total internal volume of the system V_0 by means of shortening pipes, hoses, etc. Any effort to increase stiffness by design means would lead to some increase in the natural frequency and thus to an increase in γ and, as a result, to an additional effective stiffness increase (since natural frequencies of pneumatic actuators are frequently located in a 0.5 to 2.5

Hz range, which corresponds to a transition between isothermic and adiabatic conditions).

The compliance of a flexible hose walls does not play a significant role in the overall compliance (provided that reinforced hoses are used) because of the high compressibility of air.

The effective compliance of pneumatic actuators can be substantially reduced (or stiffness increased) by using additional mechanical reduction stages at the output (similarly to the modifications evaluated in Section 6.3).

It is worth noting that the compliance of a pneumatic system is not constant and varies along the stroke together with the changing volume in the pressurized part of the cylinder.

6.2.3 Compliance of Hydraulic System Components

Hydraulic actuators are characterized by the highest force (torque)-to-size ratios and thus, potentially, are the most responsive ones. Accordingly, they find a wide application for heavy payload actuators, although their applications for medium payloads are diminishing in favor of electromechanical systems. One of the substantial disadvantages of hydraulic actuators is their relatively low effective stiffness. Although the compressibility of oil is many orders of magnitude less than that of air, very high pressures in hydraulic systems lead to a significant absolute compression of oil as well as to deformations of containing walls (in cylinders, pipes, hoses, etc.).

The compliance of the compressed fluid in a cylinder or a pipe can be analyzed using the schematic in Fig. 6.13. By definition, compliance is the ratio between the incremental displacement Δx and the incremental force ΔP that causes the displacement.

Displacement Δx of the piston in Fig. 6.13 under the force ΔP applied to the piston is caused by the compressibility of the fluid in the cylinder, pipe, or hose, and is also caused by the incremental expansion of their walls. Compress-

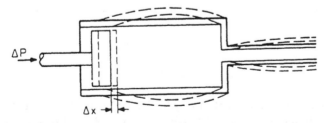

Figure 6.13 Displacement of piston in hydraulic cylinder caused by oil compressibility and by cylinder and pipes expansion.

ibility of a liquid is characterized by its volumetric compressibility modulus K_0 = $\Delta p/(\Delta V/V)$, where Δp = increment of the internal pressure; V = initial volume of liquid; and ΔV = incremental change of the volume caused by application of Δp. For a typical hydraulic oil, K_0 = $\sim 1,600$ MPa = 1.6×10^3 N/mm^2.

The combined effects of liquid compressibility and wall expansion in pipes can be conveniently characterized by an effective modulus K_{ef}. If a section of the pipe is considered [length l_p; internal diameter d; wall thickness t; Young's modulus and Poisson's ratio of the wall material E and v, respectively; cross-sectional area $A = \pi d^2/4$; internal volume $V = (\pi d^2/4)l$], an increment of internal pressure would cause an increase in the internal diameter (e.g., [9])

$$\Delta d = 2\ \Delta p(d/E)d^2/[(d + 2t)^2 - d^2][(1 + v) + (1 + v)(d + 2t/d)^2]$$
$$= \sim(\Delta p/E)(d^2/t)[1 + 2(1 + v)(t/d)] \tag{6.24}$$

since $t \ll d$. This increment of diameter is equivalent to a specific increment of pipe volume

$$\Delta V_p/V_p = \sim 2(\Delta d/d) = 2(\Delta p/E)(d/t)[1 + 2(1 + v)(t/d)] \tag{6.25}$$

An equivalent modulus associated with the volume change described by Eq. (6.25) is

$$K_p = \Delta p/(\Delta V_p/V_p) = Et/2d[1 + 2(1 + v)(t/d)] \tag{6.25'}$$

Thus, the total effective change of volume caused by the compressibility of liquid and by the expansion of pipe walls caused by a pressure increase Δp is

$$(\Delta V/V)_{ef} = \Delta p/K_0 + 2\ (\Delta p/E)(d/t)[1 + 2(1 + v)(t/d)] \tag{6.26}$$

and the effective modulus of a pipe

$$K_{ef,p} = \Delta p/(\Delta V/V)_{ef} = K_0/\{1 + 2\ (K_0/E)(d/t)[1 + 2(1 + v)(t/d)] \tag{6.26'}$$

or

$$1/K_{ef,p} = 1/K_0 + 1/K_p \tag{6.27}$$

The same formula is also applicable to hydraulic cylinders. For a steel pipe $d = 25$ mm, $t = 2.0$ mm, $E = 2.1 \times 10^5$ MPa, $v = 0.23$, $K_0 = 1.6 \times 10^3$ MPa, and $K_{ef,p} = 0.81\ K_0$. For a copper pipe of the same dimensions ($E = 1.2 \times 10^5$

MPa, $v = 0.35$), $K_{ef,p} = 0.71 \; K_0$. For a steel cylinder ($d = 63.5$ mm, $t = 6.3$ mm), $K_{ef,p} = 0.84 \; K_0$.

For flexible hoses, the values of E and v are rarely available, but in some cases data on "percent volume expansion versus pressure" is available from the hose manufacturers, from which the equivalent modulus K_h can be calculated for any pressure p. Figure 6.14 shows data provided by Rogan and Shanley, Inc., on the comparison of six hose types of $\frac{1}{4}$-in. diameter. There is a very substantial (10–20 times) scatter of the K_h value for various hose types. The best hoses are comparable with or, even, superior to metal pipes. For the pressure range $p = 3$ to 10 MPa, equivalent modulus $K_h = 0.1$ to 1.2×10^3 MPa. Since hose expansion is determined experimentally, it includes the effects of oil compress-

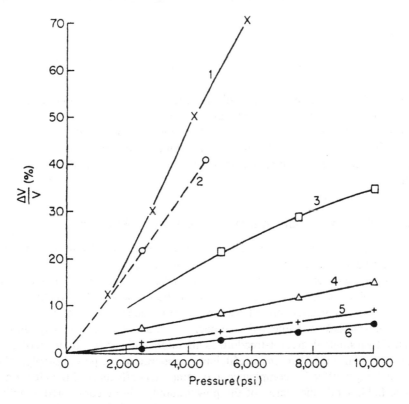

Figure 6.14 Volumetric expansion of flexible hoses ($\frac{1}{4}$-in. bore) under pressure: (1) Aeroquip 100R2, 2781-4; (2) Synflex 3R80; (3) Synflex 3V10; (4) 4006 CSA (Kevlar); (5) R&S Polyflex 2006 St; and (6) R & S Polyflex 2006 StR.

ibility, thus $K_h = K_{ef,p}$. For the hoses in Fig. 6.14, $K_{ef,p} = 0.03-0.73\ K_0$. Some hoses exhibit a very significant nonlinearity of the hardening type.

If a force ΔP_c is applied to a piston having an effective cross-sectional area A_c, it will result in a pressure increment $\Delta p_c = \Delta P_c/A_c$. This pressure increment would cause volume changes: ΔV_c in the cylinder; ΔV_p in the rigid piping; and ΔV_h in the flexible hoses, whose magnitudes are determined by their respective volumes and effective moduli. The total volume change is

$$\Delta V = \Delta V_c + \Delta V_p + \Delta V_h = \Delta X_c A_c \tag{6.28}$$

or

$$\Delta X_c A_c = \Delta P_c(V_c/K_{ef,c} + V_p/K_{ef,p} + V_h/K_{ef,h}) \tag{6.29}$$

or, finally,

$$\Delta X_c/\Delta P_c = (V_c/K_{ef,c} + V_p/K_{ef,p} + V_h/K_{ef,h})/A_c \tag{6.30}$$

where ΔX_c = incremental piston displacement caused by the volume changes.

One way to reduce the compliance originated in the hydraulic system is by the introduction of a mechanical reduction stage, as illustrated in Section 6.3.

The compliance of hydraulic systems changes along the stroke of the actuating cylinder(s) because of the changing volume of hydraulic fluid under compression.

6.2.4   Dynamic Parameters of Electric Motors (Actuators)

The electromagnetic field connecting the stator and rotor of a driving motor or actuator demonstrates quasi-elastic and/or damping properties [10]. The effective compliance and damping of the motor could noticeably influence both the compliance breakdown and the dynamic characteristic of mechanical structures (especially if mechanical reduction between the motor and the point of interest is not very substantial). A mathematical model of a motor can be, typically, approximated by a single-degree-of-freedom oscillator including rotor (moment of inertia I_r), compliance e_{em}, damping coefficient c_{em}, inertia of the power supply, where e_{em} and c_{em} = effective compliance and damping coefficients of the electromagnetic field. While the inertia of the power supply can be considered as infinity in cases of general machinery drives, for servodrives it is a complex parameter representing the dynamic characteristics of the feedback and/or feedforward systems (e.g., [11]). For conventional motors, both d.c. motors and induction a.c. motors, the amplitude-frequency characteristic of a motor while it is excited by a periodically varying torque on the shaft has a pronounced, although highly damped, resonance peak [10].

A specific feature of electric motors is their very high dynamic compliance (low stiffness). Dynamic stiffness and damping coefficients associated with the electromagnetic field of industrial a.c. induction motors and d.c. motors are given by Rivin [1,10] as

$$k_{em} = pT_{max}; \qquad c_{em} = s_{max} \, \omega_e I_r \qquad\qquad (6.31a,b)$$

for induction motors, where p = number of poles on the stator; $\omega_e = 2\pi f_e$ = angular frequency of the line voltage (Hz); T_{max} = maximum (breakdown) torque of the motor; s_{max} = slippage associated with T_{max}; and I_r = rotor moment of inertia. Equations (6.31a,b) describe parameters of an induction motor when it runs on the stable (working) branch of its torque-speed characteristic (speed is decreasing and slippage is increasing with increasing torque). However, during the start-up period, the motor initially runs along the unstable (starting) branch of its characteristic, on which increasing torque is associated with increasing rpm. As the result, induction motors may develop a very intense *negative damping*, which can produce self-exciting vibrations in the driven mechanical system and dangerous overloads [1,10]. This peculiar feature of induction motors can be utilized in special cases when reduction of damping in the driven system or even negative damping is desirable.

For d.c. motors

$$k_{em} = 1/\omega_0 v\tau_e; \qquad c_{em} = \tau_e/I_r \qquad\qquad (6.32)$$

where $v = s/T$ = slope of torque T versus slippage s characteristic; ω_0 = no-load speed of the motor (rad/s); and $\tau_e = L_r/R_r$ = electromagnetic time constant, where L_r and R_r are inductance and resistance of motor windings.

Similar properties of the electromagnetic field are characteristic also for linear motors.

Since natural frequencies of the electric motor systems are rather low, in the range of 1–10 Hz for medium horsepower motors, their dynamic behavior can be greatly influenced by the control system. It is shown [11] that effective damping of a *direct drive motor* can be substantially improved by using velocity feedback. However, because of this close coupling between the motor and the control system, the effective compliance of the motor may also be influenced by the parameters of the control system. It was shown that the effective linkage stiffness of a robot with direct drive motors was rather low [11] and corresponds to a fundamental natural frequency of 2.5–7.5 Hz; the motor natural frequency was about 3.65 Hz.

For conventional (nondirect) drives, when a motor is connected to the driven component through a transmission reducing its rpm, the effect of electric motor compliance on the effective system compliance is usually small. According to the "reduction" procedure described in Section 6.3, the motor compliance has to be divided by the square of a usually high transmission ratio. However, because of the same reasons, influence of the electromagnetic stiffness (compliance) and damping of the driving motor on dynamics of high speed mechanical drives (whose rpm are equal or exceeding the motor rpm) can be quite significant. With a proper design approach, this influence can be used to noticeably improve effective damping of the drive system [10] (see also Sect. 6.6.2).

6.3 PARAMETER REDUCTION IN MATHEMATICAL MODELS

In a complex mechanical system such as a robot, a machine tool, or a vehicle, or even in a subsystem such as a gearbox or a chain drive, there are many compliant components as well as many inertias and energy-dissipating units. If all the compliance and/or inertia and/or damping values are calculated and known, the breakdown of their distribution throughout the system cannot be done by simply adding up the numbers and calculating the fractions (percentages) of each contributor participation in the total sum. The reason for this is that contributions of partial compliances (as well as inertias and dampers) to the system's behavior depend on kinematic relationships between the design components whose parameters are considered. For compliances and inertias, it is reflected in the roles that specific components play in the overall potential and kinetic energy expressions.

Accordingly, before the breakdown can be constructed and analyzed, all the partial values have to be *reduced* (referred, reflected) to a selected point (or to a selected part) of the system. If such a reduction is properly done, neither natural frequencies nor modes of vibration are affected. The overall compliance reduced (referred) to a certain component of the system would be the same as the compliance value measured by the application of force or torque to this component and then recording the resulting deflection. For the reduction algorithm to be correct, the condition should be satisfied that magnitudes of both potential and kinetic energy are the same for mathematical models of the original and the reduced systems.

Several typical and important examples of the reduction procedure are considered here. They are intended to serve as computational tools to be used in analyzing real systems as well as to illustrate general concepts. For generality, the reduction of both elastic constants and inertias is addressed.

Figure 6.15 shows beam 1 pivoted to support structure 2 and also connected to it through revolute spring 3 having angular compliance e_ϕ (or angular stiffness $k_\phi = 1/e_\phi$). If force F is acting at distance l from the pivot, the angular deflection

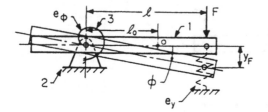

Figure 6.15 Reduction of angular compliance to equivalent translational compliance.

ϕ of the beam can be easily calculated by first calculating moment M of force F relative to the center of rotation

$$\phi = e_\phi M = e_\phi Fl \tag{6.33}$$

However, in many cases it would be beneficial to describe the compliance of this system in terms of linear compliance e_y at the force application point (e.g., the compliance at the end of arm for a robot). By definition, such a compliance is

$$e_y = y_F/F \tag{6.34}$$

where y_F = linear displacement at the force application point caused by force F. Using Eq. (6.33),

$$e_y = y_F/F = \phi l/F = e_\phi Fl^2/F = e_\phi l^2 \tag{6.35}$$

Since the potential energy of an elastic system is

$$V = \tfrac{1}{2}k\Delta^2 = \tfrac{1}{2}\Delta^2/e \tag{6.36}$$

where k and e = *generalized stiffness and compliance* and Δ = *generalized deflection*. The equivalency of the "initial" compliance e_ϕ and "reduced" compliance e_y can be easily proven as

$$V = \tfrac{1}{2}\Delta^2/e = \tfrac{1}{2}y_F^2/e_y = 1/2 \ \phi^2 l^2/e_\phi l^2 = \tfrac{1}{2}\phi^2/e_\phi \tag{6.37}$$

Of course, the reduction can be performed to an intermediate point as well, such as to the point o at a distance l_o from the center (Fig. 6.15). Obviously, in this case

$$e_o = e_\phi l_o^2 \tag{6.35'}$$

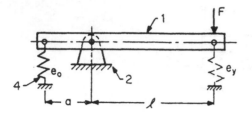

Figure 6.16 Reduction of translational compliance to a different location.

Figure 6.16 shows the same beam as in Fig. 6.15, but instead of a revolute spring at the pivot, there is a linear spring 4 at the opposite side of the beam. Compliance of spring 4 is e_o (or, stiffness $k_o = 1/e_o$).

First, let us reduce compliance e_o to the force application point (reduced compliance e_y). Force F is transformed by the leverage action of beam 1 into force $F_o = F(l/a)$ acting on spring 4. The deformation of spring 4 is

$$y_o = e_o F_o = e_o F(l/a) \tag{6.38}$$

This deformation is transformed by the leverage effect of beam 1 into deformation $y = y_o l/a$ at the force application point. Accordingly

$$e_y = y/F = y_o(l/a)/F = e_o F(L/a)(l/a)Fl = e_o(l^2/a^2) \tag{6.39}$$

Compliance e_o can also be reduced to angular compliance e_ϕ at the center of beam rotation as follows:

$$e_\phi = \phi/M = (y_o/a)/Fl = e_o F(l/a)(l/a)/Fl = e_o/a^2 \tag{6.40}$$

Naturally, Eq. (6.40) is identical to Eq. (6.35). It is easy to verify that the potential energy is invariant for both reduction procedures in Eqs. (6.39) and (6.40).

It is convenient to correlate the reduction formulas with transmission ratios in the respective cases. Transmission ratio $i_{b,c}$ between components or points b or c is defined as the ratio between the velocities of component or point b and component or point c

$$i_{b,c} = v_b/v_c \tag{6.41}$$

In the case of Eq. (6.35), the ratio between the linear velocity of the force application point $v_F = l(d\phi/dt)$ and the angular velocity of the beam $d\phi/dt$ is

$$i_{y,\phi} = l(d\phi/dt)/(d\phi/dt) = l \qquad (6.42)$$

In the case of Eq. (6.39), the transmission ratio between the velocities of the force application point (v_F) and the spring attachment point (v_o) is

$$i_{F,o} = l/a \qquad (6.43)$$

in the case of Eq. (6.40), the transmission ratio is

$$i_{F,o} = (v_o/a)/v_o = l/a \qquad (6.44)$$

Thus, in all of these cases, the reduction formula can be written as

$$e_k = e_n/i^2_{n,k} = e_n i^2_{k,n} \qquad (6.45)$$

or

$$k_k = k_n i^2_{n,k} = k_n/i^2_{k,n} \qquad (6.46)$$

where e_n and k_n = original compliance and stiffness of component or point n; e_k, k_k = compliance, stiffness reduced to component or point k; $i_{n,k}$ = transmission ratio between these components or points; and $i_{k,n}$ = transmission ratio between the same points in the opposite direction. Expressions (6.40) and (6.46) cover reduction both between the same modes of motion (linear-linear, angular-angular) and between the different modes of motion (linear-angular, angular-linear).

If a system is composed of several compliant components moving with different velocities, all the partial compliances can be reduced to a selected component (usually to the input or the output component). After the reduction procedure is performed, an analysis of the relative importance of various design components in the compliance breakdown could easily be done. This is illustrated below on the example of a system in Fig. 6.19.

Similar procedures can be developed for the transformation (reduction) of inertias. Figure 6.17 shows the same beam as Fig. 6.15, but in this case the beam is assumed to be massless and it carries a concentrated mass M at its end.

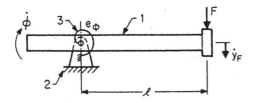

Figure 6.17 Reduction of mass to a moment of inertia.

If a translational motion of mass M is considered (in association with the deformation of the linear spring e_y), the velocity is v_F, mass is M and kinetic energy $T = \frac{1}{2}Mv_F^2$. If it is more convenient to deal with revolute motion (in association with the deformation of revolute spring 3), the angular velocity $d\phi/dt = (1/l) v_F$, the moment of inertia

$$I = Ml^2 \tag{6.47}$$

and kinetic energy

$$T = \tfrac{1}{2}I\dot{\phi}^2 = \tfrac{1}{2}Ml^2[(1/l)v_F]^2 = \tfrac{1}{2}Mv_F^2 \tag{6.48}$$

Analogously to Eq. (6.46), it can be written that

$$I_y = I_\phi i_{\phi,y}^2 \tag{6.49}$$

where $I_y = M$ and $I_\phi = I$ designate inertias associated with coordinates y, φ; and $i_{\phi,y} =$ transmission ratio between these coordinates as defined in Eq. (6.41).

In the case of Fig. 6.18, it might be desirable to consider motion in the coordinate y_F associated with the right end of the beam while both the actual spring e_o and the actual mass M_o are located at the left end. Reduction of the spring constant was discussed before; mass reduction can be performed from expressions for kinetic energy for the initial M_o and reduced M_F mass positions,

$$T_o = 1/2M_o v_o^2; \qquad T_1 = 1/2M_F v_F^2 = \tfrac{1}{2}[M_F(v_F/v_o)^2]v_o^2 \tag{6.50}$$

Thus

$$M_F(v_F/v_o)^2 = M_o; \qquad M_F = M_o(v_o/v_F)^2 = M_o i_{o,F}^2$$

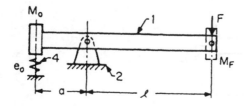

Figure 6.18   Reduction of mass to a different location.

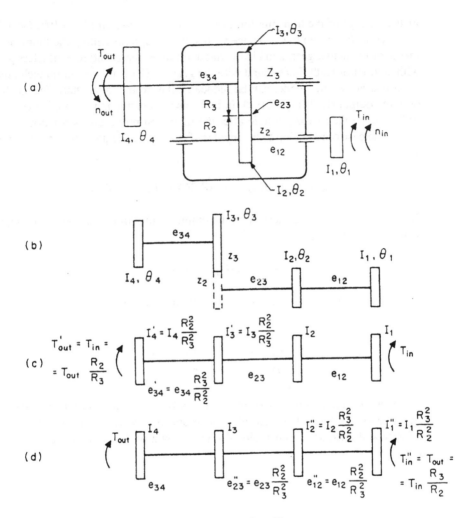

Figure 6.19 Reduction procedure for a gearbox-like system.

For example, a typical gear reducer system is shown in Fig. 6.19. It consists of two massive discs with moments of inertia I_1 and I_4 connected through massive gears with moments of inertia I_2 and I_3. The compliance of the system is caused by torsional compliances of shafts 1–2 and 3–4 as well as the bending compliance of these shafts. The latter, according to Eqs. (6.5) and (6.6), is referred to one of the shafts depending on the gear whose radius is used as R_1 in Eqs. (6.6) and (6.7). Let us assume that in the case of Fig. 6.19 the bending compliance had been computed with reference to shaft 1–2. Accordingly, the first stage mathematical model of the transmission in Fig. 6.19a is shown in Fig. 6.19b. Gear Z_2 is shown in Fig 6.19b twice—once with solid lines, representing the moment of inertia of the actual gear, and also with dotted lines, representing a massless gear with number of teeth Z_2, engaged with the gear Z_3. Using the reduction technique described above, the model can be reduced either to the input shaft (Fig. 6.19c) or to the output shaft (Fig. 6.19d). In both cases, constructing a compliance and/ or inertia breakdown is possible after the reduction procedure is performed.

In the case of reduction to the input shaft the output torque would be modified as

$$T'_{out} = T_{out}(R_2/R_3) = T_{out}(Z_2/Z_3) = T_{out}i_{3,2} = T_{out}/i_{2,3}$$

All the compliances and inertias on the output shaft will be reduced to the input shaft (Fig. 6.19c) as

$$e'_{34} = e_{34}/i^2_{2,3}; \qquad I'_4 = I'_4 i^2_{3,2}$$

If the reduction is performed using the output shaft as a reference, then (Fig. 6.19d)

$$e''_{12} = e_{12}/i^2_{2,3}; \qquad e''_{23} = e_{23}/i^2_{2,3}$$
$$I''_1 = I_1 i^2_{2,3}; \qquad I''_2 = I_2 i^2_{2,3}$$

After the reduction is performed, a breakdown of the compliance and/or inertia can be performed because all the components are now referred to the same velocity and thus, are comparable. The breakdown of the compliance is written as

$$e_{in} = e_{12} + e_{23} + e'_{34} \qquad\qquad (6.52a)$$

or

$$e_{out} = e''_{12} + e''_{23} + e_{34} \qquad\qquad (6.52b)$$

The breakdown of inertia is written as

$$I' = I_1 + I_2 + I'_3 + I'_4$$

or

$$I'' = I''_1 + I''_2 + I_3 + I_4 \qquad\qquad (6.53b)$$

As it is clear from the reduction procedure, a change in the kinematic configuration (e.g., a gear shift for a different transmission ratio, a change in the transmission ratio of a variable transmission, or a change in a linkage configuration) would completely change the compliance and inertia breakdowns. Thus, to thoroughly understand the role of the various design components in the overall compliance and/or inertia breakdown, all the critical kinematic configurations have to be analyzed.

CASE STUDY. The relative importance of the various contributors in the compliance breakdown of a geared transmission as well as the influence of a change in the kinematic configuration is illustrated in Fig. 6.20 [1,2]. Figure 6.20a shows the initial composition of the mathematical model for an actual gearbox (of a vertical knee milling machine) with the actual values of torsional, contact (keys

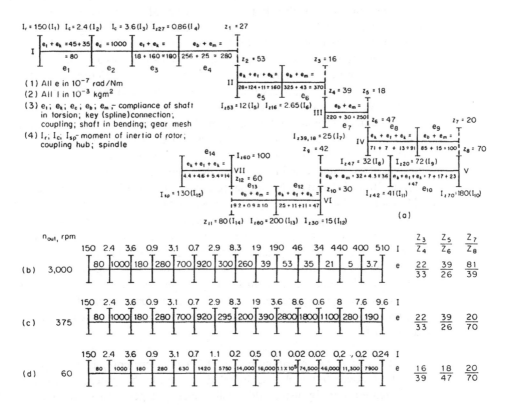

Figure 6.20 (a) Initial composition of mathematical model for a machine gearbox and final mathematical models for (b) high, (c) medium, and (d) low output (spindle) rpm.

and splines), and bending compliances as well as the compliance of a rubber coupling. All these parameters were computed in accordance to the formulas presented above in this chapter. Figures 6.20b, c, and d show changes in the compliance and inertia breakdown with all the components reduced to the motor shaft ($n_m = 1,460$ rpm, $N = 14$ KW). Different transmission ratios in the three shifting gear stages lead to dramatic changes in the breakdowns. For slow rotational speed of the output shaft (spindle), $n_{out} = 60$ rpm, the overall compliance is totally determined by the components close to the output shaft (slow-moving components whose reduction to the motor shaft involves very large multipliers) (Fig. 6.20d). The overall inertia, on the other hand, is totally determined by the fast-moving components close to the motor since reduction of inertias of the slow-moving components to the motor shaft involves very large dividers. The compliance of the rubber coupling (the largest actual compliance of a single component in the diagram in Fig. 6.20a) does not have any noticeable effect. The totally reversed situation occurs for high rotational speeds of the output shaft ($n_{out} = 3000$ rpm; Fig. 6.20b). In this case, the overall compliance is about 100 times less and is totally determined by the relatively slow-moving components close to the motor (specifically, by the rubber coupling), while the overall inertia is largely determined by the fast-moving components close to and including the output shaft (spindle). An intermediate configuration ($n_{out} = 375$ rpm; Fig. 6.20c) shows a more uniform breakdown of both compliance and inertia. Influencing either the overall compliance or the overall inertia of the gearbox by design changes would, thus, require rather different design changes depending on which system configuration is considered. However, in each configuration there are components dominating the breakdown (for $n_{out} = 60$ rpm, e_{10} represents 39% of the overall compliance and I_1 is 90% of the overall inertia; and for $n_{out} = 3000$ rpm, e_1 is 26% of the overall compliance and I_{15} is 30% of the overall inertia). Modification of these dominating components must be a starting point in the system improvement process. Design approaches to such a modification can be developed after analyzing the physical origins of the dominating components by looking into an appropriate segment of the original model in Fig. 6.20a.

As a general rule, compliances of the slowest-moving components of a system tend to be the largest contributors to the compliance breakdown, while inertias of the fastest-moving components tend to be the largest contributors to the inertia breakdown. Two important practical conclusions from this general rule are as follows:

1. To reduce the contribution of a physically very compliant component to the overall compliance, its relative speed in the system has to be increased (e.g., by using additional reducing stages after such compliant devices as harmonic drives or hydraulic actuators; see practical cases in Section 6.4). An important case is use of flexible shafts in stiffness-

critical power transmission systems. Although the actual stiffness of flexible shafts is low, the reduced stiffness values might be acceptable if the shaft is used at the high speed stage of the system.

2. To reduce the contribution of a physically massive component to the overall inertia, its relative speed in the system has to be reduced (e.g., in order to reduce the inertia of the moving linkage as seen by the driving motor, the speed of the linkage relative to the motor has to be reduced or, for a given linkage speed, the speed of the motor shaft has to be increased and the appropriate reduction means have to be introduced).

These practical conclusions should be judiciously balanced to avoid overall negative effects (e.g., an increase in the overall compliance caused by the introduction of excessive reducing means between the motor and the linkage and loss of cost-effectiveness because of the introduction of additional reduction stages). *Damping coefficients* can be reduced to system components moving with different speeds by using the same algorithms as for the reduction of compliance [Eq. (6.45)] and inertia [Eq. (6.49)] namely [2],

$$c_i^k = c_i^n / i_{n,k}^2 \tag{6.54}$$

where c_i^k = damping coefficient of unit i on shaft k and c_i^n = same but reduced to shaft n.

6.4 PRACTICAL EXAMPLES OF STRUCTURAL COMPLIANCE BREAKDOWN

Compliance breakdown for a typical gearbox is presented in Fig. 6.20. It can be done also for more diverse devices, such as a planetary transmissions [12]. Below, the breakdown of effective overall compliance for selected coordinate directions is presented for several robot manipulators (one hydraulic robot and three electromechanical robots of different structural designs [8]). Certain information about robots, such as types and dimensions of bearings, specifics of gears, etc., was not provided by the manufacturers but was taken (or assumed) from drawings and pictures available in the public domain.

6.4.1 A Hydraulically Driven Robot

The robot (Fig. 6.21) operates in spherical coordinates (radial arm extension with a maximum speed of the end effector of 0.9 m/s, rotation around the vertical axis with a maximum tangential speed of the end effector at its maximum radial extension of 2.6 m/s, and rotation in the vertical plane with a maximum tangential

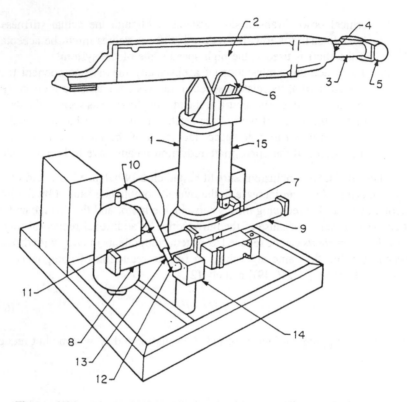

Figure 6.21 Spherical frame robot manipulator.

speed of the end effector at its maximum radial extension of 1.4 m/s). Since the rate of deceleration required for stopping in a given amount of time is higher the faster the motion, rotation around the vertical axis is the critical mode. Also, this is the most frequently used mode of motion (e.g., for all pick-and-place type applications). It is obvious that the fundamental (lowest) natural frequency determines the upper limit of the rate of acceleration/deceleration and, thus, productivity of the manipulator.

Accordingly, this mode of motion is selected to be analyzed for overall compliance and fundamental (lowest) natural frequency. The sketch of the robot in Fig. 6.21 illustrates its design features that are relevant to this mode. Column 1 carries aluminum arm carriage 2 with extending steel rods 3 supported by bronze bushings 4 and carrying end block 5 with the wrist and/or end effector (not shown). Arm carriage 2 is connected to the top surface of column 1 through pinhole connection 6.

Column 1 has rack-and-pinion unit 7 attached to its lower end with the rack driven by double cylinder systems 8 and 9. Pressurized oil is supplied to cylinders 8 and 9 from pump 10 through metal tubes 11 and 12, hose insert 13, and servo-valve 14.

The overall deflection caused by payload inertia force P is composed out of the following components (reduced to the arm end):

1. Deflection of rods 3
2. Contact deformation in bushings 4
3. Deflection of carriage 2
4. Contact deformation in pin-hole connection 6
5. Twist of column 1
6. Contact and bending deformations in rack-and-pinion unit 7
7. Compression of oil in the hydraulic cylinders
8. Compression of oil in the rigid plumbing
9. Compression of oil in the hose insert.

These components are evaluated below:

1. Deflection of a cantilever beam (Fig. 6.22) is $\delta_a = [Pl_1^2(l_1 + l_2)]/3EI$, where I = cross-sectional moment of inertia of two rods and $l_2 = 235$ mm is the distance between bushings 4; $l_1 = 800$ mm; and $\delta_a/P = 6.7 \times 10^{-4}$ mm/N.

2. One set of bushings 4, accommodating one rod, is acted upon by a force $P/2$. Reaction forces between one rod 3 and its set of bushings 4 are (Fig. 6.22)

$$F_1 = (P/2)[(l_1 + l_2)/l_2]; \qquad F_2 = Pl_1/2l_2$$

Forces F_1 and F_2 cause deformations between the rod and the bushings (caused by both structural deformation of the pin and the bushings and by deformation of surface asperities) and, accordingly an angular deflection of the rod. The contact deformations between a pin and a hole of slightly different diameters depend on the part materials, the diameter of the connection, the diametrical clearance (assumed in this case to be 20 μm or 0.0008 in.), and the magnitude of forces since the contact deformation is nonlinear (see Section 4.4.1). In this case, two

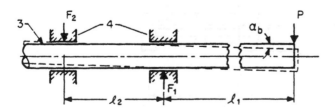

Figure 6.22 Loading schematic of sliding rods.

magnitudes of force P were considered. In the no-load case, the effective mass at the arm end only consists of the structural mass, which is equivalent to $m_1 = \sim 40$ kg, deceleration $a = 2$ g, and $P_1 = \sim 800$ N. In the maximum load case, $m_2 = 40 + 50$ kg, $a = 2$ g, and $P_2 = 1800$ N. The contact deformations in the two bushings lead to an angular deflection α_b (Fig. 6.22), which in turn causes a linear displacement $\delta_b = \alpha_b l_1$ of the end effector. Contact deformations can be easily calculated by Eqs. (4.26) and (4.32); the results for m_1 and m_2, respectively, are

$$(\delta_b/P)_1 = 0.034 \times 10^{-4} \text{mm/N}; \qquad (\delta_b/P)_2 = 0.033 \times 10^{-4} \text{mm/N}$$

3.   The arm carriage is a very massive cast aluminum part; its deformations are assumed to be negligibly small

$$\delta_c/P = \sim 0$$

4.   The arm carriage is connected to the column via the pin-bushing connection shown in Fig. 6.23. In this case, the same computational approach is used as in paragraph 2 above; the diametral clearance was again assumed to be 20 μm and the distance between the center line of the pin and the force P is equal to the full arm extension $l_3 = 1.64$ m. The force P_d in each pin-bushing connection is $P_d = P l_3/l_d$; contact deformation δ_d in each connection is calculated according to Eqs. (4.26), and (4.32). After reduction to the end effector, the result is

$$(\delta_d/P)_1 = 3.5 \times 10^{-4} \text{ mm/N}; \qquad (\delta_d/P)_2 = 1.6 \times 10^{-4} \text{ mm/N}$$

where subscripts 1 and 2 have the same meaning as they do in paragraph 2, above.

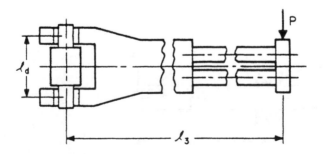

Figure 6.23   Loading schematic of extended arm.

5. The torsional compliance of the steel column that is 1.0 m high and 0.23 m diameter with 13 mm thick walls is $(\phi/T)_e = 8.3 \times 10^{-11}$ rad/N-mm, where T = torque acting on the column; accordingly,

$$\delta_e/P = (\phi/T)_e l_3^2 = 8.33 \times 10^{-11} \times 1640^2 = 2.2 \times 10^{-4} \text{ mm/N}$$

6. The compliance of the rack-and-pinion mesh (pinion diameter 254 mm, mesh width 76 mm) is calculated using Eq. (6.7). The angular compliance reduced to the pinion is $e_{rp} = 6.2 \times 10^{-11}$ rad/N-mm. This corresponds to a translational compliance reduced to the end of arm equal to

$$\delta_f/P = e_{rp} l_1^2 = 1.6 \times 10^{-4} \text{mm/N}$$

7, 8, and 9. When a force ΔP_c is applied to the piston, it moves by an increment ΔX_c because of the compression of hydraulic oil in the cylinder, in the rigid pipes, and in the hoses, and also because of the incremental expansion of their walls, as described by Eq. (6.30) in Section 6.2.3. For the considered robot, the cylinder volume in the middle rack position $V_c = 6.43 \times 10^5$ mm^3, $V_p = 5.1 \times 10^5$ mm^3, $V_h = 3.9 \times 10^5$ mm^3, $K_{ef,c} = 0.84K = 1.34 \times 10^3$ N/mm^2, $K_{ef,p} = 0.71K = 1.14 \times 10^3$ N/mm^2, $K_{ef,h} = 0.12 \times 10^3$ N/mm^2 (as calculated from data provided by a hose manufacturer, line 1 in Fig. 6.14), and $A_p = 3,170$ mm^2. Accordingly, from Eq. (6.30),

$$\Delta X_c/\Delta P_c = 4.1 \times 10^{-4} \text{ mm/N}$$

Since the action line of ΔX_c is at $l_c = 254/2 = 127$ mm distance from the column axis and the end effector is at $l_3 = 1640$ mm, the compliance of the hydraulic drive reduced to the arm end is

$$(\Delta X/\Delta P)_{ghi} = (\Delta X_c/\Delta P_c) (l_2^2/l_c^2) = 690 \times 10^{-4} \text{mm/N}$$

For better visibility of contributions of the cylinder (7), the rigid piping (8), and the flexible hose (9) segments, their compliances reduced to the end of arm can be calculated separately as $(\Delta X/\Delta P)_g = 8 \times 10^{-3}$ mm/N; $(\Delta X/\Delta P)_h = 7.5 \times 10^4$ mm/N; and $(\Delta X/\Delta P)_i = 53 \times 10^{-3}$ mm/N. Finally, the total compliance for the fully loaded ($m = 50$ kg) manipulator reduced to the arm end is as follows:

$$e = \Delta X/\Delta P = (6.7 + 0.34 + 0 + 1.6 + 2.2 + 1.6 + 80 + 75 + 530) \, 10^{-4} \text{mm/N}$$
$$= 700 \times 10^{-4} \text{mm/N} = 70 \times 10^{-6} \text{m/N}$$

Since full extension of the arm is $l_3 = 1.64$ m, it is equivalent, according to Eq. (6.40), to

$$e_\phi = 70 \times 10^{-6}/1.64^2 = 26 \times 10^{-6} \text{ rad/N-m}$$

Experimental data on static compliance of a Unimate 2000 robot is $e = 12$ to 105×10^{-6} m/N; thus the calculated data is within the realistic range of compliance.

A rough estimation of the structural arm mass m_s reduced to the end of arm is $m_s = \sim 40$ kg, thus the total inertia including the payload is $m = 90$ kg and the fundamental natural frequency is

$$f = \frac{1}{2\pi} \sqrt{\frac{1}{70 \times 10^{-6} \times 90}} \approx 2.0 \text{ Hz.}$$

This low value of natural frequency is in good correlation with the test results that have shown that it takes 0.5 s for the arm of a Unimate 2000 B robot to respond to a control input [8].

The availability of the breakdown allows a designer to find a simple means to substantially increase natural frequency by:

1. Replacement of the flexible hose with the state-of-the-art hose, line 6 in Fig. 6.14 ($K_{ef,h} = 1.4 \times 10^3$ N/mm^2) or with a metal pipe. This would result in $e_\alpha = 21.4 \times 10^{-6}$ m/N and $f_\alpha = 3.6$ Hz.
2. After item 1 is implemented, shortening the total piping length by 50%. The expected result is $e_\beta = 15.3 \times 10^{-6}$ m/N and $f_\beta = 4.3$ Hz.
3. After items 1 and 2 are implemented, introduction of a gear reduction stage with a transmission ratio $i < 1$ between the pinion and the column. Then, compliance of the hydraulic system would enter the breakdown after multiplication by i^2, according to Eq. (6.45). For $i = 0.5$, the overall compliance would become $e_\gamma = 4.8 \times 10^{-6}$ m/N and $f_\gamma = 7.6$ Hz.

It can be seen that the very easily attainable modifications 1 and 2 above would more than double the natural frequency (and reduce by half the duration of the transient period), and then the design modification 3 would additionally increase natural frequency 25–80%. Even after all these modifications are implemented, the mechanical compliances, including the most important contributor-link compliance, do not play a very significant role in the breakdown (less than 1% initially, about 15% after modifications 1, 2, and 3 are implemented). This is typical for hydraulically driven robots.

6.4.2 Electromechanically Driven Robot of Jointed Structure (Fig. 6.24a)

For a comparison, the same mode of motion is considered rotation around the vertical (column) axis at the maximum arm outreach. The identified contributors to the effective compliance are:

1. Deflection of the forearm under the inertia force
2. Contact deformations in the joint between the forearm and the upper arm (elbow joint)
3. Deflection of the upper arm

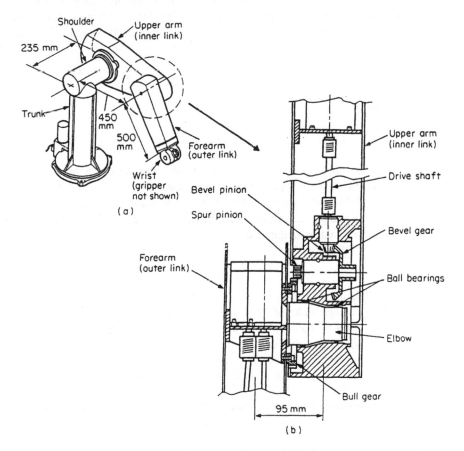

Figure 6.24 Electromechanical jointed robot.

4. Contact deformations in the joint between the upper arm and the shoulder
5. Twisting of the vertical column (waist) inside the trunk
6. Angular deformation of the gear train between the driving motor and the column (waist)

In the previous case all the components of structural compliance had been reduced to translational compliance at the end effector. In this case all the components will be reduced to torsional compliance around the vertical axis. Reduction of a linear displacement Δ caused by a force P (compliance Δ/P) to torsional compliance e around an axis at a distance a from the P and Δ vectors is performed by Eq. (6.40).

1. The forearm is an aluminum shell, $l_1 = 500$ mm long with a 1.5-mm wall thickness, having an average cross section 90×120 mm. Accordingly, $I = 8.7 \times 10^5$ mm^4, $EI = 6.1 \times 10^{10}$ Nmm^2, and with the distance between the extended end of the forearm and the waist axis

$$l_{0,\alpha} = \sqrt{1016^2 + (235 - 95)^2} = 1026 \text{ mm}$$

Thus

$$e_a = (1/l^2)(l_1^3/3EI) = 6.5 \times 10^{-10} \text{ rad/N-mm} = 6.5 \times 10^{-7} \text{ rad/N-m}$$

2. The elbow joint between the forearm and the upper arm (Fig. 6.24b) operates with two ultralight ball bearings that accommodate the moment from the payload transmitted between the end effector and the drive motor. The components of the bearing deformation are calculated by Eqs. (3.15) and (3.16). The resulting torsional compliance is

$$e_b = 38.4 \times 10^{-7} \text{ rad/N-m}$$

3. The upper arm is approximated as a hollow aluminum beam 105 mm deep with average width of 227 mm and a 3 mm thick wall ($I = 2.1 \times 10^6$ mm^4 and $EI = 1.46 \times 10^{11}$ N-mm^2) and with an active length $l_2 = 450$ mm. The effective radius from the waist axis

$$l_0 = \sqrt{450^2 + 235^2} = 508 \text{ mm}$$

Accordingly, considering loading by the force P and by the moment, $M = Pl_1$,

$$e_c = (1/508^2)(1/1.46 \times 10^{11})[450^3/3 + (500 \times 45^2)/2]$$

$$= 21 \times 10^{-10} \text{ rad/N-mm} = 21 \times 10^{-7} \text{ rad/N-m}$$

4.   The shoulder joint is similar to the elbow joint with different dimensions. The overall torsional compliance of the shoulder joint is found to be

$$e_d = 13.9 \times 10^{-7} \text{ rad/Nm}$$

5.   The waist is an aluminum tubular part with an outer diameter of 165 mm, a height of $l_3 = 566$ mm, and a wall thickness of $t_3 = 7$ mm. Accordingly, its torsional stiffness is

$$e_e = l^3/GJ = 9.3 \times 10^{-7} \text{ rad/N-m}$$

6.   The torsional compliance of the steel spur gear mesh, reduced to the bull gear attached to the waist, is characterized by $k_m = 6 \times 10^{-5}$ mm^2/N in Eq. (6.7) and also by $b = 7.5$ mm, the radius of the bull gear $R = 165$ mm and $\alpha = 20$ deg. In this case, the power is transmitted to the bull gear through two preloaded pinions (antibacklash design), thus the compliance is reduced in half. Since the reduction ratio between the pinions and the bull gear is very large (about 1:12), the compliance of the preceding stages of the motor-waist transmission does not play any significant role. Accordingly,

$$e_f = 1.6 \times 10^{-7} \text{ rad/N-m}$$

As a result, the breakdown of torsional compliance of the robot in the considered mode is

$$e = (6.5 + 38.4 + 21 + 12.9 + 9.3 + 1.6)10^{-7} \text{ rad/N-m} = 89.7 \times 10^{-7} \text{ rad/N-m}$$

With the payload 2.5 kg and the effective structural mass at the end effector 3 kg (at the distance from the waist axis $R_0 = \sim 1$ m), the total moment of inertia $I = 5.5 \times 1^2 = 5.5$ kg-m^2 and natural frequency

$$f = \frac{1}{2\pi} \sqrt{\frac{1}{5.5 \times 89.7 \times 10^{-7}}} = 22.5 \text{ Hz}$$

The compliance of this electromechanical robot is much lower than that of the hydraulically driven robot considered in Section 6.4.1, which results in a much higher natural frequency (22.5 Hz vs. 2.0 Hz) and, accordingly, a much better performance (faster acceleration/deceleration, shorter settling time).

The breakdown of the compliance in this case is much more uniform but still is dominated by one component. This component, 38.4×10^{-7} rad/N-m, is associated with the elbow joint equipped with ball bearings; a significant contribution of this joint to the arm deflection was observed also during tests. This defect can be easily alleviated by not very significant modifications of the joint (e.g., a larger spread of the bearings and/or the selection of more rigid bearings). These measures could realistically reduce the elbow joint compliance by about 50%. If this is achieved, the total compliance becomes $e = 71.5 \times 10^{-7}$ rad/Nm and the natural frequency increases about 12% up to $f = 25.3$ Hz. The total (bending) compliance of the links is about 30% of the overall compliance without design modifications and becomes about 38% after the suggested modifications. Thus, in this manipulator, stiffening of the upper arm (inner link) might be warranted after stiffening of the elbow joint is performed.

6.4.3 Electromechanically Driven Parallelogram Robot with Harmonic Drives

The parallelogram structure manipulator is shown in Fig. 6.25. A purely rotational movement of the payload in the vertical plane is considered (Fig 6.26a). For the rated payload $m = 10$ kg and an assumed deceleration of 1 g, the inertia force of the payload $P = 98$ N. A free-body diagram of the forearm is shown in Fig. 6.26b, where the forearm is presented as a beam on two compliant supports. Since the rotational motion of the forearm is driven by the motor-harmonic reducer located in the joint between the crank of the rear upper arm and the base,

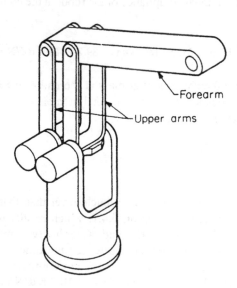

Figure 6.25 Electromechanical robot with parallelogram structure.

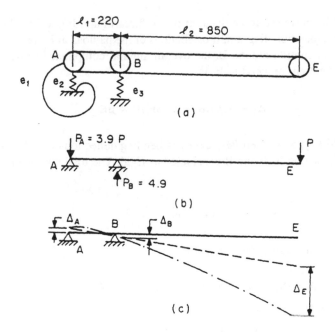

Figure 6.26 Schematics of forearm of parallelogram robot.

the torque generated by the force P around pivot B is absorbed by the motor reducer whose torsional compliance is shown as e_1. The rated torsional stiffness of the harmonic drive is $k = 92,000$ lb-in/rad $= 10,400$ N-m/rad for transmitted torques $< 0.2 \, T_r$, and $k = 515,000$ lb-in/rad $= 58,000$ N-m/rad for torques $> 0.2 \, T_r$ (manufacturer's data), thus $e_1^a = 9.6 \times 10^{-5}$ rad/N-m at $T < 0.2 \, T_r$ and $e_1^b = 1.72 \times 10^{-5}$ rad/N-m at $T > 0.2 \, T_r$. Here $T_r = 835$ lb-in. is rated torque of the reducer.

Compliance e_2, which is caused by tensile deformations of the rear upper arm AD and by deformations of two pivot bearings in A, can be represented as $e_2 = e_2' + e_2'' + e_2''' + e_2''''$, where e_2' = tensile compliance of the solid part of the rear upper arm; e_2'' = tensile compliance of two sides of the (upper) section of the rear upper arm, without the middle rib; e_2''' = tensile compliance of the sections of the arm accommodating the bearings at A; and e_2'''' = combined compliance of two pivot bearings.

Compliance $e_3 = e_3' + e_3'' + e_3''' + e_3''''$ at support B represents deformations of the front upper arm BC and consists of essentially the same components as e_2, with a possibility of e_3'''' having a different magnitude because of its load dependence. Compliance e reduced to end point E is determined by e_1, e_2, and e_3 as well as by the bending compliance of the link (forearm) ABE at E.

Figure 6.26b shows the force diagram of the forearm and Fig. 6.26c shows the translational deflection diagram. If oscillations of the payload at E are considered as oscillatory motion around B, the overall angular compliance caused by translational deflection, reduced to E, is

$$e' = [(\Delta_A + \Delta_B)/AB + \Delta_E/BE]/P \times BE$$

where $\Delta_A = 3.86\, Pe_2$; $\Delta_B = 4.86\, Pe_3$; and Δ_E = bending deflection of the forearm caused by P with supports A and B considered as rigid. The total compliance at E is

$$e = e' + e_1$$

The breakdown of the compliance is as follows:

$e'_2 = e'_3 = 0.42 \times 10^{-8}\text{m/N};$ $e''_2 = 0.11 \times 10^{-8}\text{m/N};$ $e'''_3 = 0.61 \times 10^{-8}\text{m/N}$

$e^{iv}_2 = 1.03 \times 10^{-8}\text{m/N};$ $e^{iv}_3 = 1.0 \times 10^{-8}\text{m/N}$

$\Delta_A = 8.4 \times 10^{-8}P;$ $\Delta_B = 10.2 \times 10^{-8}\, P;$ $\Delta_E = 90 \times 10^{-8}\, P$

$(\Delta_A + \Delta_B)/AB = [(8.4 + 10.2)/0.22] \times 10^{-8}P = 84.5 \times 10^{-8}P \text{ rad/N}$

$\Delta_E/BE = (90/0.85) \times 10^{-8}P = 106 \times 10^{-8}P \text{ rad/N}$

Thus, the total structural compliance is $e = 22.4 \times 10^{-7}$ rad/N-m.

Since the harmonic drive compliance $e^a_1 = 9.6 \times 10^{-5}$ rad/N-m and $e^b_1 = 1.72 \times 10^{-5}$ rad/N-m, the total compliance at the end of arm is

$$e^a = (22.4 + 960) \times 10^{-7} = 982.4 \times 10^{-7} \text{ rad/N-m}$$

$$e^b = (22.4 + 172) \times 10^{-7} = 194.4 \times 10^{-7} \text{ rad/N-m}$$

With the rated payload of 10 kg and the effective structural mass reduced to the forearm end also about 10 kg, the total moment of inertia around the pivot B is $I_B = 20 \times 0.85^2 = 14.45$ kg-m^2 and the natural frequency is

$$f^a = \frac{1}{2\pi} \sqrt{\frac{1}{14.45 \times 982.4 \times 10^{-7}}} = 4.2 \text{ Hz}$$

$$f^a = \frac{1}{2\pi} \sqrt{\frac{1}{14.45 \times 194.4 \times 10^{-7}}} = 9.25 \text{ Hz}$$

The breakdown of compliance in this case is dominated by the harmonic drive compliance (98–89% of the overall compliance). This situation can be improved by a total redesign or by a mechanical insulation of the harmonic drive from the structure through the introduction of a reducing stage (e.g., gears) or by using an oversized harmonic drive unit. The forearm bending compliance constitutes about 1–6% of the overall compliance. This percentage will rise if the drive stiffness is enhanced.

6.4.4 Electromechanically Driven Spherical Frame Robot

The stiffness of the arm in Fig. 6.27 in a vertical plane is analyzed for all ranges of its radial positions. The arm is driven by a motor with a harmonic reducer (1 in Fig. 6.27). The root segment of the arm consists of tapered tubular member 2 with counterbalance 3 and rigid partitions 4 and 5 into which two guide rods 6 are secured (guide rods 6 are reinforced with gussets 6a). Rods 6 support four open Thompson ball bushings 7 attached to intermediate tubular segment 8 with two rigid end walls to which two guide rods 9 are secured. Rods 9 support four open Thompson ball bushings 10 attached to end tubular member 11 carrying rigid end piece 12 with wrist 13. The radial motion of arm end 12 is accomplished by a motor–ball-screw–cable system drive (not shown in Fig. 6.27), which assures equal relative motion in both link connections.

The main sources of compliance in response to force *P* are as follows:

1. Bending of double-supported cantilever beam 11
2. Contact deformations in ball bushings 10 (the open sides of the bushings are directed downward in Fig. 6.27)
3. Bending of guide rods 9
4. Bending of intermediate link 8
5. Contact deformations in ball bushings 7 (the open sides of the bushings are directed upward)

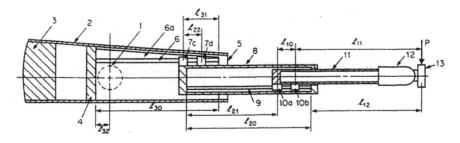

Figure 6.27 Electromechanical spherical robot arm.

6. Bending of reinforced guide rods 6
7. Bending of root link 2
8. Torsional compliance of harmonic drive 1

1. The bending compliance of link 11 is

$$e_a = l_{11}^2(l_{10} + l_{11})/3(EI)_{11}$$

2. To calculate the deflection (compliance) caused by the contact deformations in the bushings, link 11 has to be considered as a rigid beam in compliant supports. For the force P direction as shown in Fig. 6.27, the left bushing is loaded toward its open side (stiffness $k_a = k_{op}$), and the right bushing is loaded toward its closed side (stiffness $k_b = k_{cl}$). The stiffness characteristics for ball bushings in both directions are given by Rivin [8]. For the opposite direction of force P, the stiffness characteristics are reversed. Reaction forces in bushings a and b are, respectively,

$$P_a = P(l_{11}/l_{10}); \qquad P_b = P[(l_{10} + l_{11})/l_{10}]$$

Accordingly, the compliance at the arm end caused by the bushings is

$$e_b = (1/k_a)(l_{11}^2/\, l_{10}^2) + (1/k_b)[(l_{11})^2/l_{10}^2]$$

3. The forces acting on guide rods 9 from the bushings (numerically equal but opposite in direction to forces P_a and P_b) cause the deflection of the rods as shown in Fig. 6.28. Deflections y_a and y_b in points a and b are caused by both forces P_a and P_b in each, as follows, where the first subscript designates the location (point a or b) and the second subscript designates the force (P_a or P_b):

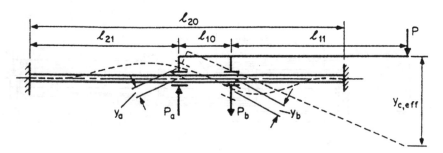

Figure 6.28 Effect of bending of guide rods on the arm-end deflection.

$$y_a = y_{aa} + y_{ab} = -\frac{Pl_{11}(l_{20} - l_{21})^2 l_{21}^2}{6l_{10}E_r I_r l_{20}^3}(2l_{21}^2 - 2l_{21}l_{20})$$

$$+ \frac{P(l_{10} + l_{11})(l_{20} - l_{21} - l_{10})^2 l_{21}^2}{6l_{11}E_r I_r l_{20}^3}(2l_{21}^2 + 2l_{10}l_{10} - 2l_{10}l_{20} + 3l_{20}l_{10})$$

$$y_b = y_{ba} + y_{bb} = -\frac{Pl_{11}l_{21}^2(l_{20} - l_{21} - l_{11})^2}{6l_{11}E_r I_r l_{20}^3}(2l_{21}^2 + 2l_{21}l_{10} - 3l_{20}l_{10})$$

$$+ \frac{P(l_{10} + l_{11})(l_{20} - l_{21} - l_{10})^2 (l_{21} + l_{10})^2}{6\,l_{10}E_r I_r \,l_{20}^3}(2l_{21}^2 + 4l_{21}l_{10} + 2l_{10}^2 - 2l_{21}l_{20} - 2l_{20}l_{10})$$

The effective deflection at the arm end because of y_a and y_b is easy to derive from Fig. 6.28

$$y_{\text{ceff}} = \frac{l_{11}}{l_{10}}y_a - \frac{l_{10} + l_{11}}{l_{10}}y_b$$

and

$$e_c = y_{\text{ceff}}/P$$

4.   The bending of intermediate link 8 is caused by the reaction force and the moment in the connection between guide rods 9 and the right end plate of link 8. Link 8 is considered as supported on two bearing blocks (ball bushings) 7c and 7d. Both the force and the moment depend on the relative position of links 11 and 8. Bending deflection y_s of link 8 end and its end slope θ_8 are causing an inclination of guide rods 9, carrying link 11, by an angle

$$\alpha_8 = y_8/l_{20} + \theta_8$$

and, accordingly, deflection and compliance at the arm end

$$y_{\text{deff}} = \alpha_8 (l_{20} + l_{12}); \qquad e_d = \alpha_8(l_{20} + l_{12})/P$$

5.   Compliance e_e at the arm end, which is caused by deformations in ball bushings 7c and 7d, is calculated similarly to e_b with the appropriate dimensional changes

$$e_e = \frac{1}{k_c}\frac{(l_{12} + l_{20} - l_{22})^2}{l_{22}^2} + \frac{1}{k_d}\frac{(l_{12} + l_{20})^2}{l_{22}^2}$$

6.   Compliance e_f at the arm end, which is caused by the bending of reinforced guide rods 6 and 6a, is calculated similarly to e_c with the appropriate
dimensional changes

$$e_f = \frac{y_{fef}}{P} = \left(\frac{l_{12} + l_{20} - l_{22}}{l_{22}} y_c - \frac{l_{12} + l_{20}}{l_{22}} y_d \right) \bigg/ P$$

7.   Compliance, which is caused by the bending of root link 2, is calculated
analogously to e_d with the appropriate dimensional changes, thus

$$\alpha_2 = y_2/l_{30} + \theta_2; \qquad y_{geff} = \alpha_2(l_{30} - l_{31} + l_{20} + l_{12}); \qquad e_g = y_{geff}/P$$

8.   The linear compliance at the arm end, which is caused by the angular
compliance e_{hd} of harmonic drive 1, is

$$e_h = e_{hd}(l_{30} - l_{31} - l_{32} + l_{20} + l_{21})^2$$

Angular stiffness $k_{hd} = 1/e_{hd}$ is specified by the manufacturer as $k'_{hd} = 45,000$
lb-in./rad for torques less than 100 lb-in. and $k''_{hd} = 234,000$ lb-in./rad for torques
above 500 lb-in.

Because of the rather cumbersome expressions for force and moment reactions and for deflections, and also because of the complex relationships between
the compliance components and the arm outreach, it is convenient to use a computer to construct compliance breakdown as a function of the arm outreach.

The breakdown is shown in Fig. 6.29. Although again the breakdown is
dominated by compliance e_{hd} of the harmonic drive, especially at low motor torques, compliance component e_c, which is caused by guide rods 9, is comparable
with the component e_{hd} and even exceeds it at high torques and short arm configurations. Ball bushings between link 11 and guide rods 9 also make a noticeable
contribution.

In this case, the reduction of the harmonic drive contribution to the overall
compliance can be achieved by the same approach, as was suggested in Section
6.4.3 (additional reduction stage or an oversized transmission). Components e_c
and e_b can be reduced by several design approaches, such as reinforcing of the
guide rods, increasing the ball bushing size (i.e., rod diameters), and using other
types of linear bearings [8].

The overall contribution of links bending ($e_a + e_d + e_g$) is 3.5–8.5% for
17.9 in. outreach, diminishing to 1.4–6.1% at 42.5 in. outreach.

The natural frequency of the arm with the rated payload of 5 lb (2.25 kg)
for a motor torque < 100 lb-in is 10.7 Hz at 17.9 in. outreach and 4.95 Hz at
42.5 in. outreach. The latter number correlates well with the experimental data.

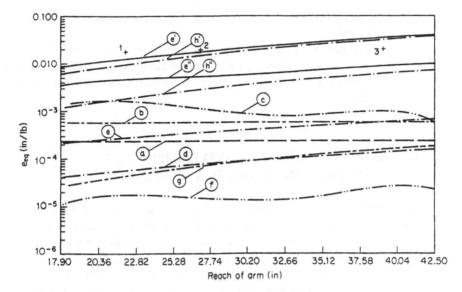

Figure 6.29 Compliance breakdown for robot arm in Fig. 6.27 vs. its extension; letters at the lines designate sources of compliance as listed in the text; e' and e'' = overall compliance for $T < 100$ lb-in. and $T > 100$ lb-in.; experimental points are measured at: (1) $T = 70$ lb-in., (2) $T = 90$ lb-in., and (3) $T = 127$ lb-in.

Measured data for static compliance are shown in Fig. 6.29 and indicate a reasonably good correlation with the computed values.

6.4.5 Summary

Examples in Sections 6.4.1 through 6.4.4 show the following:

1. Computation of the static compliance breakdown allows, with a relatively small effort, the evaluation of the design quality and selection of the most effective way to make improvements while in the blueprint stage.
2. Compliance breakdown of the considered robotic structures is usually dominated by one component whose identification allows for a subsequent significant improvement of the system performance.
3. In many instances, a reasonably minor design effort aimed at stiffening the identified dominating compliant component would lead to a major improvement in the structural stiffness.
4. Accordingly, special efforts directed toward the development of basic

mechanical structural units, especially transmissions and joints are warranted.

5. Linkage compliance does not play a critical role in the compliance breakdown of the considered robotic structures; as a result, increased arm outreach is not always accompanied by increased compliance at the arm end.

6.5 MORE ON STIFFNESS AND DAMPING OF ANTIFRICTION BEARINGS AND SPINDLES

Antifriction bearings, both ball and roller bearings, may play a special role in balance of stiffness and damping of important mechanical systems. For example, an overwhelming majority of machine tool spindles are supported by antifriction bearings that determine stiffness and damping of the spindle system, which, in turn, frequently determines performance quality of the machine tool. Parameters affecting stiffness and damping of antifriction bearings include rotational speed, magnitudes of external forces and of preload forces, clearances/interferences of fits on the shaft and in the housing, amplitudes and frequencies of vibratory harmonics, viscosity and quantity of lubricants, and temperature. Some of these factors are addressed below.

6.5.1 Stiffness of Spindles

Spindles are frequently supported by roller bearings, which are sensitive to angular misalignment between their rollers and races. To maintain a tolerable degree of misalignment, it is recommended [13] that the distance between the bearings does not exceed 4–5 diameters of the spindle and the length of its overhang in the front (for attachment of tools or part-clamping chucks) is kept to the minimum. To provide for a normal loading regime of the roller bearings, structural stiffness of the spindle itself for light and medium machine tools should be at least 250 N/μm (1.4 × 10^6 lb/in.). This stiffness value is computed/measured if the spindle is considered as a beam simply supported at the bearings' locations and loaded by a concentrated force in the midpoint between the supports. Spindles of high precision machines tools, while usually loaded by small magnitudes of cutting forces, should have at least two times higher stiffness, 500 N/μm (2.8 × 10^6 lb/in.), in order to provide adequate working conditions for the roller bearings [13].

Radial stiffness of a spindle unit can be enhanced by using more rigid bearings and/or by increasing the spindle diameter. The latter approach is less attractive since it reduces the maximum rpm of the spindle. Effectiveness of using stiffer bearings depends on the role that the bearings play in the overall stiffness of the spindle unit. Smaller spindles (50–60 mm diameter) contribute so much to the overall compliance at the spindle flange/nose (70–80%) that enhancement of the bearings' stiffness would not be very noticeable (see Fig. 6.30).

Stiffness of the spindle at its flange can be significantly enhanced by using short spindles of large diameters. Such an approach resulted in stiffness 450–2500 N/μm (2.5 − 14 × 10^6 lb/in.) at the spindle flange for lathes with maximum part diameter 400 mm [13].

It is important to assure by the design means that stiffness of the spindle is the same in all directions. Anisotropic stiffness may result in an elliptic shape of the machined part on a lathe, in a distorted surface geometry during milling operations, etc. On the other hand, properly oriented vectors of the principal stiffness may result in enhanced chatter resistance of the machine tool (e.g., [14]).

Optimal dimensioning of the spindle to obtain its maximum stiffness depends on its design schematic. Spindles of high speed/high power machining centers and of some other machine tools are frequently directly driven by an electric motor whose rotor is the spindle. In such cases, the spindle is loaded only by the cutting forces (Fig. 6.31a) (and also by the driving torque from the motor). However, many machine tools have a driving gear or a pulley located between its bearings. In such cases two external forces are acting on the spindle: cutting force P and the radial force from the pulley or the resultant of the tangential and radial forces from the gear (Q; Fig. 6.31b). Radial deflection of the spindle at the application point of the cutting force P, considering deformation of the spindle itself and its bearings and assuming one-step structure of the spindle (constant

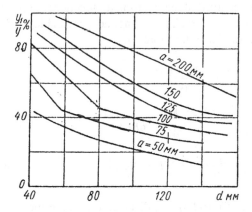

Figure 6.30 Ratio of the spindle proper deformation (y_1) to total deflection of the spindle (y) unit as function of diameter d and overhang a.

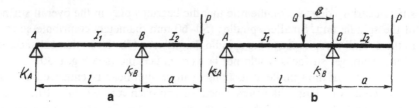

Figure 6.31 Typical loading schematics of spindles.

diameter between the bearing supports and another constant diameter of the overhang section), is [13]

$$\Delta = P\left[\frac{la^2}{3EI_1} + \frac{a^3}{3EI_2} + \frac{1}{k_B}\frac{(l+a)^2 + a^2(k_B/k_A)}{l^2}\right] \tag{6.55}$$

and angular deflection at the front bearing from the cutting force P is

$$\theta_B = Pal/3EI \tag{6.56}$$

Here I_1 and I_2 = cross-sectional moments of inertia of the section between the bearings, of the overhang section, respectively; and k_A and k_B = stiffnesses of the rear and front bearings units.

If the driving gear is placed between the bearings as in Fig. 6.31b, the total radial force Q acting on the spindle from the gear mesh or from the pulley is proportional to the cutting force P but has a different direction than P. Accordingly, both P and Q should be resolved along coordinate directions x (vertical) and y (horizontal). The components of radial and angular deformations of the spindle at the application point of the cutting force are

$$\Delta_{x,y} = P_{x,y}\left[\frac{la^2}{3EI_1} + \frac{a^3}{3EI_2} + \frac{1}{k_B}\frac{(l+a)^2 + (k_B/k_A)}{l_2}a^2\right]$$
$$+ Q_{x,y}\left[\frac{1}{k_B}\frac{(l+a)(l-b) - (k_B/k_A)ab}{l^2} - \frac{a}{6EI_1l}(b^3 + 2l^2b - 3lb^2)\right] \tag{6.57}$$

$$\theta_B = \frac{1}{3EI_1}\left[P_{x,y}al - \frac{Q_{x,y}}{2l}(b^3 - 2l^2b - 3lb^2)\right] \quad \text{rad} \tag{6.58}$$

It can be seen from Eqs. (6.55)–(6.58) that distance l between the spindle bearings is critical for its overall stiffness (deformation Δ for given cutting force

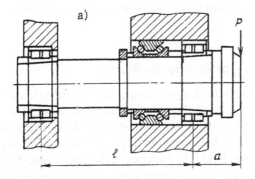

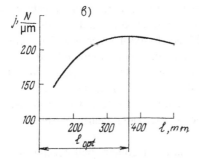

Figure 6.32 Radial stiffness as function of distance between bearing supports (*b*) for spindle unit *a*.

P). Fig. 6.32b [7] illustrates this statement for a typical spindle unit in Fig. 6.32a supported by two double-row roller bearings of the 3182100 series and having a given overhang length *a*. In a general case, optimization of ratio $K = l/a$ is important for improving performance of spindle units. Similar calculations for spindles having nonuniform cross sections can be easily performed by computer [15].

Special attention must be given to reducing the overhang length *a*. It can be achieved by a careful design of both the spindle and the tool holding/part holding devices, which constitute an extension of the spindle and determine the "effective overhang" (defined as the distance between the application point of the cutting force and the front bearing). It can be done by making tool holders of a solid design rather than a hollow one in order to reduce overhang of the tool and ultimately by integrating the tool with the spindle or by integrating the part holding chuck with the spindle (there are some commercial designs of lathes using this approach), among others.

6.5.2   Stiffness and Damping of Antifriction Bearings

While stiffness of bearings for critical applications, such as for machine tool spindles, is a critical parameter, it is not the only critical parameter to be considered by the designer. The other very important parameters are the allowable speed (rpm or dn where d = diameter of the bore in mm, and n = rotational speed in rpm) and the heat generation. Fig. 6.33 [16] compares radial stiffness of different types of antifriction bearings with d = 100 mm at two radial loads. While the angular contact ball bearing (contact angle $\beta = 12°$) has the lowest stiffness, it is widely used for spindles of high precision and/or high-speed machine tools.

Radial stiffness of both roller and ball bearings depends significantly on their preload (see also Chapter 3). Stiffness of double-row spindle roller bearings of the 3182100 series is shown in Fig. 6.34 as a function of preload. The preloading is achieved by axial shifting of the inner race, having a tapered bore, along the tapered spindle. When the deformation e caused by the preload exceeds $2\delta_{ro}$, where δ_{ro} = initial radial clearance in the bearing, its stiffness does not change significantly. The reason for this phenomenon is a rather weak nonlinearity of the radial load-deflection characteristic of the roller bearings (see Chapters 3 and 4).

Determination of radial stiffness of angular contact ball bearing is a more involved process due to their more pronounced nonlinearly and also due to complexity of their geometry. It has been suggested [16] that the following algorithm provides reasonably accurate data:

(a)   Axial compliance factor K_A characterizing the relative axial shift of the races under the axial preload force A_0 is determined from Fig. 6.35 as a function of relative preload A_o/Q_{lim} and the contact angle β; Q_{lim} = static load capacity of a given bearing.

(b)   Radial compliance coefficient K_r is determined from Fig. 6.36, where P = radial loading on the bearing.

(c)   Elastic mutual approaching of the races is calculated as

$$\delta_r' = 0.03 d_b K_R \qquad (6.59)$$

where d_b = diameter of balls in the bearing.

(d)   Radial stiffness of the bearing itself is

$$k_R = P/\delta_r' \qquad (6.60)$$

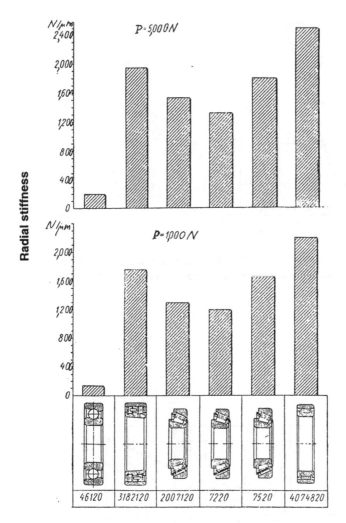

Figure 6.33 Comparison of radial stiffness of several types of bearings ($d = 100$ mm).

stiffness of the support is

$$k_s = P/(\delta'_r + \delta''_r) \tag{6.61}$$

where δ''_r = total of contact deformations between the bearing and the housing and between the bearing and the shaft from Eq. (3.15). It is important to remember that thermal deformation of the spindle and cen-

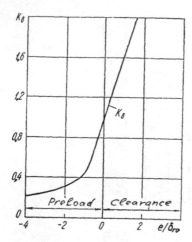

Figure 6.34 Compliance coefficient K_δ for double-row spindle roller bearings of series 3182100 and 4162900 as a function of clearance/preload.

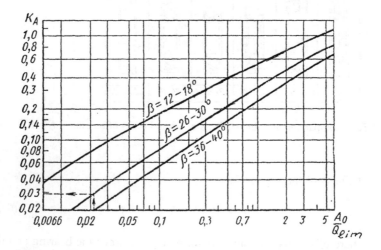

Figure 6.35 Axial compliance factor K_A for determining compliance of angular contact ball bearings.

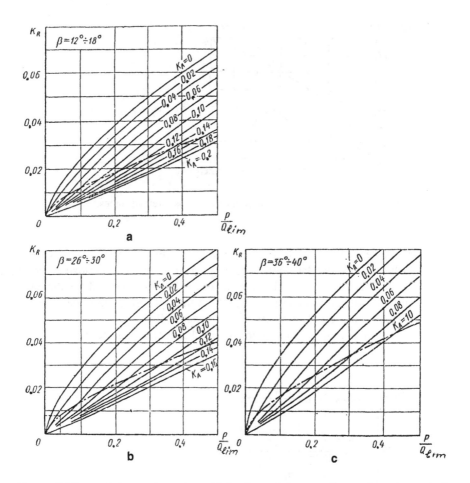

Figure 6.36 Radial compliance factor K_R for determining compliance of angular contact ball bearings: (a) $\beta = 12 \div 18°$; (b) $\beta = 26 \div 30°$; and (c) $\beta = 36 \div 40°$.

trifugal forces cause changes of the preload forces and of the effective contact angles β.

Preloading of angular contact ball bearings causes increasing of the actual contact angle. Figure 6.37 [17] shows incremental increases $\Delta\beta$ of the contact angle for angular contact bearings with nominal contact angles $\beta_0 = 15°$ and $26°$ as a function of $Q_p/z\, d_b^2$, where Q_p = preload force, z = number of balls, and d_b = diameter of the ball. It can be seen that the larger contact angle changes are developing for smaller nominal contact angles. Selection of the preload mag-

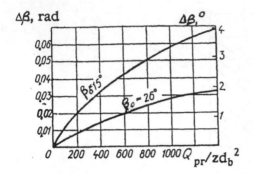

Figure 6.37   Change of actual contact angle vs. preload forces.

nitude depends on the load acting on the bearing. Figure 6.38 [7] shows radial stiffness k_r of angular contact bearings as a function of preload force Q_{pr} and external radial force P. The line for $Q_{pr} = 440$ N is typical for the case when all balls are accommodating force P. Falling pattern of the lines is due to the gradual reduction of numbers of the loaded balls.

An important phenomenon for angular contact ball bearings at high-speed

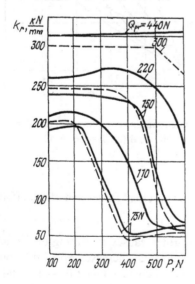

Figure 6.38   Radial stiffness of angular contact ball bearings 36100 (solid line) and 36906 (broken line) with contact angle $\beta = 15°$ as function of external force P and preload force Q_p.

regimes ($dn > \sim0.5 \times 10^6$ mm-rpm) is development of high centrifugal forces and gyroscopic moments acting on the balls. These forces are becoming commensurate with the external forces and the preload forces. The centrifugal forces are pressing the balls towards the outer races, thus changing the effective contact angles and the kinematics of the balls as well as redistributing contact loads in the bearing. These factors lead to reduction of stiffness. Figure 6.39 [7] gives radial deflection of the flange of the spindle with the journal diameter 110 mm supported by two pairs of angular contact ball bearings (in the front and in the rear bearing supports) and loaded by force $P = 1000$ N. It can be seen that the deflection is increasing (thus, stiffness is decreasing) for the same preload magnitude with increasing rpm. The rate of deflection increase is steeper when the preload force is lower.

The centrifugal forces can be significantly reduced by using lighter ceramic balls instead of steel balls. This alleviates the stiffness reduction at high rpm and rises the speed limit of the bearings. Figure 6.40 [18] shows dependence of radial stiffness of angular contact bearings ($d = 50$ mm) on rotational speed for (a) steel and (b) ceramic balls. The initial (static) stiffness of the bearings with ceramic (silicon nitride) balls is higher than for the bearing with steel balls due to higher Young's modulus of the ceramic. A more gradual stiffness reduction with increasing rpm is due to much lower (~2.5 times) density and thus magnitude of centrifugal force for the ceramic balls. A similar effect can be achieved by using bearings with hollow rollers (Section 8.1.1) or balls.

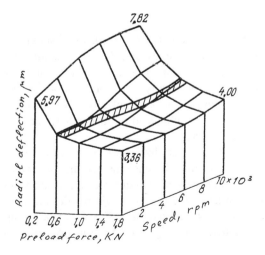

Figure 6.39 Radial deflection of spindle flange (journal diameter $d = 110$ mm) under radial load 1000 N as function of rpm and preload force.

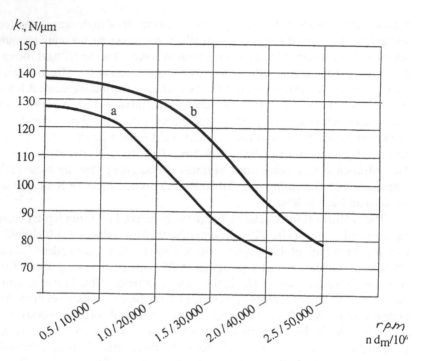

Figure 6.40 Radial stiffness of angular contact ball bearing (d = 50 mm) vs. speed: (a) steel balls; (b) silicon nitride balls.

An important factor in some applications is *axial stiffness* of spindle bearings, which may significantly influence accuracy and dynamic stability of the machining system. Although the share of radial compliance of the spindle bearings in the overall spindle compliance is usually not more than 40–60%, the axial stiffness (compliance) of the spindle unit is completely determined by the bearings. The axial stiffness is noticeably influenced by inaccuracies of components of the bearings, especially at low loads.

Since the finishing operations generate very low loads, influence of inaccuracies (such as nonuniformity of race thickness, dimensional variation and nonideal sphericity of balls, and nonperpendicularity of spindle and headstock faces to their respective axes) must be considered. Figure 6.41 [16] provides information on axial stiffness of high precision thrust ball bearings (Fig. 6.41a) and angular contact ball bearings with contact angle 26° (Fig. 6.41b) as functions of preload force Q_p.

It is important to understand that stiffness of the spindle unit is determined not only by stiffness of the bearings but also by seemingly minor design issues. Figure 6.42 [16] shows influence of positioning of the bearing relative to the

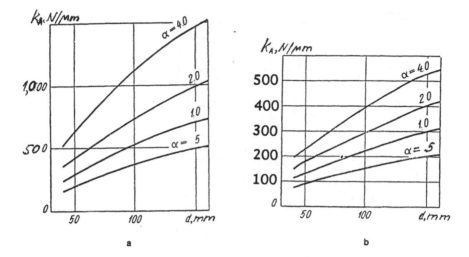

Figure 6.41 Axial stiffness k_a of ball bearings: (a) thrust bearings of 8100 and 8200 series; (b) angular contact bearings of 46100 series. Preload force $Q_p = \alpha d$, N, where d = bore diameter in mm.

supporting wall on its stiffness. The highest resulting stiffness develops when the outer race of the bearing is fit directly into the housing, without any intermediate bushing, and is symmetrical relative to the median cross section of the wall (1). Use of an intermediate bushing (2), and especially an asymmetrical installation with a small cantilever of the bearing (3) lead to significant deterioration of stiffness as well as service life of the bearing.

Damping in bearings for spindles and other critical shafts plays an important role in dynamic behavior of the system. As in many other mechanical systems, various design and performance factors, such as fits between the bearing and the shaft/housing, preload, lubrication, and rotational speed, influence both stiffness and damping. Frequently, these factors result in very different, sometimes oppositely directed, stiffness and damping changes. A survey of three major studies of stiffness and damping of rolling element bearings was presented by Stone [19].

It was found that stiffness and damping of both roller and ball bearings are significantly dependant on their rotational speed. Figure 6.43 shows stiffness, damping coefficient, and steady state temperature for a grease-lubricated tapered roller bearing for two mounting conditions (0 μm and 5 μm clearance). It can be seen that stiffness k_L is increasing with rpm, more for the bearing with 5 μm initial clearance. Although the initial stiffness for the bearing with 5 μm clearance was ~50% lower than for 0 μm clearance, this difference is decreasing with increasing rpm, probably due to thermal expansion causing preload development in both bearings. On the other hand, damping coefficient c_L is decreasing with

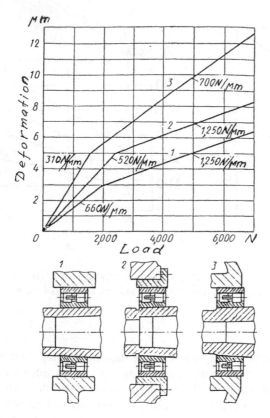

Figure 6.42 Influence of location of the front spindle bearing on compliance of the support.

rpm up to 1600–1700 rpm. Actual reduction of damping is even more pronounced than the plot in Fig. 6.43b illustrates, since the damping parameter important for applications is not c_L but log decrement δ related to c_L by Eq. (A.3.8) in Appendix 3. For the same mass m of the vibrating parts and same c_L, δ is decreasing with increasing k_L. This effect results in a smaller increase in δ above 1600 rpm than shown for c_L in Fig 6.43b.

Similar results are shown in Fig. 6.44 for a heavily preloaded angular contact ball bearing (internal deformation by the preload force 25 μm) with low viscosity (AWS 15) and high viscosity (AWS 32) lubricating oil. Damping is expressed in degrees of phase shift β between the exciting force and the resulting deformation under vibratory conditions. Since $\delta = \pi \tan \beta$ and β does not exceed 7°, the relative damping changes are proportional to the phase angle readings. Figure 6.44 demonstrates the same correlations as Fig. 6.43: stiffness is increasing while

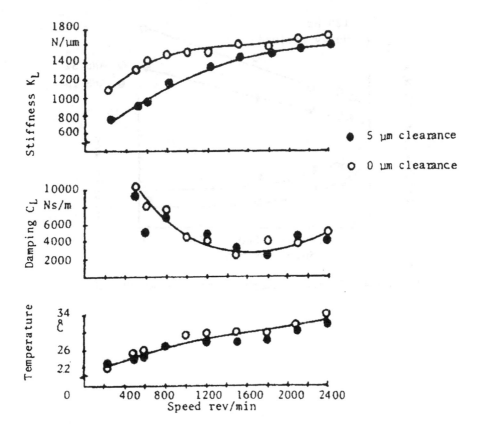

Figure 6.43   Variation of stiffness and damping with speed for tapered roller bearing NN3011 KUP lubricated with grease Arcanol L78.

damping is decreasing with increasing rpm. While the dynamic stiffness is increasing with increasing viscosity of lubricating oil (due to more effort required for squeezing the viscous oil from contact areas under load), damping is increasing with decreasing viscosity (due to more intense motion of oil in the contacts, thus higher friction losses).

Figure 6.45 demonstrates that stiffness of an angular contact ball bearing is decreasing while damping is increasing with increasing amplitude of vibratory force. These effects are in full compliance with the dependencies derived in Section 4.6.2.

It is interesting to note that angular (tilt) stiffness k_m and damping coefficient for angular vibration c_m for tapered roller bearing are both increasing with increasing preload (Fig. 6.46). This effect is due to a different mechanism of the tilt vibratory motions. These are mostly tangential motions between the rollers and

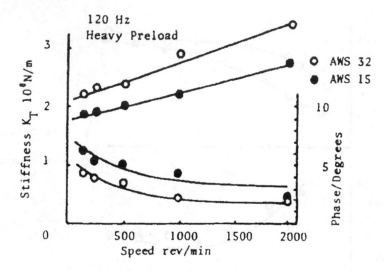

Figure 6.44   Stiffness and damping of angular contact ball bearings.

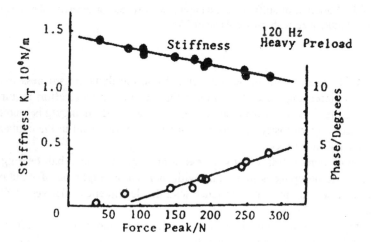

Figure 6.45   Variation of stiffness and phase angle vs. amplitude of excitation force.

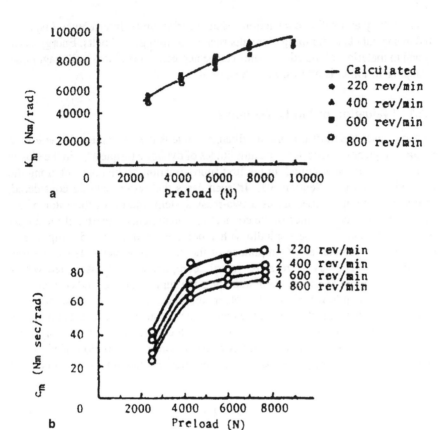

Figure 6.46 Variation of (a) tilt stiffness and (b) damping coefficient with preload force and speed.

the races, and it is obvious that both resistance to displacement (stiffness) and friction losses (damping) are increasing when the normal forces between the rollers and the races are increasing.

6.6 DAMPING IN POWER TRANSMISSION SYSTEMS [20]

Damping in power transmission systems is determined by damping in the working medium of the driving motor [e.g., by damping of the electromagnetic field between stator and rotor; see Eqs. (6.31) and (6.32)] and by energy dissipation in joints (key and spline connections, shaft/bearing units, interference fits, etc.), and in special elastodamping elements such as couplings and belts. Energy dissipation in materials of the components can be neglected due to its small magnitude. Damping in joints is addressed in Section 4.7.

6.6.1 Evaluation of Modal Damping

In many cases the most important vibratory mode is the fundamental (the lowest natural frequency) mode at which amplitudes of dynamic torque M_{dyn} in the elastic connections can be assumed, in the first approximation, to be constant along the reduced system (see Section 6.3). In such cases the system can be considered, also as a first approximation, as a two-mass system: inertia of the rotor of the driving motor—compliance of the connecting components—inertia of the output member. The latter can be a spindle with a tool in a machine tool, driving wheels of a wheeled vehicle, etc. Compliance of the connecting elements can be presented as $e = e_1 + e_2$. Here e_1 is "elastic compliance" not associated with a significant amount of energy dissipation; it is the sum of the reduced compliances of torsion and bending of shafts. The other component $e_2 = ae$ is "elastodamping compliance," which is the sum of the reduced compliances of components whose compliance is due to contact deformations and due to deformations of polymeric materials (couplings, belts, etc.). Energy dissipation for one period of vibratory process (relative energy dissipation) ψ for systems having not a very high damping

$$\psi = 2\,\delta \qquad\qquad (6.62)$$

where δ = log decrement of the vibratory process.

In a transmission system composed of shafts, gears, key and spline connections, bearings, etc., but not containing special damping elements such as couplings or belts, the damping is determined only by energy dissipation in joints δ_j. Relative energy dissipation in joints is $\psi_j = 2\,\delta_j > 0.5\text{--}0.7$ (see Section 4.7.3), and for the whole transmission system

$$\psi = a\psi_j \qquad\qquad (6.63)$$

In a typical gearbox, compliance of key and spline connections is about 35%; considering also contact compliance of gear meshes and of bearing units, $a = \sim 0.45$. Thus, $\psi > 0.45\ \psi_j = 0.22\text{--}0.31$ and log decrement for a gearbox is $\delta = \psi/2 = 0.11\text{--}0.15$. These values were confirmed experimentally.

In more general cases the fundamental mode is not as simple as it was assumed above, or there is a need to evaluate damping of higher vibratory modes. In such cases, differences of a and ψ for different segments of the system must be considered, as well as distribution of dynamic elastic moment M_{dyn} along the system. Distribution of M_{dyn} along the system is, essentially, a mode shape that can be computed using a variety of available software packages. Parameters a and ψ for each segment (elastic element) of the system can be easily determined from the original model such as shown in Fig. 6.20a, in which the composition of compliance for each component is shown. For example, compliance e_5 in Fig. 6.20a is composed of compliances of two key connections (28 and 11×10^{-7} rad/N-m) and one torsional compliance of a shaft, 124×10^{-7} rad/N-m. Thus, $a_5 = (28 + 11)/(28 + 11 + 124) = 0.24$. Since the key connections are not tightly fit in this gearbox, from Section 4.7.3 the value for δ can be taken as an average between the tight fit key (0.05–0.1) and the sliding spline (0.3–0.4), or $\delta = {\sim}0.18$ and $\psi = {\sim}0.36$. Values of a and ψ for other components can be determined in a similar way. After M_{dyn}, a, and ψ are known for each ith segment of the computational model, the damping parameters for this mode can be calculated as

$$\psi = 2\delta = \frac{\sum_i M_{dyn_i}^2 \psi_i a_i e_i}{\sum_i M_{dyn_i}^2 e_i} \tag{6.64}$$

After these parameters are calculated, resonance amplification factor for this mode can be determined as

$$Q = \pi/\delta = 2\pi/\psi \tag{6.65}$$

Instead of M_{dyn_i}, relative values M_{dyn_i}/M_{dyn_1} can be used in Eq. (6.64).

EXAMPLE We now compute damping characteristics for two lower vibratory modes for gearbox of the vertical milling machine whose mathematical models are given in Fig. 6.20d, for $n_{sp} = 60$ rmp. The computation is performed for a simplified system shown in Fig. 6.47a, in which values of a_i are also given for each elastic segment. Simplification is performed using the method described by Rivin [1,2] (see Sect. 8.6.1) that allows to reduce the number of inertias and elastic elements (i.e., degrees of freedom) of a dynamic model without distorting its lower natural frequencies and modes. Vibratory modes of M_{dyn} for two lowest modes corresponding to natural frequencies f_{n_1} and f_{n_2} are given in Fig. 6.47b in a logarithmic scale. From the plots in Fig. 6.47b, values of $M_{dyn i}/M_{dyn 1}$ can be easily computed. These values are given in Table 6.2 both in dB (upper lines for each f_n) and in absolute numbers (lower lines). Using data from Table 6.2 and values of a from Fig. 6.47a, damping can be easily calculated. Two cases are compared below. In the first case (A), ψ_i are assumed constant for each segment

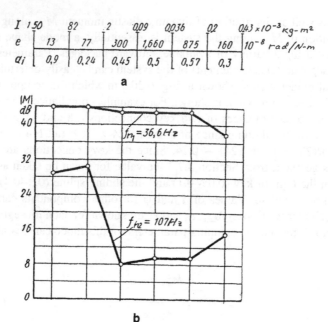

Figure 6.47 (a) Simplified dynamic model and (b) amplitudes of elastic torque in the elastic elements at two lower natural frequencies for the dynamic system in Fig. 6.20d.

of the model in Fig. 6.47a, $\psi_i = 0.6$; in the second case (B), damping of coupling ($e_1 = 13 \times 10^{-8}$ rad/N-m in Fig. 6.43a) is increased to $\psi_1 = 1.0$. After substituting all the parameters into Eq. (6.64):

(A) At natural frequency f_{n_1}, $\psi_{f_1} = 0.3$; and at f_{n_2}, $\psi_{f_2} = 0.21$
(B) At natural frequency f_{n_1}, $\psi_{f_1} = 0.306$; and at f_{n_2}, $\psi_{f_2} = 0.24$

These are very valuable results. They show that for this system the overall system

Table 6.2 Relative Dynamic Elastic Moment for Fig. 6.43

Natural frequency	No. of elastic segments					
	1	2	3	4	5	6
f_{n_1}	0	0	−1.3	−1.3	−1.3	−6.0
	1.0	1.0	0.86	0.86	0.86	0.5
f_{n_2}	0	+0.5	−21	−19.5	−19.5	−14
	1.0	1.2	0.09	0.106	0.106	0.2

damping at the fundamental natural frequency is significantly higher than at the second natural frequency. They also show that a substantial increase of damping in the coupling does not noticeably influence damping at the fundamental mode (2% increase) while increasing damping at the second mode by about 15%. These results would be very different for other configurations of the dynamic system. For example, for higher n_{sp} for the system in Fig. 6.20, the coupling compliance constitutes a larger fraction of the overall compliance, thus increase of damping in the coupling would increase the system damping at the fundamental mode much more significantly. This statement was validated by experimental results for high damping coupling tested at $n_{sp} = 375$ rpm, which are described in Article 2.

6.6.2 Utilization of Driving Motor Damping

The electromagnetic field between the stator and rotor of electric motors can be described as a highly damped elastic connection whose compliance and damping coefficients for some motor types are described by expressions (6.31) and (6.32). Similar expressions can be derived for other types of rotational motors as well as for linear motors, which are becoming more and more popular for machine tool feed drives. The damping capacity of the electromagnetic field in the driving motors is very substantial. Damping of the "natural" vibratory system representing the motor and consisting of an inertial member (rotor) connected to the stator by a connection having elastic and damping parameters described by eqs. (6.31) and (6.32) is characterized by relative damping values as high as $c/c_{cr} = 0.2$–0.4 (log decrement $\delta = 2\pi(c/c_{cr}) = 1.3$–2.5), much higher than damping of the mechanical part of the power transmission system ($\delta = 0.1$–0.3). However, since usually the natural frequency of the driving motor is much lower than the lowest natural frequency of the mechanical system, the motor does not noticeably influence damping of the mechanical system.

It was shown in Rivin [10] that it is possible to utilize high damping of the driving motor for damping enhancement of the whole power transmission system. This can be achieved by developing a design of the system in which dynamic coupling between the motor and the mechanical system is enhanced.

Fig. 6.48 shows a simplistic model of power transmission in which I_r–e_t–I_t is a single-degree-of-freedom two-mass model of the mechanical part of the transmission (e.g., obtained by the reduction method described in Section 8.6.1) and I_e–e_{em}–I_r is a dynamic model of the motor. Here, I_r = moment of inertia of the rotor (modified by the inertias of the mechanical system components reduced to the rotor during the reduction process described in Section 8.6.1); I_t = (modified) moment of inertia of the mechanical system; I_e = inertia of the electric power supply system ($I_e = \infty$ for motors plugged into a major electric network); and e_t and c_t = compliance and damping coefficients of the electromagnetic sys-

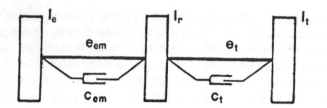

Figure 6.48 Two-degrees-of-freedom model of transmission with driving motor.

tem of the motor. Dynamic coupling between the systems I_r–e_m and I_r–e_t–I_t is characterized by the dynamic interconnection factor [10]

$$\sigma = 2\gamma \frac{\omega_m \omega_t}{\omega_m^2 - \omega_t^2} \qquad (6.66)$$

where

$$\gamma = \sqrt{\frac{I_e I_t}{(I_e + I_r)(I_r + I_t)}}$$

or for $I_e = \infty$

$$\gamma = \sqrt{\frac{I_t}{I_r + I_t}}, \quad \omega_m^2 = \frac{I_e + I_r}{e_m I_e I_r}, \quad \omega_t^2 = \frac{I_r + I_t}{I_r + I_t}$$

The larger σ, is the more interrelated are the motor and the mechanical tranmission systems. It is shown in Rivin [1, 10] that damping values for the natural modes of the system in Fig. 6.48 are

$$\Delta_1 \cong \frac{1}{2} \frac{(1 + \sqrt{1 + \sigma^2})\Delta_m - (1 - \sqrt{1 + \sigma^2})\Delta_t}{\sqrt{1 + \sigma^2}}$$

$$\Delta_2 \cong \frac{1}{2} - \frac{(1 - \sqrt{1 - \sigma^2})\Delta_m + (1 + \sqrt{1 + \sigma^2})\Delta_t}{\sqrt{1 + \sigma^2}} \qquad (6.67)$$

Here $\Delta_i = \omega_i (c/c_{cr})_i$, where $(c/c_{cr})_i$ = the relative damping value for the ith mode ($i = 1,2$), and ω_i = angular natural frequency of the ith mode. Usually $\omega_1 = 2\pi f_{n1} \approx \omega_m$, where ω_m = angular natural frequency of the motor system. If σ is small, e.g., $\sigma \ll 1$, then $\Delta_1 \approx \Delta$ and $\Delta_2 \to \Delta_t$. It means that the damping values of the natural modes of the complete system in Fig. 6.48 at the first and the second modes are very close to the damping values of partial systems (motor

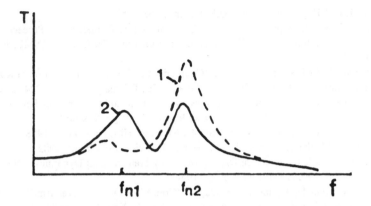

Figure 6.49 Amplitude-frequency characteristics of a motor-driven transmission: 1, weak coupling between mechanical system and motor field; 2, strong coupling. f_{n_2} = mechanical-system related natural frequency; f_{n_1} = motor-related natural frequency.

system and mechanical transmission system). However, if σ is large, the situation is different. For the limiting case when σ is very large, $\Delta_1 \approx \Delta_2 \approx (1/2)(\Delta_m + \Delta_t)$. It means that damping at the second (usually the most important for applications) mode ω_2 is significantly increasing at the expense of damping reduction at the less important mode ω_1.

Increasing value of the factor σ can be achieved by bringing partial natural frequencies ω_m and ω_t closer together and/or by increasing σ. The latter can be achieved, for example, by installing flywheel in an appropriate part of the mechanical transmission.

Figure 6.49 [10] compares amplitude-frequency characteristics of two transmissions—one with a weak coupling between dynamic systems of the motor and of the transmission (a), and another with a strong coupling (b).

REFERENCES

1. Rivin, E.I., Dynamics of Machine Tool Drives, Mashgiz Publishing House, Moscow, 1966 [in Russian].
2. Rivin, E.I., "Compilation and Compression of Mathematical Model for a Machine Transmission," ASME Paper 80-DET-104, ASME, New York, 1980.
3. Levina, Z.M., "Ball-spline connections," In: Components and Mechanisms for Machine Tools, Vol. 2, ed. by D.N. Reshetov, Mashinostroenic Publishing House, Moscow, 1972, pp. 334–345 [in Russian].

4. Rivin, E.I., "Key Connection," U.S. Patent 4,358,215.
5. Rivin, E.I., and Tonapi, S., "A Novel Concept of Key Connection," Proceedings of 1989 International Power Transmission and Gearing Conference, ASME, New York, 1989.
6. Iwatsubo, T., Arri, S., and Kawai, R. "Coupled Lateral-Torsional Vibration of Rotor System Trained by Gears," Bulletin of the JSME, February 1994, pp. 224–228.
7. "Computational analysis of deformations and dynamic and temperature characteristics of spindle units," ENIMS, Moscow, Russia, 1989 [in Russian].
8. Rivin, E.I., Mechanical Design of Robots, McGraw-Hill, New York, 1988.
9. A.H. Burr, Mechanical Analysis and Design, Elsevier, New York, 1982.
10. Rivin, E.I. "Role of Induction Driving Motor in Transmission Dynamics," ASME Paper 80-DET-96, ASME, New York, 1980.
11. Asada, H., Kanade, T., and Takeyama, I., "Control of Direct-Drive Arm," In: Robotics Research and Advanced Applications, ASME, New York, 1982.
12. Tooten, K., et al, "Evaluation of torsional stiffness of planetary transmissions," Antriebstechnik, 1985, Vol. 24, No. 5, pp. 41–46 [in German].
13. Sokolov, Y.N., "Spindles," In: Components and Mechanisms for Machine Tools, ed by D.N. Reshetov, Mashinostroenie Publishing House Moscow, 1972, Vol. 2, pp. 83–90 [in Russian].
14. Tobias, S.A., Machine Tool Vibration, Blackie, London, 1965.
15. Levina, Z.M., and Zwerev, I.A., "Computation of static and dynamic characteristics of spindle units using finite elements method," Stanki i Instrument, 1986, No. 10, pp. 7–10 [in Russian].
16. Figatner, A.M., "Antifriction bearing supports for spindles," In: Components and Mechanisms for Machine Tools, Vol. 2, ed. by D.N. Reshetov, Mashinostroenie Publishing House, Moscow, 1972, pp. 192–277 [in Russian].
17. Levina, Z.M., "Analytical Expressions for Stiffness of Modern Spindle Bearings," Stanki i Instrument, 1982, No. 10, pp. 1–3 [in Russian].
18. "Bearings for High Speed Operation," Evolution, 1994, No. 2, pp. 22–26.
19. Stone, B.J., "The State of the Art in the Measurements of the Stiffness and Damping of Rolling Element Bearings," Annals of the CIRP, 1982, Vol. 31/2, pp. 529–538.
20. Rivin, E.I., "Calculation of Dynamic Loads in Power Transmission Systems," In: Components and Mechanisms of Machine Tools [Detali i mekhanismi metallorezhuschikh stankov], Vol. 2, ed. by D.N. Reshetov, Mashinostroenie Publishing House, Moscow, 1972, pp. 30–82 [in Russian].

7

Design Techniques for Reducing Structural Deformations and Damping Enhancement

Previous chapters have addressed various correlations between design features of structures and their stiffness and damping. Although it is rather obvious how to use these correlations for enhancement of the structural stiffness and damping, it is useful to emphasize some very effective design techniques for stiffness and damping enhancement. This usefulness justifies, in the author's opinion, some inevitable repetitions. In some cases, such as described in Section 7.5, issues of stiffness, damping, and mass cannot be separated.

7.1 STRUCTURAL OPTIMIZATION TECHNIQUES

In many structures, there are critical directions along which deflections must be minimal (i.e., stiffness must be maximized). Frequently these critical directions relate to angular deformations, usually caused by bending. Angular deformations are dangerous since even small angular deformations may result in large linear deformations if the distance from the center of rotation is significant.

Angular deformations are naturally occurring in non-symmetrical structures, such as so-called C-frames (Fig. 7.1a). C-frame structures are frequently used for stamping presses, drill presses, welding machines, measuring systems, and more. They have an important advantage of easy access to the work zone. However, C-frame machines exhibit large deformations ("opening of the frame") under the work loads. In drill presses such deformations may cause nonperpendicularity of the drilled holes to the face surfaces; in stamping presses they may cause nonparallelism of upper and lower dies and other distortions that result in a fast wear of stamping punches and dies as well as in poor quality of parts. To

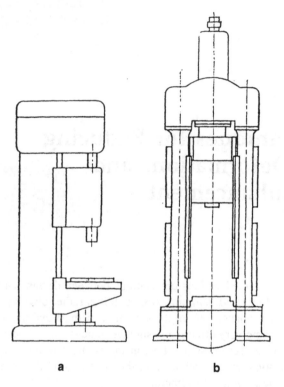

Figure 7.1 Typical schematics of machine frames.

reduce these undesirable effects, the cross sections of the structural elements (thus, their weights) are "beefed up" and the rate of performance is reduced.

A very effective way of enhancing stiffness of C-frame machine structures is to replace them with two-column ("gantry") or three- to four-column systems (Fig. 7.1b). Such symmetrical architecture minimizes both the deformations and their influence on part accuracy and on tool life. The difference between maximum deformations of the overhang cross beam of a C-frame structure (radial drill press; Fig. 7.2a) and of the cross beam of a gantry machine tool is illustrated in Fig. 7.2b.

Stiffness is a very important parameter in machining operations since relative deformations between the part and the tool caused by the cutting forces are critical components of machining errors. In the case of a cylindrical OD grinder in Fig. 7.3a, both the part 1 and the spindle of the grinding wheel 2 can deflect under the cutting forces. On the other hand, in the centerless grinding process in Fig. 7.3b part 1 is supported between grinding wheel 2, supporting wheel 4, and sta-

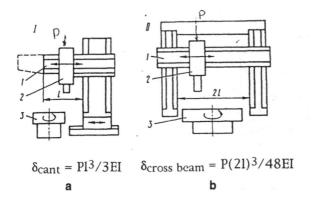

$$\delta_{cant} = Pl^3/3EI \qquad \delta_{cross\ beam} = P(2l)^3/48EI$$

        a   **b**

Figure 7.2 Deformations of (a) C-frame and (b) gantry machine frames.

tionary steady rest 3. As a result, the part deformations are reduced, thus greatly improving its cylindricity.

In many designs, critical deformations are determined by combinations of several sources. For example, deformation at the end of a machine tool spindle is caused by bending deformations of the spindle as a double-supported beam with an overhang, and by contact deformations of the bearing supports. Changing geometry of the system may influence these sources of deformations in totally different ways. For example, changing the span between bearings of the spindle in Fig. 7.4 with a constant cross section would change its bending deformations but would also change reaction forces in the supports (thus, their deformations), and modify the influence of the support deformations on the deflections at the end. The situation is even more complex for actual spindle designs, which are characterized by greatly different cross sections along the axis. These cross sections and the bearing sizes would be changed if the overall geometry (e.g., dis-

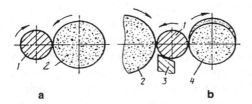

        a   **b**

Figure 7.3 Support conditions of the machine parts on (a) an OD grinder and (b) a centerless grinder.

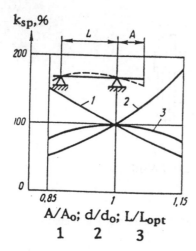

k_{sp},%

200

100

0

0.85   1   1,15

A/A$_0$; d/d$_0$; L/L$_{opt}$

1   2   3

Figure 7.4 Influence of design parameters on spindle stiffness.

tance between the supports) changes. Thus, optimization of the spindle stiffness becomes a complex multiparametric interactive problem. Although there have been many attempts to solve it [1], the optimization of the spindle designs still requires a combination of the computational results and the expertise of the designers (see also Section 6.5.1).

A very powerful design technique resulting in enhancement of the effective structural stiffness is the use of rational loading patterns. One approach to rationalization of the loading pattern (load distribution) is the use of supporting/load-bearing devices providing a continuous load distribution rather than concentrated forces. For example, hydrostatic guideways provide more uniform load distribution than rolling friction guideways, thus reducing local deformations. Another advantage of hydrostatic guideways is their self-adaptability; the positioning accuracy of heavy parts mounted on hydrostatically supported tables can be corrected by monitoring the oil pressures in each pocket.

Another design technique resulting in a desirable load distribution is illustrated in Fig. 7.5 [2], showing a setup for machining a heavy ring-shaped part 1 on a vertical boring mill. If the part were placed directly on the table 3, it would deform as shown by the broken line. The deformation can be significantly reduced by use of intermediate spacer (supporting ring) 4 with extending arms 2. The ring 4 applies the weight load between the guideways 5 and 6, thus eliminating the moment loading of the table 3 on its periphery and reducing its bending deformations.

Figure 7.6 shows a setup for machining (turning and/or grinding) of a crank-

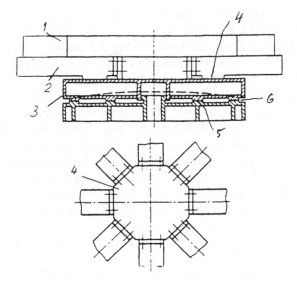

Figure 7.5 Enhancement of effective bending stiffness of rotating round table 3 of a vertical boring mill by intermediate spacer 4 with arms 2 for machining oversized part 1.

shaft having relatively low torsional stiffness. The *effective stiffness* of the crankshaft is enhanced (twisting of the shaft caused by the cutting forces is reduced) by driving the part from two ends using synchronized drivers (bevel gears and worm reducers). This technique was suggested and patented in the United States in the beginning of the century and is still widely used on crankshaft grinding machines.

Other systems in which changes in support conditions and load distribution

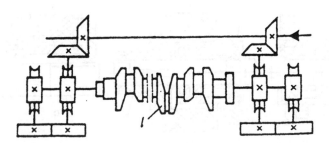

Figure 7.6 Compensation of low torsional stiffness of crankshaft 1 by driving from both ends.

result in reduced deformations, and thus in enhancement of effective stiffness, are shown in Figs. 7.7 and 7.8 [3]. Both figures show components of power transmission systems. Figure 7.7, which is similar to Fig. 5.5, compares methods of mounting a pulley on the driven shaft (spindle or other output shaft). In a conventional embodiment in Fig. 7.7a, the pulley 3 is mounted on the cantilever extension of the shaft 4; the deformations are relatively large and stiffness at the pulley attachment point is low. The design in Fig. 7.7b shifts the loading vector, which in this case passes through the shaft bearing. Thus, although the pulley 1 is mounted on the cantilever shaft segment 2, the deformations are significantly reduced as if it were not cantilever. The effective stiffness is further dramatically increased in the case of Fig. 7.7c, in which the pulley 5 is supported not by the shaft but by the housing 6. The shaft is not subjected to any radial forces but only to the torque, thus the radial stiffness is determined not by the shaft deformations but by much smaller deformations of the bearings.

Figure 7.8 emphasizes the importance of avoidance, or at least shortening, of overhang segments under load. In case aII, load from the thrust bearing is transmitted through the massive wall whose deformation is negligible, while in aI this load inflicts bending of a relatively thin overhang ring. Figure 7.8b(II) shows how a minor design change—machining the bevel gear as a part of the shaft—significantly reduces overhang of the gear as compared with the design in Fig. 7.8b(I), in which the gear is fit onto the shaft. Design II is more expensive but is characterized by a much higher stiffness.

Optimization of complex mechanical structures in order to reduce their deformations caused by performance-related forces requires, first of all, identification of the dominant sources of the objectionable deformations. Such identification can be made by application of static loading to the system and measuring of the ensuing deformations (*compliance breakdown*) or by dynamic excitation of the structure and measuring its responses (*modal identification*).

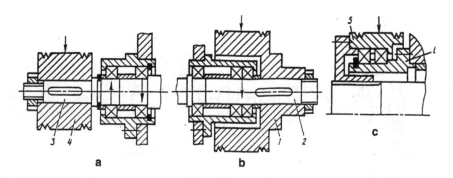

Figure 7.7   Alternative designs of cantilever shafts.

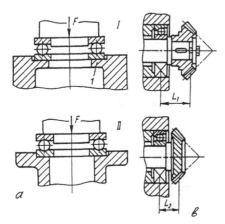

Figure 7.8 Design techniques for (a) avoiding and (b) reducing cantilever loading.

Contact deformations play important roles in compliance breakdown of many mechanical systems. Since contact deformations are highly nonlinear, the dynamic testing results have to be treated very cautiously. Currently, dynamic testing (evaluation of amplitude-frequency characteristics and of modal shapes) are usually performed by impact excitation of the system by an instrumented hammer, and processing the responses by computing Fourier transforms of the input and output signals. Although this powerful technique is much more convenient than direct (by static loading) stiffness/compliance evaluation of mechanical systems, it is sometimes forgotten that it is applicable only to linear systems.

It is also very important to remember that the hammer excitation applies relatively small forces to the system. Thus, if the operational forces in the system are large (e.g., static and dynamic cutting forces in machine tools), the testing has to be performed by applying static forces having the appropriate magnitudes, and/or the dynamic testing has to be performed while the system is preloaded by the specified static forces.

Static "compliance breakdown" can be constructed computationally (e.g., see Section 6.4) or experimentally. Appendix 2 describes an experimental study of the static compliance breakdown for a precision OD grinder.

7.2 COMPENSATION OF STRUCTURAL DEFORMATIONS

Structural deformations caused by weight of the components as well as by payloads (e.g., cutting forces) can be reduced by passive and/or active (servocon-

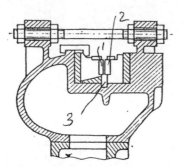

Figure 7.9 Work area of a thread-rolling machine.

trolled systems) means. Reduction of the structural deformations is equivalent to enhancement of structural stiffness of the system.

Figure 7.9 shows the work zone of a thread-rolling machine in which two roll dies, 1 and 2, are generating thread on blank 3. To achieve a quality thread, dies 1 and 2 must be parallel. However, due to high process loads, initially parallel dies would develop an angular misalignment due to structural deformations of the die-holding structure, thus resulting in the tapered thread. To prevent this undesirable effect, the structure is preloaded/predeformed during assembly in order to create an oppositely tapered wedged clearance between the roll dies. The rolling force would make the dies parallel, thus resulting in an accurate cylindrical thread.

In tapered toolholder/spindle interfaces (Fig. 7.10), the standard tolerances on the toolholder angle and on the angle of the spindle hole (International Standard ISO 1947) are assigned in such a way that the angle of the spindle hole is always smaller than the toolholder angle. Pulling the toolholder into the spindle reduces the difference in the angles and guarantees clamping at the front part of

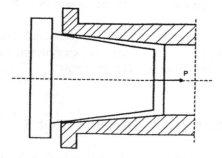

Figure 7.10 Standardized fit of tapered interfaces (exaggerated).

the connection, thus increasing stiffness by reducing the tool overhang (also see Chapter 4 and Section 8.2.2).

Many large machine tools and other production machines develop large deformations under weight forces. Since relative positions of the heavy units are not constant (moving tables and carriages, sliding rams, etc.), the weight-induced deformations frequently disrupt the normal operation. It is important to note that, generally, the weight-induced deformations cannot be reduced by just "beefing up" the parts. In a radial drill press in Fig. 7.11, deformation of the cantilever cross beam 1 caused by weight W of the spindle head 2 is

$$\delta = Wl^3/3EI \tag{7.1}$$

If all dimensions of the machine are scaled up by factor K, then:

$$l_1 = Kl; \quad I_1 = K^4I; \quad W_1 = K^3W; \quad \delta_1 = K^3WK^3l/3EK^4I = K^2\delta \tag{7.2}$$

Thus, a straightforward "beefing up" did not reduce deformations—just the opposite—and more sophisticated design approaches are necessary to achieve the required stiffness.

Deformations in large machines can be very significant. Deformations in a heavy vertical boring mill in Fig. 7.12 [2] under weight loads are $\delta_1 = 1.25$ mm (0.05 in.) and $\delta_2 = 1.0$ mm (0.04 in.). Deformation of the table under the combined load of cutting force and weight load, $F = 3,000$ KN (650,000 lb), is $\delta_3 = 0.05$ mm (0.002 in.).

Many techniques are used to reduce these deformations and, consequently, to enhance the effective stiffness of large machines. A simple and effective technique is "forward compensation" of the potential deformations by intentional distortion of shape of the guideways (Fig. 7.13) [3].

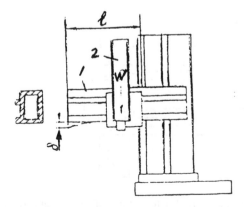

Figure 7.11 Schematic of a radial drill press.

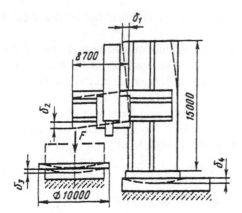

Figure 7.12 Deformations in a large vertical boring mill under weight and cutting forces.

Tolerances on straightness of the long guideways are always assigned to be towards a convex shape in the middle of the guideway. Guideways on cross beams have to be machined while the crossbeam is in place, is attached to columns, and is preloaded. Then, horizontal guideways would stay flat under the gravity force; otherwise they would sag.

There are many fabrication techniques for creating ''predeformed'' frame parts. The most straightforward technique is scraping. A highly skilled operator can relatively easily scrape the required convex or concave profile, only slightly deviating from a flat surface. Another technique is illustrated in Fig. 7.14. It involves intentional creation of residual thermal deformations; the dots in Fig. 7.14 indicate areas of local heating with a gas torch to 150–200°C [2].

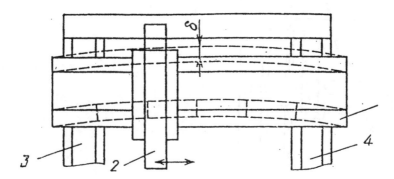

Figure 7.13 Compensation of weight-induced deformations of cross beam by intentional distortion of its original shape.

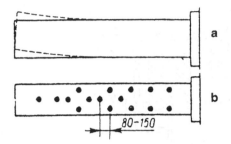

Figure 7.14 Forward compensation of (a) ram deformation by (b) generating thermal distortions by localized heating.

Although these techniques are "rigid" ones that do not allow for adjustment of the degree of compensation, the techniques illustrated in Figs. 7.15–7.18 [2,4] are more "flexible" since they provide for the adjustments. In Fig. 7.15a, the structural member 1 is made hollow. Inside the member 1 there are two auxiliary beams 2 of a significant rigidity. Each beam 2 can be deformed in the positive and negative z directions by tightening bolts 3 or 4, respectively. Deformation of the beams 2 is causing an oppositely directed deformation of the structural member 1. A similar system is shown in Fig. 7.15b. It has only one auxiliary beam 2, but the adjusting bolts 3, 4, and 5 can introduce relative deformations between the structural member 1 and the beam 2 in two directions x and z and thus the member 1 can be "predistorted" in two directions.

Predistortion of the cantilever structural member 1 in Fig. 7.15c is achieved by tensile loading of the auxiliary rod 2, which causes compensation of the weight-induced deflection δ.

A similar system in Fig. 7.15d [4] enhances effective stiffness of ram 3 of a coordinate measuring machine. Ram 3 carries quill 6 with measuring probe 7, thus the measurement accuracy is directly dependent on its deformations. Bending of ram 3 caused by its weight is changing with changing its overhang. The changing deformation is compensated by correction bar 9, which is placed above the neutral plane of ram 3 and attached to the latter next to quill 6 location. Tension of bar 9 causes an upward bending of ram 3. To generate the required tensile force, the opposite end of bar 9 is attached to the small arm 10 of lever 12, which is pivoted on the back side 11 of ram 3. Strained wire 14 is attached to the large arm 13 of lever 12 and wrapped on the large diameter 16 of double roller 15 fastened to housing 1. Another strained wire 17 is attached to ram 3 and wrapped on the smaller diameter 18 of roller 15. When the overhang of ram 3 is increasing, wire 17 is rotating roller 15 thus forcing wire 14 to pull lever 12, which in turn applies tension to correction bar 9 and straightens ram 3.

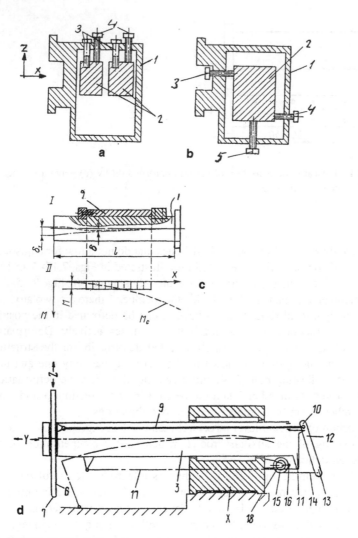

Figure 7.15 Adjustable systems for application of opposing bending moments.

While the systems in Fig. 7.15 are close-contour mechanically prestressed structures, there are frequently used hydraulic compensating systems applying adjustable forces to the structural members. The system in Fig. 7.16 has a hydraulic cylinder-piston unit 2 applying counterbalancing force to the heavy spindle head 3. Pressure regulator 1 can be set to compensate the weight load of the spindle head 3 or can vary the counterbalancing force depending on the specified

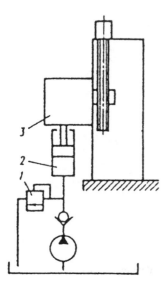

Figure 7.16 Hydraulic counterbalancing of spindle head 3.

parameters, such as the cutting force. A similar system for a gantry machine tool is shown in Fig. 7.17. In this case the hydraulic cylinder 1 "transfers" the weight load of the moving heavy carriage 3, which contains the spindle head, from the crossbeam 2 to the auxiliary frame 4.

The system in Fig. 7.18 is using the counterbalancing weight 4 connected with the heavy cantilever ram 1 moving in and out via cables 2 and 3. Cam 6

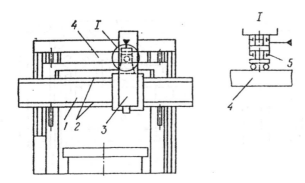

Figure 7.17 Hydraulic system for transference of spindle head 3 weight to auxiliary frame 4.

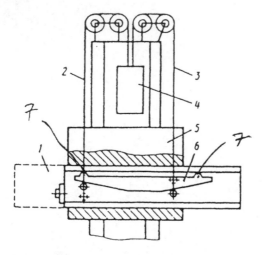

Figure 7.18 Position-dependent weight counterbalancing device.

attached to the ram 1 is engaged with the cables 2 and 3, and transmits load from the counterweight 4 to the ram 1 via two supporting legs 7. Motion of ram 1 is accompanied by a precalculated redistribution of reaction forces between the legs 7 thus preventing sagging of the ram 1 under weight loads.

A rational design of bearing supports for power transmission shafts can beneficially influence their computational models thus resulting in significant stiffness increases. Figures 7.19a and b show two design embodiments of a bearing support for a machine tool spindle. In both cases, two identical bearings are used, the

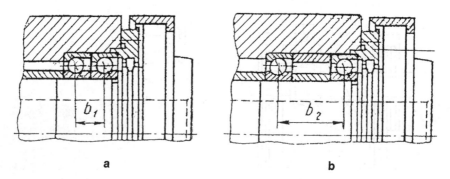

a   b

Figure 7.19 Stiffness enhancement of (a) front spindle bearing unit by (b) a minor design change.

only difference being the distance between the bearings. In the design b, the distance is larger ($b_2 > b_1$), which would create a "built-in" effect for bending of the spindle by the cutting forces, while the design a should be considered as a simple support (low angular stiffness restraint). Thus, design b will exhibit a higher stiffness and a better chatter resistance.

While the design changes reducing both eccentricity of loading conditions and overhang are beneficial for stiffness enhancement, they have to be done cautiously. Figure 7.20 [5] shows three designs of the front spindle bearing. The double row roller bearing is preloaded by tightening bolts 1 acting on preloading cover 2 and through cover 2 on the front bearing. Stiffness at the spindle end in design a is higher than in design b due to closeness of the preloading bolts to the axis, $C < C_1$. It is also higher than in design c due to the shorter overhang, $a < a_1$. However, due to closeness of preloading bolts to the bore for the bearings, bolt tightening in design a creates large distortions Δ of the housing and outer races of the bearings (Fig. 7.20d), which are detrimental for accuracy. Distortion of the outer race in design c is absent since the bolt 1 is removed from the bearing by a significant distance l_3.

The techniques described above allow one to compensate for deforma-

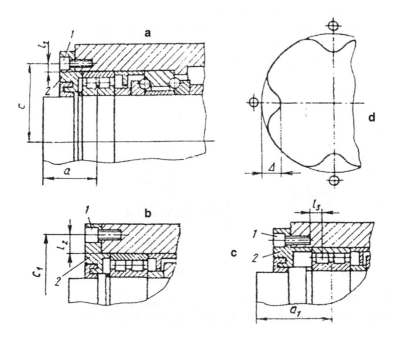

Figure 7.20 Alternative design embodiments of a front spindle bearing unit.

tions by design means. Although these techniques result in better, more robust systems, compensation of structural deformations in computer controlled (CNC) machines can also be achieved by modifications of control signals sent to work organs.

7.3 STIFFNESS ENHANCEMENT BY REDUCTION OF STRESS CONCENTRATIONS

Stress concentrations in structural components may result in large local deformations and thus in increasing overall deformations and reduced structural stiffness. Special attention should be given to alleviation of stress concentrations in stiffness-critical systems. It can be achieved by a judicious selection of design components and/or their shape and dimensions.

Figure 7.21 [3] shows a bearing support for shaft 1. If forces acting on the shaft and the resulting shaft deflections are relatively small, the tapered roller bearing in Fig. 7.21a exhibits high stiffness since rollers 2 are contacting bearing races along their whole length. However, at larger shaft deformations, the length of contact between the rollers and the races is reduced and the contact stresses are concentrated at the roller edges a, thus creating large local deformations. In this case, it can be advantageous to replace the single tapered roller bearing with tandem angular contact ball bearings which share the total reaction force $R_1 + R_2$, as shown in Fig. 7.21b. The shaft deformation is reduced (and its effective stiffness increased) due to increased angular stiffness of the support (close to

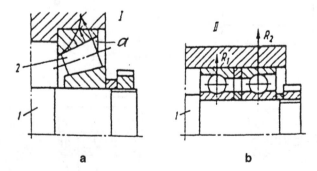

a   b

Figure 7.21 Enhancement of stiffness of shaft bearing unit by replacing (a) rigid but misalignment-sensitive tapered roller bearing with (b) two spread-out angular contact ball bearings.

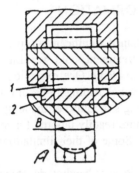

Figure 7.22 Stress concentration in roller guideways.

built-in conditions in Fig. 7.21b vs. single support conditions in Fig. 7.21a), and due to elimination of the stress concentrations.

Figure 7.22 is a roller guideway in which width B of the rollers 1 is less than width of the guideway 2. It results in a nonuniform pressure distribution as illustrated by diagram A. The stress concentrations and the deformations are reduced if widths of the rollers and of the guideway are the same (dashed line in the diagram). Since deformations in the guideway are proportional to stresses, reduction of stress concentrations leads to a stiffness enhancement.

Sharp stress concentrations are also characteristic for interference fits (press fit and shrink fit) connections. These stress concentrations and the resulting deformations can be significantly reduced by judicious shape modifications of one or both connected parts. Influence of shape of the bushing 1 on the contact stress distribution in a shrink-fit connection is shown in Fig. 7.23.

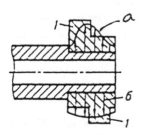

Figure 7.23 Reduction of stress concentration in interference fit connections.

7.4 STRENGTH-TO-STIFFNESS TRANSFORMATION

While strength of metals is relatively easy to enhance significantly (up to 3–7 times for steel and aluminum) by alloying, heat treatment, cold working, etc., their stiffness (Young's modulus) is essentially invariant. One exception from this universal rule are aluminum-lithium alloys, in which adding 3–4% of lithium to aluminum results in a 10–15% increase of Young's modulus of the alloy (see Table 1.1). Fiber-reinforced composites can be more readily tailored for higher stiffness, but they cannot be used in many cases. Some of their limitations are discussed in Chapter 1.

The most critical mode of loading is bending because bending deformations can be very large even for not very high forces; thus bending stiffness can be very low, and in many cases this determines the effective stiffness of the structure. Bending stiffness can be enhanced by reducing spans between supports of the components subjected to bending, by reducing the overhang length of cantilever components, and by increasing cross sectional moments of inertia ("beefing up"). The first two techniques are frequently unacceptable due to design constraints, but "beefing up" of cross sections can even be counterproductive since it inflates dimensions and increases weights of the components, as illustrated above in Section 7.1.

7.4.1 Buckling and Stiffness

"Buckling" of an elongated structural member loaded with a compressive axial force P is loss of stability (collapse) of the structural member when the compressive force reaches a certain critical magnitude P_{cr}, which is also called the *Euler force*. Usually, the buckling process is presented as a discrete situation: stable/ unstable. However, the process of development of instability is a gradual continuous process during which bending stiffness of the structural member is monotonously decreasing with increasing axial compressive force P. The member collapses at $P = P_{cr}$, when its bending stiffness becomes zero.

This process can be illustrated on the example of a cantilever column in Fig. 7.24a. Bending moment M causes deflection of the column. If the axial compressive force $P = 0$, bending stiffness

$$k_0 = M/x \qquad (7.3)$$

where x = deflection at the end of the column due to moment M. If $P \neq 0$, it creates additional bending moment Px, which further increases the bending deformation and thus reduces the effective bending stiffness. The overall bending moment becomes [6]

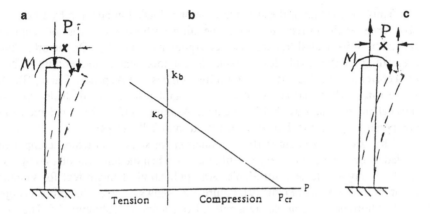

Figure 7.24 Stiffness change (b) of a cantilever beam loaded by (a) a compressive and (c) a tensile force.

$$M_{ef} = M/(1 - P/P_{cr}) \tag{7.4}$$

and the resulting bending stiffness is approximately

$$k_b = (1 - P/P_{cr})k_o \tag{7.5}$$

Figure 7.24b illustrates dynamics of the stiffness change for the column in Fig. 7.24a with increasing compressive force P. Equation (7.5) was experimentally validated in Jubb and Phillips [7]. The stiffness-reducing effect presented by this equation must be considered in many practical applications. For example, supporting of a machined part (e.g., for turning or grinding) between the centers involves application of a substantial axial force, which may result in a significant reduction of bending stiffness for slender parts having relatively low P_{cr} (see Article 3).

If the axial force is a tensile force $(-P)$ instead of compressive force P, Fig. 7.24c, then the effect is reversed since the tensile force creates a counteracting bending moment $(-Px)$ on the deflection x caused by moment M. The effective bending moment is thus reduced,

$$M_{ef} = M(1 + P/P_{cr}) \tag{7.6}$$

and the effective stiffness is increased

$$k_b = k_o(1 + P/P_{cr}) \tag{7.7}$$

Thus, preloading of the structural member loaded in bending by a compressive force results in reduction of its bending stiffness (and in a corresponding reduction of its natural frequencies; see Appendix 3 and [7]). On the other hand, preloading by the tensile force results in enhancement of its bending stiffness (and corresponding increase in natural frequencies; see Appendix 3 and [7]). This is the same effect that allows one to tune guitar strings (and strings of other musical instruments) by their stretching. A similar effect for two-dimensional components (plates) is described in Ivanko and Tillman [8].

The effect of bending stiffness reduction for slender structural components loaded in bending can be very useful in cases when the stiffness of a component has to be adjustable or controllable. An application of this effect for vibration isolators is proposed in Platus [9]. Figure 7.25 shows a device that protects object 14 from horizontal vibrations transmitted from supporting structure 24. The isolation between 14 and 24 is provided by several isolating elements 18 and 16. Each isolator 18 and 16 is a thin stiff (metal) post, 60 and 32, respectively. The horizontal stiffness of the isolation system is determined by bending stiffness of posts 60 and 32. This stiffness can be adjusted by changing compression force applied to posts 60 and 32 (in series) by loading bolt 66. The horizontal stiffness can be made extremely low (even negative if the compressive force applied by bolt 66

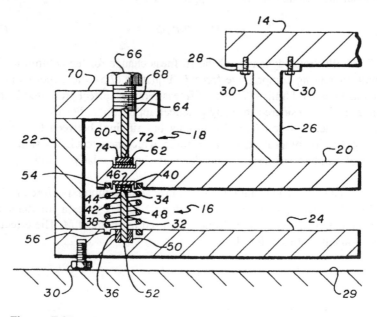

Figure 7.25 Vibration isolator for horizontal motion, based on using buckling effect for stiffness adjustment.

exceeds the critical force). In the latter case, the stability can be maintained by springs 38.

7.4.2 "Reverse Buckling" Concept

Very effective stiffness enhancement of structural components subjected to bending can be achieved by axial preloading in tension of the component loaded in bending for reducing its bending deformations. This technique can be called "reverse buckling" or the "guitar string" effect.

As it was shown in Chapters 3 and 4, preloading of a belt or chain transmission, ball bearings, joints, etc., also results in enhancement of stiffness of the preloaded system. However, belts, chains, balls and races, etc., are subjected to higher forces than the same components in nonpreloaded systems. Thus, part of the load-carrying capacity of the system (*strength*) is traded for increase of its effective *stiffness*. The reduction of strength is also the price to be paid for using the "reverse buckling" effect. Thus, all stiffness enhancement techniques based on applying a preload to the system can be named as *strength-to-stiffness transformation*.

The tensile preload of a beam not only results in enhancement of its bending stiffness, but also increases its buckling stability, especially for axial compressive forces applied within the span of the beam. The beam in Fig. 7.26 is pretensioned between its supporting points 0 and 2 by force P_0. It is also loaded by axial force P_1 applied at point 1 within its span. The total (tensile) force acting on the upper section of the beam (l_2) is $P_0 + P_1$, while the total (tensile) force acting on the lower section (l_1) is $P_0 - P_1$. Thus, application of the compressive force would not generate compressive stresses until the compressive force $P_1 > P_0$. It means that the critical (Euler) force is effectively enhanced by the amount of P_0.

Both bending stiffness enhancement and buckling stability enhancement are very important for long ball screws that are frequently used in precision positioning systems, e.g., for large machine tools. Although bending deformation due to weight and thermal expansion–induced forces can adversely affect the accuracy, high payload forces applied through the nut within the span of the screw could cause collapse of the screw. Both effects can be alleviated by stretching the ball screw.

Figure 7.27b shows design of the bearing support for a ball screw positioning system schematically shown in Fig. 7.27a. Two thrust bearings 4, 5 and a long roller radial bearing 3 generate "built-in" supports that can accommodate a high magnitude of the axial preload. It has to be noted though, that an inevitable temperature increase during intensive operation of the ball screw unit would lead to thermal expansion of the screw and, thus, to reduction of the axial tensile force.

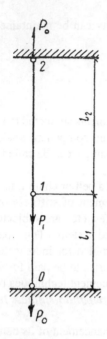

Figure 7.26 Pretensioned rod loaded with an axial load within its span.

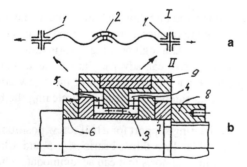

Figure 7.27 Schematic of (a) ball screw translational drive and (b) end support design: 1, end supports; 2, ball nut; 3, radial bearing; 4 and 5, thrust bearings; 6 and 7, spacers; 8, preloading nut; and 9, stationary frame.

This effect can be compensated by using springs (for example, Belleville springs) between spacer 7 and tightening nut 4.

Application of this approach (stretching a slender structural member to enhance its bending stiffness) to machining (turning or grinding) of long slender parts without steady rests is described in Article 4. It is important to note that stretching of the part during machining results in residual compressive stresses on the part surface after machining is completed and the tensile force is withdrawn. These compressive stresses are very beneficial for fatigue endurance of the part while in service. The residual compressive stresses would reduce or even completely cancel residual tensile stresses that develop in the part surface during turning operation and are detrimental for the fatigue endurance of the part. Also, there is information that application of tensile force reduces effective hardness of the part surface (surface layer 15–20 μm deep) and thus improves machinability. Vickers hardness reduction can be 5–10% for steel specimens at 400 MPa tensile stress, and 15–20% for titanium specimens at 600 MPa tensile stress.

Application of this effect to two-dimensional components is shown in Fig. 7.28 [10]. Grinding wheels having internal cutting edge (Fig. 7.28a) are frequently used for slicing semiconductor crystals into wafers. The abrasive ring 1 (the internal part of the wheel) is held by the metal disc 2 clamped in the housing (not shown). Axial (bending) stiffness of the wheel determines its chatter resistance as well as thickness accuracy of the wafers. Since the axial stiffness is proportional to Young's modulus of the holding disc 2 and to the third power of its thickness, and is inversely proportional to the square of its internal radius, introduction of large wafer diameters reduces the axial stiffness.

Since thickness and/or its variation lead to increased losses of the expensive crystal material, it was suggested [10] to generate two-dimensional stretching of the holding disc by introducing a temperature gradient along the disc radius. Figure 7.28b shows dependence of compliance coefficient e on the temperature gradient $T^* = T_a - T_b$. Bending stiffness of the wheel is

$$k = E_2\, h_2^3/ea^2 \tag{7.8}$$

where E_2 = Young's modulus of the holding disc material.

The above analysis [expressions (7.3)–(7.7)] assumed that the axial forces always have the same directions regardless of bending deformations of the beam. While this assumption holds for the cases of Figs. 7.25–7.27 and Article 4, in many cases, such as in the self-contained systems discussed below, directions of the axial forces are changing with the beam deflection, and follow the beam axis inclination at the beam ends (the "following force"). It can be shown (e.g., [11]) that in such case the critical force is significantly higher than for the case of constant directions of the axial forces. For example, for a cantilever column as in Fig. 7.29a, the buckling force for compression by a vertical axial force P is

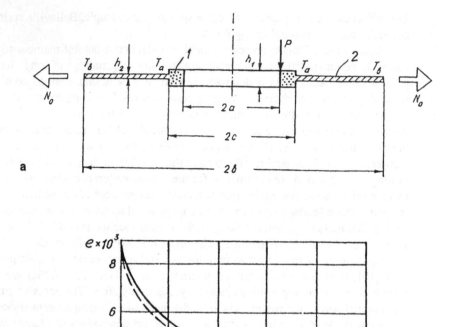

Figure 7.28 Effect of temperature gradient–induced tensile prestressing of grinding wheel with internal cutting edge (a) on its compliance coefficient e (b).

$$P_{cr} = \pi^2 EI/4l^2 = 2.47 \; EI/l^2 \qquad (7.9a)$$

while for the case of Fig. 7.29b it is

$$P_{cr}^f = 20.05 \; EI/l^2 \qquad (7.9b)$$

or about 8.2 times higher.

Expressions for bending stiffness enhancement of beams with the tensile axial preload in Fig. 7.30 are given in Rivin [12]. For double-supported beams

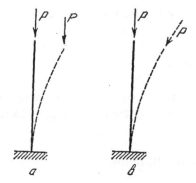

Figure 7.29 Typical cases of compression loading of a column: (a) vertical force; (b) "follower" force.

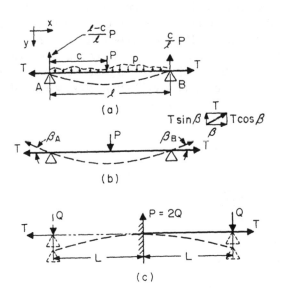

Figure 7.30 Beams under combined bending and axial force loading: (a) double-supported beam, axial force; (b) double-supported beam, following axial force; and (c) cantilever beam.

loaded in the middle by force P, deflection under the force while the axial force T is applied is

$$y_{max} = K(y_o)_{max} \tag{7.10}$$

where $(y_o)_{max}$ = maximum deflection without application of the tensile force $(T = 0)$. The *deflection reduction coefficient* K for the case in Fig. 7.30a (axial tensile force) is

$$K' = [\alpha l/2 - \tanh (\alpha l/2)]/1/3(\alpha l)^3 \tag{7.11a}$$

and for the case in Fig. 7.30b (following tensile force)

$$K'' = [\alpha l/2 - \tanh(\alpha l/2)]/1/3(\alpha l)^2 \tanh(\alpha l/2) \tag{7.11b}$$

Here parameter α is defined as $\alpha^2 = T/EI$, where EI = bending rigidity of the beam.

A built-in cantilever beam loaded at the end (solid line in Fig. 7.30c) can be considered as one half of a double-supported beam loaded in the middle (dotted line in Fig. 7.30c). Instead of force P and length l, as in Figs. 7.30a and b, force $2Q$ and length $2L$ are associated with the simulated double-supported beam in Fig. 7.30c. Thus, the deflection reduction coefficient for cantilever beams can be calculated from Eqs. (7.11a) and (7.11b) if L is substituted for $l/2$.

The stiffening effect of the tensile force as given by Eq. (7.11) can be assessed from Fig. 7.31, in which $\mu = \alpha l/2 = \alpha L$. To better visualize the stiffening effect of the tensile force, a beam of a solid round cross section (diam-

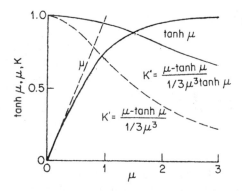

Figure 7.31   Reduction of bending deflection caused by applied tensile force.

eter *D*) can be considered. For such beam $I = \pi(D^4/64)$, cross-sectional area $A = \pi(D^2/4)$, $I/A = D^2/16$, and

$$\alpha l = l\sqrt{\frac{T}{EI}} = l\sqrt{\frac{T}{A}\frac{1}{E}\frac{A}{I}} = l\sqrt{\frac{\sigma_T}{E}\frac{16}{D^2}} = 4\frac{l}{D}\sqrt{\frac{\sigma_T}{E}} = 4\frac{l}{D}\sqrt{\varepsilon_T} \qquad (7.12)$$

where σ_T = tensile stress from the tensile force T and ε_T = relative elongation caused by T.

Several important conclusions can be made from Eqs. (7.11) and (7.12) and from Fig. 7.31:

1. The stiffening effect from the tensile force is substantially higher for a cantilever beam than for a double-supported beam.
2. The stiffening effect is substantially higher for the case of axial tensile force than for the case of the following force.
3. The higher stiffening effect can be achieved if higher tensile stress σ_T or strain ε_T were tolerated. The allowable tensile stress is, in many cases, determined by the yield stress of the beam material.
4. For the same given allowable σ_T, the technique is more effective for lower modulus materials (higher $\sigma_T/E = \varepsilon_T$).
5. The effectiveness quickly increases with increasing l/D (slenderness ratio of the beam). It explains the very high range of stiffness (or pitch) change during tensioning a guitar string.

7.4.3 Self-Contained Stiffness Enhancement Systems

Structural use of the beam-like components preloaded in tension, as described above in Section 7.4.2 and in Article 3, is limited. There is a need for external force application devices, such as in musical string instruments, in a translational drive in Fig. 7.27, or in a machining arrangement for precision turning of slender parts described in Article 4. A self-contained device in Fig. 7.32 [12] is a composite beam having an external tubular member 1 and an internal core (bolt) 4. Since the cross-sectional moment of inertia of a round beam is proportional to the fourth power of its diameter, both strength and stiffness of the composite beam in bending are determined by the external tubular member whose stiffness is usually 85–90% of the total stiffness. Outside member 1 can be stretched by an axial tensile force $-P$ applied by tightening bolt 4. This tensile force is compensated (counterbalanced) by the equal compression force P applied to internal core 4. These forces would cause stiffening of the external member as indicated by Eq. (7.7), and reduction of stiffness of the internal core as indicated by Eq. (7.5). Since value of the buckling force for the external member P_{cr_e} is much higher than value of P_{cr_i} for the internal core, the relative stiffness change is more pronounced for the internal core. However, due to the insignificant initial contribution of the internal core to the overall stiffness (10–15%), even a relatively small increase in stiffness of the external member would result in an improvement of the overall stiffness even if the internal core stiffness is reduced to a negligible value.

The problem with the design in Fig. 7.32 is the danger of collapsing of the long internal core due to the low value of its Euler's force. The collapse could be prevented if the internal core were supported by the inner walls of the external tubular member. However, a precise fitting of a long bolt serving as the internal core is very difficult, especially for slender beams for which this approach is especially effective.

It seems that the best way of assuring the required fit between the external and internal components of the composite beam is to form the internal core in place by casting [13,14]. If the core expands and the expansion is restrained, e.g., by covers or caps on the ends of the external tubular member, this would apply a tensile force to the external member and an equal compressive force to the internal core. Although this would generate the overall stiffness enhancement effect as described above, collapse of the internal core would be prevented. Even the stiffness reduction of the internal core per Eq. (7.5) would be alleviated, since its Euler's force would be very high due to the supporting effect of the inner walls of the external tubular member.

There are many materials that are expanding during solidification. The most common material is water; the volume of ice is about 9% larger than the volume of water before freezing. Another such material is bismuth, which expands 3.3% in volume during solidification. Although work with ice requires low temperatures, work with bismuth requires relatively high temperatures (its melting point is 271°C). However, some alloys of bismuth, so-called fusible alloys, have relatively low solidification temperatures and also expand during solidification. Some experiments [14] were performed with a fusible alloy Asarco LO 158 from Fry Co. (50% Bi; 27% Pb; 13% Sn; 10% Cd). It is used for soldering applications;

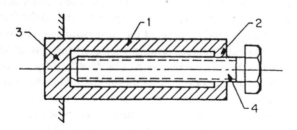

Figure 7.32 Enhancement of bending stiffness by internal axial preload.

its melting point is at 70°C (158°F), below the water boiling temperature. This alloy has about 0.6% volumetric growth, which can be modified by varying the bismuth contents.

If the core is cast inside a tubular rod and is restrained by plugging the internal bore of the rod, then expansion of the core would be constrained. The process of longitudinal expansion would cause stretching of the tubular rod and compression of the core with equal forces.

The force magnitude can be determined from the model in Fig. 7.33. The initial configuration of the tubular rod (length L_2) and the cast nonrestrained internal core after the expansion (length L_1) are shown by solid lines. The equilibrium configuration of the assembly (the case when the internal bore of the rod is plugged after casting) is shown by dashed lines. The final length of the assembly is L_3. Magnitude of the resulting force compressing the core and stretching the rod is P_0. In the final condition, the core is compressed by Δ_1 and the rod is stretched by Δ_2.

Obviously,

$$\Delta_1 + \Delta_2 = L_1 - L_2 = \Delta \tag{7.13}$$

$$\Delta_1 = \frac{PL_1}{E_1 A_1}; \qquad \Delta_2 = \frac{PL_2}{E_2 A_2} \tag{7.14}$$

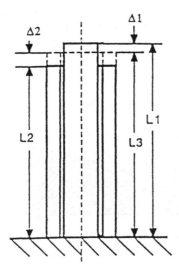

Figure 7.33 Model of tubular rod with expanding core.

where E_1 and A_1 = Young's modulus and cross-sectional area of the core, respectively, and E_2 and A_2 = corresponding parameters for the rod. Increments Δ_1, Δ_2, and Δ are usually very small quantities in comparison with L_1 and L_2, and $L_1 \approx L_2 = L$ in Eq. (7.14). Then,

$$\frac{PL}{E_1A_1} + \frac{PL}{E_2A_2} = \Delta \qquad (7.13')$$

or

$$P = \Delta/L\left(\frac{1}{E_1A_1} + \frac{1}{E_2A_2}\right) = \Delta/\left(L\frac{\dfrac{E_2A_2}{1 + \dfrac{E_2A_2}{E_1A_1}}}{}\right) \qquad (7.15)$$

In this expression Δ represents expansion of the core in its unrestrained condition. When the expansion is restrained, the material can expand in radial direction and its behavior under pressure might be different from behavior during the free expansion process. Accordingly, Eq. (7.15) must be considered as only an approximate one.

Experimental study was performed [14] using aluminum tubes $L = 0.71$ m (28 in.), $D_o = 12.7$ mm (0.5 in.), and $D_i = 9.3$ mm (0.37 in.) with threaded plugs at both ends, as shown in Fig. 7.34. The inside volume of the tube was filled with the molten Asarco LO alloy. It was found that frictional conditions on the tube wall have a very profound effect on expansion during solidification. Lubrication of the wall with a silicon grease results in a much more consistent behavior.

Changes in both static and dynamic stiffness were monitored during the so-

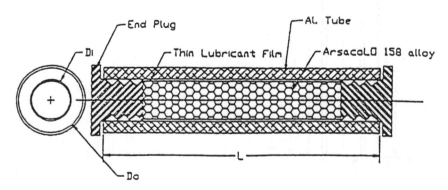

Figure 7.34 General layout of aluminum beam with AsarcoLO 158 alloy.

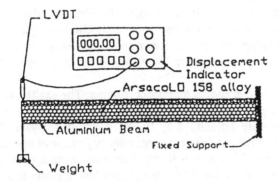

Figure 7.35 Experimental setup for static deflection test.

lidification process. Fig. 7.35 shows the setup for monitoring static stiffness change of the tube filled with the constrained fusible alloy during the solidification process. The beam is clamped at its right end, while the left end was loaded with a weight. Deflection of the left end is monitored by an LVDT probe. Table 7.1 shows results of the static test. Deflection of the tube under 1.1 N (0.25 lb) weight was monitored during the solidification process. After first 10 min during which the initial shrinkage of the fusible alloy is gradually replaced by its expansion, the deflection is decreasing. Effective stiffness k_{ef} of the beam was calculated by dividing the weight value by the deflection. During the 30 min of monitoring, k_{ef} increases about 20%.

The static stiffness test results reflect not only changes in the beam stiffness due to the internal preload caused by expansion of the core, but also deformations in the clamp. Since deformations in the clamp are not influenced by internal preloading of the clamped beam, the actual stiffness increase may be underesti-

Table 7.1 Time History of Deflection of Test Tube Under 0.25 lb Weight (Lubricated)

Time T (min.)	Deflection (in.)	Static stiffness (lb/in.)
0	0.05125	4.87
5	0.04915	5.08
10	0.0508	4.92
15	0.04755	5.25
25	0.065	5.37
30	0.0433	5.77

mated. In the dynamic test setup shown in Fig. 7.36, the free-free boundary conditions were used for the beam being tested, thus completely eliminating influence
of compliance in the clamping devices. The beam (tube) was suspended to the
supporting frame by means of a thin string (wire) whose influence on the tube
vibrations is negligible.

The tube was excited at one end by an impact hammer. The excitation was
applied several times during the cooling/solidification process of the fusible alloy,
and the transfer functions were determined by an FFT spectrum analyzer. Plots
of the transfer functions recorded at different times after beginning of solidification ($t = 0$) are given in Fig. 7.37. At $t = 0$ only four lower natural modes are
pronounced in Fig. 7.37; higher natural modes are of such low intensity that they
are lost in the noise, possibly due to increase of damping caused by friction
between the liquid or semiliquid fusible alloy and the walls. The damping is more
pronounced at the higher modes since the distance between the nodes is small.
With increasing $t \geq 10$ min, the higher modes appear. Even the lower modes are

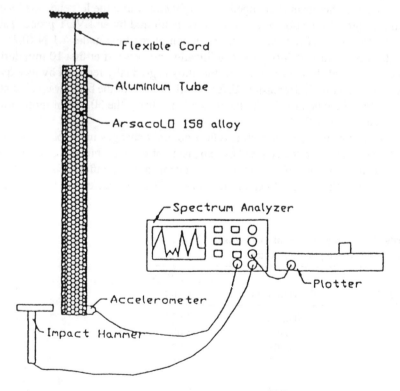

Figure 7.36 Experimental setup for free–free vibration test.

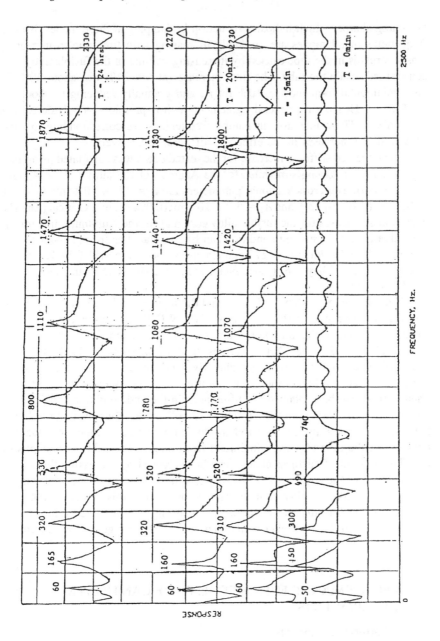

Figure 7.37 Response spectra at different times; 10 dB/div, 0–2500 Hz range.

becoming sharper which indicates reduction in damping at these modes. Natural frequencies of all modes are gradually increasing with time due to increasing tensile stresses in the tube caused by increasing volume of the fusible alloy core constrained inside the tube. The final increase (at $t = 24h$) was about 20% for the fundamental frequency f_1, 10% for f_2, and gradually reducing to about 4% for f_6 to f_9. This is in a general agreement with expression (A.3.2) derived in Appendix 3. These increases in natural frequencies are corresponding to 44%, 21%, and 8% increases in the effective stiffness, respectively.

The expected stiffness increase for the tested beam was calculated using Eqs. (7.7) and (7.13). Young's modulus for the Asarco LO 158 alloy is $E_1 = 42$ GPa (6×10^6 psi). The cross-sectional parameters are $I = 9.1 \times 10^{-6}$ m^4; $A_1 = 5.6 \times 10^{-5}$ m^2 (0.105 sq. in.); and $A_2 = 5.9 \times 10^{-5}$ m^2 (0.094 sq. in.). Critical (Euler's) loads and stiffnesses for the fusible alloy core and for the aluminum tube for the dynamically tested free-free condition can be calculated as for the clamped-free columns whose length is $L_c = L/2$ and that are subjected to the "follower force" [see Eq. (7.10)]. They are, respectively, P_{cr_1}, = 2510 N (530 lb), and P_{cr_2} = 10,250 N (2170 lb).

The relative expansion of the core and the internal preload force were determined [14] to be $\Delta/L = 0.1\%$ (0.001), 1700 N (377 lb), respectively, and the stiffness increase of the aluminum tube is determined from Eq. (7.7) to be $1 + 1700/10,250 = 1.17$ times. Actual stiffness increase was measured to be 1.2–1.44 times, which exceeds the predicted value. Even larger deviations in the "positive direction" occur for higher modes of vibration, as can be seen in Fig. 7.37. This phenomenon can be explained by differences in the end conditions, which may be not exactly representing the "follower force" model, thus resulting in lower magnitudes of the Euler force. Another reason might be deviations of actual parameters of the tube (E_2, wall thickness) from ones used in the calculations of P_{cr_2}. The amount of stiffness change can be increased by increasing the degree of expansion of the cast core. High strength aluminum can tolerate strains up to 0.003–0.005 within its elastic range. At least one half of this range, ~0.002, can be used for transforming into stiffness, thus resulting in a computed value of 1.35 times increase and even higher actual increases, in the range of 1.4–1.8 times.

7.5 PERFORMANCE ENHANCEMENT OF CANTILEVER COMPONENTS

7.5.1 General Comments

Cantilever structures are frequently critical parts of various mechanical systems (boring bars and internal grinding quills, smokestacks, towers, high-rise buildings, booms, turbine blades, robot links, etc.). Some of these structures are stationary, like boring bars for lathe use, smokestacks, etc.; some rotate around their longitudinal axes, like boring bars and other tools for machining centers, quills; yet others perform a revolute motion around a transverse axis at one end, like turbine blades or robot links. A specific feature of a cantilever structural component is its naturally limited stiffness due to lack of restraint from adjacent structural elements that usually enhance stiffness of noncantilever components. This feature, together with low structural damping, causes intensive and slow decaying transient vibrations as well as low stability margins for self-excited vibrations. As in many other cases, both stiffness and damping considerations are critical and are interrelated in designing the cantilever components.

The overall dimensions of cantilever components are usually limited. External diameter of a boring bar or cross-sectional dimensions of a turbine blade are limited by application constraints. Stiffness enhancement by selecting a material with higher Young's modulus E is limited by Young's moduli of available materials and by excessive prices of high-modulus materials. To enhance natural frequency of a component or to reduce deflection caused by inertia forces of a component rotating around its transverse axis, there is a possibility of shape optimization with the cross section gradually diminishing along the cantilever. This is, however, limited by other design constraints. For a smokestack, the main constraint is diameter of the internal passage; for a boring bar, some minimum space at the end is necessary to attach a cutting tool and there is a need for an internal space to accommodate a dynamic vibration absorber (DVA); for a robotic link, there is a need to provide space for cables, hoses, power transmission shafts, etc.

Damping enhancement, critical for assuring stability of the structure, is usually achieved by DVAs. Effectiveness of a DVA is determined by the ratio of its inertia mass to the effective mass of the component being treated (the mass ratio μ). However, the size of the inertia mass in the cantilever systems is limited: in free-standing systems, like towers and high-rise buildings, by economics of huge inertia mass units; in application-constrained systems, like boring bars and grinding quills, by the available space inside the structure. The effective mass of the component, on the other hand, can be relatively high if the component is made from a high Young's modulus material, such as steel. Inertia masses made from high specific density materials (e.g., tungsten alloys, $\gamma = \sim 18$) allow for relatively high $\mu = \sim 1.0$ for solid steel boring bars, but even higher ratios are needed to stabilize boring bars with length L to diameter D ratio $L/D > 8$–9.

According to performance requirements, there are two groups of cantilever design components: (a) stationary and rotating around the longitudinal axis; and (b) rotating around the transverse axis. Components of group (a) require high static stiffness (e.g., boring bars for precision machining whose deflections under varying cutting forces result in dimensional errors), as well as high damping to enhance dynamic stability (to prevent self-excited vibrations) and to reduce high-frequency microvibrations causing accelerated wear of the cutting inserts and the resulting deterioration of accuracy (generation of tapered instead of cylindrical bores, gradual change of bore diameters). The loss of dynamic stability can be caused by cutting forces in cutting tools (chatter), by wind in smokestacks and towers, etc. It can be shown that dynamic stability improves with increasing value of criterion $K\delta$, where K is effective stiffness of the system and δ is log decrement of the fundamental vibratory mode. Components of group (b) require high stiffness to reduce deflections caused by angular acceleration/deceleration, and high natural frequency in order to reduce time required for vibrations caused by transient (usually start/stop) motions to decay (settle) below the specified amplitudes.

7.5.2 Stationary and Rotating Around Longitudinal Axis Cantilever Components

To increase stiffness K, the component of given dimensions must be made of a material with high E. Specific weight (density) of metals is usually increasing with the increasing E (with the exception of beryllium, too expensive for general purpose applications) (see Table 1.1). To increase damping by using DVA, its mass ratio μ must be increased. Since the dimensions and specific gravity of the inertia mass are limited, the effective mass of the cantilever component has to be reduced, thus the "system contradiction" [15] of this system is the following: the component (e.g., boring bar) must be at the same time rigid (thus, heavy) and light. This contradiction was resolved by separation of the contradictory properties in space [15]. A simple analysis [16] shows that stiffness of a cantilever beam is determined by its root segment (7/8 of the total potential energy in bending is concentrated in the root half of a uniform built-in/free cantilever beam having a constant cross section along its length). The effective mass is determined by the overhang segment (3/4 of the total kinetic energy is concentrated in the end half of the cantilever beam vibrating in its fundamental mode). Thus, making the root segment from a high modulus but heavy material would increase stiffness while not influencing significantly the effective mass at the end. On the other hand, use of a light material for the overhang segment assures a low effective mass but does not affect the stiffness significantly. Such a design, in which the rigid and light segments are connected by a preloaded joint 6 (Fig. 7.38), was

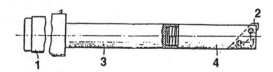

Figure 7.38 Combination boring bar: 1, clamp; 2, cutting tool; 3, rigid root segment; 4, light overhang segment.

suggested in Rivin and Lapin [17]. Figures 7.39b, c and d [16] show stiffness, effective mass, and natural frequency as functions of segment materials and position of the joint between the segments as marked on the model in Fig. 7.39a. It can be seen that while the effective stiffness of the combination structure is only insignificantly lower than the stiffness of the structure made of the solid high E material, the natural frequency (and effectiveness of the DVA mounted at the cantilever end of the structure) are significantly improved. Optimization of the segment lengths (position of the joint) can be performed using various criteria, such as maximum natural frequency, maximum effectiveness of DVA, etc., and their combinations with different weighting factors. The in-depth analytical and experimental evaluation of this concept in application to boring bars (the root segment made of sintered tungsten carbide, the overhang section made of aluminum), as well as optimization based on maximization of $K\delta$ criterion, are given in Article 5. The optimized boring bars (both stationary and rotating) performed chatter-free at $L/D = 15$.

7.5.3 Components Rotating Around Transverse Axis

Solid Component

Components of group (b) require high natural frequency (in order to accelerate decay of transient vibrations) and reduction of end point deflections caused by inertia forces associated with acceleration/deceleration of the component. The basic component (robotic link), made of a material with Young's modulus E, having a uniform cross section with the cross-sectional moment of inertia I and mass per unit length γ, is shown in Fig. 7.40a. The link moves with a constant angular acceleration $\ddot{\theta}_0$. The inertia force acting on an element having an infinitesimal length du and located at a distance u from the link end is

$$dF_u = \gamma du \, [\ddot{\theta}_0(L - u)] \qquad (7.16)$$

The intensity of inertia forces described by Eq. (7.16) is illustrated by the diagram in Fig. 7.40b. The inertia force acting on each element du generates bending moments in all cross sections of the link to the left of the element. At a cross section that is situated at a distance $x > u$ from the end of link, the incremental bending moment generated by dF_u is equal to

$$dM_x = dF_u(x - u) = \gamma\ddot{\theta}_0(L - u)(x - u)du \qquad (7.17)$$

The total magnitude of moment M_x in cross section x due to inertia forces can be computed by the integration of the moment increments generated by all elements du to the right of the cross section x

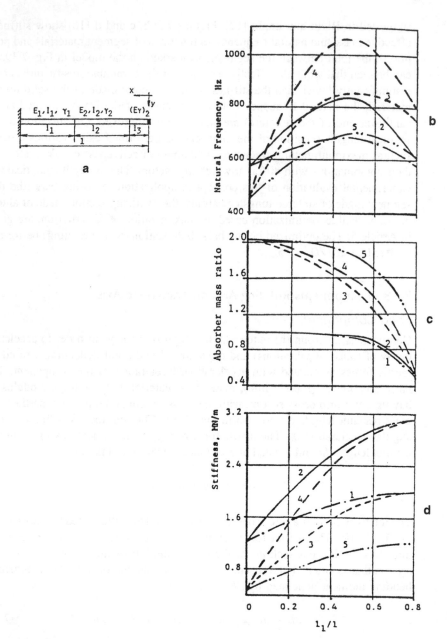

Figure 7.39   (a) Combination boring bar; (b) its natural frequency; (c) absorber mass ratio; and (d) effective stiffness. 1, Tungsten/steel; 2, sintered tungsten carbide/steel; 3, tungsten/aluminum; 4, sintered tungsten carbide/aluminum; 5, steel/aluminum.

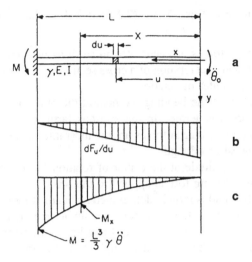

Figure 7.40 Schematic of (a) a revolving massive beam and (b, c) its loading.

$$M_x = \int_0^x \gamma\ddot{\theta}_0(L - u)(x - u)du = \gamma\ddot{\theta}_0\left(\frac{L}{2}x^2 - \frac{x^3}{6}\right) \qquad (7.18)$$

Moment distribution is illustrated by the diagram in Fig. 7.40c. If shear deformations are neglected (which is justified by the slenderness of the component), the bending deflection is due to the changing curvature of the link caused by the bending moments. A moment M_x acting on a cross section x causes a change in curvature that is equivalent to angular displacement between the end faces of an infinitesimal element dx [18]

$$d\psi = (M_x/EI)dx \qquad (7.19)$$

This angular deformation is projected into a deformation of the end of the link in the y-direction

$$dy = xd\psi = (M_x/EI)xdx \qquad (7.20)$$

The total deflection of the link end is the sum of increments dy generated by all the cross sections along the link or

$$y = \int_0^L \frac{M_x}{EI}xdx = \frac{7}{6}\frac{\gamma\ddot{\theta}_0}{EI}L^5 \qquad (7.21)$$

Analysis of the derivation steps, Eqs. (7.16)–(7.21), as well as diagrams in Fig. 7.40b and c shows that:

1. The most intense inertia forces are generated in sections near the free end of the link since linear acceleration of a cross section is proportional to its distance from the center of rotation.
2. The greatest contributors to the bending moment at the built-in end of the link are inertia forces generated in the sections farthest from the center of rotation since these inertia forces are the most intense, and are multiplied by the greatest arm length.
3. The bending moment magnitude at the center of rotation (at the joint) is the magnitude of the driving torque applied to the link.
4. The deflection at the free end is largely determined by bending moments near the center of rotation since these moment magnitudes are the greatest, and angular deformations near the center of rotation are transformed into linear displacements at the free end through multiplication by the largest arm length.

These observations can be rephrased for design purposes as the following: a well-designed link would be characterized by a light end segment and a rigid segment near the pivot, with not very stringent requirements as to rigidity of the former and/or to the specific weight of the latter.

Combination Link

The above specifications present the same contradiction [15] as discussed above in Section 7.5.2 for the cantilever components, which are stationary or rotate about the longitudinal axis: the component has to be heavy (rigid) and light at the same time. Accordingly, it can also be resolved by a combination link design. Such a design is relatively easy to implement by a reliable joining of several segments of the same cross-sectional shape that are made of different materials.

The optimization procedure for such a combination link, however, is very different from the described above procedure for a component that is stationary or rotating about its longitudinal axis. To perform the optimization procedure, the equations derived above have to be modified. First, E and γ are no longer constant along the length of the link; they are constant only within one segment. Second, the link end may carry a lumped mass payload for the robot arm or the effective mass of the preceding links for the intermediate links, and can be acted upon by the reaction torque from a preceding link. In even further approximation, the end-of-link mass may also possess a moment of inertia and be variable depending on the system configuration. Third, actual links cannot be modeled as built-in beams since (angular) compliance of the joints can be of comparable magnitude with the bending compliance of the link [12].

The model in Fig. 7.41 incorporates a lumped mass and a moment at the link end, as well as an elastic joint (pivot). For this model, Eq. (7.18) will be written as follows [12]:

$$M_x = M_0 + m_e \ddot{\theta}_0 x + \int_0^{l_1} \gamma_1 \ddot{\theta}_0 (L - u)\,du + \int_{l_1}^{x} \gamma_2 \ddot{\theta}_0 (L - u)(x - u)\,du \quad (7.22)$$

for $L > x > l_1$. If x belongs to another segment ($x < l_1$), only the first integral would remain, with the integration performed from 0 to x. From Eq. (7.22) for $x < l_1$

$$M_{x_1} = M_0 + m_e \ddot{\theta}_0 + \gamma_1 \ddot{\theta}_0 \left(\frac{L}{2} x^2 - \frac{x^3}{6} \right) \quad (7.23)$$

and for $x > l_1$

$$M_{x2} = M_0 + m_e \ddot{\theta}_0 x + \ddot{\theta}_0 \left\{ \gamma_1 \left[x \left(Ll_1 - \frac{1}{2} l_1^2 \right) - \frac{1}{2} Ll_1^2 + \frac{1}{3} l_1^3 \right] \right.$$
$$\left. + \gamma_2 \left[-\frac{1}{6} x^3 + \frac{1}{2} Lx^2 + \left(\frac{1}{2} l_1^2 - Ll_1 \right) x + \frac{1}{2} Ll_1^2 - \frac{1}{3} l_1^3 \right] \right\} \quad (7.24)$$

The moment at the joint ($x = L$) is

$$M_L = M_0 + \ddot{\theta}_0 \left[m_e L + \gamma_1 \left(L^2 l_1 - Ll_1^2 + \frac{1}{3} l_1^3 \right) \right.$$
$$\left. + \gamma_2 \left(\frac{1}{3} L^3 - L^2 l_1 + l_1^2 L - \frac{1}{3} l_1^3 \right) \right] \quad (7.25)$$

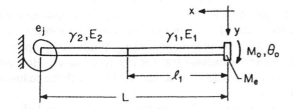

Figure 7.41 A combination link.

Analogously, Eq. (7.21) must also be integrated in a piecemeal fashion:

$$y = \int_0^{l_1} \frac{M_{x1}}{E_1 I} x\,dx + \int_{l_1}^{L} \frac{M_{x2}}{E_2 I} x\,dx + M_L e_j L = \frac{M_0}{2I}\left(\frac{l_1^2}{E_1} - \frac{l_1^2}{E_2} + \frac{L^2}{E_2}\right)$$

$$+ \frac{\ddot{\theta}_0}{I}\left[\frac{m_e L}{3}\left(\frac{l_1^3}{E_1} - \frac{l_1^3}{E_2} + \frac{L^3}{E_2}\right) + \frac{\gamma_1}{E_1}\left(\frac{Ll_1^4}{8} - \frac{l_1^5}{30}\right)\right.$$

$$+ \frac{\gamma_1}{E_1}\left(\frac{L^4 l_1}{3} - \frac{5}{12} L^3 l_1^2 + \frac{L^2 l_1^3}{6} - \frac{1}{12} Ll_1^4\right) \tag{7.26}$$

$$+ \frac{\gamma_2}{E_2}\left(\frac{11}{120} L^5 - \frac{1}{3} L^4 l_1 + \frac{5}{12} L^3 l_1^2 - \frac{1}{6} L^2 l_1^3 - \frac{1}{24} Ll_1^4 + \frac{1}{30} l_1^5\right)\right]$$

$$+ e_j L\left\{M_0 + \ddot{\theta}_0\left[m_e L^2 + \gamma_1\left(L^2 l_1 - Ll_1^2 + \frac{1}{3} l_1^3\right) + \gamma_2\left(\frac{1}{3} L^3 - L^2 l_1 + l_1^2 L - \frac{1}{3} l_1^3\right)\right]\right\}$$

Equations similar to Eqs. (7.25) and (7.26) can be derived if the component (link) comprises more than two segments.

If $\gamma_1 < \gamma_2$ and/or $E_2 > E_1$, at a certain l_1 deflection at the free end would be minimal. The extremum will be more pronounced if the mass load at the free end (m_e) is small and/or compliance e_j of the joint is small. If the link was initially fabricated from a rigid but relatively heavy material, minimization of its deflection would be accompanied by reduction in its inertia and thus by reduction in the required joint torque M_L required to achieve the prescribed acceleration $\ddot{\theta}_0$.

Results of computer optimization of the link structure in Fig. 7.41 made of aluminum ($E_1 = 0.7 \times 10^5$ MPa, $\gamma_1 = 2.7$) and steel ($E_2 = 2.1 \times 10^5$ MPa, $\gamma_2 = 7.8$) with different mass loads and joint compliances are given in Rivin [12]. Figure 7.42 illustrates these results for the case of rigid joint ($e_j = 0$).

The plots of relative deflection values versus l_1/L in Fig. 7.42 are calculated for various mass loads (payloads) characterized by factor $K = m_0/m_{st}$, where $m_{st} =$ mass of the link if it were made of steel. For $K = 0$ (no payload), the optimization effect is the most pronounced, with reduction of deflection of 64%, and reduction in driving torque of 66% for $(l_1/L)_{opt} = \sim0.5$. This case represents circumstances for which reduction of deflections is especially desirable, such as in precision measurements or high-speed laser processing. When a larger payload is used, the effect of the reduced structural mass is less pronounced. However, even at $K = 1$ the end deflection is reduced a noticeable 13%, with the reduction of the required torque of 15%.

An important additional advantage of the optimized combination link is an increase in the fundamental natural frequency of bending vibrations of the link.

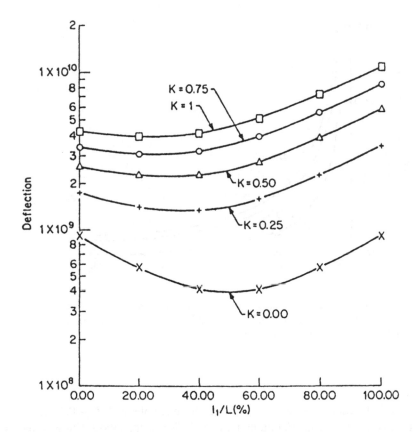

Figure 7.42 Deflection of a combination steel-aluminum link with various payloads versus length ratio (rigid joint).

The increase in natural frequencies (for the case $m_o = 0$, $e_j = 0$) for the deflection-optimized combination versus a steel link is 59% for $K = 0$, 9% for $K = 0.5$, and 5% for $K = 1.0$. The relationships between the fundamental natural frequency and various parameters of a steel-aluminum combination link are shown in Fig. 7.43, in which $Q = \Delta_{st}/\Delta_{ej}$, where $\Delta_{ej} = PL_{ej}^2 =$ deflection at the free end loaded by a force P due to the joint compliance alone, and $\Delta_{st} = PL^3/3E_2I =$ deflection of a solid steel link loaded by the same load due to bending alone.

Although a combination of more than two segments could be beneficial on some occasions, computational analysis of an aluminum-titanium-steel combination versus an aluminum-steel combination has not shown a significant improvement.

Use of a light or low-modulus material for the overhang portion of the combi-

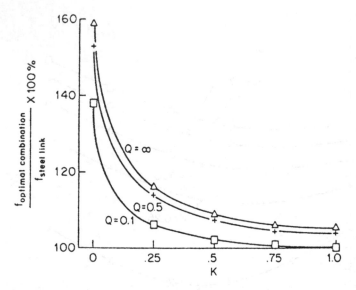

Figure 7.43 Fundamental natural frequency of an optimized combination steel/aluminum link with various payload and joint compliance parameters as a percentage of fundamental natural frequency for a solid steel link.

nation link reduces its bending stiffness. For example, at $l_1/L = 0.4$ the stiffness of the combination steel/aluminum link is 25% less than the stiffness of the steel link (for a case of loading by a concentrated moment at the link end). In most cases, such a reduction is more than justified by the reduction of the acceleration-induced deflection, reduction in the required driving torque, and by increase in natural frequency. In many cases, use of a heavy material, such as steel, is not even considered due to extremely large magnitude of the required driving torque. In such circumstances, the bending stiffness increase in comparison with the solid aluminum link (more than twofold) would be very significant.

Realization of this concept in actual robots depends on the development of an inexpensive and reliable joining technique for metallurgically different materials comprising the combination component without introducing compliance or extra mass at the joint. One of the approaches is based on using nonlinearity of contact deformations discussed in Chapter 4. Two components pressed together with high contact forces would behave as a solid system. A very light and small tightening device can be (and has been) realized by using a wire made of a so-called shape memory alloy [19].

7.6 ACTIVE (SERVOCONTROLLED) SYSTEMS FOR STIFFNESS ENHANCEMENT

While significant improvements in stiffness characteristics of a mechanical system can be achieved by optimizing its structural design in accordance with the concepts and analyses presented above, there are certain limits for the structural stiffness of mechanical components and systems. For example, the "combination structure" cantilever boring bars described in 7.5.2 and Article 4 allow for a chatter-free machining of long holes requiring length-to-diameter ratios up to $L/D = 15$. However, while there is no chatter, static deflections of such a long cantilever beam by cutting forces are very significant and may require numerous passes to achieve the required precision geometry of the hole if the original hole had significant variation of the allowance to be removed by the boring operation [see Chapter 1, Eq. (1.3)].

In such cases, active or servocontrolled systems for compensation of mechanical deflections can be very useful. Figure 7.44 [2] shows an active system for reducing deformations of the boring bar 1 for machining precision holes. The axis of the machined hole in part 2 should coincide with the beam of laser 5. This beam is reflected from prism 6 rigidly attached to boring bar 1. The reflected beam is reflected again from semireflective mirror 7 and its position is compared by photosensor 8 with the reference beam (the primary laser beam refracted by semireflective mirror 7). Controller 10 processes the difference between the desired and the actual positions of the beam and generates voltage to activate piezo actuator 4 for a corrective action on boring bar 1. Although this system only maintains the axis of the boring bar in a precise position and does not respond to changes in cutting tool 3 due to its wear, it is also possible to use the machined surface as a reference. In such case, both deflections of the boring bar and the tool wear would be taken into consideration.

A similar but more complex laser-based system for compensation of boring bar deflections is described in Catskill et al. [20]. The system is using a laser

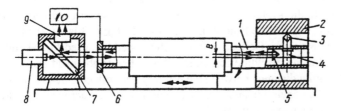

Figure 7.44   Active axis alignment system for a boring bar.

simultaneously producing two beams with different wavelengths (blue and green beams). One beam is used to control the radial position, and another to control the inclination of the tool. The measured guiding accuracy of piezo actuators is 10 μm.

There are many designs of active boring bars (e.g. [21,22]) as well as active systems for other stiffness and vibration-critical components in which the active systems are used to suppress self-excited vibrations in the system or response of the system to external vibratory excitations. Since the frequency range of the undesirable vibrations can be very broad, up to 300–500 Hz, the servocontrolled vibration suppression system becomes very complex and is frequently too delicate to adequately perform in the field or shop floor conditions. It seems to be more reasonable and in many cases definitely more feasible to perform the vibration-suppression tasks by passive means. Such means include rational designs such as the combination structure for the cantilever components, use of high damping materials, of dynamic vibration absorbers like in Article 5, etc. The active systems are more effective if used only for correction of static or other slow-developing deviations of the critical components from the desired geometrical positions. If an active system is used to control vibrations of a precision low-stiffness system, such as a long boring bar, compensation of its static deflections is still necessary. Thus, while the vibration sensor for such system could be very small, a deflection-measuring (or, better, output accuracy measuring) system is still necessary. It could be laser-based, as described above, or based on other principles.

Active systems can be and are successfully used also to correct positions and to enhance effective stiffness of massive frame parts. Figure 7.45 [2] shows a correction system for angular positioning of spindle head 1 mounted on horizontal ram 2. Electronic angular sensor (level) 1 measures inclination of ram 2. The signal from level 3 is conditioned and amplified, and used to actuate pilot valve 4 changing the balance of flow between two hydrostatic pockets 5 and 6. This, in turn, changes gaps h_1 and h_2, respectively, in these pockets thus restoring the desired precise orientation of the spindle head.

Hydraulic actuating systems have advantages versus piezoelectric systems in cases when large forces with sizeable deformations have to be realized. The high pressure hydraulic system similar to one in Fig. 7.45 is used for vertical ram 1 carrying milling spindle head 2 in Fig. 7.46 [2]. Laser 3 is attached to the main frame of the machine. Deflection δ of milling head 2 caused by the cutting forces results in shifting of cylindrical mirror 4 attached to the head and, consequently, in a changing distance between the primary and reflected beams on array of photo diods 5. The deviation signal is then conditioned and activates a hydraulic servosystem similar to one in Fig. 7.45.

Many large machines have inadequate stiffness of their frames and thus require stiffening by massive foundation blocks (see Chapter 5). However, building

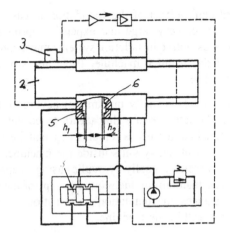

Figure 7.45   Active deflection correction system for traveling spindle head.

large foundation blocks is very expensive and time consuming. Use of such blocks impairs flexibility of the production lines since the machine cannot be easily relocated in response to changing product lines. Sometimes there is a need to install a heavy and not adequately rigid machine on the shop floor directly, in a location not allowing building of a large foundation. In many instances, heavy machines require low-frequency vibration isolation, which calls for low

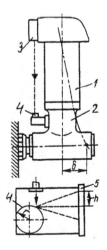

Figure 7.46   Active deflection correction system for a cantilever milling spindle.

stiffness mounts excluding the stiffening effect of the foundation in case of a
direct installation on the isolators or for very large and expensive spring-sus-
pended foundation blocks. In such cases, active installation systems may economically enhance effective stiffness of the installed machine. There are two basic
types of active installation systems [23]:

1. Low-frequency vibration isolators (usually pneumatic) having built-in
    level-maintaining devices; the latter assure that the distance between
    the machine bed and the foundation surface remains constant for slow
    changing forces, such as travel of heavy units inside the machine.
2. Comprehensive installation systems (usually hydraulic) using a separate
    reference frame and actuators maintaining the required shape of the ma-
    chine regardless of any changes in foundation deformations and/or
    changing weight distribution within the machine.

1. A typical self-leveling vibration isolator is shown in Fig. 7.47 [23,24].
The isolator comprises two basic units: passive damped pneumatic spring A and
height controller B. Both units are mounted on the base 1. Top supporting platform 7 is guided by ring 2 and sealed by O-ring 8. Shaped partition 4 separates the
cavity under supporting platform 7 into damping chamber 3 and load-supporting
chamber 5, which are connected by a calibrated orifice. Weight load on mounting
foot 6 of the installed machine is supported by platform 7, which is resting on
pressurized air in chamber 5. Compressed air 17 enters damping chamber 3 and
then load-supporting chamber 5 via controller B. Control rod 16 carries closing
surfaces of both inlet and bleeding valves. Bleeding capillary 13 is attached to
elastic membrane 10 separating lower and upper volumes of the controller.

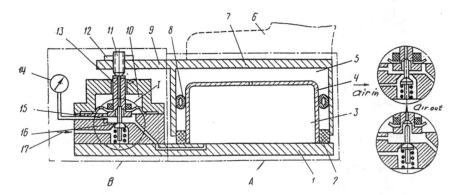

Figure 7.47 Vibration isolation mount Serva-Levl SE (Barry Controls Co.): (A) resilient element; (B) height controller.

Feedback is realized by extension 9 of platform 7. When the weight load on platform 7 increases, e.g., due to travel of a heavy unit inside the installed machine, the equilibrium between the weight load on platform 7 and air pressure in chamber 5 is distorted, platform 7 is sagging down, and feedback extension 9 is pushing capillary 13 down. Control rod 16 also moves down and opens the inlet valve. Since air pressure in the compressed air line is higher than in chambers 3 and 5, air pressure inside the isolator is increasing until the higher weight load on platform 7 is counterbalanced. When the weight load on platform 7 is reduced, the excess air pressure in chamber 15 lifts capillary 13 thus opening the bleeding valve until platform 7 returns to equilibrium.

There are many commercially available leveling isolators similar to the one in Fig. 7.47. Although they may provide a high leveling accuracy (the height of the isolator can be maintained within ~1.0 µm), the settling period is usually quite long, up to 10–20 s. The settling period is increasing with reduction of isolator stiffness (of natural frequency of the isolation system).

Some designs of the leveling isolators use fluidics control systems instead of the valve-based control systems [23,25]. Both systems have similar overall performance characteristics, but fluidics systems have a smaller "dead zone."

Since the system does not respond to fast changes in the relative position, it can have a low dynamic stiffness (low natural frequency) while it has a very high static stiffness (level maintenance).

2. A comprehensive installation system in which positions of the mounting points are precisely maintained in relation to an unloaded reference frame was suggested and tested [26] (Fig. 7.48). Bed 1 of a large machine tool is installed on foundation 4 by means of leveling mounts 8 located at points A_1, A_2, and A_3. Distances between the mounting points are made so small that deflections of the frame parts between the mounts are negligible. Reference frame 3 is constructed around the machine and is supported at three points P_1, P_2, and P_3, which are located in relatively undeformed areas of the foundation. Level 2 can be used to adjust and correct, if necessary, the horizontality of frame 3. Height sensors 9 are hydraulic pilot valves regulating flow of hydraulic fluid to mounts 8. Line pressure is generated by pump 6 and adjusted by chock valve 7. When the weight load on the mounts increases, bed 1 is bending, the slider in the corresponding pilot valve 9 is shifting, and hydraulic fluid pressure in the affected mount 8 is also increasing until the set distance between the supporting plane of bed 1 and reference frame 3 in that location is restored. Analogously, when the weight load on a mount decreases, the hydraulic pressure in the respective mount also decreases until the distance between the bed and the frame is restored. The tests demonstrated that the level in relation to the reference plane is maintained within ~10 µm, and the settling time is about 0.2 s (for the vertical natural frequency of the machine tool on the mounts 41.5 Hz).

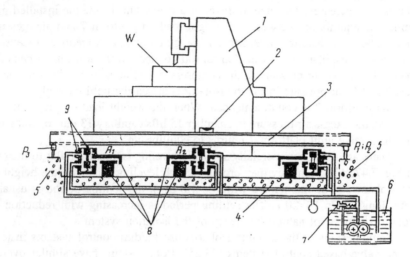

Figure 7.48 Active installation system with a reference metrology frame.

Performance of the system depends on design of the reference frame. Large frames may have excessive thermal gradients and ensuing deformations unless special measures (cooling, temperature stabilization) are undertaken. The frame design can be simplified if a "virtual" frame is used which employs system 8 of interconnected fluid chambers (Fig. 7.49) [26]. Isolated object 1 is attached to the foundation via mounts 7. There is a chamber over each mount, and sensors 2 fastened to the object, are following the fluid level in the respective chambers. To reduce influence of the contact force exerted by sensors 2 on the fluid surface, sensors are interacting with the fluid surface via buoys 3. Such system guaranties horizontality of the reference plane, but may be sensitive to air streams, temperature changes, variations in the contact forces of the sensors, etc. It was suggested [27] to modify the "virtual frame" by filling the system of the interconnected chambers with a low melting point material that is periodically melted to restore the ideal leveling, but is solid during the operation of the system. Another approach is to use a "pseudo fluid," i.e., to fill the system with a bulk material or with small balls that are cemented by friction forces in normal condition and thus the surface in the chambers is, effectively, rigid. The system is periodically "liquefied" by applying high frequency vibrations in order to restore the accurate leveling of all chambers.

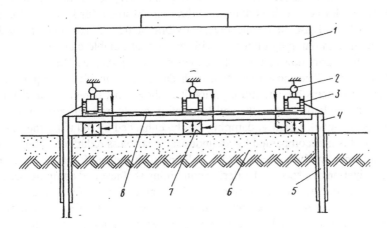

Figure 7.49 Active installation system with reference frame comprising interconnected fluid vessels.

7.7 DAMPING ENHANCEMENT TECHNIQUES

Damping enhancement allows one to reduce vibrations and/or dynamic loads in mechanical systems prone to dynamic effects. There are three main approaches for reducing vibrations/dynamic loads:

1. A major change in the system design (enhancement of structural stiffness and/or modification of natural modes of vibration (see Section 8.6); change parameters of transient processes, such as starting and stopping of driving motors; etc.).

2. A change in the working process parametes in order to avoid a resonance or a self-excitation regime (change of speed of the work organ, e.g., of a cutter in a milling or mining machine; change of work organ design, e.g., change of cutting angles for a tool, of a number or positioning of cutting teeth; speed modulation of major motions, such as spindle rpm in machine tools; etc.).

3. Use of special antivibration devices, usually dampers and dynamic vibration absorbers, which allow one to enhance dynamic stability, reduce levels of forced vibrations and intensity of dynamic loads, and protect the machine or measuring device from harmful vibrations, all without major changes in the basic design.

The antivibration devices may be exposed to significant payload forces (e.g., power transmission couplings, see Article 2) or weight forces (e.g., vibration isolators, see Article 1), or can be installed outside of the force-transmitting path (dynamic vibration absorbers, dampers). With the exception of low-damped precisely tuned devices, such as dynamic vibration absorbers for systems vibrating with well-defined discrete frequencies, antivibration devices are characterized by presence of high damping elements with or without flexible elements.

Structural damping can also be enhanced by a judicious "tuning" of the existing system. The most important high damping "tunable" components of mechanical systems are structural joints. It is shown in Chapter 4 that damping of joints depends on surface/fit and lubrication conditions, but also on contact pressures (preload). Usually, damping can be enhanced by reducing the preload force, but this is accompanied by an undesirable stiffness reduction. Thus, "tuning" of a joint involves selecting and realizing such preload force, which results in maximization of a specified criterion combining stiffness and damping parameters. Most frequently, such a criterion is $K\delta$, where K = stiffness and δ = log decrement associated with the joint. Other examples of "tunable" structural elements include driving motor, whose stronger dynamic coupling with the driven mechanical systems results in damping enhancement (see Section 6.6.2) and tuning of high damping tool clamping systems (Section 8.3.1).

Any antivibration device has at least one flexible element and/or one damping (energy-dissipating) element. Frequently, it is more desirable to combine flexibility and damping properties in one element. It can be achieved by using pneumatic, hydraulic, electromagnetic, electrodynamic, and piezoelectric systems, among others, which can relatively easily combine required elastic and damping characteristics, or by using elastodamping materials.

7.7.1 Dampers

Special damping elements (*dampers*) are used when damping in the elastic element of an antivibration device is inadequate. Most frequently used are viscous dampers, Coulomb friction dampers, magnetic and electromagnetic dampers, impact dampers, and passive piezoelectric dampers, among others. An extensive survey of damping systems is given in Article 6 [28]. Effects of damping on vibration isolation systems are discussed in Article 1, and on power transmission systems in Article 2. Some practical comments given below may be useful.

1. While *viscous dampers* are not very popular with designers since they usually require a rather expensive hardware (fitted moving parts, seals) and maintenance, they have to be seriously considered in special circumstances. First of all, viscous dampers can be natural in lubricated rotating and translational systems (see Section 5.3). Secondly, recent progress in development of time-stable electrorheological and magnetorheological fluids, as well as of electronically controlled hydraulic valves, made variation of viscous resistance in the viscous

dampers a reasonably easy task. Such damping control frequently represents the easiest implementation of active vibration control systems [29].

    2. *Coulomb friction dampers* use friction between nonlubricated surfaces. Since the friction force is proportional to pressure between the friction surfaces, there is a need for preloading. Since wear of the nonlubricated friction surfaces can be significant, it may influence the preload force magnitude if the preloading springs have high stiffness. A low stiffness preloading spring should be deformed for a large amount to create the preload force, thus a small change of deformation due to wear of the frictional surfaces would not noticeably influence the preload force.

    The Coulomb friction dampers are not effective for small vibration amplitudes. The smaller the vibration amplitude, the stiffer the damper should be. If the stiffness is inadequate, small vibration amplitudes would not result in a relative motion between the frictional surfaces.

    No damping action would occur in a Coulomb friction damper if the relative motion between the frictional surfaces is a combination of steady and vibratory motions, and the velocity amplitude of the vibratory motion is less than the velocity of the steady motion.

    Figure 7.50 shows a Coulomb friction damper providing damping action in three directions—one vertical and two horizontal directions.

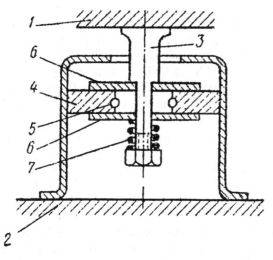

Figure 7.50 Three-dimensional Coloumb friction damper: 1 and 2, components experiencing relative vibratory motion; 3, ram; 4, frictional sectors (e.g., high friction components); 5, ring-shaped spring; 6, frictional plates for horizontal motions; 7, preloading spring for horizontal motions.

3. *Elastodamping materials* usually have relatively low elastic moduli (with some exceptions, see 7 below), allow for large deformations, and have high energy dissipation under vibratory conditions. The most widely used elastodamping materials are elastomers (rubbers), fiber-based materials (felt), volumetric wire mesh materials, plastics, and some composites. Frequently, both static and dynamic characteristics of such materials are nonlinear. In many cases this is a beneficial factor (see Chapter 3; Articles 1 and 2). Stiffness of elastodamping materials for dynamic/vibratory loading (dynamic modulus or dynamic stiffness) is usually greater than stiffness at slow (static) loading with frequency below 0.1 Hz. The ratio of dynamic-to-static stiffness is characterized by dynamic stiffness coefficient K_{dyn} (see Chapter 3), which can be up to 5–10.

Relative energy dissipation in elastodamping materials in the low frequency range, below $\sim$200 Hz, does not depend strongly on frequency. Accordingly, damping of a material can be characterized by log decrement δ which can be as high as 1.5–3.0. Usually, both K_{dyn} and δ are amplitude-dependent (see Article 1). Thus, such materials can be described as having hysteretic damping with $r = 1$ (see Appendix 1). This fact is very important for applications, especially for designing vibration isolators and vibration isolating systems, since increasing damping of isolators, while reducing undesirable resonance amplitudes, does not lead to a significant deterioration of isolation effectiveness (transmissibility) in the after-resonance frequency range (see Appendix 1).

4. *Volumetric wire mesh materials* can be very effective for dampers that have to be used in aggressive environments and under very intense vibrations. They are made from stainless steel cold-drawn wire, possibly nonmagnetic. In one production process, a net is made from wire 0.1–0.6 mm in diameter on a knitting machine. Then the net is wrapped into a "pillow," that is cold pressed in a die under pressure up to 100 MPa (15,000 psi). In another process, a tight spiral is made of 0.15–1.0 mm diameter wire, then the spiral is stretched to 5–600% elongation and placed into a die where it is compressed to attain the required shape.

Wire mesh elements are usually loaded in compression. With increasing load, the number of contacts between the wires is increasing, resulting in stiffness increase. Doubling of the compressive force results in $\sim$1.5 times stiffness increase. The allowable compressive loading $p_{max} = 3$–20 MPa (450–3000 psi) depending on the wire diameter. Allowable dynamic (shock) overloads may reach 8–$10 \ p_{max}$.

Angles between the contacting wires and contact forces are independent random parameters. Any lubrication is squeezed out of the contacts and the friction is essentially of dry (Coulomb) type. Due to high friction forces, the volumetric mesh is quite rigid, and its static deformation is qualitatively similar to elastofrictional connections like one analyzed in Section 4.8.1. For such connections both stiffness and energy dissipation are frequency-independent but strongly amplitude-dependent (energy dissipation increasing, stiffness decreasing with increasing amplitude).

Under dynamic (vibratory) loading the contacting wires are slipping against each other. In the slipping contacts the vibratory energy is dissipated and the lengths of wire segments that can deform in bending and/or tension are increasing. At small realtive vibration amplitudes a (amplitude divided by thickness of the element), dynamic loads between the wires are small and slippage occurs only in those contacts where initial contact pressures are low or angles of contacting wires are beneficial for inducing slippage. With increasing a, the number of slipping contacts increases, and amplitudes of the slippage motions also increase. This results in increasing energy dissipation (log decrement δ) and decreasing K_{dyn}. With further increase of a, both compliance and energy dissipation approach their limiting values when all possible slippages are realized. Since the total energy associated with the vibratory motion is proportional to a^2, the relative energy dissipation $\psi \approx 2\delta$ has a maximum. Figure 3.2 shows typical correlations $\delta(a)$ and $K_{dyn}(a)$ for one type of wire mesh element from recordings of free (decaying) vibrations at $a = 0.4 - 15 \times 10^{-3}$. At low a, δ is small, $\delta = 0.15$–0.2, and $K_{dyn} = 8$–10. With increasing a, δ quickly rises, reaching $\delta = 1.5$–2.0 and peaking at $a = 7 - 10 \times 10^{-3}$, and then slowly decreasing. K_{dyn} is monotonously decreasing with increasing a, asymptotically approaching $K_{dyn} = 1$.

It can be concluded that the wire mesh materials are very effective at large vibration amplitudes since they have very high δ at the large amplitudes, can withstand high temperatures and aggressive environments, have low creep rate, and can be loaded with significant forces. However, at low amplitudes they are very stiff and have low damping. A quantitative comparison of wire mesh and elastomeric materials applications for vibration isolation at various vibration amplitudes is given in Article 1.

5.   *Felt* is a fabric produced by combining fibers by application of mechanical motions, chemicals, moisture, and heat, but without waving or knitting. Felt is usually composed of one or several grades of wool with addition of synthetic or plant fibers. The best grades of felt are resistant to mineral oils, greases, organic solvents, cold/dry environment, ozone, and UV light. Felt structure is similar to that of the wire mesh (chaotic interaction between the fibers) but the felt fibers are much more compliant than the steel wires and have their own material damping, while material damping of the steel wires is negligible. As a result, the amplitude dependencies for both K_{dyn} and δ are less steep (Fig. 3.2). The allowable compression loads on felt pads are much lower than for wire mesh, $p_{max} = 0.05$–0.35 MPa (7.5–53 psi), and up to 0.8–2.0 MPa (12–30 psi) for thin pads.

6.   *Rubber* is composed of a polymeric base (gum) and inert and active fillers. Active fillers (mostly, carbon black) are chemically bonded to the polymeric base and develop complex interwoven structures that may break during

the deformation process and immediately re-emerge in another configuration. The character of these breakage–reconstruction events is discrete, similar to Coulomb friction, whereas a body does not move until the driving force reaches the static friction force magnitude. When the body stops, the static friction force is quickly reconstructed. Thus, the deformation mechanism of the active carbon black structure in a filled rubber is somewhat analogous to the deformation mechanism of the volumetric wire mesh structures. Accordingly, dynamic characteristics of rubber are similar to those of felt (discussed above). They are composed of dynamic characteristics of the active carbon black structure (K_{dyn} and δ are strongly amplitude-dependent and frequency-independent), and of dynamic characteristics of the polymer base (K_{dyn} and δ are amplitude-independent; frequency-dependency in 0.01–150 Hz range is different for different types of rubber). The relative importance of these components depends on contents of active fillers; for lightly filled rubbers the amplitude dependencies are not noticeably pronounced. Some rubber blends have ingredients preventing building of the carbon black structures [30]; these blends do not exhibit the amplitude dependencies even with heavy carbon black content.

Figure 7.51 [30] shows amplitude dependencies for shear modulus G and log decrement δ for butyl rubber blends which differ only in percent contents of

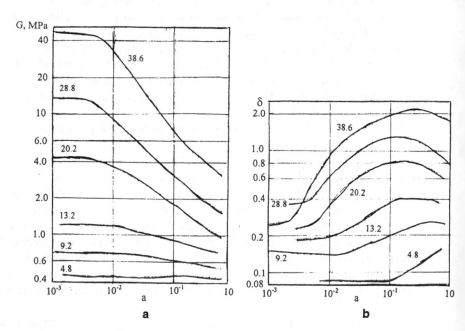

Figure 7.51 Amplitude dependencies of (a) shear modulus G and (b) log decrement δ for butil rubbers vs. percentage content of carbon black.

carbon black. These dependencies are very steep in their areas of change (up to 15:1 change in G; up to 8:1 change in δ) and demonstrate "peaking" of δ at a certain amplitude.

A lack of consideration for the amplitude dependencies of the dynamic characteristics of elastodamping materials while designing damping-enhancement means may result in a very poor correlation between the expected and the realized system characteristics, as well as in an inadequate performance.

7. *High modulus elastodamping materials* are also available. They are represented by metals having high internal energy dissipation. The high damping metals group includes lead, some magnesium alloys, nickel-titanium (NiTi) alloys ("shape memory" or "superelastic" alloys), etc. The highest modulus and strength are associated with NiTi alloys, which demonstrate high damping (log decrement $\delta = 0.5$) when prestressed to $\sim$70 MPa (10,000 psi) [31] (see Section 8.3.1).

Due to high modulus/strength, the high damping metals can be used as structural materials for critical parts, such as clamping devices (see Section 8.3.1).

8. Optimization of dynamic behavior of dynamically loaded or vibrating mechanical systems can be helped by using, wherever possible, *criterial expressions connecting stiffness and damping* parameters. In some cases, an indiscriminate increase in damping is desirable. Such cases are usually associated with resonating systems. The resonance can occur in a structure experiencing translational vibrations or in a torsional (e.g., power transmission) system. Such cases in which δ can be considered as a criterion are represented by power transmission couplings as described in Article 2.

In many cases, which are addressed in several chapters here, the criterion connecting stiffness and damping is $K\delta$. The criterion is very important for optimization of dynamic stability of a system (its resistance to development of a self-excitation process). Although direct applications of this criterion have been addressed in several sections, a more involved practical case of using this criterion for structural optimization of cantilever boring bars is described in Article 5. In this case, shifting the joint between the "stiff" and the "light" segments of the bar is changing both effective stiffness and effective mass of the bar. Increase in the length of the "light" segment reduces the stiffness but can be used to enhance effectiveness of the dynamic vibration absorber, thus enhancing the damping capacity. However, the problem is complicated since the dynamic system has more than one degree of freedom.

It is shown in Article 1, that quality of vibration isolation of precision objects can be characterized in some typical cases by the criterion K/δ. Use of this criterion allows one to optimize selection of materials for vibration isolators depending on amplitude and frequency, which are prevalent in the specific cases.

The influence of mount parameters on chatter resistance of the mounted machine is shown in Article 1 to be characterized by criterion $K^{3/2}\delta$. Knowledge of this criterion allows one to optimize material selection for mounts (vibration isolators), which are used for installation of chatter-prone machines and equipment.

Stiffness/damping criteria are derived in Article 1 for some other cases of machinery installation.

7.7.2 Dynamic Vibration Absorbers

A *dynamic vibration absorber* (DVA) is a dynamic system (frequently a single-degree-of-freedom mass/damped spring oscillator) attached to a body whose vibrations have to be reduced. An appropriate tuning of a DVA results in reduction of vibration amplitudes of the vibrating body, while the mass of the absorber ("inertia mass") may exhibit high vibration amplitudes. The basics of operation of a DVA are provided in textbooks on vibration. The state of the art in design and application of DVAs is extensively surveyed in Sun et al. [31]. Several practical comments on design of effective DVAs are given below.

1.   Tuning parameters needed for the most effective performance of a DVA are derived in vibration handbooks for the case of a sinusoidal (harmonic) vibration of the vibrating body. However, it is usually not explained that these *tuning parameters* are not universal and *depend on characteristics of the vibration* that need to be suppressed. It is important to understand that if vibrations of the body whose vibrations are to be suppressed by attaching a DVA are not sinusoidal (e.g., random vibrations), or if the absorber has to enhance dynamic stability of the system rather than to suppress vibration amplitudes, the tuning parameters may significantly change. These two cases are addressed in Article 5. Another special case that may require a special dynamic analysis to develop optimal tuning conditions is reduction (acceleration of settling process) of transient vibrations.

2.   Effectiveness of a properly tuned DVA is determined mostly by its *mass ratio* μ, which is the ratio between the inertia mass of the absorber and the effective mass of the body whose vibration characteristics have to be modified. Although the inertia mass size is limited by packaging constraints, as in the case of boring bar (see Section 7.5.2), or by economics, it is often forgotten that the mass ratio has two components. Weight/mass reduction of the vibrating body is frequently feasible and, if realized, it allows to significantly enhance effectiveness of the absorber. An example of such an approach is given in Sections 7.5.2 and 7.5.3 and in Article 5.

3.   In many applications, it is desirable to suppress vibrations of the vibrating body in a wide frequency range. This can be achieved by using high damping connection between the inertia mass and the vibrating body. However, elastomeric or polymeric materials possessing high damping capacity have their damping and stiffness parameters influenced by many factors: amplitude and frequency of vibrations; temperature; process variation in making the material; etc. As a

result, computational optimization of the absorber can hardly be realized at the "first try." It is beneficial to use tunable connections for a DVA that allow one to correct the tuning imperfections. The "tunability" can be easily attained by using nonlinear elastomeric elements, for example, as shown in Figs. 3.17 and 3.18 or in Article 5.

REFERENCES

1. Levina, Z.M., and Zwerev, I.A., "Finite element analysis of static and dynamic characteristics of spindle units," Stanki i instrument [Machines and Tooling], 1986, No. 10, pp. 7–10 [in Russian].
2. Bushuev, V.V., "Compensation of elastic deformations in machine tools," Stanki i instrument, 1991, No. 3, pp. 42–46 [in Russian].
3. Bushuev, V.V., "Design and loading patterns," Stanki i instrument, 1991, No. 1, pp. 36–41 [in Russian].
4. Wieck, J., "Compensator for coordinate-measuring machine," Patent of German Democratic Republic (East Germany), No. 133,585, 1979.
5. Bushuev, V.V., "Paradoxes of design solutions," Stanki i instrument, 1989, No. 1, pp. 25–27 [in Russian].
6. Timoshenko, S.P., and Gere, J.M., Theory of Elastic Stability, McGraw-Hill, New York, 1961.
7. Jubb, J.E.M., and Phillips, I.G., "Interrelation of structural stability, stiffness, residual stress and natural frequency," Journal of Sound and Vibration, 1975, Vol. 39, No. 1, pp. 121–134.
8. Ivanko, S., and Tillman, S.C., "The natural frequencies of in-plane stressed rectangular plates," Journal of Sound and Vibration, 1985, vol. 98, No. 1, pp. 25–34.
9. Platus, D.L., "Vibration isolation system," U.S. Patent 5,178,357, 1993.
10. Petasiuk, G.A., and Zaporozhskii, V.P., "Enhancement of axial stiffness of grinding wheels," Sverkhtverdie materiali [Superhard Materials], 1991, No. 4, pp. 48–50 [in Russian].
11. Feodosiev, V.I., Selected Problems for Strength of Materials, Nauka Publishing House, Moscow, 1973 [in Russian].
12. Rivin, E.I., Mechanical Design of Robots, McGraw-Hill, New York, 1988.
13. Rivin, E.I., "Method and means for enhancement of beam stiffness," U.S. Patent 5,533,309, 1996.
14. Rivin, E.I., and Panchal, P., "Stiffness enhancement of beam-like components," ASME, New York, 1996.
15. Fey, V.R., and Rivin, E.I., The Science of Innovation, The TRIZ Group, Southfield, MI, 1997.
16. Rivin, E.I., "Structural optimization of cantilever mechanical elements," ASME Journal of Vibration, Acoustics, Stress and Reliability in Design, 1986, Vol. 108, pp. 427–433.

17. Rivin, E.I., and Lapin, Yu.E., "Cantilever tool mandrel," U.S. Patent 3,820,422, 1974.
18. Timoshenko, S.P., and Gere, J.M., Mechanics of Materials, Van Nostrand Reinhold Co., New York, 1972.
19. Rivin, E.I., et al., "A high stiffness/low inertia revolute link for robotic manipulators," In: Modeling and Control of Robotic Manipulators and Manufacturing Processes, ASME, New York, 1987, pp. 253–260.
20. Catskill, A., et al., "Development of a high-performance deep-hole laser-guided boring tool: Guiding characteristics," Annals of the CIRP, 1997, Vol. 46.
21. Glaser, D.J., and Nachtigal, C.L., "Development of a hydraulic chambered, actively controlled boring bar," ASME Journal of Engineering for Industry, 1979, Vol. 101, pp. 362–368
22. Dornhöfer, R., and Kemmerling, K., "Boring with long bars," VDI Z., 1986, Vol. 128, pp. 259–264 [in German].
23. Rivin, E.I., Active Vibration Isolators and Installation Systems, NIIMASH, Moscow, 1971 [in Russian].
24. Harris, C. (ed.), Shock and Vibration Handbook, 3rd Edition, McGraw-Hill, New York, 1988.
25. Push, V.E., Rivin, E.I., and Shmakov, V.T., "Vibration isolator," USSR Certificate of Authorship 261,831 (1970).
26. Hailer, J., "A self-contained leveling system for machine tools—an approach to solving installation problems," Maschinenmarkt, 1966, Vol. 72, No. 70 [in German].
27. Rivin, E.I., "A device for automatic leveling," USSR Certificate of Authorship 335,448 (1970).
28. Johnson, C.D., "Design of passive damping systems," Trans. of ASME, Special 50th Anniversary Design Issue, 1995, Vol. 177, pp. 171–176.
29. Karnopp, D., "Active and semi-active vibration isolation," Trans. of ASME, Special 50th Anniversary Design Issue, 1995, Vol. 117, pp. 177–285.
30. Davey, A.B., and Payne, A.R., Rubber in Engineering Practice, McLaren & Sons, London, 1964.
31. Sun, J.Q., Jolly, M.R., and Norris, M.A., "Passive, adaptive and active tuned vibration absorbers—A survey," Trans. of ASME, Special 50th Anniversary Design Issue, 1995, Vol. 117, pp. 234–242.

8

Use of "Managed Stiffness" in Design

While in most cases mechanical systems benefit from enhancement of their stiffness, there are many important cases where stiffness reduction is beneficial or where there is an optimal range of stiffness values. Some examples of such cases are as follows: generation of specified constant in time forces for preloading bearings, cam followers, etc.; vibration isolators (e.g., see Appendix 1); force measuring devices (load cells) in which a compromise must be found since the reduction of stiffness improves sensitivity of the device but may distort the system and/or the process being measured; compensating resilient elements for precision overconstrained systems; reduction of stress concentration; improvement of geometry and surface finish in metal cutting operations by intentional reduction of stiffness of the machining system; use of elastic elements for limited travel bearing and guideways; use of anisotropic components having significantly different stiffness in different directions; and trading off stiffness for introduction of damping into the system. This chapter addresses some of these issues.

8.1 BENEFITS OF INTENTIONAL STIFFNESS REDUCTION IN DESIGN COMPONENTS

Our common sense tells us that solid components can tolerate higher loads than the same components that are ''weakened'' by holes or cuts. However, this belief does not consider very complex interactions between the components, which involve contact deformations on very small surface areas, stress concentrations, and nonuniform load sharing between several contact points (e.g., in power transmission gears with contact ratios exceeding 1.0). Quite frequently, stiffness reduction for some components may result in increasing of the overall stiffness and/ or in improved performance due to changing load and deformation distribution patterns.

8.1.1 Hollow Roller Bearings

A very important mechanical component is roller in contact with a flat surface (guideways) or with a round surface of much smaller curvature (bearings). It was shown [1,2] that a hollow roller (roller with an axial bore) develops significantly lower contact (Hertzian) stresses due to the larger contact area resulting from deformation of the roller's body. With increasing parameter $a = r/R$, where R is radius of the roller and r is radius of the bore, the wall is getting so thin that its bending stresses may become a critical parameter. Figure 8.1 [2] displays dimensionless plots of σ_{max}/E_1 and P/RLE_1 as functions of angular width $2\phi_o$ of the contact rectangle. Here σ_{max} = maximum contact stress; P/RLE_1 = dimensionless load on the roller, E_1 = Young's modulus of the roller; P = force acting on the roller; L = length of the roller; and ϕ_o is measured in angular minutes. The same width of the contact area $2\phi_o$ for a roller with larger a would develop at a lower force P. These plots can be used to determine σ_{max} using the following

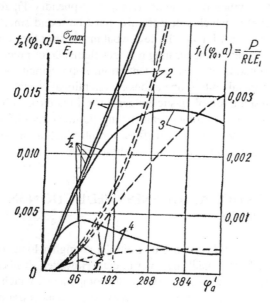

Figure 8.1 Dimensionless stresses in contact of hollow rollers with a flat surface: 1, a = 0; 2, a = 0.5; 3, a = 0.75; 4, a = 0.875.

procedure: calculate dimensionless load on a roller P/RLE_1; draw the line parallel to abscissa from this value of dimensionless load on the right ordinata to the line representing $f_1(\phi_o, a) = P/RLE_1$ for the given a; this determines ϕ_o, thus intersection of vertical line from this ϕ_o with a plot for $f_2(\phi_o, a) = \sigma_{max}/E_1$ for the given a solves the problem.

Table 8.1 [1] compares magnitudes of P that cause a certain maximum contact stress (σ_{max} = 590 MPa) for a steel roller R = 50 mm, L = 200 mm at various a. A thin-walled roller (a = 0.875) can absorb 14% higher load while being about four times lighter. Of course, the thin-walled roller is more compliant overall due to its bending deformation. But even relatively small holes, $a \approx$ 0.5, which do not significantly influence the overall deformation, could be very beneficial for roller bearing and guideways applications.

In a roller bearing for an aircraft gas turbine rotor [1], rollers with R = 10 mm and L = 20 mm are placed around the circle with diameter D_o = 200 mm. The typical cage design allows one to place 24 rollers with angular pitch α = 15°. The bearing is loaded with radial force P = 5,000 N at n = 10,000 rpm; it is not preloaded, and the load on the most loaded roller can be calculated, using Eq. (5.24a) as

$$P_o = P \bigg/ \left(1 + 2 \sum_{i=1}^{12} \cos^2 i\alpha \right) = P/6.01 = 832 \, N \tag{8.1}$$

Centrifugal forces press roller to the outer race. Mass of one roller is $M = \gamma \pi R^2 L$ = 0.05 kg where γ = 7.8 × 10^{-3} kg/m³ is density of steel. Linear velocity at the center of the roller is V = 0.5(10,000/60)$2\pi(D_o - R)$ = 47.1 m/s and centrifugal force on one roller is $P_{c.f.} = mv^2/(D_o/2)$ = 1,110 N, which is more than the payload per one roller. Thus, the maximum radial force on the most loaded roller is

$$P_{max} = P_o + P_{c.f.} = 1,942 \text{ N}$$

Table 8.1 Compressive Load P_{max} Causing Stress 590 MPa in Contact of Roller with Flat Surface Depending on Degree of Hollowness a of Steel Roller R = 50 mm, L = 200 mm

a	P_{max} (N)	P_{max} (%)	Weight (%)
0	95,100	100	100
0.8	98,000	102.9	36
0.875	108,800	119.2	23.4

If $a = 0.5$, then the contact stresses are, practically, not affected (the maximum contact stress σ_{max} would decrease, but rather insignificantly; see Fig. 8.1). However, the presence of the hole reduces mass of the roller by 25%, and $P'_{c.f.}$ = 0.75 $P_{c.f.}$ = 833 N. The payload on the most loaded roller in this case is

$$P'_o = P_{max} - P'_{c.f.} = 1,109 \text{ N}$$

and the corresponding allowable external force

$$P = P'_o \times 6.01 = 6,665 \text{ N}$$

Thus, the rated load on the bearing is 33% higher while its weight is reduced by ~ 5%. In many cases, increase in the rated load is not as important as increase of the life span of the component. The correlation between the load and the length of life for roller bearings is

$$Ph^{0.3} = \text{const.} \tag{8.2}$$

where h = number of hours of service. Thus, the length of service h' of the bearing with hollow rollers can be found from

$$(h'h)^{0.3} = P'/P = 1.34$$

or $h' = h \times 1.34^3 = $ ~2.6 h, and the life resource of the bearing is 2.6 times longer.

There is an important secondary effect of this design change. Use of hollow rollers allows to change design of the cage by using the holes to accommodate pins of the modified cage. This allows one to increase the number of rollers to 30 ($\alpha = 12°$), and also reduces friction between the rollers and the cage. Using Eq. (8.1), it is easy to find that the allowable external force can be increased to $P'' = 8,580$ N for the same maximum load on the roller (1,942 N). This would increase the rated load by 67% as compared with the original solid roller bearing, or prolong its life by a factor of 4.5.

Similar roller bearings (Fig. 8.2) [3–6] were tested (and are now marketed) for high speed/high precision machine tool spindles. In this case, values of a = 0.6–0.7 are considered optimum since it was found that at such values of a the balance between contact and bending deformations of the rollers is the best. The bearings are uniformly preloaded by using slightly oversized rollers; they are used without cages. It was established [4] that if the rollers are initially tightly packed, they become separated by small clearances after a few revolutions. Such bearings have only about 50% of the maximum load capacity of the solid roller bearings (partly due to use of a part of the strength to enhance stiffness by pre-

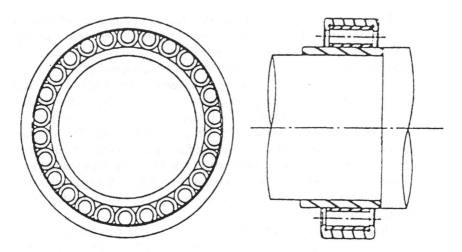

Figure 8.2 Spindle bearing with hollow rollers.

load). However, it is compensated by the combination of high stiffness, high rotational accuracy (runout less than 1 μm is reported), and high speed performance up to $dn = 3.5 \times 10^6$ mm-rpm (due to reduced weight and centrifugal forces from the rollers). It is interesting to note that similar results both in effective mass and in stiffness can be achieved by using high Young's modulus, low density ceramic balls, or hollow steel rollers with intentionally reduced stiffness.

8.1.2 Stiffness Reduction in Power Transmission Gears

Two principal challenges in designing power transmission gears are: increasing payload for given size/weight and reliability, and reduction of vibration and noise generation for high speed transmissions. The payload capacity is determined by contact stresses in the mesh; by bending stresses in the teeth (especially, stress concentrations in the fillets connecting the teeth with the rim); and by dynamic loads generated due to deviations from the ideal mesh kinematics, especially at high speeds. These deviations are caused by: imperfect uniformity of the pitch; deviations of involute tooth surfaces from their ideal shapes; deformations of shafts, bearings, and connections; changing mesh stiffness during one mesh cycle; and non-whole number contact ratio causing abrupt changes in number of engaging teeth in each mesh cycle. Vibration and noise generation are also closely correlated with the dynamic loads in the mesh.

Conventional techniques for handling the above-listed factors are modification of the tooth geometry (intentional deviation from the ideal involute profiles by flanking and/or crowning the tooth surfaces, optimization of fillet shapes, etc.); improvements in gear material (use of highly alloyed steels with sophisticated selective heat treatments, use of high purity steel, hard coatings, etc.); and tightening manufacturing tolerances for critical high speed gears. While significant improvements of the state-of-the-art gears were achieved by these approaches, new developments along these lines are bringing diminishing returns for ever-increasing investments. For example, while higher accuracies in pitch and profile generation lead to reduction of dynamic loads, thus to increasing payloads and to noise reduction, costs of further tightening of tolerances for already high precision gears are extremely high. However, even ideal gears would deviate from the ideally smooth mesh due to deformations of teeth that vary during the mesh cycle and due to distortions caused by deformations of shafts, bearings, and connections.

Because of these complications, another approach is becoming more and more popular—intentional reduction of stiffness of the meshing gears (this development is in compliance with the universal Laws of Evolution of Technological Systems formulated in the Theory of Inventive Problem Solving (TRIZ); e.g., see [7]).

Introduction of elastic elements into power transmission gears is a subject of many patents, starting from the last century [8]. Some of the typical approaches are shown in Fig 8.3 [9,10]. Five design groups shown in Fig. 8.3 as a–e achieve different effects. In the designs in Fig. 8.3a, compliant teeth experience smaller contact (Hertzian) stresses, similarly to the hollow roller described in Section 8.1.1. Since the highest bending stresses develop in the fillets between the teeth and the rim, they are not significantly affected by a slot in the "upper body" of the tooth. However, by enhancing the tooth compliance by means of introduction of groove between the teeth ("artificial stress concentrator") [Fig 8.3a(3)], peak stresses (in the fillet) can be reduced for the optimal dimensioning of the groove by 20–24% [10]. Use of compliant teeth may also result in a very dramatic reduction of dynamic load amplitude P_d, which is expressed as follows:

$$P_d = \psi v_o \sqrt{km} \qquad (8.3)$$

Here v_o = tangential velocity of the meshing gears, which determines impact velocity; ψ = coefficient reflecting gear parameters, accuracy, etc.; k = stiffness of impacting pair of teeth; and m = effective mass of the gears.

Designs in Fig. 8.3b are characterized by the same contact stresses, the same bending stresses in designs b1 and b4, and somewhat increased bending stresses (due to increased effective height of the teeth) in cases b2 and b3. The effect of these designs is reduction of dynamic loads due to the reduced stiffness in accordance with Eq. (8.3). Similar effects are realized by designs in Figs. 8.3d and e,

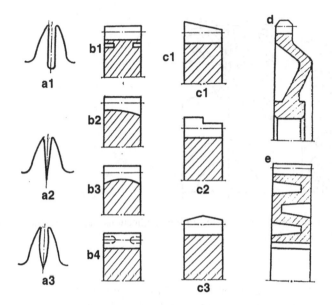

Figure 8.3 Gears with enhanced compliance: (a) compliant teeth; (b) compliant interface between teeth and gear hub; (c) shaped teeth to compensate stiffness nonuniformity of mesh; (d) compliant gear hub; (e) rim elastically connected to hub.

in which the enhanced radial compliance of hubs is equivalent to enhanced tangential compliance of teeth in the designs of Figs. 8.3b and c.

Periodic variation of stiffness of the meshing teeth, both due to the constantly changing radial position of the contact point on each tooth and due to the non-whole value of the contact ratio, is a powerful exciter of parametric vibrations, especially in high speed gears. The stiffness variation can theoretically be compensated by modification of the tooth shape in the axial direction, e.g., as in Fig. 8.3c.

Although designs in Fig. 8.3a–e are rather difficult for manufacturing, designs in Fig. 8.4 are the most versatile. These designs use conventionally manufactured gear rims connected with the gear hub (which in this case can be made from a low-alloyed inexpensive steel or even from cast iron) by a special flexible connection. Both contact and bending strength of the teeth under a static loading are the same as for a conventional (solid) gear. However, self-aligning of the rim in relation to the hub results in a more uniform load distribution along the teeth. Another important effect is reduction of dynamic stresses due to reduction of both k and m in Eq. (8.3). Such designs have been extensively tested in Berestnev [9]. These tests demonstrated that ratio of maximum-to-minimum stresses along

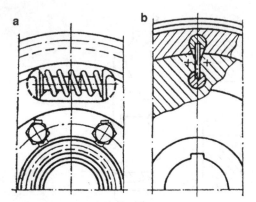

Figure 8.4 Composite gears with elastic rim-hub connections: (a) coil spring connection; (b) flat spring connection.

the tooth were 1.1–1.4 (depending on the total load magnitude) for the composite gear with the flexible connection vs. 1.65–1.85 for the solid gear. This difference was maintained both in static tests and in the working transmissions. It results in much slower wear rates of teeth profiles in the composite gears. Figure 8.5 shows a typical comparison of rms of vibration acceleration for the solid and composite gears. A very significant difference (three times or 10 dB) is characteristic for dynamic loading in the mesh.

The flexible rim-hub connection presents a very simple and economical way to reduce dynamic loads in the mesh as well as noise levels of power transmission gears. However, conventional flexible connectors have flexibility not only in the circumferential (desirable) direction, but also in the radial direction. An excessive radial rim-hub compliance may distort the meshing process and is undesirable. Frequently, the hub and the rim have a sliding circumferential connection, as in Fig. 8.4. The inevitable friction and clearances may introduce performance uncertainty. This situation can be corrected by introducing elements with anisotropic stiffness, which have a low stiffness value in one direction and a high stiffness value in the orthogonal direction. Such stiffness characteristics are typical, for example, for thin-layered rubber-metal laminates described in Section 3.2 and in more detail in Rivin [11] (Article 3).

Although the anisotropic elements may be designed to fully satisfy requirements to the stiffness values in two orthogonal directions, it is often useful to separate these functions between two distinctly different design components. It was suggested in Rivin [12,13] to use high compression stiffness of the thin-layered rubber-metal laminates for the radial restraint, while designing them with very low shear (circumferential) stiffness. The required circumferential stiffness

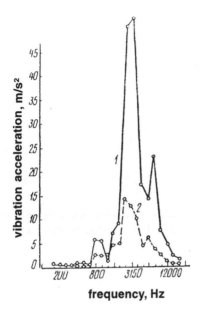

Figure 8.5 Spectra of rms vibration acceleration of the gear housing for (1) solid gears and (2) composite gears tested on a four-square test rig.

is provided by a flexible torsional connector (coupling) (Fig. 8.6). Superior performance characteristics of the torsional connection (coupling) using multiple rubber cylinders compressed in the radial direction as described in Rivin [12,13] allows one to package the connection into the available space between the rim and the hub of even heavy duty power transmission gears.

A similar effect can be achieved by a totally different approach—modification of the meshing system by using elastic elements in order to separate sliding and rolling in the mesh. Two such systems were proposed in Rivin [14,15].

The first system [14] is most applicable to gear teeth characterized by a constant curvature of the tooth profile. This feature is typical for so-called conformal or Wildhaber/Novikov (W/N) gears (e.g., [16]). Slider 1 in Fig. 8.7 has the same shape as tooth profile 2 and is attached to it by rubber-metal laminate 3. During the mesh process, counterpart tooth profile 4 engages without sliding with slider 1. Sliding between the meshing profiles, which is necessary for the meshing process, is accommodated by shear deformation of laminate 3 while the tangential load is transmitted by compression of laminate 3. The contact pressures between the meshing profiles, which are relatively low for the W/N gears (about 6–10 times lower than for the contact between conventional involute profiles) is further

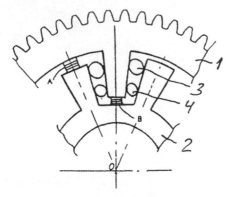

Figure 8.6 Torsionally flexible rim-hub connection for a power transmission gear with radial restraint by rubber-metal laminates: 1, rim; 2, hub; 3 and 4, flexible coupling elements; A and B, rubber-metal laminates

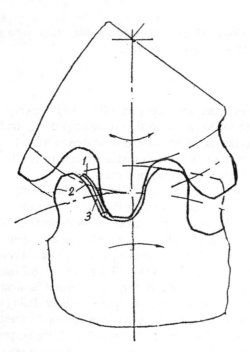

Figure 8.7 Gears in which sliding in the mesh is accommodated by internal shear in rubber-metal laminate coating.

reduced due to a higher compliance of slider 1 and laminate 3 in compression than the compliance of the steel-to-steel contact. This, combined with the high load-carrying capacity of thin-layered rubber-metal laminates in compression (see Chapter 3 and [11]) allows one to improve load capacity of the gears while at the same time compensates for center distance inaccuracies and reduces noise generation (15–20 dB; see [17]).

The second system [15] resolves the combined rolling/sliding motion between the two meshing involute profiles into separate rolling and sliding motions. A pure rolling motion takes place between involute profile 1 of one gear and the specially designed profile 2 of slider 3 on the counterpart gear (Fig. 8.8). Slider 3 is attached to tooth core 4 of the counterpart gear with a possibility of sliding relative to core 4. The sliding is realized by connecting slider 3 with cylindrical or flat surface of tooth core 4 via rubber-metal laminate 5. The sliding motion of slider 2 relative to core 4 is accommodated by shear deformation of laminate 5. Again, the incremental compression compliance of rubber-metal laminate 5 results in substantial reduction of dynamic loads and noise generation as well as in very low sensitivity of the mesh to manufacturing inaccuracies [18].

8.1.3 Stiffness Reduction of Chain Transmissions

While reduction of dynamic loads is important for power transmission gears in order to enhance their load-carrying capacity and reduce noise generation, it is even more important for power transmission chains. Operation of the chain drives

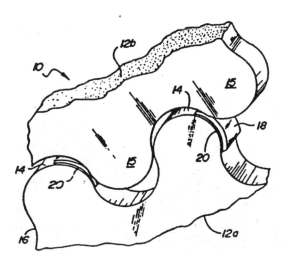

Figure 8.8 Composite gears with separation of sliding and rolling motions.

is naturally associated with generation of relatively high dynamic loads when a new chain link is engaging with the sprocket [19]. The dynamic loads are responsible for high noise levels of the chain drives; they limit the load-carrying capability and, especially, maximum speeds of chains. It was suggested [20,21] to reduce dynamic loads inherent to chain drives by compliant attachment of sprocket teeth to the sprocket body.

A normal engagement between the chain and the sprocket would take place if deformation of the sprocket tooth is not excessive and does not prevent engagement of the next tooth with the next link of the chain. An experimental study [21] was performed with the driving sprocket design shown in Fig. 8.9. Each tooth 1 of the sprocket is connected with hub 2 by pivots 5 and is supported by rim 3. The sprocket with rigid teeth (reference) had a metal rim, while the sprocket with compliant teeth had the rim made of rubber, as shown in Fig. 8.9. Teeth 1 are held in contact with rim 3 by rubber bands 4.

The test results demonstrated a three- to sixfold reduction of dynamic loads associated with entering new links into engagement with the sprocket teeth for the compliant sprocket. It was also concluded that the payload is more uniformly distributed between the teeth of the compliant sprocket.

8.1.4 Compliant Bearings for High-Speed Rotors

Rotational speeds of machines such as turbines and machine tool spindles are continuously increasing. While balancing, both static and dynamic, has become

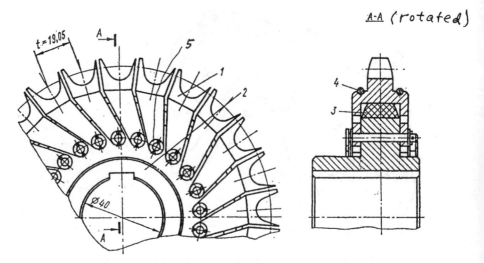

Figure 8.9 Chain sprocket with compliant teeth.

a routine procedure, there are situations where balancing of rotors before their assembly with bearings is not adequate for assuring low vibration levels of the machine and low dynamic loads on the bearings. In turbines, the balancing conditions of high-speed rotors may change due to thermal distortions, especially for horizontal rotors, which sag when stopped while their temperature is still elevated [22]. Machine tool spindles carry tools whose balance is changing due to wear of cutting inserts or grinding wheels, variations in clamping conditions resulting in slight eccentricities, change of mass distribution during dimensional adjustments, etc. The resulting unbalance exhibits itself in high levels of vibration and high dynamic loads transmitted through the bearings to housings and other frame parts. These dynamic loads reduce the life span of the bearings and also result in undesirable temperature increments, which, in turn, increase highly undesirable thermal deformations of spindles.

Although the conventional approach to bearing designs is to increase their stiffness, significant benefits can be often obtained by an *intentional reduction of the bearing stiffness*. It is well known (e.g., see [22]) that after a rotor passed through its first critical speed, its center of mass tends to shift in the direction of its rotational axis. If some masses attached to the rotor have mobility, a self-balancing effect can be realized. Since the first critical speed of a rigid rotor is usually very high (e.g., for machine tool spindles), it is artificially reduced by using compliant bearings. The existing autobalancing devices use special bearing systems, which sustain high stiffness of bearings at working conditions and reduce stiffness of bearings when the balancing is required.

In many cases, reduction of dynamic loads on high-speed bearings is the most important. It was demonstrated in Kelzon et al. [23] that if a rotor is supported by compliant bearings (Fig. 8.10), the dynamic forces between the rotor and the bearings disappear if the following conditions are satisfied:

$$m_1 - k_1/\omega^2; \qquad m_2 = k_2/\omega^2 \tag{8.4}$$

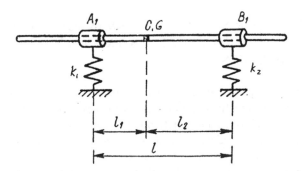

Figure 8.10 Rigid rotor rotating in two compliant bearings.

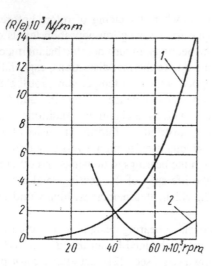

Figure 8.11 Dynamic pressure between rigid rotor and bearing vs. rpm: 1, rigid bearings; 2, compliant bearings tuned for 60,000 rpm.

Here m_1 and m_2 = masses (nonrotating) of bearings A and B, respectively; k_1 and k_2 = stiffness coefficients of the bearings; and ω = rotational speed of the rotor (rad/sec). Figure 8.11 shows the load per unit length of the bearing for a high speed rigid rotor in rigid bearings (line 1) and rotor in compliant bearings for which the conditions in Eq. (8.4) are satisfied at $n = \omega/2\pi = 60,000$ rpm (line 2). It can be seen that the force acting on the bearing is greatly reduced at rotational speeds around 60,000 rpm. Use of this approach to turbine rotors and to machine tool spindles (e.g., [24]) demonstrated significant reductions in vibration levels as well as temperature reduction of the bearings.

8.2 COMPENSATION FOR STATIC INDETERMINACY AND/OR INACCURACIES IN MECHANICAL SYSTEMS AND TAPERED CONNECTIONS

Static indeterminacy and/or imperfect dimensional accuracy can adversely effect performance characteristics of mechanical systems, as illustrated in Chapters 4 and 5. In many cases, effects of both static indeterminacy and of inaccuracies can be alleviated or completely eliminated by introducing low stiffness compensating elements, e.g., such as shown in Fig. 5.24. While the system in Fig. 5.24 is over-constrained by an excessive number of positive (hardware) restraints, there are numerous cases when the excessive restraints are due to friction forces causing excessive loads and position uncertainties in the system. Use of the compensating elements can also be illustrated on tapered (conical) connections analyzed above in Chapter 4.

8.2.1 Use of Managed Stiffness Connections to Reduce Friction-Induced Position Uncertainties

Friction-induced position uncertainties in mechanical connections develop due to inevitable variation of the friction coefficients and of the normal forces in the frictional contacts. These uncertainties are especially important (and objectionable) in precision devices, where in many cases position uncertainties even within fractions of 1 μm cannot be tolerated. The important case of such a device is the so-called kinematic coupling, which is used as a repeatable connection between the tool and the tool carriage in a precision lathe [25].

The kinematic coupling concept is used for providing statically determined connection between two mechanical components. The statically determined connection provides six restraints for six degrees of freedom of the component. Frequently, it is realized by using a three-grooves/three-balls connection (Fig. 8.12). Each ball has two contacts (one with each side of the respective groove). Although, theoretically, these contacts are points, actually they are contact areas,

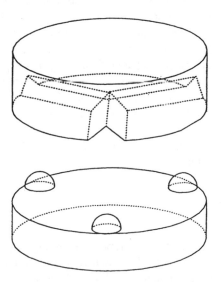

Figure 8.12 Kinematic coupling with three V-grooves and three balls.

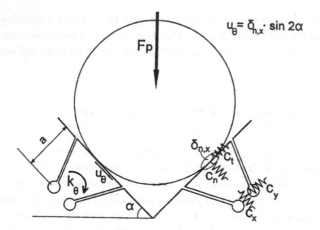

Figure 8.13 Ball in V-groove with "elastic hinges."

which become relatively large if the connection is preloaded for enhancing its stiffness. As a result of the finite sizes of the contact areas, friction forces are developing along these areas that cause uncertainty of relative positioning between the two connected components. In fact, the system becomes a statically indeterminate one on the microlevel. It was found in Schouten et al. [25] that this positional uncertainty (hysteresis) could be 0.4–0.8 µm. The solution for this problem proposed in [25] is to introduce compliance in the design by making the contact areas "floating" (Fig. 8.13). The compliance was introduced on each side of each groove by making two longitudinal holes along each groove and by machining slots, which partially separate the contact areas from the grooved plate. In such a design, tangential forces generated by the preload force F_p due to inclination of the contact surface do not generate dynamic friction forces and microdisplacements, but slightly elastically deform the contact segments of the groove wall under the static friction forces.

This approach resulted in reduction of the hysteresis down to 0–0.1 µm, albeit with reduction of the overall system stiffness.

8.2.2 Tapered Connections

Definition of the Problem

Tapered connections are frequently used for fast and repetitive joining of precision parts without pronounced radial clearances. The alternative to the tapered connection is connection by fitting precisely machined cylindrical surfaces. However, for easy-to-assemble cylindrical connections, the diameter of the male component must be somewhat smaller than the diameter of the female component. This results in an inevitable radial clearance. Even with the clearance, especially when its magnitude is kept small, assembly of a cylindrical connection requires precision positioning of the components before the assembly. A tapered connection fabricated with the same degree of precision is easy to assemble and, when preloaded with an axial force, it is guaranteed to have a contact at the large or small diameter of the connection depending on the tolerance assignment. As the result, there would be no radial clearance along the whole length of the connection but there would always be some angular play between the connected components. This angular play can be eliminated if a very expensive individual matching of the components is performed. The angular play results in an indeterminate angular position within the tolerance of the male component in relation to the female component and thus in an undesirable runout and stiffness reduction of the extending part of the former. Since frequently the components have to be interchangeable within a large batch (such as spindles and tool holders for machine tools), the matching is only rarely practical.

Although it is not very difficult to maintain very stringent tolerances on angles of the taper and the hole in the tapered connection, it is extremely difficult (unless expensive individual matching is used) to maintain at the same time a high dimensional accuracy on diameters and axial dimensions of the connected parts. These difficulties are due to a statically indeterminate character of the connection since the simultaneous taper and face contact is associated with an excessive number of constraints. As a result, a simultaneous contact of the tapered and the face (flange) surfaces is economically impossible to achieve for interchangeable parts. Because of this, standards on tapered connections (such as U.S. National Standard ASME B5.50-1994 for " 'V' Flange Tool Shanks for Machining Centers With Automatic Tool Changers") specify a guaranteed clearance between the face surfaces of the connected parts. Figure 8.14 shows the standard arrangement for the most frequently used 7/24 taper connectors (taper angle 16° 36' 39"). The axial clearance between the face of the spindle and the flange of

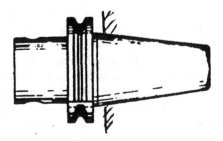

Figure 8.14 Positioning of standard 7/24 tapered connection.

the male taper is specified to be about 3 mm (0.125 in.). Axial position of the male taper in relation to the tapered hole for the tool holder/spindle connections can vary for the same components within 15–20 μm due to variations of the dimensions, of the friction conditions of the contacting surfaces, and of the magnitude of the axial force [26,27]. It is even more uncertain within a batch of interchangeable components. One of the most important applications of the tapered connections is their use for machine tools, especially CNC machining centers, since they have to combine very high accuracy, accommodation of large forces, and interchangeability between large inventories of spindles and toolholders

Tapered Toolholder/Spindle Interfaces for Machine Tools: Practical Sample Cases

Until recently, the steep taper interfaces satisfied basic requirements of machining operations. However, fast development of high accuracy/high speed/high power machine tools caused by increased use of hard-to-machine structural materials, by proliferation of high performance cutting inserts, and by required tight tolerances, posed much more stringent requirements to machining systems. Modern machine tools became stiffer while the interfaces did not change. Now, the toolholder and the toolholder/spindle interface are the weakest links in the machining system.

Standard 7/24 Connection

The standard steep taper connections are discussed in detail in Section 4.4. They have many positive features for machine tool applications. They are not self-locking thus allowing fast connections and disconnections with a simple drawbar design not requiring a "kick out" device. The taper is secured by tightening the toolholder taper in the tapered hole of the spindle. The solid taper allows one to use the tapered part for supporting the tool, thus reducing its overhang from the spindle face. Only one dimension, the taper angle, has to be machined with a high degree of precision. As a result, the connection is inexpensive and reliable.

Many shortcomings of the 7/24 taper interface are due to the fact that the taper surface plays two important roles simultaneously: precision location of the toolholder relative to the spindle, and clamping in order to provide adequate rigidity to the connection. It is practically impossible to make interchangeable tapers that have both face and taper contact with the spindle, thus the standard 7/24 tool/spindle interface has a large guaranteed clearance between the face of the spindle and the toolholder flange. Radial location accuracy is not adequate since standard tolerances specify a "minus" deviation of the hole angle and a "plus" deviation of the toolholder angle, resulting in a clearance in the back of the connection between the male taper and the spindle hole (see Section 4.4). A typical AT4 qualitet (ISO IS1947) has 13 angular sec tolerance for each angle, which may result in radial clearance as great as 13 μm (.0005 in.) at the back end of the connection. This radial clearance may lead to a significant runout and to worsening balancing condition of the tool. The absence of face contact between toolholder and spindle leads also to micromotions between the male and female tapers within the radial clearance under heavy cutting forces, leading to accelerated wear of the front part of the spindle hole ("bell mouthing") and in the extreme cases to fretting corrosion of the spindle, which results in even faster wear. It also results in indeterminacy of axial positioning of the tool within 25–50 μm (.001–.002 in.).

It is important to discriminate between two major shortcomings of conventional 7/24 taper interfaces:

1. Radial clearance in the back part of the tapered connection due to the taper tolerancing, which reduces stiffness, increases runout, introduces unbalance, and creates the potential for micromotions causing fretting corrosion and wear.

2. Mandated axial clearance between the flange of the toolholder and the face of the spindle, which creates uncertainty in the axial positioning of the toolholder. The axial uncertainty is also very undesirable when the machined part is measured on the machining center using touch probes; in such cases, a special procedure is required for axial calibration of the probe. This shortcoming can be eliminated by assuring the face contact between the toolholder and spindle flanges. However, due to the steepness of the 7/24 taper, providing simultaneous contact at both taper and face requires fractional micrometer tolerances. This is difficult but possible to achieve for a given toolholder/spindle combination, and some toolholder manufacturer and user companies provide such precision fitting for critical machining operations. However, it is impractical for the huge inventory of toolholders and spindles, since variations in the spindle gage diameter are in the range of tens of micrometers.

Requirements for Toolholder/Spindle Interfaces

Modern machining centers combine high accuracy, high installed power of the driving motor, and high maximum spindle rpm. However, these three categories of machining conditions are not frequently used in the same time and pose different requirements to the toolholder/spindle interface.

High accuracy machining operations require high accuracy of tool positioning both in radial and in axial directions. However, both radial and axial accuracy of standard 7/24 interfaces are adequate for many operations, such as drilling and many cases of end milling.

High power cutting is frequently performed at not very high spindle rpm. The most important parameters of the interface for such regimes are high stiffness and/or damping assuring good chatter resistance, and high stability of the tool position. For standard 7/24 interfaces, the latter requirement is not satisfied due to the small clearance in the back of the tapered connection. This allows for small motions under heavy cutting forces, and consequently to fretting corrosion and fast bell mouthing of the spindle.

High speed operations require assurance that the interface conditions are not changing with the increasing spindle rpm. The front end of high speed spindles expands due to centrifugal forces, as much as 4–5 µm at 30,000 rpm for the taper #30 [28]. Since the toolholder does not expand as much, the spindle expansion increases the effective length of the cantilever tool, reduces its stiffness, and changes the axial position of the toolholder. Maintaining a reliable contact in the connection under high rpm requires a very significant initial interference, as much as 15–20 µm for the taper #40. The clearance in the connection caused by the taper tolerances can also be eliminated by introduction of preloading (interference) along the whole length of the tapered connection. However, for the standard solid steel taper the interference has to be at least 13 µm (for AT4 grade) to eliminate the clearance. Such magnitudes of interference are impractical for conventional connections since this would require extremely high drawbar forces and disassembly of the connection with the high interference for the tool changing operation would be close to impossible. In addition, high magnitudes of interference for solid toolholders would result in bulging of the spindle, similar but to a larger degree than bulging shown in Fig. 4.24 and resulting from the interference fit of a *hollow* toolholder. This may have a detrimental effect on spindle bearings. A better alternative technique for holding toolholder and spindle together at high rpm is to generate axial flange contact between them with high friction forces, exceeding centrifugal forces. In such case, the spindle expansion would be prevented.

High speed machining requires very precise balancing of the toolholder. Precision balancing of standard 7/24 taper interfaces can be disrupted by the inherent unbalance of the large keys and keyslots, and also by the aforementioned clearance in the back of the connection causing radial runout and thus unbalance. The keys are required to transmit torque (in addition to the tapered connection) and to orient the angular position of the toolholder relative to the spindle, which may be desirable for some operations (such as fine boring). The need for the torque-transmitting keys can be eliminated if enhanced friction forces were generated in the connection. The desired orientation, if needed, can be maintained by other means.

Alternative Designs of Toolholder/Spindle Interfaces

The Theory of Inventive Problem Solving (TRIZ) [7] teaches that if a technological system becomes inadequate, attempts should be made to resolve its contradictions by using its internal resources, without major changes in the system (i.e., to address a so-called *miniproblem*). Only if this approach has proven to be unsuccessful should major changes in the technological system be undertaken, i.e., a *maxiproblem* should be addressed. However, in solving the toolholder/spindle interface problem, the miniproblem was not addressed until recently. Due largely to competitive considerations, several manufacturers undertook drastic changes (see Section 4.2.2), thus making existing spindles and the present huge inventory of 7/24 toolholders obsolete. All 7/24 taper tools of the same size are now interchangeable, but adding one machine tool with a different spindle or interface to a machine shop requires an expensive duplication of a large inventory of toolholders.

The following advanced toolholder designs with 7/24 taper resulted from addressing the miniproblem related to the toolholder/spindle interface. This goal can be formulated as development of an interface system exhibiting higher stiffness, better accuracy, and better high speed performance than conventional interfaces while being fully compatible with existing toolholders and spindles, and not prohibitively expensive. This effort took two directions in accordance with two sets of requirements formulated above. The first direction was development of a toolholder providing for taper/face interface with the spindle. This eliminates indeterminacy of axial position of the conventional 7/24 taper interface, improves its stiffness and radial runout, and creates better balancing conditions for high rpm use due to elimination of the need for torque-transmitting keys. The second direction was modification of existing toolholders in order to improve their stiffness and runout characteristics. The second direction is important, since it has the potential for upgrading performance characteristics of millions of existing toolholders with absolutely minimal changes in the system. Since the problems of the standard connections are mostly due to the static indeterminacy of the system, there is a definite need for an elastic link in the system. However, the elastic elements must have an acceptable ("managed") stiffness in order to satisfy the performance requirements, and must be very accurate in order to satisfy the accuracy requirements for the connection.

After an extensive survey of the state of the art [27] and analysis based on TRIZ [7], seven designs were developed and tested [29]. These designs are based on a "virtual taper" concept, in which there are discrete points or lines defining a tapered surface and contacting the tapered spindle hole. All the design elements providing the contacts are elastic; thus axial force from the drawbar brings the toolholder into face contact with the spindle by deforming only the elastic elements and not the whole shank of the toolholder as in other designs, e.g., described in Section 4.4.2. Such an approach allows one to achieve the desired interference but without absorbing a significant portion of the drawbar force and also without creating a high sensitivity to contamination. The latter is characteristic for the designs having hollow deforming tapers contacting with the spindle hole along the whole surface. Most of the designs were rejected because they introduced new critical design dimensions in addition to the taper angle, thus requiring expensive precision machining.

The final version (WSU-1), shown in Fig. 8.15 [29], requires the same accuracy of machining as what is needed to manufacture the standard 7/24 toolholders. It has a tapered (7/24) shank, 60d, whose diameter is smaller than the tapered shank of a standard toolholder with the same diameter of flange 60b. A metal or plastic cage 62c containing a number of precision balls 68 of the same diameter is snapped on (or attached to) shank 60d. In the free condition, the gage diameter of the virtual taper defined by the balls is 5–10 μm larger than the gage diameter of the spindle. Under the axial force applied by the drawbar 22, the balls undergo elastic deformation and the toolholder moves inside the spindle, stopping after flange face 60c of the toolholder touches face 16 of the spindle. The cage 66 has lips 62b and 62c interacting with grooves 60f and 60g, thus protecting its inside area from dirt.

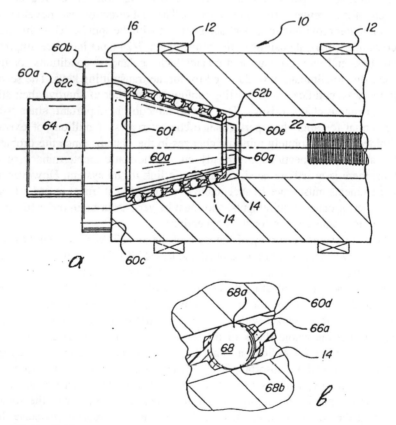

Figure 8.15 Taper/face interface WSU-1 with spherical elastic elements: (a) general layout; (b) ball/tapered surfaces interaction.

This system uses precision balls as elastic elements. Balls made of various metals (e.g., steel, titanium, aluminum), glass, plastics, etc., are available with sphericity and diameter accuracy within fractions of 1 μm, at very reasonable prices. The high dimensional accuracy of the balls guarantees the high accuracy of their elastic characteristics. The deformation of the ball-taper connection in Fig. 8.15 is composed of Hertzian deformations between the balls and two tapered (male and female) surfaces, and the solid body deformation of the ball (see Table 4.2). It was found that plastic balls do not have an adequate stiffness to support a long and heavy tool during its insertion into the spindle, thus steel and titanium balls are being used.

If steel balls are used (e.g., 6 mm diameter balls for a #50 spindle taper), then with a safety factor of 2.0 for contact stresses, axial motion δ_{ax} of the toolholder for up to 35 μm (~.0015 in.) is allowable (see Table 8.2). If precision titanium or glass balls are used, then axial motion up to 70 μm (~.003 in.) is allowable. After the face contact is achieved, it is tightened by the drawbar force. Since a relatively low force is required to deform the balls, the axial force required for the axial motion to achieve the face contact is significantly lower than in the "shallow hollow taper" designs. Thus, a larger part of the drawbar force can be used for the face clamping. There were concerns expressed about denting of the spindle and toolholder tapered surfaces in the contact areas with the balls. It was shown in Braddick [30] that a small permanent deformation (1/4 wavelength of green light) of a flat steel plate in contact with a steel ball develops at the load $P = SD^2$, where D is diameter of the ball. Coefficient S is 2.4 N/mm^2 for the plate made from hardened steel with 0.9%C, 7.2 N/mm^2 for the plate made from superhardened steel with 0.9%C, and 5 N/mm^2 for the chromium-alloyed ball-bearing steel plate tempered at 315°C (spring temper). For a 6 mm ball and $S = 2.4$ N/mm^2, $P = 86$ N. Actual testing (6 mm steel ball contacting a hardened steel plate) did not reveal any indentation marks at loads up to 90 N (20 lb). This

Table 8.2 Deformation and Stress Parameters of WSU-1 Toolholder with 6 mm Balls

	Load per ball (N)											
	4.5				36				121			
Ball material	δ	δ_{ax}	k	σ	δ	δ_{ax}	k	σ	δ	δ_{ax}	k	σ
Steel: $\sigma_{al} = 5.3$ GPa	1.3	9	6.8	1.1	5.0	32	13.6	2.2	12	82	20.4	3.3
Glass: $\sigma_{al} = 4$ GPa	2.5	16	1.5	0.6	9.5	66	3.0	1.2	21	150	4.6	1.9

δ = radial deformation of ball (μm); δ_{ax} = axial shift of toolholder (μm); k = stiffness per ball (N/mμ); σ = contact stress (GPa); and σ_{al} = allowable contact stress (GPa).

force corresponds to a deformation about 10 μm and axial displacement of the toolholder of about 70 μm (∼.003 in.). This assures a normal operation of the system even with steel balls. No danger of denting exists when the balls are made from a lower Young's modulus material, such as titanium.

Since only a small fraction of the drawbar force is spent on deformation of balls, friction forces in the flange/face contact are very high. Thus, in many cases the keys are needed only as a safety measure since an instantaneous sliding can occur at dynamic overloads during milling. However, it would be beneficial to further enhance friction at the face contact by coating the toolholder face in order to eliminate the need for keys altogether. Measured static friction coefficients f for one type of coating are given in Fig. 8.16 vs. normal pressure in the contact. It can be seen that f does not depend significantly on the presence of oil in the contact area. Due to the high friction, the connection can transmit very high torques without relying on keys. For a #50 taper, axial force 25 KN (5,600 lb), and $f = 0.35$, such a connection can transmit 360 Nm torque which translates into 180 KW (250 HP) at $n = 5,000$ rpm. Even higher f, up to $f = 0.9$, are realizable by a judicious selection of the coating [31].

The measured runout of the toolholder in Fig. 8.15 was less than that of any standard toolholder, since there is no clearance between the toolholder and the spindle hole. Balls with sphericity within 0.25 μm (medium accuracy grade) were used. A significant stiffness enhancement due to face contact was measured in line with the data on stiffness of the shallow hollow taper connections with the face contact.

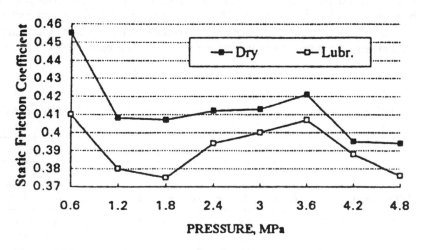

Figure 8.16 Static friction coefficient between coated flat surfaces.

The design in Fig. 8.15 solves the same problems as the other alternative designs listed in Section 4.4.2 (axial registration, high stiffness, insensitivity to high rpm), but without their shortcomings. No changes in the spindle design are required; manufacturing of the new shank is no more complex than manufacturing of the conventional standard shanks; the cost differential is minimal; shrink-fit tools or other tool clamping devices can be located deep inside the holder, thus reducing the necessary tool overhang and enhancing the effective stiffness; and no bulging of the spindle occurs.

Performance testing of the WSU-1 design was performed on a milling machine equipped with a manually operated drawbar that was instrumented with strain gages to measure the axial force. Face and slot milling operations were performed at the regimes creating maximum allowable load on the cutters. Significant improvements in flatness and surface finish of the machined surfaces over conventional interfaces have been observed [32].

As it was noted in Section 8.2.2, not all applications require the axial indexing of the toolholder. If the axial indexing is not required, the main problems of the standard 7/24 taper toolholders are reduced stiffness, large runout, and fretting. All these shortcomings are due to the clearance at the back of the taper connection. Successful testing of the design in Fig. 8.15 led to the design of interface WSU-2 (Fig. 8.17) [33], which greatly alleviates these problems.

The toolholder in Fig. 8.17 has taper 1 of the standard dimensions. At the back side of the taper a coaxial groove 2 is machined. Inside this groove one or more rows of precision balls 3 are packed. The balls protrude out of the groove by an amount slightly exceeding the maximum possible clearance between the male and female tapers in the connection, and are held in place by rubber or plastic filling 4. The balls deform during the process of inserting the toolholder

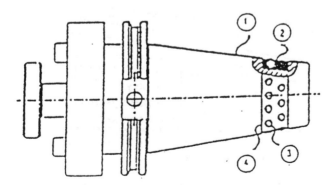

Figure 8.17 Modified 7/24 taper interface WSU-2.

thus "bridging" the clearance. This deformation assures the precise location of the toolholder in the tapered hole, and also provides additional stiffness at the end of the tool since it prevents "pivoting" of the toolholder about its contact area at the front of the connection. The micromotions of the shank inside the spindle hole and the resulting wear are greatly reduced.

This modification can be applied both to existing and to newly manufactured toolholders. Although the modification is very simple and inexpensive, it results in a very significant increase in effective stiffness and reduction of runout of the interface. The stiffness increase is especially pronounced at low drawbar forces. Figure 8.18 presents typical plots of stiffness vs. axial force for conventional and modified #50 toolholders. Stiffness was measured under vertical (Y-direction) load applied 40 mm (~1.5 in.) in front of the spindle face in the direction of the keys, and in the perpendicular direction. While at low axial forces the stiffness increases as much as threefold, even at high axial forces the increase is still very significant. The stiffness of WSU-2 is comparable to (and in some cases exceeds) the stiffness of the taper/face interfaces. As a result, this simple modification may be sufficient for many machining operations without the need for radical changes in the toolholder design.

Runout reduction was measured by comparing runout of the toolholder with the machined groove with and without balls. Reductions in the range of 10–50% have been observed, depending on the fabrication quality of the toolholder. Flatness and surface finish for face milling with the WSU-2 interface are significantly better than with the conventional 7/24 interface, although not as good as with the taper/face WSU-2 interface [32].

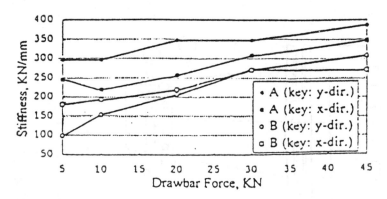

Figure 8.18 Stiffness of 7/24 #50 toolholders with machined groove: (A) with two rows of 6 mm balls (WSU-2; Fig. 8.14); (B) without balls.

8.3 TRADING OFF STIFFNESS FOR IMPROVING OVERALL PERFORMANCE: PRACTICAL EXAMPLES

Performance of mechanical systems is determined by many parameters. While stiffness is one of the critical parameters, it is definitely not the only one. Clear understanding of roles of various parameters allows to achieve significant overall performance improvements if some degrading of one parameter is accompanied by substantial gains in other parameters. This approach is illustrated below by several practical examples.

8.3.1 Improving Chatter Resistance of Machining Systems by Using Reduced Stiffness/High Damping Tool Clamping Systems

It is universally accepted that performance of a machining system (chatter resistance, accuracy, tool wear) is improving with increasing stiffness of its components. Many efforts (frequently very expensive and only marginally useful) arc directed to increasing stiffness of machining systems (machine tool frames, tooling and fixturing structures, structural connections, etc.). A blind effort to enhance stiffness may represent a simplistic approach as it is demonstrated in the survey of research publications on the role of stiffness in machining systems (Article 7). These are some reasons why the issue of stiffness in machining is not a straightforward one:

1. The chatter resistance is dependant not only on stiffness but also on damping.
2. State-of-the-art high speed machine tools are characterized by high power but low cutting forces, especially at finishing regimes, thus an extremely high stiffness is not needed for reduction of the cutting force-induced deformations below the required tolerance.
3. Intentional stiffness reduction of high speed bearing units may allow to improve bearing conditions (see Section 8.1.4).

Since the chatter resistance of a machining system improves with increasing value of the criterion $K\delta$, where K is effective stiffness of the systems and δ is a measure of its effective damping (e.g., log decrement δ), it can be concluded that some stiffness reduction can be tolerated if it is accompanied by a more significant increase in damping. This approach was, indirectly, used in designing the composite boring bar shown in Fig. 7.39a and discussed in detail in Article 5. Making the overhang part of the boring bar from a light material resulted in some reduction of its stiffness. This reduction is about 15% for a tungsten carbide–aluminum bar as compared with a solid tungsten carbide bar. However, since the mass ratio μ of DVA was disproportionally increased and subsequently the effective damping was at least doubled, the value $K\delta$ and, accordingly, chatter resistance was also significantly increased.

The simplest application of this approach to machining systems can be achieved by modifying tool clamping devices. Two generic cases were reported in Rivin and Kang [34] and Rivin and Xu [35].

Turning of Low Stiffness Parts

In Rivin and Kang [34], turning of a long slender part clamped in the chuck and supported by the tailstock is considered. Dynamic behavior of the machine frame does not significantly influence stability of the cutting process when a slender bar is machined. The equivalent stiffness of the work piece and its end supports (chuck, spindle, tailstock) are considerably lower than the structural stiffness of the machine, thus the effective equivalent stiffness, which is determined by the weakest element in the force transmission path, is also relatively low. Under the chatter conditions the system spindle–work piece–tailstock is vibrating. The conventional approach in machining such parts is to provide additional support means, such as steady rests, which are bulky, expensive, and do not perform well for stepped or asymmetric shafts. Another technique is described in Article 4, in which stiffness of the part is enhanced by application of a tensile force. While effective, this technique requires special means for applying the tensile force. In some cases, it is desirable to achieve the stable no-chatter cutting as well as improved accuracy and surface finish, with minimum changes in the machining system. Since the work piece has low stiffness and damping, and since these parameters are difficult to modify (unless external devices like steady rests or the tensioning means are used), a natural way to improve the stability is from the cutting tool side. An effective approach to doing it is by adding damping to the cutting tool. However, stiffness of the tool is very high as compared with the work piece stiffness, and its vibratory displacements are very small. As a result, damping enhancement of the tool would not have a noticeable effect on the overall damping of the machining system since the energy dissipation is proportional to vibratory velocity and/or displacement of the damping element. Thus, to achieve enhancement of damping in the machining system, enhancement of the cutting tool damping must be accompanied by reduction of its stiffness. With a proper tuning of the dynamic system, an additional damping then would be pumped into the work piece subsystem and stability of the cutting process would be increased.

The following factors are influencing stability of the cutting process:

1. Work piece material and geometry.
2. Tool geometry and stiffness.
3. Cutting regimes such as cutting speed, feed, and depth.

All these factors can be studied in the model in Fig. 8.19, where k_w and c_w are stiffness and damping coefficients of the work piece, k_t and c_t are the same for the tool, and k_c and c_c are effective stiffness and damping of the cutting process (see Section 1.4). These are dependent on tool and work piece materials, tool geometry, and cutting regimes. The most important parameter is the damping associated with the cutting process, which is related to the cutting conditions. Under certain conditions the damping of the cutting process becomes negative, and if damping of the work piece and tool subsystems is not high enough, this negative damping overcomes the positive damping in the system and instability occurs. The system stability can be analyzed for any given set of parameters in the model in Fig. 8.19 [34]. Optimum values of the combination of tool parameters k_t and c_t (or δ_t, log decrement of the tool system) can be obtained by such analysis.

Although the tool stiffness in the direction radial to the work piece has to be reduced, it must stay high in all other directions. It can be achieved by mounting the tool in a sleeve made of the thin-layered rubber-metal laminate material described in Section 3.2 and in Article 3 [11] (Fig. 8.20). The radial component P_x of the cutting force causes shear deformation of the laminate (low stiffness direction), while the components P_y and P_z, orthogonal to P_x, load this laminated sleeve in compression, which is characterized by a very high stiffness. The radial dynamic stiffness of the ''reduced stiffness'' cutting tool tested in Rivin and Kang [34] was $K_t = 10,350$ lb/in. while the original tool stiffness was $k_t = 162,000$ lb/in. (about 15 times reduction). The rubber in the laminates had high damping corresponding to $\delta_t = 1.8$, which is about 10–15 times higher than damping of the conventional tool. It is important to note that the overall stiffness of the machining system was reduced to a much lesser degree, as shown in Fig. 8.21, since the low work piece stiffness was the determining factor in the original system. The work piece was a bar 0.7 in. (18 mm) in diameter and 15 in. (376 mm) long. An important feature of the plots in Fig. 8.21 is a dramatic improvement in

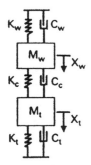

Figure 8.19 Mathematical model of the cutting system.

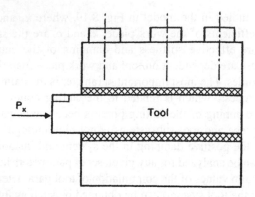

Figure 8.20 ''Reduced stiffness'' tool in the laminate sleeve.

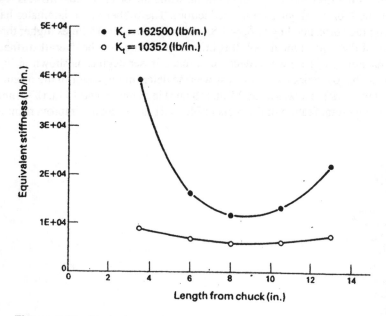

Figure 8.21 Equivalent static stiffness between work piece and tool along the work piece length.

uniformity of stiffness along the work piece, which resulted in good cylindricity of the machined bar, within 0.07 mm (0.0028 in.) vs. 0.125 mm (0.005 in.) with the conventional tool. Relative vibration amplitudes between the tool and the work piece have diminished about two to four times for cutting with the reduced stiffness tool as compared with cutting with the conventional tool. Surface finish with the reduced stiffness tool was acceptable, $R_a = 2.2$ μm vs. $R_a > 10$ μm for the conventional tool, and chatter resistance of the process was significantly improved.

Modification of Tool Clamping Systems for Cantilever Tools [35]

High stiffness of tool-clamping systems for cantilever tools (boring bars, end mills, etc.) is specified in accordance with two requirements. High *static stiffness* is required in order to reduce deformations of the machining system under cutting forces (which result in dimensional inaccuracies of the machined surfaces), and high *dynamic stiffness* is needed in order to reduce self-excited (chatter) and forced vibrations resulting in a poor surface finish and reduced tool life. Static stiffness of the tool clamping devices is, in most cases, higher than is needed for machining with the required tolerances, since at the roughing regimes characterized by high cutting forces the requirements to geometric accuracy are not very stringent, while at the finishing regimes the cutting forces are very low. It is not always understood that high stiffness is frequently combined with very low damping. This combination has a negative impact on surface finish and tool life. Since high damping is usually characteristic for polymeric materials whose very low Young's moduli make them unacceptable for the tool clamping devices, the absolute majority of the clamping devices are made of steel having very high Young's modulus but very low loss factor $\eta = \tan \beta = 0.001{-}0.003$ ($\delta = 0.003{-}0.009$).

It was discovered in Rivin and Xu [35] that an alloy of ~50% Ni and ~50% Ti (NiTi, or Nitinol), which is known for its "shape memory effect," has very high damping when prestressed in tension or compression. Figure 8.22 presents the loss factor of the NiTi specimen as a function of the prestress magnitude and the cyclic stress amplitude. One can see from the plot that the test specimen has extremely high damping even at small cyclic stress amplitudes when subjected to the optimal prestress of 10,500 psi (10 MPa). The loss factor of $\eta = \tan \beta = 0.06{-}0.1$ ($\delta = 0.19{-}0.3$) is 20–100 times higher than the loss factor of steel. It is important that such a high loss factor develops even at low cyclic stress amplitudes (1000 psi or 6.5 MPa) since vibrations of tools, both self-excited and forced, are associated with relatively small stresses.

Cross sections of a three-dimensional plot (loss factor vs. prestress and vs. cyclic stress, respectively) are shown in Figs. 8.23 and 8.24. Figures 8.22–8.24 show that at certain combinations of the prestress and cyclic amplitudes the internal damping of this material is becoming very high (log decrement δ up to 0.33).

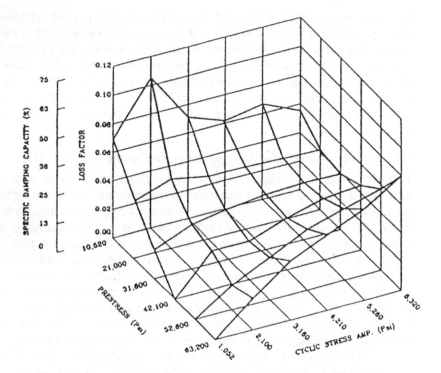

Figure 8.22 Loss factor vs. prestress and cyclic stress for NiTi specimen 0.5 in. diameter, 2.5 in. long under compression.

Figure 8.25 shows that the dynamic modulus of elasticity (stiffness) is decreasing with the increasing stress amplitude, similarly to dependence of stiffness on vibration amplitude for other high-damping materials, see Chapter 3 and Section 4.6.2. The dynamic modulus is in the range of 5.5–10.5 $\times$ 10^6 psi (0.37–0.7 $\times$ 10^5 MPa), or three to six times lower than Young's modulus of steel. Since damping of this material is 20–100 times higher than steel, criterion $K\delta$ can be higher than for steel. However, in a structural use of NiTi both stiffness and damping of a device using this material would certainly be modified by other components of the device, by joints, etc.

The described effect was used in [35] for making a clamping device for a small cantilever boring bar from the high damping prestressed NiTi alloy. The approach described in Section 7.5.2 and Article 5 allows one to enhance effectiveness of dynamic vibration absorbers, which are used to increase damping of cantilever structures, such as boring bars. However, the proposed techniques require

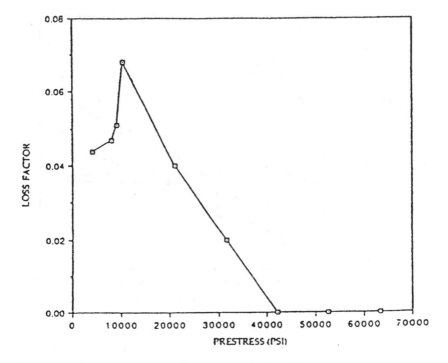

Figure 8.23   Loss factor at cyclic stress amplitude 1,050 psi as function of prestress.

design modifications which are feasible only for tools having external diameter not less than 15–20 mm (0.625–0.75 in.). There is no available technique to improve chatter resistance of small cantilever tools rather than making them from a high Young's modulus material (e.g., sintered tungsten carbide).

Since small boring bars cannot be equipped with built-in vibration absorbers, the damping enhancement should be provided from outside, e.g. by making a clamping device from the high-damping material. The tool, a boring bar with nominal diameter 0.25 in. (6.35 mm) was clamped in a NiTi bushing (Fig. 8.26).

The prestress was adjusted by changing amount of interference in the bar–NiTi bushing connection. The reference (conventional) boring bar was mounted in the steel bushing. These NiTi and steel bushings were clamped in the standard steel slotted square sleeve with a 0.25 in. bore, which was in turn clamped in the tool holder of a 12 in. lathe. Dynamic characteristics (natural frequency and log decrement) were determined by a free vibration test.

The test results are shown in Table 8.3. It can be seen that at high prestress

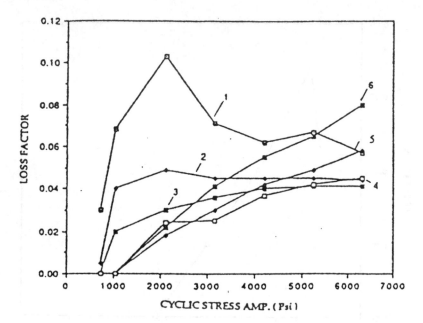

Figure 8.24 Loss factor at various prestress magnitudes as function of cyclic stress amplitude: 1, prestress 10,500 psi; 2, 21,000 psi; 3, 31,600 psi; 4, 42,100 psi; 5, 52,600 psi; 6, 63,200 psi.

levels, damping of the boring bar clamped in the SMA bushing is 5.5 times higher than for the similar bar clamped in the steel bushing. However, clamping in a steel bushing results in ~15% higher natural frequency (about 30% higher effective stiffness).

As it was mentioned before, dynamic stability of cantilever tools is determined by the criterion

$$K_{ef}\delta = (2\pi)^2 m_{ef} f_n^2 \delta = A f_n^2 \delta \qquad (8.5)$$

where m_{ef} = effective mass of the system; f_n = natural frequency; $\delta = \pi\eta$ = log decrement; and η = loss factor. Since m_{ef} is constant for all the tests summarized in Table 8.3, A is also constant and values of $f_n^2\delta$ listed in Table 8.3 are proportional to $K_{ef}\delta$. It can be seen that at the high prestrain the dynamic stability criterion with the NiTi holder bushing is about four times higher than with the steel holder. While in the machine tool setting this difference may be reduced

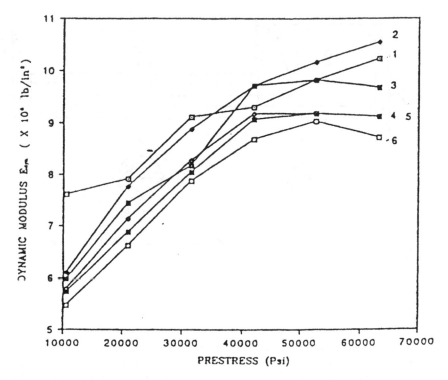

Figure 8.25 Dynamic modulus vs. prestress for various values of cyclic stress amplitude: 1, cyclic stress amplitude 1050 psi; 2, 2100 psi; 3, 3160 psi; 4, 4120 psi; 5, 5260 psi; 6, 6320 psi.

due to influence of stiffness (compliance) and damping in the attachment of the tool holder to the machine tool, a significant improvement of cutting conditions can still be expected.

Cutting tests were performed on a lathe with a non-rotating boring bar (diameter $D = 0.25$ in., overhang length $L = 1.9$ in., $L/D = 7.5$). Due to the shorter length (and higher structural bending stiffness of the boring bar) as compared with the setup in Fig 8.26, influence of the clamping bushing material was even more pronounced. This resulted in a larger difference between natural frequencies of the boring bar mounted in steel (f_{st}) and in NiTi (f_{NiTi}) bushings. Natural frequencies measured on the lathe were $f_{st} = 1315$ Hz; $f_{NiTi} = 855$ Hz (average between two directions) and log decrement, respectively; $\delta_{st} = 0.025$; and $\delta_{NiTi} = 0.113$. Accordingly, $(f^2\delta)_{st} = 43,230$ and $(f^2\delta)_{NiTi} = 80,500$, about two times difference.

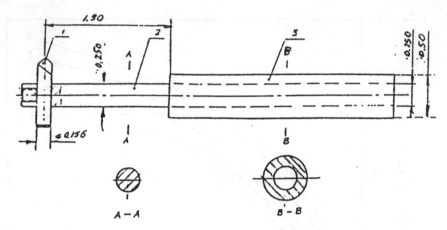

Figure 8.26 Boring bar with high-damping NiTi clamping bushing: 1, tool bit; 2, boring bar; 3, NiTi bushing.

Cutting tests were performed on work pieces made of 1045 medium carbon steel and 316 stainless steel. Surface finish (R_a) was compared for the cutting tests when the boring bar was clamped in the NiTi (stressed to the strain value $\varepsilon = 1,300 \times 10^{-6}$) and in the steel bushings. All tests were performed with feed 0.0015 in./rev (0.038 mm/rev) by boring an 1.8 in. (46 mm) diameter hole. The tests have demonstrated that the boring bar clamped in the prestressed NiTi bushing is more chatter resistant. This resulted in significant improvements in surface finish, up to two times reduction in R_a [35].

A similar approach can be used for clamping end mills of both small and large size (since it is very difficult to insert dynamic vibration absorbers in end mills), as well as for other cantilever systems.

Table 8.3 Free Vibration Test Results for Boring Bar in SMA and Steel Clamping Bushings

Prestrain		NiTi bushing			Steel bushing	
ε, 10^{-6}	δ	f_n (Hz)	$f_n^2\delta$	δ	f_n (Hz)	$f_n^2\delta$
850	0.058	515	15,400	0.016	602	5800
1050	0.088	531	24,800	0.016	638	6500

8.4 CONSTANT FORCE (ZERO STIFFNESS) VIBRATION ISOLATION SYSTEMS

Effectiveness of vibration isolation systems in filtering out unwanted vibrations can be improved by reducing natural frequencies of the isolated objects on vibration isolators. If the isolators support the weight of the isolated object, reduced natural frequency in vertical directions and the corresponding reduced stiffness of the isolator lead to an unacceptably large static deflection of the isolator(s) and to their static instability and packaging problems. Static deflection in centimeters of a vertical spring caused by the weight of the supported object can be expressed as

$$\Delta = \frac{25}{f_n^2}; \qquad f_n = \frac{5}{\sqrt{\Delta}} \tag{8.6}$$

where f_n = vertical natural frequency, Hz. Very low natural frequencies in the range of 1–4 Hz, which are often desirable especially for vibration protection of humans, are associated with static deflections of 1.5–25 cm (0.6–10.0 in.). Such deflections are hardly attainable since they require very large dimensions of the isolators. For example, rubber isolators of conventional design loaded in compression have to be at least 6.5–10 times taller than the required static deformation, which results in highly unstable systems. The stability can be enhanced by introduction of a bulky and expensive "inertia mass" (foundation block).

In some cases, the problems associated with isolating systems characterized by such and even lower natural frequencies can be alleviated by using so-called *constant force* (CF) elastic systems [35]. The constant force F as a function of deflection is equivalent to "zero stiffness"

$$K = dF/dz = 0 \tag{8.7}$$

CF systems can provide zero stiffness at one point of their load-deflection characteristic or on a finite interval of the load-deflection characteristic. Since the stiffness values vary, the CF systems are always nonlinear. Besides vibration isolation systems, CF systems are very effectively used for shock absorption devices were the CF characteristic is the optimal one (e.g., see [37]).

Figure 8.27a illustrates a CF system having a stiffness compensator. Main spring 1 having constant stiffness k_1 (linear load-deflection characteristic in Fig. 8.27b) cooperates with compensating springs 2 having total stiffness k_2. Due to geometry of the device, load-deflection characteristic of springs 2 is nonlinear as shown in Fig. 8.27c. The effective load-deflection characteristic of the system is a summation of load-deflection characteristics for springs 1 and 2 (Fig. 8.27d).

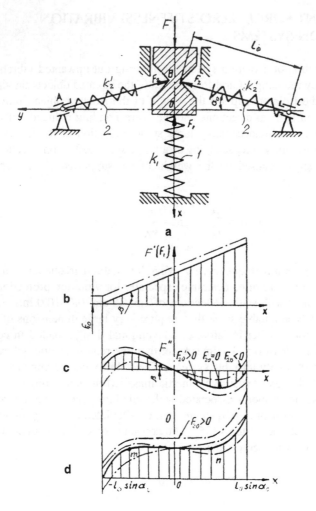

Figure 8.27 Zero-stiffness suspension with stiffness compensator.

The basic effective load-deflection characteristic shown as the solid line in Fig. 8.27d has zero stiffness at deflection $x = 0$ and very low stiffness on the interval m–n (working interval). Weight of the supported object (not shown in Fig. 8.27a) is compensated by an initial preload F_{10} of the main spring 1. Change of the weight can be accommodated by a corresponding change of the preload, as shown by a chain line in Fig. 8.27d. Effective stiffness of the device can be adjusted by changing preload F_{20} of compensating springs 2 (two preload magnitudes are shown by broken lines). All initial values are indicated by subscript zero.

Figure 8.28a shows application of the concept illustrated by Fig. 8.27a for a vibration-protecting handle for hand-held impact machines (jack hammer, concrete breaker, etc.). The handle consists of handle housing 1 and two links 2 and 3, which are connected via pivots 0' and 0" with the machine 4, and via rollers 5 and 6 with the handle. Links 2 and 3 are engaged by gear sectors 6 and 7, which assure their proper relative positioning. The elastic connection is designed as two pairs of springs 8, 9 and 10, 11, with springs 10, 11 (stiffness k_s) being shorter than springs 8, 9 (stiffness k) and the latter having progressively decreasing pitch (like in Fig. 3.3)

Figure 8.28b shows the measured load-deflection characteristic of the handle. An increase of stiffness of the elastic connection (due to "switch-out" of some coils in the progressively coiled springs 8 and 9 and also due to "switching in" of springs 10 and 11 after some initial deformation of springs 8 and 9), results in a two-step characteristic in Fig. 8.28b having two CF (low stiffness) sections

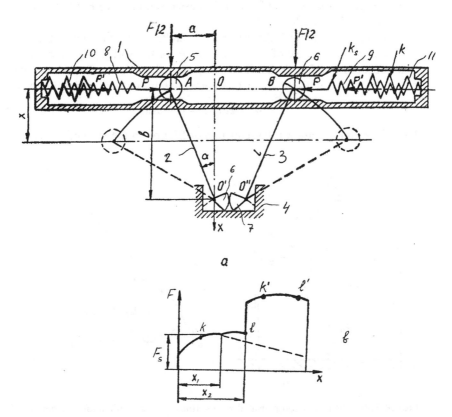

Figure 8.28 Handle for zero-stiffness vibration protection system for jack hammer.

k–l and k'–l'. The second CF section serves as a safety device in case the specified operator pressure F_s on the handle is exceeded. The test results confirmed effectiveness of this system for vibration protection.

Figure 8.29a shows another embodiment of a similar system that is self-adjusting for changing static (weight) loads. The main difference between the systems in Figs. 8.27 and 8.29 is a frictional connection between load-carrying bar 1 attached to main spring 2 and sleeve 3 contacting with poles 4 transmitting forces from compensating flat springs 5. Sleeve 3 is preloaded on bar 1 but when the load F is increasing beyond a preassigned increment, sleeve 3 is slipping along bar 1 and stops in a new position. Figure 8.29b illustrates the load-deflection characteristic of the system.

Another type of CF devices is based on using a specially shaped elastic element. The elastic element in Fig. 8.30 is a complex shape spring in which the side parts AM and CH act as the main spring 1 in Fig 8.27a, and top part ABC acts as compensating spring 2 in Fig. 8.27a. Interaction between the top and the side parts generates reaction forces F'' whose resulting force F' has a characteristic similar to Fig. 8.27c.

The third group of CF devices involves linkage-based compensating devices as in Fig. 8.31. Increase of restoring force F' of the spring with increasing displacements of handles A and B is compensated by decreasing (shortening) of arm b associated with the force F. Approximately, in the working interval

$$W = 2F(b/a) \cong \text{constant} \tag{8.8}$$

Use of the linkage allows the reduction of friction in the system.

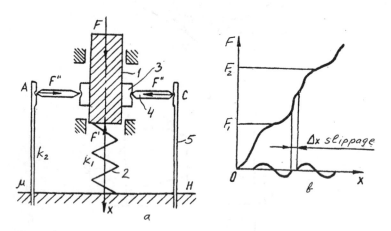

Figure 8.29 Stiffness-compensated zero-stiffness system with automatic height adjustment.

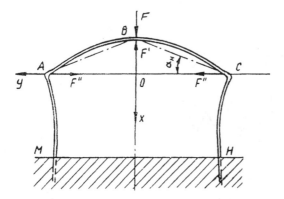

Figure 8.30 Zero-stiffness complex shape spring.

The fourth group is characterized by using cams for the compensating devices (Fig. 8.32). By selecting an appropriate profile, any shape of the load-deflection characteristic can be realized. However, the characteristic is very sensitive to relatively minor errors of the cam profile.

The fifth group of CF devices employs the buckling phenomenon. When a mechanical system buckles, its resistance to external forces (stiffness) ceases to exist (see Chapter 7). Special shapes of elastomeric devices, such as an "inverted flower pot" shock absorber shown in Fig. 8.33a [37] can be used for a large travel while exhibiting the constant resistance force (Fig. 8.33b). Although the total height of the rubber element is 130 mm, deflection no less than 100 mm can be tolerated. A similar effect within a somewhat smaller range, can be achieved by axial [37] or radial (Chapter 3) compression of hollow rubber cylinders.

A low-stiffness system with widely adjustable load-deflection characteristic

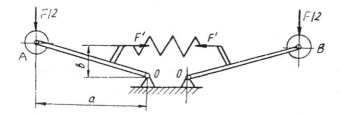

Figure 8.31 Zero-stiffness combination spring/linkage system.

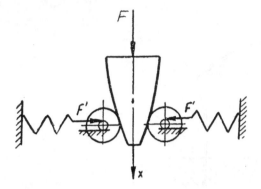

Figure 8.32 Cam-spring variable stiffness system.

is shown in Fig. 8.34a [38]. It comprises housing 1 and T-shaped drawbar 2. The "shelf" of drawbar 2 applies force to leaf springs 3 having an initial curvature that results in their reduced resistance to buckling. Between the convex surfaces of springs 3 and housing 1, auxiliary elastic elements 4 are placed. Presence of elements 4 with stiffness k leads to increase in critical buckling force P_{cr} for leaf springs 3 as [38]

$$P_{cr} = P_{cr_o} + 2lk/\pi^2 \tag{8.8}$$

where l = length of leaf springs 3 and P_{cr_o} = critical (buckling) force without auxiliary elements 4. It is convenient to use an annular pressurized pneumatic

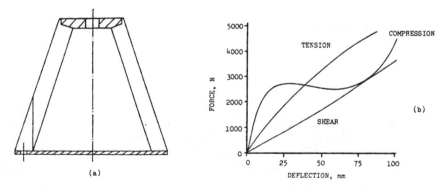

Figure 8.33 (a) Buckling-type shock mounting and (b) its force-deflection characteristic.

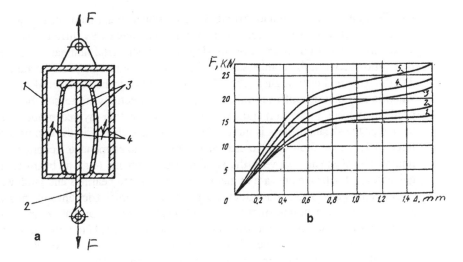

Figure 8.34 (a) Adjustable low stiffness elastic system and (b) its load-deflection characteristics: 1, without pneumatic sleeve; 2, with pneumatic sleeve, air pressure $p = 0$; 3, $p = 0.2$ MPa (30 psi); 4, $p = 0.4$ MPa (60 psi); 5, $p = 0.6$ MPa (90 psi).

sleeve between springs 3 and housing 1 as auxiliary element 4. In this case, change of pressure in the sleeve would change its stiffness and consequently, load-deflection characteristic of the system. Figure 8.34b shows a family of load-deflection characteristics for $l = 238$ mm and cross section of each spring 3 (width $\times$ thickness) 30×4.5 mm^2 at various magnitudes of air pressure in the annular sleeve.

CF or low stiffness characteristic can be obtained also from specially designed fully pneumatic and hydraulic devices. Such systems can be equipped with servocontrol systems accommodating the static (weight) loads, thus the low stiffness for small incremental dynamic motions (low dynamic stiffness) is not accompanied by large deformations (e.g., see Fig. 7.47).

8.5 ANISOTROPIC ELASTIC ELEMENTS AS LIMITED TRAVEL BEARINGS

Usually, high stiffness of a structural component in some direction is associated with a relatively high load-carrying capacity in the same direction. For example, both stiffness and load-carrying capacity in compression for the thin-layered rubber-metal laminates are increasing with decreasing thickness of the rubber layers (see Section 3.2; Fig. 3.9 and Article 3). The rubber-metal laminates as well as some types of metal springs (e.g., flat springs) may have very different stiffness values in different directions. For thin-layered rubber-metal laminates, the stiffness ratios between the compression and shear directions can be in the range of 3,000–5,000 and even higher [11]. These ratios have to be considered together with the fact that the stiffness values in different directions are reasonably independent of loads in the perpendicular directions. For thin-layered rubber-metal laminates the shear stiffness is increasing ~15% for a change in the compression force 100:1. This allows one to use anisotropic elastic elements to accommodate limited displacements between the structural components (e.g., as bearings and guideways). Such elements are in many ways superior to bearings and guideways of conventional designs since they do not have external friction and thus are responsive to even infinitesmal forces and displacements. On the contrary, the friction-based bearings do not respond to the motive forces below the static friction force. Another advantage of the elastic guideways is absence of clearances (backlashes). The elastic connections can even be preloaded.

Desirable performance for some mechanical systems depends not only on stiffness values of the stiffness-critical components, but also on a properly selected stiffness ratio(s) in different linear and/or angular directions. While the chatter resistance of a metal cutting machining system can be improved by increasing its stiffness (as referred to the cutting tool), it also depends significantly on orientation of the principal stiffness axes and on the ratio of the maximum and minimum principal stiffness (e.g., see [39]). A proper selection of stiffness ratios in the principal directions is important for assuring a satisfactory performance of vibration isolation systems. Such systems are often subjected to contradictory requirements: on one hand they have to provide low stiffness for the high quality isolation, and on the other hand they should have a relatively high stiffness in order to assure stability of the isolated object from the rocking motion caused by internal dynamic loads, spurious external excitations, etc. A typical example is a surface grinder that requires isolation from the floor vibration to produce high accuracy and high surface finish parts, but that also generates intense transient loads due to acceleration/deceleration of the heavy table. An effective way of achieving both contradictory goals (good isolation and high stability) is to use isolators with judiciously selected stiffness ratios (see Article 1).

Optimal stiffness ratios for the situations described in the above paragraph usually do not exceed 0.3–3.0. A much higher stiffness ratio may be called for in cases when a mechanical connection is designed to accommodate some relative motion between the connected components. Significant displacements (travel) between the components are usually accommodated by continuous motion (sliding or rolling friction) guideways. However, limited travel motions are better accommodated by elastic connections having distinctly anisotropic stiffness values in the principal directions. These connections have such positive features as:

"solid state" design not sensitive to contamination and not requiring lubrication; absolute sensitivity to even minute forces and displacements that some continuous motion guideways lack due to effects of static friction (on a macroscale for sliding guideways, on a microscale for rolling friction guideways); absence of backlash and a possibility of preloading in the direction of desirable high stiffness without impairing the motion-accommodating capabilities; and usually more compact and lightweight designs.

Two basic types of the anisotropic guideways/bearings are elastic "kinematic" suspensions for precision (e.g., measuring) devices, in which the forces are very small and the main concern is about accuracy, and rubber-metal laminated bearings/guideways that are used for highly loaded connections.

8.5.1 Elastic Kinematic Connections

Elastic kinematic connections using metal, usually spring-like, constitutive elements provide low stiffness (low resistance to motion) in one direction while restraining the connected components in the perpendicular direction. Their main advantage is extremely high sensitivity to small magnitudes of forces and displacements and very low hysteresis. Such devices are most suitable to be used in precision instruments where accommodation of high forces is not required. Depending on the application, the designer may select out of a huge variety of designs for accommodating rotational (revolute) motion (pivots or hinges), translational motion, or for transforming from translational to rotational motion. Some typical devices are described below [40].

Elastic Connections for Rotational Motion

The simplest revolute connection is shown in Fig. 8.35a. It consists of frame 1, elastic (spring) strip 2, and connected (guided) link 3. Usually, strip 2 is initially flat (leaf spring), although it can be initially bent. Moving link 3 may rotate by angle $\theta \leq 15°$ about the axis parallel to the long side of the cross section of strip 2. However, position of this axis may shift depending on the displacement of link 3 and on the active forces.

The double-band pivot in Fig. 8.35b is comprised of frame 4, elastic bands 1 and 2, and connected (guided) link 3. Bands 1 and 2 can be flat or bent (shaped). Position of the intersection axis I–I and the angle between bands 1 and 2 can vary. This pivot can also accommodate rotational angle $\theta \leq 15°$.

Torsional guideways in Fig. 8.35c comprise frame 1 and moving element 3 connected by several slender, torsionally elastic rods 2. Rotation of element 3 relative to frame 1 is accommodated by torsional deformations of rods 2. This device is very rugged and may accommodate large forces/torques.

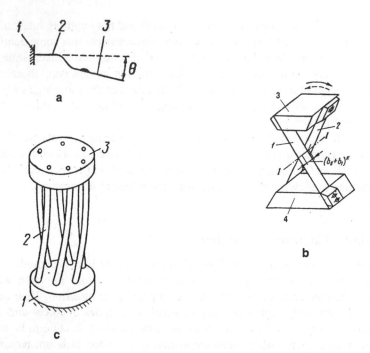

Figure 8.35 Elastic revolute motion guideways: (a) single strip pivot; (b) cross-strip pivot; (c) torsional connection with several elastic rods.

Elastic Connections for Translational Motion

The spring parallelogram in Fig. 8.36a, consists of frame 1 connected with moving element 3 by two flat or shaped (as shown) springs 2 and 4. In the neutral (original) condition springs 2 and 4 are parallel and lengths AB = CD. The main displacement s_1 of element 3 is always accompanied by a smaller undesirable displacement s_2 in the perpendicular direction. The maximum allowable displacement is usually $s_1 < 0.1L$. The mechanism is very simple but s_2 may be excessive, and its performance is sensitive to variations in forces applied to moving elements 3.

The more stable design is the reinforced spring parallelogram shown in Fig. 8.36b. Its springs 2 and 4 are reinforced by rigid pads 5. Pads 5 substantially increase buckling stability of the device under compressive forces applied to element 3, and also improve consistency of the motion parameters under moments/torques applied to element 3. The devices in both Fig. 8.36a and Fig. 8.36b have a disadvantage of not providing for a straight path of the moving element 3 due to a noticeable magnitude of the "parasitic" motion s_2. A "double

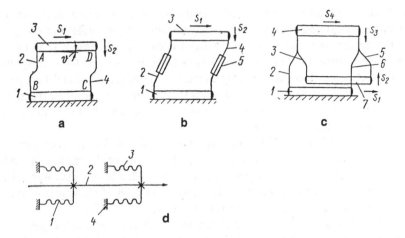

Figure 8.36 Elastic translational motion guideways: (a) spring parallelogram; (b) spring parallelogram with reinforcements; (c) double (series) spring parallelogram; (d) guideways with bellows.

parallelogram" in Fig. 8.36c allows for a relatively long travel of the moving element (slider) 7, $s_1 \leq 0.25\ L$, with a greatly reduced s_2. The elastic connection comprises frame 1, elastic bands 2, 3, 5, and 6, moving element 7, and intermediate moving element 4. The resulting motion of slider 7 is very close to the straight motion since its "parasitic" motion s_2 is a sum of transfer motion s_3 of intermediate element 4 and oppositely directed motion s_{74} of slider 7 relative to element 4

$$|s_2| = |s_3 + s_{74}| < |s_3|$$

Usually, magnitudes of s_3 and s_{74} cannot be made exactly equal due to different longitudinal forces acting on elastic bands 2, 3, 5, and 6. However, their differences are not very substantial and s_2 is greatly reduced.

A very accurate motion direction can be realized by using axisymmetrical systems, such as double-bellows system in Fig. 8.36d. This device has moving element 2 attached to the centers of two bellows 1 and 3, which are fastened to frame 4.

Elastic Motion Transformers

Elastic motion transformers are used in precision devices in which backlashes and "dead zones" (hysteresis) are not tolerated. The motion transformer in Fig.

8.37a can be called a *"double reed"* mechanism. It transforms a small rectilinear motion s_1 of driving element 5 into a significant rotation angle θ of driven link 3. The device comprises two leaf springs (reeds) 2 and 4, frame 1, and driving and driven elements 5 and 3. Reed 2 is fastened to frame 1; the corresponding end of reed 4 is connected to driving element 5. Initially, reeds 2 and 4 are parallel with a small distance (which determines the transmission ratio) between them. This simple device may have a transmission ratio up to 0.1°/μm with the range of rotation of element 3 up to 5°.

The device in Fig. 8.37b is utilizing postbuckling deformation of leaf spring 2. Driving element 1 pushes leaf spring 2 in the longitudinal direction, and driven element 3 is attached to the opposite end of spring 2. Although the transmission ratio is not constant (a nonlinear mechanism), position of the center of rotation is reasonably stable. The range of rotation angle is $\theta < 20°$.

The greatest transmission ratio together with large rotation angles can be realized by a "twisted strip" motion transformer in Fig. 8.37c. A small linear displacement of the ends of a prestretched elastic strip composed of two identical segments 1 and 3 with opposing (left and right) twist direction is transformed into large rotation of driven element 2. The relative displacement $s_1 = s_{11} - s_{12}$ of the ends of elastic strip 1.3 is transformed into rotation angle θ of driven element 2 with transmission ratio $i_\theta = d\,\theta/ds_1 = 0.8°/\mu m - 10°/\mu m$.

8.5.2 Accommodation of Limited Travel Using Thin-Layered Rubber-Metal Laminates

While the elastic kinematics connections are mostly used in precision instruments, thin-layered rubber-metal laminates can be and are used in heavy-duty devices subjected to very high preloads and/or payloads.

The most heavy-duty applications of rubber-metal laminates are in civil engi-

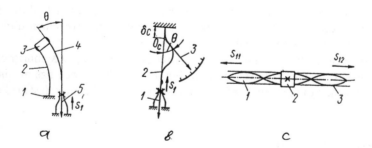

Figure 8.37 Elastic motion transformers: (a) double reed device; (b) buckled strip device; (c) twisted strip device.

neering—for supporting bridges and for protecting buildings from earthquakes [37]. Bridge bearings have to accommodate expansion and contraction of the structure caused by temperature and humidity variations, and also allow small rotations caused by bending of the bridge span under heavy vehicles. Although relative changes of the longitudinal dimensions due to variations in temperature and humidity are small, the absolute displacements can be significant due to large spans of modern bridges. The bridge bearings must have high vertical stiffness to prevent excessive changes of pavement level caused by traffic-induced loads, and low horizontal stiffness in order to minimize the forces applied to the bridge supports by expansion and contraction of the span. Previously, rolling or sliding bearings were universally used for bridges. However, the rollers are not performing well for very small displacements due to a gradual development of small dents ("brinelling") causing increases in static friction and wear of the contact surfaces. Although sliding bearings (usually, Teflon–stainless steel combinations) are competing with rubber-metal laminated bearings, they require more maintenance since the sliding zone must be protected from contamination. They may also require special devices for accommodating small angular motions, which can be naturally accommodated by properly dimensioned laminates.

The laminates used for supporting bridges (as well as for earthquake protection of buildings) have rubber layers 5–15 mm thick with the aspect ratio (ratio of the smaller dimension in the plane view to thickness) usually not less than 15. While the specific load-carrying capacity of the rubber-metal laminates with such relatively thick rubber layers is much smaller than that of the thin-layered rubber-metal laminates described in Chapter 3 and in Article 3, specific compressive loads as high as 15–30 MPa (2250–4500 psi) can be easily accommodated. Shear deformations of the bridge bearings exceeding 100 mm do not present a problem for the rubber-metal laminates, while the compression deformation does not exceed 3 mm.

Use of rubber-metal laminates for mechanical devices is based on the same properties that made them desirable for the bridge and building supporting bearings. These properties include combination of high stiffness in one (compression) direction and of low stiffness in the orthogonal (shear) directions. Usually, high stiffness is associated with high allowable loading in the direction of high stiffness (see Article 3). Thin-layered rubber-metal laminates are used in three basic applications: (1) anisotropic elastic elements; (2) bearings or guideways for limited travel; and (3) compensation elements.

Use of Rubber-Metal Laminates as Anisotropic Elastic Elements

This application is illustrated in Fig. 8.38 on the example of vibration-stimulated gravity chute for conveying parts and scrap from the work zone of stamping presses [41]. If inclination of a chute cannot be made steeper than 15–20°, part

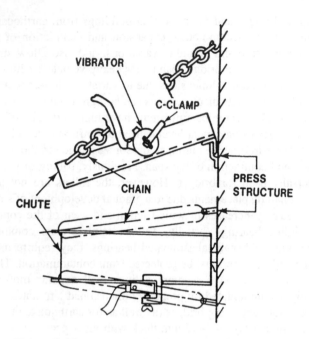

Figure 8.38 Vibration-stimulated gravity chute.

and scrap pieces can be stuck on the sliding surface of a gravity chute. In such cases, vibration stimulation of the chute is used to assure easy movement of the part and scrap pieces. A pneumatic vibration exciter (vibrator) having a ball forced around its raceway by compressed air or an unbalanced impeller is attached to the side of the chute (Fig. 8.38). The useful stimulating effect of vibration is accompanied by excessive noise levels due to resonances between the chute structure and high frequency harmonics of the intense vibratory force from the vibrator. Reduction of the noise levels was attempted to achieve by enhancing effectiveness of the vibration–stimulation effect and, consequently, reducing the magnitude of the required vibratory force.

It is known that the most effective vibration-assistant conveyance of particles along a flat surface develops when the rotating force vector describes an elliptical trajectory. However, ball and turbine vibrators generate circular trajectories of the vibratory force vector. To transform the circular trajectory into an elliptical trajectory of the vibratory force, a force vector transformer in Fig. 8.39a was proposed. Vibrator 1 is attached to mounting bracket 3 via two rubber strips (gaskets) 2 with a large aspect ratio. Due to a significant anisotropy of strips 2 (high compression stiffness in the direction of holding bolt and low shear stiffness in the orthogonal directions), the respective natural frequencies f_c and f_s of vibra-

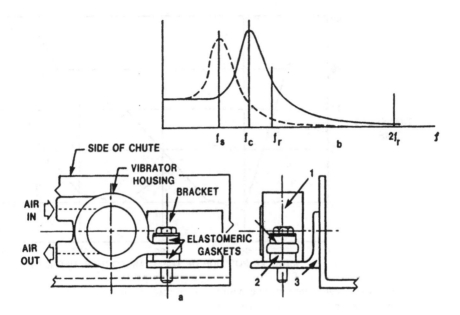

Figure 8.39 (a) Vibration force vector transformer and (b) its transmissibility curves.

tor 1 in these two directions are very different (Fig. 8.39b). The mounting device (i.e., its stiffnesses in the compression and shear directions) is designed in such a way that the fundamental frequency of vibratory force (rpm of the ball or the turbine) f_r is correlated with f_c and f_s as shown in Fig. 8.39b. Then the compression component of the vibratory force is amplified, and the shear component is attenuated, thus creating the required elliptical trajectory of the vibratory force vector; orientation of this trajectory relative to the chute surface can be adjusted by positioning bracket 3. Figure 8.40 shows the time of part travel along the chute as a function of orientation of the force vector transformer. The minimum time is 1.8 s vs. 20 s for the same vibrator without the force vector transformer (12 times improvement). Such improvement allows one to reduce pressure of the compressed air and reduce noise by 5–6 dBA, additional noise reduction (also about 5 dBA) is achieved due to isolation of higher harmonics (such as $2f_r$ in Fig. 8.39b) of the vibratory force from the chute structure. This sizeable noise reduction is accompanied by a significant reduction in energy consumption (savings of compressed air).

Use of Rubber-Metal Laminates as Limited Travel Bearings

The most typical application of thin-layered rubber-metal laminates is for accommodation of small displacements ("limited travel bearings"). Advantages of such

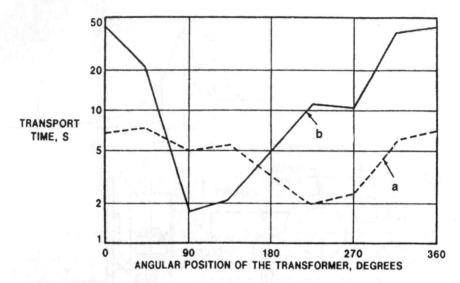

Figure 8.40 Part transport time with vibration force vector transformer: (a) counter-clockwise rotation; (b) clockwise rotation.

bearings are their solid state design not sensitive to contamination; no need for lubrication; absolute sensitivity to small forces and displacements; possibility of preloading due to virtual independence of their shear resistance from compressive forces; low energy losses; and generation of restoring force due to elastic character of the connection. Some such applications are described in Section 8.1.2 and Figs. 8.7 and 8.8, in which generation of the restoring force is important for returning the sliders to their initial positions.

Important mechanical components whose performance is based on accommodation of small displacements are U-joints and misalignment compensating couplings (such as Oldham coupling).

The U-joint (or Cardan joint) allows transmitting rotation between two shafts whose axes are intersecting but not coaxial. Figure 8.41 shows a U-joint with rubber metal laminated bushings serving as the trunnion bearings. It is interesting to analyze efficiency of such a U-joint. For angle α between the connected shafts, each elastic bushing is twisted $\pm\alpha$ once per resolution of the joint. With angular stiffness k_α of each bearing, maximum potential energy stored in one bushing during the twisting cycle is

$$V = k_\alpha \frac{\alpha^2}{2} \qquad (8.10)$$

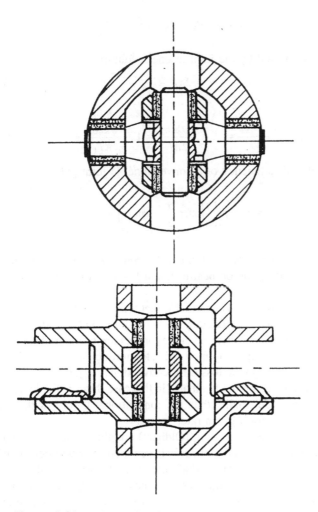

Figure 8.41 U-joint with elastomeric (rubber-metal laminated) trunnion bearings.

and the energy dissipation during one twisting cycle in one bushing is

$$\Delta V_1 = \psi k_\alpha \frac{\alpha^2}{2} \tag{8.11}$$

where $\psi = \delta/2$ = relative energy dissipation of the rubber blend used for the laminate and δ is its log decrement. The total energy dissipation in four bushing during one revolution is

$$\Delta V = 4\Delta V_1 = 2\psi k_\alpha \alpha^2 \tag{8.12}$$

The total energy transmitted by the joint in one revolution is

$$W = 2\pi T \tag{8.13}$$

and efficiency is

$$\eta = \frac{W - \Delta V}{W} = 1 - \frac{2\psi k_\alpha \alpha^2}{2\pi T} = 1 - \frac{\psi k_\alpha \alpha^2}{\pi T_1} \tag{8.14}$$

It can be compared with efficiency of a conventional U-joint [11]

$$\eta = 1 - f\frac{d}{R}\frac{1}{\pi}\left(2\tan\frac{\alpha}{2} + \tan\alpha\right) \tag{8.15}$$

where d = effective diameter of the runnion bearing; $2R$ = distance between the centers of the opposite trunnion bearings; and f = friction coefficient in the bearings.

It can be seen from Eqs. (8.14) and (8.15) that although efficiency of a conventional U-joint is a constant, efficiency of the U-joint with elastic bushings increases with increasing load (when the energy losses are of the highest importance). The losses in the elastic U-joint at the rated torque can be 1–2 decimal orders of magnitude lower than the losses for conventional U-joints. Due to high allowable compression loads on the laminate (in this case, high radial loads), the elastic U-joints can be made smaller than the conventional U-joint with sliding or rolling friction bearings for a given rated torque.

Figure 8.42a shows a compensating (Oldham) coupling that allows one to connect shafts with a parallel misalignment between their axes without inducing nonuniformity of rotation of the driven shaft and without exerting high loads on the shaft bearings. The coupling comprises two hubs, 1 and 2, connected to the respective shafts and an intermediate disc 3. The torque is transmitted between driving member 1 and intermediate member 3, and between intermediate member 3 and driven member 2, by means of two orthogonal sliding connections a–b and c–d. Because of the decomposition of a misalignment vector into two orthogonal components, this coupling theoretically assures ideal compensation while being torsionally rigid. The latter feature may also lead to high torque/weight ratios. However, this ingenious design finds only an infrequent use, usually for noncritical low speed applications. Some reasons for this are as follows:

1. Since a clearance is needed for the normal functioning of the sliding connections, the contact stresses are nonuniform with high peak values (Fig. 8.42b). This leads to a rapid rate of wear.
2. The lubrication layer in the highly loaded contact areas is squeezed out,

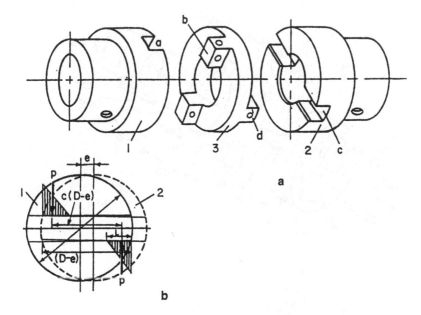

Figure 8.42 (a) Oldham coupling and (b) contact stress distribution in its sliding connections.

thus the effective friction coefficient is high, $f = 0.1–0.2$. As a result, the coupling exerts high forces

$$F = Pf = 2fT/D \qquad (8.16)$$

on the connected shafts. Here P = tangential force acting in each sliding connection and D = external diameter of the coupling.

3. The coupling does not compensate misalignments below $0.5–1.0 \times 10^{-3}$ D. At smaller misalignments hubs 1 and 2 and intermediate disc 3 stay cemented by the static friction forces and sliding/compensation does not occur.
4. The coupling component must be made from a wear-resistant material (usually heat-treated steel) since the same material is used for the hub and disc structures and for the sliding connections.

Since displacement in the sliding connections a–b and c–d in Fig. 8.42a are small (equal to the magnitude of the shaft misalignment), the Oldham coupling is a good candidate for application of the thin-layered rubber metal laminates. Figure 8.43 [42] shows such application, which was extensively studied at Wayne State University.

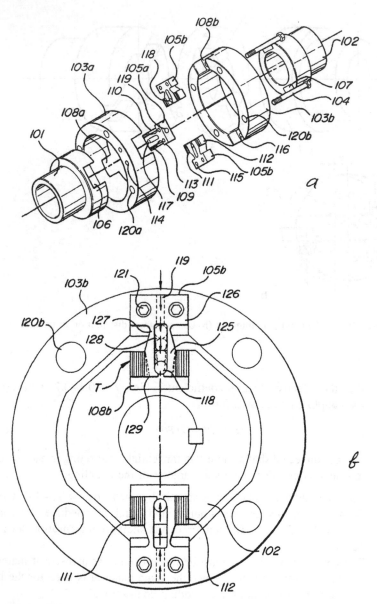

Figure 8.43 Oldham coupling with elastomeric (rubber-metal laminated) connections.

In Fig. 8.43a, hubs 101 and 102 have slots 106 and 107, respectively, whose axes are orthogonal. Intermediate disc can be assembled from two identical halves, 103a and 103b. Slots 108a and 108b in the respective halves are also orthogonally orientated. Holders 105 are fastened to slots 108 in the intermediate disc and are connected to slots 106 and 107 via thin-layered rubber-metal laminated elements 111 and 112 as detailed in Fig. 8.43b. These elements are preloaded by sides 125 of holders 105, which spread out by moving preloading roller 118 radially toward the center.

This design provides for the kinematic advantages of the Oldham coupling without creating the above-listed problems associated with the conventional Oldham couplings. The coupling is much smaller for the same rated torque than the conventional one due to the high load-carrying capacity of the laminates and the absence of the stress concentrations shown in Fig. 8.42b. The intermediate disc (the heaviest part of the coupling) can be made from a light strong material, such as aluminum. This makes the coupling suitable for high-speed applications.

The misalignment compensation stiffness and the rated torque can be varied by proportioning the laminated elements (their overall dimensions, thickness and number of rubber layers, etc.). The loads on the connected shafts are greatly reduced and are not dependent on the transmitted torque since the shear stiffness of the laminates does not depend significantly on the compression load.

The efficiency of the coupling is similar to efficiency of the elastic U-joint described above.

Use of Rubber-Metal Laminates as Compensators

Use of the laminates as compensators is exemplified by the above example of bridge bearings. Similar applications are important also for precision mechanical devices, such as long frames for machine tools and measuring instruments. It is required that the frame is always parallel to the supporting structure and is connected with the supporting structure by very rigid (in compression) elements. However, these rigid elements must exhibit very low (in fact, as low as specified) resistance for in-plane compensatory movements caused, for example, by temperature changes and/or gradients.

Another interesting application of the rubber-metal laminates is for spherical compensators (washers). Conventional spherical compensating seats have their convex and concave spherical surfaces connected by a frictional contact. If the compensation of an angular misalignment has to be performed while the seat is loaded by a significant force (e.g., the weight load for compensators of machinery mounts, or the bolt preload for compensators used in dynamically loaded bolted connections), then the frictional connection results in a "dead zone" whereas small misalignments are not compensated.

The spherical compensator shown in Fig. 8.44 [11] is free from this shortcoming and has an infinite sensitivity.

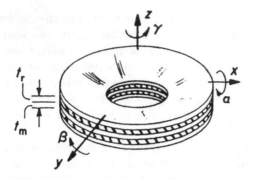

Figure 8.44 "Solid state" frictionless spherical compensator with rubber-metal laminate connection.

8.6 PARAMETERS MODIFICATION IN DYNAMIC MODELS

Dynamic performance of a mechanical system having more than a single degree of freedom depends on stiffness values of its components in a nontrivial way. The same is true for masses and damping parameters of the components. Thus, even a substantial change (e.g., increase) of stiffness (as well as of mass or damping) of some components and/or their connections in order to achieve a desirable shift in values of natural frequencies and/or vibration amplitudes might be ineffective if the "wrong" stiffness were modified. However the expenses associated with stiffness and other parameter modifications are similar for both "right" and "wrong" components being modified. To maximize effectiveness of the modifications, the role of the stiffness/inertia/damping component slated for modification must be clearly understood. This is similar to modifications of static compliance breakdowns (see Chapter 6), whereas the effectiveness of the design modification is determined by the importance of the stiffness component to be modified in the compliance breakdown. Two techniques, briefly described here, allow to modify dynamic performance of the system more effectively by "managing" the parameters being modified.

8.6.1 Evaluation of Importance of Stiffness and Inertia Components in Multi-Degrees-of-Freedom Systems

Dynamic models of real-life mechanical systems usually have many degrees of freedom, sometimes up to 100, and a corresponding number of natural frequencies and vibrating modes. However, the practically important are in most case only two to three lowest natural frequencies and modes. When these lowest natural frequencies/modes have to be modified, it is important to know which stiffness and/or inertia components of the model significantly influence the frequencies/ modes of interest and which ones do not. These conclusions are not obvious and can not be arrived at by analyzing the full model.

However the stiffness and inertia components of the dynamic model which do not influence the selected lowest natural frequencies and modes can be quickly and easily identified by using Rivin's Compression Method developed for chain-like dynamic models [43,44]. The method is based on the fact that any chain-like model (e.g., see Figs. 6.20b, c, and d) can be broken into single-degree-of-freedom partial subsystems of two types as shown in Fig. 8.45: two-mass systems (Fig. 8.45a) and single-mass systems (Fig. 8.45b). If a chain-like dynamic model is broken into such partial dynamic systems, the latter become *free body diagrams*, with the remainder of the model on both sides replaced by torques T and position angles ϕ (for a transmission system), forces and linear coordinates for a translational systems, etc.

It was shown in Rivin [43,44] that within a specified frequency range $0-f_{\lim}$ the complexity of the dynamic model can be reduced (its number of degrees of freedom reduced) without introducing significant errors in the natural modes of vibration. To achieve such "compression," some partial subsystems of type a in Fig. 8.45 have to be replaced with subsystems of type b, and vice versa. There are several conditions that must be observed for such transformation. First of all, the natural frequency of the partial subsystem to be replaced must be much higher than the higher limit $f_{\lim}$ of the frequency range of interest. Natural frequencies of the systems in Fig. 8.45 are respectively,

$$n_{ak} = \sqrt{\frac{I_k + I_{k+1}}{e_k I_k I_{k+1}}}; \qquad n_{bk} = \sqrt{\frac{e_{k-1} + e_k}{I_k e_{k-1} e_k}} \qquad (8.17)$$

If the vibratory modes below $f_{\lim}$ have to be maintained within ± 2 dB in the compressed system, then subsystems being transformed should have natural fre-

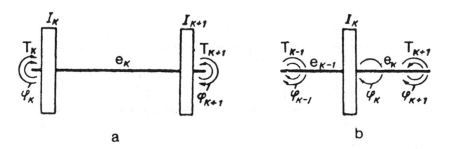

Figure 8.45   Chain-like dynamic model broken into two single-degree-of-freedom partial subsystems: (a) two-mass and (b) single-mass systems.

quencies $n \geq 3.5$–$4.0\ \omega_{\lim}$, where $\omega_{\lim} = 2\pi f_{\lim}$. In this case the natural frequencies of the compressed system will be within 2–3% of the corresponding natural frequencies of the original system. If only values of the natural frequencies in the frequency range 0–$f_{\lim}$ are of interest, they will remain within 5–10% of the corresponding natural frequencies of the original system if the partial subsystems having $n \geq 2$–$2.5\ \omega_{\lim}$ are transformed.

Figure 8.46 illustrates this compression algorithm. The initial dynamic model in Fig. 8.46a is broken into partial subsystems type a (Fig. 8.46b), and partial subsystems type b (Fig. 8.46c) (first step). Then, for each partial subsystem in Figs. 8.46b and c the value $n^2 = 1/I_k^* e_k^*$ is calculated, where $I_k^* = I_k I_{k+1}/(I_k + I_{k+1})$, $e_k^* = e_k$ for a kth subsystem type a, and $I_k^* = I_k$, $e_k^* = e_{k-1}e_k/(e_{k-1} + e_k)$ for a subsystem type b. As a next step, subsystems having $n^2 \gg \omega^2_{\lim}$ (i.e., $I^*e^* \ll \omega^2_{\lim}$) are replaced with *equivalent subsystems of the opposite type*. The

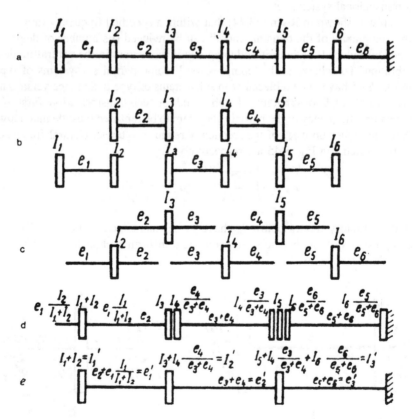

Figure 8.46 Compression algorithm: (a) initial dynamic model; (b) partial subsystem a; (c) partial subsystem b; (d) intermediate stage of the transformation; (e) final version of the "compressed" dynamic model.

equivalent subsystem of type b for the subsystem in Fig. 8.45a would have its parameters as

$$I_k' = I_k + I_{k+1}; \qquad e_{k-1}' = [I_{k+1}/(I_k + I_{k+1})]e_k; \qquad e_k' = [I_k/(I_k + I_{k+1})]e_k \qquad (8.18)$$

The equivalent subsystems of type a for the subsystem in Fig. 8.45b would have

$$I_k'' = \frac{e_k}{e_{k-1} + e_k}; \qquad I_{k+1}'' = \frac{e_{k-1}}{e_{k-1} + e_k}\ I_k; \qquad e_k'' = e_{k-1} + e_k \qquad (8.19)$$

After all possible substitutions are performed; the dynamic model is rearranged. In the Fig. 8.46 example, the first type a subsystem and the fourth and sixth type b subsystems have been replaced with the equivalent subsystems of the opposite types. The intermediate stage of the transformation is shown in Fig. 8.46d and the final version of the "compressed" dynamic model is shown in Fig. 8.46e. Instead of six degrees of freedom as in the initial model, the final model has only three degrees of freedom. However, natural modes and/or natural frequencies of both systems in the $0-f_{lim}$ range are very close if the stated above conditions for the subsystems transformation had been complied with.

If the transformed subsystem is at the end of the chain, e.g., $I_1-e_1-I_2$ or $e_5-I_6-e_6$ in Fig. 8.46a, then there appears a "residue" after the transformation. Such "residues" are represented by a "free" compliance $e_1[I_1/(I_1 + I_2)]$ or a "free" inertia $I_6[e_5/(e_5 + e_6)]$ in Fig.8.46d. These components do not participate in the vibration process and have to be abandoned. This means that the importance of the corresponding parameters in the original system (compliance e_1, moment of inertia I_6) for the dynamics of the model in Fig. 8.46a in the specified frequency range $0-f_{lim}$ is limited. It is especially so when the abandoned segment represents a substantial part of the original component.

The transformation process can be performed very quickly, even with a pocket calculator. If it demonstrates that a certain elastic (compliance) or inertia component does not "survive" the transformation, then its modification in the original system would not be effective for modification of dynamic characteristics in the specified frequency range. This compression algorithm was extended to generic (not chain-like) dynamic systems in Banakh [45].

8.6.2 Modification of Structural Parameters to Control Vibration Responses

Vibratory behavior of structures and other mechanical systems can be modified by the so-called modal synthesis based on experimental modal analysis. This approach can provide desirable resonance shifts, reduction of resonance peaks, shifting and optimal placement of nodal points, etc. However it can not identify values of such structural parameters as mass or damping, which have to be added to or subtracted from values of these parameters of the current structural components in order to achieve desired vibratory responses. This task can be realized by using the modification technique presented in Sestieri and D'Ambrogio [46].

This modification technique is based on experimentally determined frequency response functions (FRF) between selected points in the structure. After the matrix $H_0(\omega)$ of all FRF is measured, it can be modified by structural modifications at the selected points (ω is angular frequency). These modifications include adding and or subtracting stiffness, mass (inertia) and/or damping at the selected points. The matrix of such modifications can be written as $\Delta B(x \cdot \omega)$, where x = a vector of values of these modifications. The matrix of FRF of the modified system is $H(x, \omega)$ and it can be expressed as

$$H(x, \omega) = [I - H_0(\omega) \Delta B (x \cdot \omega),]^{-1} H_0(\omega) \qquad (8.20)$$

This technique [46] allows one to determine the required $\Delta B(x \cdot \omega)$, i.e., to identify the required changes in the structural parameters, in order to realize the required modified FRF matrix $H(x, \omega)$. The latter may represent the system modification in which magnitudes of selected FRF are limited (thus, vibration responses are constrained) or natural frequencies are changed in a specified manner (some increased, some reduced), etc. The importance of this technique is the fact that it results in well-defined requirements for changes in stiffness, mass, and damping parameters.

This can be demonstrated on an example of two-degrees-of-freedom system in Figure 8.47. It was assumed that the structural parameters of this system (stiffness values k_1, k_2, k_3, and mass values m_1, m_2) are not known, but only its FRF matrix (inertance matrix) is known, which also identifies its angular natural frequencies to be $\omega_1 = 364$ rad/s and $\omega_2 = 931$ rad/s. It was required to modify

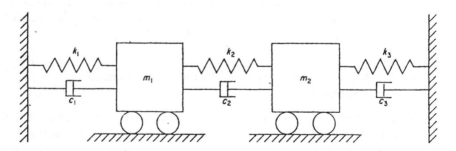

Figure 8.47   Two-degrees-of-freedom system.

this system in order to shift the natural frequencies to $\omega_1' = 300$ rad/s and $\omega_2' = 1{,}000$ rad/s. Application of this frequency-response modification technique resulted in recommending the following modifications: $\Delta m_1 = 0$; $\Delta m_2 = 1.46$ *kg*; $\Delta k_1 = 0$; $\Delta k_2 = 2\times10^5$ *N/m*; $\Delta k_3 = 0$. Thus, the solution demonstrated very different influence of various structural parameters on the structural dynamic characteristics of interest. Both stiffness and mass values must be managed to achieve the desired dynamics effects.

This technique was applied to modification of a high-speed machining center in order to reduce magnitude of amplitude-frequency characteristic of the spindles [47]. *It was found that adding stiffness, mass, or damping at the selected points would be useless.* The only recommendation from the evaluations using the algorithm from Sestieri and D'Ambrogio [46] was to install a dynamic vibration absorber on the spindle sleeve. The effectiveness of this approach was proven experimentally.

REFERENCES

1. Grigoriev, A.M., and Putvinskaya, E.I., "Rational geometric parameters of hollow supporting rollers," In: *Detali Mashin*, Tekhnika Publishing House, Kiev, 1974, No. 19, pp. 72–78 [in Russian].
2. Grigoriev, A.M., and Putvinskaya, E.I., "Contact between hollow cylinder and flat surface," In: Detali Mashin, Tekhnika Publishing House, Kiev, 1974, No. 19, pp. 79–83 [in Russian].
3. Bhateja, C.P., "A hollow roller bearing for use in precision machine tools," Annals of the CIRP, 1980, Vol. 29/1, pp. 303–308.
4. Bowen, W.L., and Bhateja, C.P., "The hollow roller bearing," ASME Paper 79-Lub-15, ASME, New York, 1979.
5. Bhateja, C.P., and Pine, R.D., "The rotational accuracy characteristics of the preloaded hollow roller bearings," ASME J. of Lubrication Technology, 1981, Vol. 103, No. 1, pp. 6–12.
6. Holo-Rol Bearings, Catalog of ZRB Bearings, Inc., Connecticut, 1997.
7. Fey, V.R., and Rivin, E.I., The Science of Innovation, The TRIZ Group, Southfield, Michigan, 1997.
8. Sawer, J.W., "Review of interesting patents on quieting reduction gears," Journal of ASNI, Inc., 1953, Vol. 65, No. 4, pp. 791–815.
9. Berestnev, O.V., Self-Aligning Gears, Nauka i tekhnika Publishing, House, Minsk, 1983 [in Russian].
10. Berestnev, O.V., et al, "Study of stress distribution in gears with artificial stress concentrators," In: *Mashinostroenie*, Vysheishaya Shkola Publishing House, Minsk, 1982, No. 7, pp. 96–100 [in Russian].
11. Rivin, E.I., "Properties and prospective applications of ultra-thin-layered rubber-metal laminates for limited travel bearings," Tribology International, Vol. 16, No. 1, pp. 17–26.

12. Rivin, E.I., "Torsional connection with radially spaced multiple flexible elements," U.S. Patent 5,630,758, 1997.
13. Rivin, E.I., "Conceptual developments in design components and machine elements," ASME Transactions, Special 50th Anniversary Design Issue, 1995, Vol. 117, pp. 33–41.
14. Rivin, E.I., "Gears having resilient coatings," U.S. Patent 4,184,380, 1980.
15. Rivin, E.I., "Conjugate gear system," U.S. Patent 4,944,196, 1990.
16. Chironis, N., "Design of Novikov gears," In: Chironis, N., ed., *Gear Design and Application*, McGraw-Hill, New York, 1967.
17. Rivin, E.I., and Wu, R.-N., "A novel concept of power transmission gear design," SAE Tech. Paper 871646, 1987.
18. Rivin, E.I., and Dong, B., "A composite gear system with separation of sliding and rolling," Proceedings of the 3rd World Congress on Gearing and Power Transmission, Paris, 1992, pp. 215–222.
19. Shigley, J.E., Mechanical Engineering Design, 3rd Edition, McGraw-Hill, New York, 1977.
20. Zvorikin, K.O., "Engagement of chain links with compliant sprocket teeth," In: *Detali Mashin*, No. 40, Tekhnika Publishing House, Kiev, 1985, pp. 3–8 [in Russian].
21. Bondarev, V.S., et al., "Study of chain drive sprockets with compliant teeth," In: *Detali Mashin*, No. 40, Tekhnika Publishing House, Kiev, 1985, pp. 8–13 [in Russian].
22. Den Hartog, J.P., Mechanical Vibrations, McGraw-Hill, New York, 1956.
23. Kelzon, A.S., Zhuravlev, Yu. N., and Yanvarev, N.V., Design of Rotational Machinery, Mashinostroenie Publishing House, Leningrad, 1977 [in Russian].
24. Kelzon, A.S., et al, "Vibration of a milling machine spindle housing with bearings of reduced static rigidity," Vibration Engineering, 1989, Vol. 3, pp. 369–372.
25. Schouten, C.H., Rosielle, P.C.J.N., and Schellekens, P.H.J., "Design of a kinematic coupling for precision applications," Precision Engineering, 1997, Vol. 20, No. 1, pp. 46–52.
26. Tsutsumi, M., et al., "Study of stiffness of tapered spindle connections," Nihon Kikai gakkai rombunsu [Trans. of the Japan. Society of Mechanical Engineers], 1985, C51(467), pp. 1629–1637 [in Japanese].
27. Rivin, E.I., "Trends in tooling for CNC machine tools: tool-spindle interfaces," ASME Manufact. Review, 1991, Vol. 4, No. 4, pp. 264–274.
28. Meyer, A., "Werkzeugspannung in Hauptspindeln für hohe Drehfrequenzen" [Holding tools in spindles rotating with high speeds], Industrie-Anzeiger, 1987, Vol. 109, No. 54, pp. 32–33 [in German].
29. Rivin, E.I., "Tool holder-spindle connection," U.S. Patent 5,322,304, 1994.
30. Braddick, H.J.J., "Mechanical Design of Laboratory Apparatus," Chapman & Hall, London, 1960
31. Gangopadhyay, A., "Friction and wear of hard thin coatings," In: *Tribology Data Handbook*, ed. by E.R. Booser, CRC Press, Boca Raton, FL, 1997.
32. Agapiou, J., Rivin, E., and Xie, C., "Toolholder/spindle interfaces for CNC machine tools," Annals of the CIRP, 1995, Vol. 44/1, pp. 383–387.
33. Rivin, E.I., "Improvements relating to tapered connections," U.S. Patent 5,595,391, 1996.

34. Rivin, E.I., and Kang, H. "Improving machining conditions for slender parts by tuned dynamic stiffness of tool," International J. of Machine Tools and Manufacture, 1989, Vol. 29, No. 3, pp. 361–376.
35. Rivin, E.I., and Xu, L., "Damping of NiTi shape memory alloys and its application for cutting tools," In: *Materials for Noise and Vibration Control*, 1994, ASME NCA-Vol. 18/DE, Vol. 80, pp. 35–41.
36. Alabuzhev, P.M., et al., Use of Constant Force Elastic Systems for Vibration Protection Devices, Vibrotechnika, Mintis Publishing House, Vilnius, 1971, No. 4(13), pp. 117–127 [in Russian].
37. Freakley, P.K., and Payne, A.R., Theory and Practice of Engineering with Rubber, Applied Science Publishers, London, 1978.
38. Rogachev, V.M., and Baklanov, V.S., Low Frequency Suspension with Stabilization of Static Position of the Object, Vestnik Mashinostroeniya, 1992, No. 5, pp. 10–11 [in Russian].
39. Tobias, S.A., Machine Tool Vibration, 1965, Blackie, London.
40. Tseitlin, Elastic Kinematic Devices, 1972, Mashinostroenie Publishing House, Leningrad [in Russian].
41. Rivin, E.I., "Noise abatement of vibration stimulated material-handling equipment," Noise Control Engineering, 1980, No. 3, pp. 132–142
42. Rivin, E.I., "Torsionally rigid misalignment compensating coupling," U.S. Patent 5,595,540, 1997.
43. Rivin, E.I., Dynamics of Machine Tool Drives, Mashinostroenie Publishing House, Moscow, 1966 [in Russian].
44. Rivin, E.I., "Computation and compression of mathematical model for a machine transmission," ASME Paper 80-DET-104, ASME, New York, 1980.
45. Banakh, L., "Reduction of degrees-of-freedom in dynamic models," Mashinovedenie, 1976, No. 3., pp. 77–83 [in Russian].
46. Sestieri, A., and D'Ambrogio, W., "A modification method for vibration control of structures," Mechanical Systems and Signal Processing, 1989, Vol. 3, No. 3, pp. 229–253.
47. Rivin, E.I., and D'Ambrogio, W., "Enhancement of dynamic quality of a machine tool using a frequency response optimization method," Mechanical Systems and Signal Processing, 1990, Vol. 4, No. 3, pp. 495–514.

Appendix 1

Single-Degree-of-Freedom Dynamic Systems with Damping

A.1 VISCOUS DAMPING

This brief description attempts to illustrate differences and special features of some of the various damping mechanisms typical for mechanical systems. A classic single-degree-of-freedom (SDOF) mechanical system in Fig. A.1.1 comprises mass m, spring k, and viscous damper c. The equation of motion of this system at free vibration condition (no external forces, $F_0 = 0$, $a_f = 0$) can be written as

$$m\ddot{y} + c\dot{y} + ky = 0 \qquad (A.1.1a)$$

where $y = $ displacement of mass m. When viscous friction in the damper is not very intense, $c < 2\sqrt{km}$, then the solution of (A.1.1) is

$$y = e^{-nt}(C_1 \sin \omega^* t + C_2 \cos \omega^* t) \qquad (A.1.2)$$

where

$$n = \frac{c}{2m}, \qquad \omega^* = \sqrt{\omega_0^2 - n^2} \qquad (A.1.3)$$

where $\omega_0 = $ natural frequency of the system without damping ($c = 0$), and constants C_1 and C_2 are determined from initial conditions $y(0) = y_0$, $\dot{y} = \dot{y}_0$ as

398

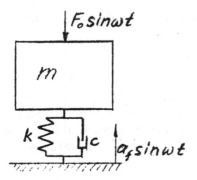

Figure A.1.1 Classic single-degree-of-freedom (SDOF) mechanical system.

$$C_1 = (\dot{y}_0 + ny_0)/\omega^*; \qquad C_2 = y_0 \tag{A.1.4}$$

Another format of the solution is

$$y = e^{-nt} \sin(\omega^* t + \beta) \tag{A.1.5}$$

where

$$A = \sqrt{\frac{(\dot{y}_0 + ny_0)^2}{\omega_0^2 - n^2} + y_0^2}; \qquad \tan \beta = \frac{y_0 \sqrt{\omega_0^2 - n^2}}{\dot{y}_0 + ny_0} \tag{A.1.6}$$

It can be seen from Eqs. (A.1.2) and (A.1.5) that variation of y in time (motion) is a decaying oscillation with a constant frequency ω^* and gradually declining amplitude (Fig. A.1.2). Parameter β is called the *loss angle* and $\tan \beta$ is the *loss factor*.

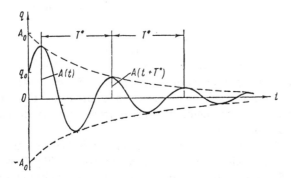

Figure A.1.2 Decaying oscillation with a constant frequency ω^* and gradually declining amplitude.

Envelopes of the decaying time history of y are described by functions

$$A = \pm A_0 e^{-nt} \tag{A.1.7}$$

where A_0 = ordinate of the envelope curve at $t = 0$.

The ratio of two consecutive peaks $A(t):A(t + T^*)$, which are separated by time interval $T^* = 2\pi/\omega^*$ (period of the vibratory process), is e^{nT^*} = constant. The natural logarithm of this ratio is called *logarithmic* (or *log*) *decrement*. It is equal to

$$\delta = nT^* = \frac{2\pi c}{\sqrt{4mk - c^2}} \approx \frac{\pi c}{\sqrt{mk}} \tag{A.1.8}$$

Thus for the same damper (*damping coefficient c*), log decrement of a system with viscous damper depends on stiffness k and mass m of the system.

For forced vibrations (excitation by a harmonic force $F = F_0 \sin \omega t$ applied to mass m), the equation of motion becomes

$$m\ddot{y} + c\dot{y} + ky = F_0 \sin \omega t \tag{A.1.1b}$$

and the amplitude of response is

$$A = \frac{F_0}{k\sqrt{\left(1 - \frac{\omega^2}{\omega_0^2}\right)^2 + \left(\frac{c}{\sqrt{mk}}\frac{\omega}{\omega_0}\right)^2}} = \frac{F_0/k}{\sqrt{\left(1 - \frac{\omega^2}{\omega_0^2}\right)^2 + \left(\frac{\delta}{\pi}\right)^2\left(\frac{\omega}{\omega_0}\right)^2}} \tag{A.1.9a}$$

Expression (A.1.9a) is plotted in Fig. A.1.3 for various δ.

Frequently, there is a need for vibration isolation. Two basic cases of vibration isolation are: (1) protection of foundation from force $F = F_0 \sin \omega t$ generated within the object (machine) represented by mass m; and (2) protection of a vibration-sensitive object (machine) represented by mass m from vibratory displacement of the foundation $a = a_f \sin \omega t$.

In the first case, the force transmitted to the foundation is $F_f \sin \omega t$ and the quality of vibration isolation is characterized by *force transmissibility* $T_F = F_f/F_0$. In the second case, the displacement transmitted to mass m is characterized by *displacement transmissibility* $T_a = a_m/a_f$. For the single-degree-of-freedom isolation system, which can be modeled by Fig. A.1.1

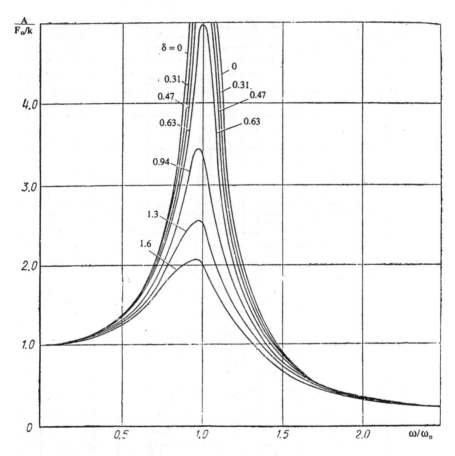

Figure A.1.3 Hysteresis loop illustrating change of deformation y for the processes of increasing (loading) and decreasing (unloading) force P.

$$T_F = T_a = \frac{\sqrt{1 + \left(\frac{\delta}{\pi}\right)^2 \left(\frac{\omega}{\omega_0}\right)^2}}{\sqrt{\left[1 + \left(\frac{\omega}{\omega_0}\right)^2\right]^2 + \left(\frac{\delta}{\pi}\right)^2 \left(\frac{\omega}{\omega_0}\right)^2}} \qquad (A.1.9b)$$

Expression (A.1.9c) is plotted by solid lines in Fig. A.1.4 for several values of δ. It is important to note that while higher damping results in decreasing transmissibility T around the resonance ($\omega = \omega_0$), it also leads to a significant deterioration of isolation (increased T) above $\omega = 1.41\,\omega_0$ (isolation range).

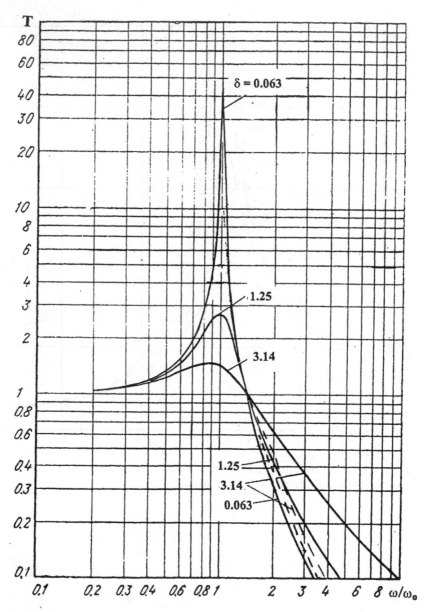

Figure A.1.4 Transmissibility of single-degree-of-freedom vibration isolation system in Fig. A.1.1 vs. frequency and damping. Solid lines indicate viscous damping; broken lines indicate hysteresis damping.

A.2 HYSTERESIS-INDUCED DAMPING

Deformation of mechanical components and joints between them (contact deformation) is not perfectly elastic. This means that deformation values during the process of increasing the external force (loading) and decreasing the external force are not the same for the same magnitudes of the external force. This effect results in developing of a *hysteresis loop* (Fig. A.1.5), which illustrates change of deformation y for the processes of increasing (loading) and decreasing (unloading) force P. The area of the hysteresis loop represents energy lost during one loading/unloading cycle. It is established by numerous tests that for the majority of structural materials, as well as for joints between components, the area of the hysteresis loop *does not strongly depend on the rate of force change* (i.e., on frequency of the loading process), but *may depend on amplitude of load/ deformation*. This statement can be formalized as an expression for the energy Ψ lost in one cycle of deformation as function of amplitude of deformation A

$$\Psi = \alpha A^{r+1} \qquad\qquad (A.1.10)$$

where α and r = constants.

To derive the law describing the decaying oscillatory process when the damping is due to hysteresis, the loss of energy during one cycle can be equated to change of system energy during one cycle of vibration. Let's consider one period of the process in Fig. A.1.2 and start the period when the displacement peaks at amplitude $A(t)$. At this moment the *kinetic energy* of the mass is $K = 0$, and all energy is stored in the spring k (Fig. A.1.1) as the *potential energy V.* In the beginning of the period

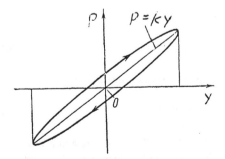

Figure A.1.5 Hysteresis loops associated with a deformation process.

$$V_0 = \frac{1}{2} kA^2(t) \tag{A.1.11}$$

At the end of the period

$$V_{T*} = \frac{1}{2} kA^2(t + T*) \tag{A.1.12}$$

The increment of the potential energy is

$$\Delta V = V_{T*} - V_0 = \frac{1}{2} k[A^2(t + T*) - A^2(t)]$$
$$= \frac{1}{2} k[A(t + T*) + A(t)][A(t + T*) - A(t)] \tag{A.1.13a}$$

The sum inside the first square brackets on the right-hand side is $\sim 2A(t) = 2A$ if the energy loss during one cycle is not very large. The difference inside the second bracket is $-\Delta A$, and

$$-\Delta V = kA\Delta A \tag{A.1.13b}$$

This increment or the potential energy is equal to the energy loss (A.1.10), or

$$-\alpha A^{r+1} = kA\Delta A \tag{A.1.14a}$$

or

$$\Delta A = -(\alpha/k)A^r \tag{A.1.14b}$$

This expression defines the shape of the upper envelope of the oscillatory process. Considering this envelope as a continuous curve $A = A(t)$, approximately

$$\Delta A = T*(dA/dt) = (2\pi/\omega*)(dA/dt) \tag{A.1.15}$$

From (A.1.15) and (A.1.14b)

$$dA/dt = -(\alpha\omega*/2\pi k)A^r \tag{A.1.16}$$

If $r = 1$, the solution of Eq. (A.1.16) is an exponential function

$$A(t) = A_0 e^{-(\alpha\omega*/2\pi k)t} \tag{A.1.17}$$

The ratio of two peak displacements $A(t)/A(t + T*) = e^{\alpha/k}$ is again constant, as in the case of viscous damper, but in this case the log decrement does not depend on mass m

$$\delta = \alpha/k \tag{A.1.18}$$

Although for $r = 1$ the log decrement does not depend on amplitude of vibration, for many real-life materials and structures value of r may deviate from 1, and then the log decrement would be changing in time with the changing vibration amplitude. The character of the change is illustrated in Fig. A.1.6 by envelopes for the vibratory process at $r = 0$ and $r = 2$, in comparison with the exponential curve (amplitude-independent log decrement) for $r = 1$. Dependence of log decrement on amplitude for fibrous and elastomeric materials is illustrated in Fig. 3.2. For many elastomeric materials (rubber blends) $r = \sim 1$.

The response amplitude for the forced vibrations for a system with hysteresis damping

$$A = \frac{F_0}{k\sqrt{\left(1 - \dfrac{\omega^2}{\omega_0^2}\right)^2 + \left(\dfrac{\alpha A^{r-1}}{\pi k}\right)^2}} \tag{A.1.19a}$$

For $r = 0$ (Coulomb friction–induced damping), for $r = 2$, and for other $r \neq 1$, the amplitude of the forced vibrations can be found after solving equation (A.1.19). However, for $r = 1$, which is typical for many rubber blends,

$$A = \frac{F_0}{k\sqrt{\left(1 - \dfrac{\omega^2}{\omega_0^2}\right)^2 + \left(\dfrac{\alpha}{\pi k}\right)^2}} = \frac{F_0/k}{\sqrt{\left(1 - \dfrac{\omega^2}{\omega_0^2}\right)^2 + \left(\dfrac{\delta}{\pi}\right)^2}} \tag{A.1.19b}$$

Equation (A.1.19b) is similar to the expression for the response to external harmonic excitation by a system with viscous friction described by Eq. (A.1.9a),

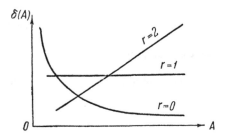

Figure A.1.6 Amplitude dependencies of log decrement δ for different damping mechanisms.

with the only difference being the second (damping) term under the radical sign. For a system with viscous friction this term is frequency dependent, while for a system with hysteresis damping it does not depend on frequency.

Transmissibility T between mass m and foundation for the force excitation of the mass $F = F_0 \sin \omega t$ or between foundation and mass m for the "kinematic" excitation of the foundation $a = a_f \sin \omega t$ for the system with the hysteretic damping when $r = 2$ is

$$T_F = T_a = \frac{\sqrt{1 + \left(\dfrac{\delta}{\pi}\right)^2}}{\sqrt{\left[1 + \left(\dfrac{\omega}{\omega_0}\right)^2\right]^2 + \left(\dfrac{\delta}{\pi}\right)^2}} \tag{A.1.19c}$$

The differences between (A.1.19c) and (A.1.9b) are due to presence in the latter and absence in the former of the frequency ratio multipliers in the second terms under the radical signs both in the numerator and in the denominator. Effect of these seemingly subtle differences is practically indistinguishable in the frequency range $\omega = 0 - 1.41 \, \omega_0$, as well as for small damping (δ) values. However, for larger damping values the difference in the isolation effectiveness (at $\omega > 1.41 \, \omega_0$) is very pronounced, as shown by broken lines in Fig. A.1.4. Although vibration isolators having significant viscous damping demonstrate deteriorating performance in the "isolation range," $\omega > 1.41 \, \omega_0$, vibration isolators with the hysteretic damping characterized by $r = 1$ provide effective performance in the "isolation range" while limiting transmissibility around the resonance frequency.

A system with hysteresis damping with the hysteresis damping can be described by Eqs. (A.1.1a) and (A.1.1b) if the damping coefficient c is replaced by a frequency- and amplitude-dependent coefficient

$$c_h = \alpha A^{r-1}/\pi\omega \tag{A.1.20a}$$

which for $r = 1$ becomes only frequency-dependent

$$c_h = \alpha/\pi\omega \tag{A.1.20b}$$

The loss factor for a system with hysteresis damping is

$$\tan \beta = \alpha A^{r-1}/\pi k(1 - \omega^2/\omega_0^2) \tag{A.1.21}$$

A.3   IMPACT DAMPING [1]

Impact interactions between mechanical components cause energy losses. Such impact interactions can occur in clearances between components in cylindrical joints or guideways, or in specially designed impact dampers. During a typical vibratory process in a system with clearances, impacts occur every time the system passes its equilibrium configuration. The amount of energy loss during impact can be presented as a function of relative velocity of the coimpacting components between the impact moment

$$\Psi = bv^2 \tag{A.1.22}$$

where b = a constant having dimension of mass.

Let's consider a half-period of oscillation that commences at a maximum displacement $A(0)$. In the first quarter of the period the mass is moving with a constant energy $\frac{1}{2}kA^2(0)$ and the square of velocity at the end of this quarter of the period is

$$v^2 = (k/m)A^2(0) \tag{A.1.23}$$

At this moment an impact occurs and, consequently, a loss of energy by the amount of Eq. (A.1.22). After the impact, the system moves with energy

$$kA^2(0)/2 - (bk/m)A^2(0) = \frac{1}{2}kA^2(0)(1 - 2b/m) \tag{A.1.24}$$

This energy remains constant during the second quarter of the period. Accordingly, at the end of the second quarter the potential energy is equal

$$\frac{1}{2}kA^2(T/2) = \frac{1}{2}kA^2(0)(1 - 2b/m) \tag{A.1.25}$$

Thus, the ratio of maximum displacements is

$$\frac{A(0)}{A\left(\dfrac{T}{2}\right)} = \frac{1}{\sqrt{1 - \dfrac{2b}{m}}} \tag{A.1.26}$$

The same ratio would materialize for the next half-period. Thus, for the whole period

$$A(0)/A(T) = 1/(1 - 2b/m) = \text{constant} \tag{A.1.27}$$

Since the ratio of sequential maximum displacement is constant, the envelope of the time plot of the decaying vibratory process is an exponential curve

$$A = A_0 e^{-nt} \qquad \text{(A.1.28)}$$

This process is associated with the log decrement

$$\delta = nt = \log_e[1/(1 - 2b/m)] = \sim\log_e(1 + 2b/m) = \sim 2b/m \qquad \text{(A.1.29)}$$

if $2b/m$ is small.

REFERENCE

1. Panovko, Ya.G., Introduction to Theory of Mechanical Vibration, Nauka Publishing House, Moscow, 1971 [in Russian].

Appendix 2

Static Stiffness Breakdown for Cylindrical (OD) Grinders

Although modal analysis of complex structures is very useful for determining weak links in complex mechanical structures, it can be complemented by measuring static deformations of various components and their connections under loads simulating working conditions of the system. Evaluation of the static compliance breakdowns allows us to better understand the role of small but important components, more readily simulate diverse working conditions/regimes, detect and study nonlinear deformation characteristics of some components that can confuse the modal analysis procedure, and more. Since measuring the compliance breakdown under static loading is much more time-consuming than the dynamic evaluation, the former should be undertaken in cases of critical importance or when the nonlinear behavior is strongly suspected. This appendix is a summary of a detailed analysis of static deformations for cylindrical OD grinders for grinding parts up to 140 mm in diameter and up to 500 mm long [1]. However, this study is rather generic, and its techniques and conclusions can be very useful for other mechanical systems.

The machine is sketched in Fig. A.2.1. The most stiffness-critical units are identified as follows. Wheelhead 3 houses spindle 1 supported in hydrodynamic bearings 2 and is mounted on a carriage comprising upper 4 and lower 8 housings. Hydraulic cylinder 6 and ball screw 15 with nut 16 are mounted in housing 8. The machined part is supported by headstock 21 and by tailstock (not shown in Fig. A.2.1) installed on angular (upper) table 19, which is attached to longitudinal (lower) table 18 moving in guideways along bed 11.

One end of ball screw 15 is supported by bracket 17 fastened to housing 4; the other end is driving pusher 13. The force between pusher 13 and ball screw 15 is adjusted by compression spring 12 and threaded plug 10, which are housed

409

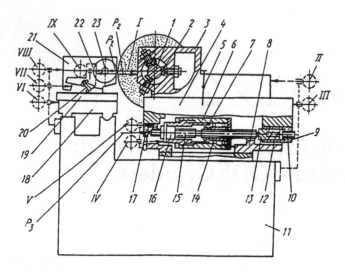

Figure A.2.1 Test forces P_1, P_2, and P_3 and measurement positions/deformation sensors I–IX for cylindrical (OD) grinder.

in bracket 9 fastened to housing 4. Thus, spring 12 generates preload (up to 1,000 N) in connection: bracket 17–ballscrew 15.

Wheelhead 3 may perform setup motions along guideways on housing 4, with its final position secured by set screw 5; fast motion (before and after machining) by piston 7 moving in hydraulic cylinder 6; and feed motion by ballscrew 15 driven by worm gear 14. During the feed motion, piston 7 is touching the left face of cylinder 6.

The principal contributors to stiffness/compliance breakdown are: spindle 1 in bearings 2; joint between wheelhead 3 and housing 6; joint between bracket 17 and screw 15; connection between ball screw 15 and ball nut 16; joint between table 18 and guideways; headstock 21 (or tailstock); and joint between supporting center 23 and housing of headstock or tailstock.

In some similar grinders the wheelhead can be installed directly on rolling friction guideways of the lower housing of the carriage, not on the upper housing as shown in Fig. A.2.1. This does not change the compliance breakdown; it is shown below that displacement of the wheelhead relative to the upper housing is only about 0.5 μm under 600 N load.

Test forces were applied (through load cells) in several locations (Fig. A.2.1): between nonrotating wheel and part (P_1); between bracket 22 attached to table 19 and wheel head (P_2); and to end face of screw 15 (P_3). The forces were varied in two ranges 0–300 N and 0–600 N. Displacement transducers were located in positions I–IX: I and II measured displacement of wheel head 3 relative to table

19 and bed 11; III measured displacement of housing 4 relative to bed 11; IV measured the joint between end face of nut 16 and housing 8; V measured the contact between end of screw 15 and housing 8; VI and VII measured displacements of tables 18 and 19 relative to bed 11; VIII measured displacement of tailstock (headstock) relative to bed 11; and IX measured displacement of supporting center relative to table 19.

Figure A.2.2 shows measuring setups for displacements of the part as well as of the headstock and tailstock and their components. Transducers X–XIII are on the tailstock side and measure displacements of part 3, supporting center 4, holder 5, mounted in sleeve 7 on balls 6, and tailstock 8 relative to upper table 9. Transducers XIV–XVII are on the headstock and measure displacement of part 3, supporting center 2, central bushing 1, and headstock 10 relative to table 9, respectively.

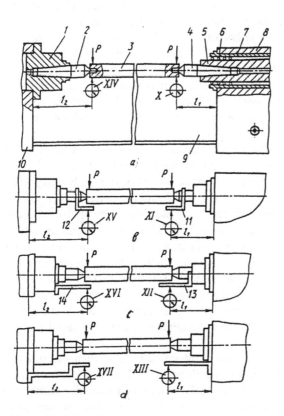

Figure A.2.2 Part support schematics used for measuring deformations of various components under radial load P; X–XVII measurement positions/deformation sensors.

Stiffness of the part as well as of components of headstock and tailstock were determined under force P applied to an end of the part from the headstock or tailstock (12 and 11 are, respectively, arm lengths of load application). To determine influence of stiffness of various components of headstock and tailstock on the total part stiffness relative to table 9, measurements were performed not only in setup of Fig. A.2.2a, but also as shown in Figs. A.2.2b, c, and d. For example, to identify influence of supporting centers stiffness relative to table 9 on the part displacements, bracket 11 or 12 was attached to front or rear center, respectively, and the load was applied to the part. Displacements of ends of these brackets were measured at distances l_1 and l_2 from the head-/tailstock. Influence of stiffness of center holder 5 relative to the table on effective stiffness of the part was measured in a similar way. To perform this measurement, bracket 13 was attached to the holder (Fig. A.2.2c) and displacements of the bracket end were recorded.

Supporting centers 2 and 4 have Morse taper #4. Depending on grinding conditions the centers could be short or long (the cylindrical part is 20 mm longer). Overhang of the center holder of the tailstock was varied during grinding within 20 mm. Distance $l_1 = 85$ mm for the short center and minimum holder overhang; 105 mm for the short center and the maximum overhang as well as for the long center and minimum overhang maximum overhang; and 125 mm for the long center and maximum overhang. Bushing 1 is fixed stationary in the headstock; thus l_2 depends only on length of center 2: $l_2 = 160$ mm for the short center and 180 mm for the long center.

Plots in Fig. A.2.3 show displacements δ vs. radial load for wheel head, tailstock and headstock, and other components. Spindle stiffness (245 N/µm) was determined from computed stiffness of its hydrodynamic bearings since the measurements were performed without spindle rotation. The broken line in Fig. A.2.3a is plotted by adding computed spindle displacements to the measured (transducer II) displacement of the wheelhead. This line represents the total effective compliance of the spindle due to compliance of all components sensing the forces from the wheelhead. These plots show that the largest contributor to compliance of the wheelhead is the ball screw/nut transmission (the distance between lines IV and V). The next contributor is spindle (the distance between the broken line and line II), the third is joint screw 15–bracket 17 (the distance between lines V and III). Other displacements/compliances are very small and can be neglected.

Stiffness of the upper and lower tables is different when measured at the tailstock and at the headstock (Fig. A.2.3b and c). This is mostly due to manufacturing imperfections of the guideways resulting in their nonuniform fit along the length and thus in stiffness variations. Since the radial force is applied at the centerline level, it also generates angular displacements of the lower table in

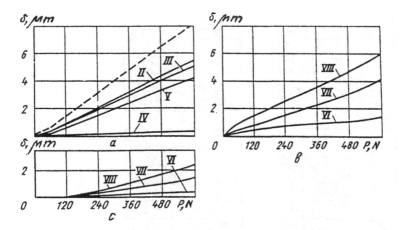

Figure A.2.3 Deformations δ (μm) of components of (a) wheelhead, (b) tailstock, and (c) headstock relative to bed under radial force *P* at different measuring points. The dashed line is the calculated displacement of spindle.

the guideways in the transverse vertical and horizontal planes. As a result, displacements of tables as well as the tailstock and headstock become uneven.

Stiffness of the headstock and tailstock depends on the attachment method of each unit to the table. The headstock is fastened by two short bolts 20 (Fig. A.2.1). The tailstock is fastened by one long bolt in the middle of its housing. This arrangment simplifies resetting of the tailstock bit reduces its stiffness. It can be concluded from lines VIII in Fig. A.2.3b and II in Fig. A.2.3a that compliance of the tailstock (taking into account also deformations of lower and upper tables) is close to compliance magnitude of the wheelhead. Figure A.2.4 gives deformations of components of both headstock and tailstock for various combinations of lengths of the centers and overhang of the sleeve. Lines X–XIII in Figs. A.2.4a–d represent deformations of the machined part, supporting center, sleeve, and tailstock, relative to the upper table measured in positions indicated in Figs. A.2.2a–d, respectively. Lines VIII in Fig. A.2.4a–d show the total displacement of the part in relation to the bed; lines II show the part in relation to the wheelhead; and the broken lines show the part in relation to the wheel spindle. Lines XIV–XVII in Figs. A.2.4d–e show displacement of the part, the supporting center sleeve, and headstock, respectively, in relation to the upper table.

These plots give an understanding of influence of deformations of each main structural component of the machine tool on the spindle-part deformations under forces up to 300 N. The data in Table 1 shows that the most compliant elements of the tailstock are the supporting center and the sleeve. Depending on the length

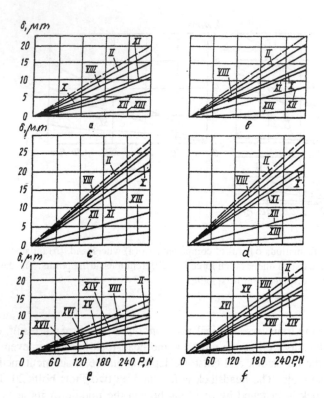

Figure A.2.4 Displacements δ (μm) of components of (a–d) tailstock and (e–f) head-stock at: (a) short support center with minimal sleeve overhang; (b) short center with maximum overhang; (c) long center with minimal overhang; (d) long center with maximum overhang; (e) short center; and (f) long center. The dashed line is spindle displacement.

Table A.2.1 Compliance Breakdown of OD Grinder

Setup of Fig. A.2.4	Spindle displacement (%)	Contribution of unit deformation (%)									
		II	VIII	X	XI	XII	XIII	XIV	XV	XVI	XVII
a	7.7	14.2	17.4	3.2	21.7	26.0	9.8	—	—	—	—
b	6.8	12.5	15.5	4.8	21.7	29.0	9.7	—	—	—	—
c	4.6	9.4	11.2	5.2	39.7	18.8	11.1	—	—	—	—
d	4.4	9.1	10.7	6.0	38.9	18.8	12.1	—	—	—	—
e	9.6	20.1	5.2	—	—	—	—	11.1	32.8	2.5	18.7
f	6.1	12.7	3.3	—	—	—	—	8.5	54.3	1.4	13.7

of the center, it is responsible for 21.7–39.7% of the total displacement of the spindle, and the sleeve is responsible for 18.8–29%.

The tailstock displacement in relation to the upper table (measuring position XIII) and in relation to the bed (position VIII), as well as displacement of the wheelhead in relation to the bed have close magnitudes. The most significant component of the headstock deformation is the supporting center (the short center accounts for 32.8%, the long center for 54.3% of the total deformation of the tailstock spindle). The spindle displacement accounts for 29.8% of the total displacement with the short center and 18.8% with the long center.

The plots in Fig. A.2.4 indicate that stiffness of the same machine tool (not considering deformations of the machined part) varies, for loading forces to 300 N, from 16.3 to 10.1 N/μm at the headstock.

Thus, the most significant component in the compliance breakdown is compliance of the supporting centers. Displacement of the loaded center is

$$\delta = \delta_b + \delta_c + \delta_a \qquad (A.2.1)$$

where δ_b = bending deformation; δ_c = radial displacement due to contact deformations; and δ_a = displacement at the load application point due to angular contact deformations. From Marcinkyavichus and Yu [1]

$$\delta_c = P(Ak^{0.5} + Bk^{0.75}); \qquad \delta_c = P(Ck^{0.25} + Dk^{0.5}); \qquad (A.2.2)$$

where P = radial force; A, B, C, and D = constants depending on the Young's modulus of the center and the sleeve, taper diameter, and distance from the load application point to the sleeve face; k = parameter depending on quality of the tapered connection sleeve = 0.1–1.0 μm-mm²/N.

For cases illustrated in Fig. A.2.4 at the force 300 N, δ_b = 1.6 μm for the short center and δ_b = 4.2 μm for the long center. Separate tests performed on the centers gave $\delta_c + \delta_a$ = 3–4 μm for the short center and 5.8–6.8 μm for the long center, and k = 0.4–0.8 μm-mm²/N. Additional tests on other machine tools and on other supporting centers resulted in values k = 0.4–1.2 μm-mm²/N. Displacements of the wheelhead relative to the upper table (measuring position I) under the force from bracket 22 to wheel head 3 (Fig. A.2.1) are close to displacement of the wheelhead relative to the bed (line II in Fig. A.2.3a) and to displacement of the tailstock relative to the bed (line VIII in Fig. A.2.3b).

To enhance the total stiffness of the grinder, the most important are stiffness values of the supporting centers, the sleeve, and the tailstock, as can be seen from Fig. A.2.4. Reducing values of parameter k can enhance stiffness of the centers. This can be achieved by improving the fit in the Morse taper connection between the center and the sleeve or by preloading this connection. The contact area in the tapered connections of the grinder on which the tests were performed was about 70–80% of the nominal contact area; it was not possible to obtain k < 0.4 µm-mm^2/N. Mutual lapping of the tapered connections was out of consideration since the centers have to be frequently replaced depending on the grinding conditions. Axial preloading of the tapered connection resulted in bulging of the sleeve (as in Fig. 4.32) and in over preloading of the balls guiding the sleeve motion. To maintain accessibility of the grinding wheel to the part, the cylindrical part and the 60° "center part" of the supporting center can be made like in the standard Morse taper #4 while the tapered seat should be dimensioned as the standard Morse taper #5. Such arrangement would reduce $\delta_c + \delta_a$ up to 2.5 times.

An analysis using data on contact deformations (see Chapter 4) has shown that 20–30% (depending on the overhang) of the sleeve deformation is due to compliance of the balls guiding the sleeve in the holder. The other 80–70% is due to bending of the sleeve inside the ball bushing. Reduction of the bending deformation can be achieved by increasing diameter of the sleeve or the number of guiding balls, or by reduction of the overhang. The first approach is unacceptable since in case of grinding of a tapered part between the centers with angular displacement of the upper table in relation to the lower table, the tailstock would interfere with the wheelhead. The third approach cannot be realized since the overhang in the present design is the minimum acceptable one. Thus, some reduction of deformations and the resulting stiffness enhancement can be achieved by increasing the number of balls and by their optimal packaging. Another approach, using rollers instead of balls, would enhance stiffness but would result in a significantly more complex and costly system.

Stiffness of the tailstock could be enhanced by using stiffer attachment to the table, e.g., by using two bolts instead of one. However, it would lengthen the setup time, which might not be desirable.

REFERENCE

1. Marcinkyavichus, A.-G. Yu, "Study of stiffness of cylindrical OD grinders," Stanki i instrument, 1991, No. 2, pp. 2–4 [in Russian].

Appendix 3

Influence of Axial Force on Beam Vibrations

The deflection equation of the beam in Fig. 7.30a subjected to a distributed load p can be written as [1]

$$\frac{d^2}{dx^2}\left(EI\frac{d^2y}{dx^2}\right) - T\frac{d^2y}{dx^2} = p \qquad (A.3.1)$$

If the distributed load p is due to inertia forces of the vibrating beam itself, then

$$p = -m\frac{\partial^2 y}{\partial t^2} \qquad (A.3.2)$$

where m = mass of the beam per unit length. If the beam has a constant cross section and uniform mass distribution, m = constant and Eq. (A.3.1) becomes the equation of free vibration

$$\frac{EI}{m}\frac{\partial^4 y}{\partial x^4} - \frac{T}{m}\frac{\partial^2 y}{\partial x^2} + \frac{\partial^2 y}{\partial t^2} = 0 \qquad (A.3.3)$$

Substituting into (A.3.3) an assumed solution with separable variables $y = X(x)U(t)$, we arrive at two equations with single variables:

$$\frac{\ddot{U}}{U} = -\omega^2 \qquad (A.3.4a)$$

$$\frac{T}{m}\frac{X^{ii}}{X} - \frac{EI}{m}\frac{X^{iv}}{X} = \omega^2 \qquad (A.3.4b)$$

The first equation is a standard equation describing vibratory motion, the second one describes the modal pattern. It can be rewritten as

$$X^{iv} - \alpha^2 X'' - k^4 X = 0 \qquad \text{(A.3.4c)}$$

where

$$\alpha^2 = T/EI; \qquad k = \sqrt[4]{\frac{m\omega^2}{EI}}$$

The solution of Eq. (A.3.4c) can be generally expressed as (e.g., [2])

$$X = C_1 \sinh r_1 x + C_2 \cosh r_1 x + C_3 \sin r_2 x + C_4 \cos r_2 x \qquad \text{(A.3.5)}$$

where

$$r_1 = \sqrt{\frac{\alpha^2}{2} + \sqrt{\frac{\alpha^4}{4} + k^4}}; \qquad r_2 = \sqrt{-\frac{\alpha^2}{2} + \sqrt{\frac{\alpha^4}{4} + k^4}} \qquad \text{(A.3.6)}$$

For the case of Fig. 7.30a of a double-supported beam, boundary conditions for $x = 0$ and $x = l$ are $X = 0$ and $X'' = 0$. Thus, $C_2 = C_4 = 0$ and

$$C_1 \sinh r_1 l + C_3 \sin r_2 l = 0; \qquad C_1 r_1^2 \sinh r_1 l - C_3 r_2^2 \sin r_2 l = 0 \quad \text{(A.3.7)}$$

Eq. (A.3.7) has nontrivial solutions if its determinant is zero or

$$(r_1^2 + r_2^2)\sinh r_1 l \sin r_2 l = 0 \qquad \text{(A.3.8a)}$$

For any magnitude of axial force $T = 0$, $r_1 l > 0$ and $r_2 l > 0$. As a result, always $\sinh r_1 l > 0$ and Eq. (A.3.8a) is equivalent to

$$\sin r_2 l = 0 \qquad \text{(A.3.8b)}$$

Since $r_2 l = 0$, Eq. (A.3.8b) is satisfied when

$$r_2 l = n\pi \ (n = 1, 2, 3, \ldots) \qquad \text{(A.3.9)}$$

Accordingly, natural frequencies ω_n can be determined from the expression

$$l\sqrt{-\frac{\alpha^2}{2} + \sqrt{\frac{\alpha^4}{4} + \frac{m\omega^2}{EI}}} = n\pi \qquad \text{(A.3.10)}$$

Transforming expression (A.3.10), an explicit expression for ω_n can be written as

$$\omega_n = \frac{n^2\pi^2}{l^2}\sqrt{\frac{EI}{m}}\sqrt{1 + \frac{Tl^2}{n^2\pi^2 EI}} \qquad \text{(A.3.11a)}$$

or

$$\omega_n = \omega_{n,\,0}\sqrt{1 + \frac{1}{n^2}\frac{T}{T_e}} \qquad \text{(A.3.11b)}$$

where $\omega_{n,\,0}$ = natural frequencies of the same beam without the axial load; T_e = Euler force for the beam; and n = order of the vibratory model. If $T < 0$ (compressive force), then all ω_n are decreasing with increasing force, and ω_1 becomes zero at T_e, when the beam buckles. If $T > 0$, then any increase in its magnitude leads to a corresponding increase in all natural frequencies of the beam. This effect is more pronounced for the lower natural frequencies, especially ω_1, due to a moderating influence of the factor $1/n^2$.

Although Eq. (A.3.11a) was derived for a double-supported beam, the generic expression (A.3.11b) seems to be valid for other supporting conditions as well.

REFERENCES

1. Rivin, E.I., Mechanical Design of Robots, McGraw-Hill, New York, 1988.
2. Craig, R.R., Structural Dynamics, John Wiley & Sons, New York, 1981.

Articles of Interest

1. Rivin, E.I., "Principles and Criteria of Vibration Isolation of Machinery," ASME Journal of Mechanical Design, 1979, Vol. 101, pp. 682–692.
2. Rivin, E.I., "Design and Application Criteria for Connecting Couplings," ASME Journal of Mechanical Design, 1986, Vol. 108, pp. 96–105.
3. Rivin, E.I., "Properties and Prospective Applications of Ultra-Thin Layered Rubber-Metal Laminates for Limited Travel Bearings," Tribology International, 1983, Vol. 18, No. 1, pp. 17–25.
4. Rivin, E.I., Karlic, P., and Kim, Y., "Improvement of Machining Conditions for Turning of Slender Parts by Application of Tensile Force," Fundamental Issues in Machining, ASME PED, 1990, Vol. 43, pp. 283–297.
5. Rivin, E.I., and Kang, H., "Enhancement of Dynamic Stability of Cantilever Tooling Structures," International Journal of Machine Tools and Manufacture, 1992, Vol. 32, No. 4, pp. 539–561.
6. Johnson, C.D., "Design of Passive Damping Systems," Transactions of the ASME, 50th Anniversary of the Design Engineering Division, 1995, Vol. 117(B), pp. 171–176.
7. Rivin, E.I., Trends in Tooling for CNC Machine Tools: Machine System Stiffness," ASME Manufacturing Review, 1991, Vol. 4, No. 4, pp. 257–263.

Principles and Criteria of Vibration Isolation of Machinery

E. I. RIVIN[1]

Principal Staff Engineer,
Ford Motor Company,
Research Staff,
Dearborn, Mich.

This paper considers a complex of problems connected with the systematic approach to vibration isolation of production machinery. After general classification of machinery and the formulation of typical features of a dynamic vibroisolation system of machinery, criteria of effective isolation for main groups of machinery are derived. Designs of isolators complying with these criteria are described.

1 Introduction

A lack of clear understanding of goals and principles of vibration isolation of real machines is often a major cause of inadequate application of machinery installation on vibration isolating mountings. The complexity of the problem is enhanced by the diversity of requirements for vibration isolation of different groups of machinery; the variety of working regimes of a particular machine as well as varying environments (e.g., the dynamic characteristics of the floor and foundation structure; the presence of vibration-producing and/or vibration-sensitive equipment close by, and so on); the degree of rigidity of the machine's bed, etc. The infinite variety of production machinery and the conditions of its use make optimal synthesis of the isolating system for a particular machine hardly practical. The appropriate way to solve this problem is to derive more or less general criteria applicable to large groups of machinery.

2 Classification of Machinery

It is useful to consider all machines by groups:

A. Vibration-sensitive machines and equipment (precision machine tools, measurement devices, etc.). The main goal of vibration isolation here is to ensure that under given external conditions relative vibrations in the working area (e.g. between the tool and work pieces) will not exceed permissible limits, e.g. to achieve desired accuracy and/or surface-finish.

B. Vibration-producing machines which apply intensive dynamic forces to the supporting structures (forging and stamping machinery, auxiliary equipment-e.g. compressors). The main goal in this case is to reduce transmission of dynamic loads to the supporting structure.

C. General machinery (e.g. machine tools of ordinary precision) which are neither very sensitive to external vibrations nor

produce excessive dynamic forces. The main goals of vibration isolation are: to facilitate installation by eliminating fastening and grouting to the floor; to protect the machine from accidental intensive external shocks and vibrations; to protect nearby sensitive equipment from occasional disturbances caused by the machine and to reduce noise and vibration level in the workshop area.

D. Machines on nonrigid structures, e.g. on upper floors or in vehicles. In this case, dynamic excitations may be amplified due to low dynamic stiffness of load-carrying structure. Thus, even ordinary machines can produce severe vibrations and likewise, the greater floor vibrations increase the need for protection even of ordinary equipment.

In all cases vibration isolation must not disturb normal performance of a machine, e.g. accuracy and production rate of machine tools.

Of course, this classification is not absolute. For example, a surface grinder is a precision machine and needs to be protected from floor vibrations. At the same time however, reversal of its heavy table produces large dynamic loads which may disturb nearby precision equipment, so the grinder is to be considered as a vibration producer.

2 Specific Features of Vibration Isolation Systems of Machinery

A. For machines installed on the floor, the natural coordinate system is one with origin at the center of gravity (c.g.) and vertical Z, transversal X and longitudinal Y axes.

The principal axes of inertia of most production machines are nearly parallel to X, Y, and Z (even for machines with asymmetrical structures). Calculations for a medium-sized lathe (asymmetrical structure) and a surface grinder (one plane of symmetry XZ) have shown that the principal axes of inertia deviate from the axes of the "natural" system only 8–10 deg. Because cos 10 deg = 0.985 ≈1.0 all products of inertia I_{xy}, I_{xz}, I_{ys} in the natural system can be assumed to be zero.

B. Spatial mass distribution in production machines is more or less identical and similar to a three-axial ellipsoid. This was confirmed by detailed computation for several machines. Thus, radii of inertia for a machine can be expressed within 10–15 per-

[1]Formerly Head of the Laboratory of Vibration Control, the All-Union Institute for Standardization in the Machine-Building Industry, Moscow, USSR).

Contributed by the Vibrations Committee of the Design Engineering Division for publication in the JOURNAL OF MECHANICAL DESIGN.

Reprinted from J. Mech. Des., Oct. 1979, Vol. 101, with permission of ASME.

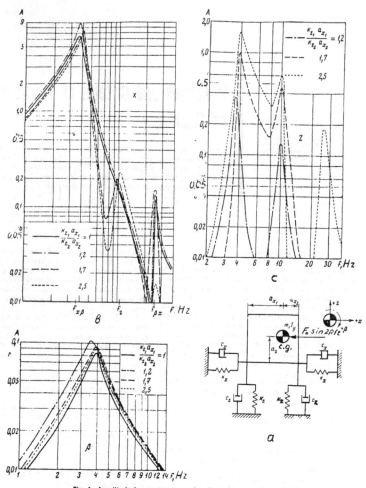

Fig. 1 Amplitude-frequency curves for displacements of c.g. in X and Z directions and rotation in X-Z plane (β) of a resiliently supported rigid body excited by harmonic force through c.g. in X-direction. (M = 2350 kg; I_y = 277kg′m²; ρ_{y_-} = $\sqrt{I_y/M}$ = 0.343 m; a_{z1} + a_{z2} = 0.7 m; a_z = 0.75 m.)

Nomenclature

a = coordinate of a mounting point (isolator) in the natural coordinate system.

a_o = averaged amplitude of floor vibrations

c/c_{cr} = damping ratio

f = frequency

$f_{\alpha\beta}; f_{y\alpha}$ = lower natural frequencies of a coupled pair of modes.

$f_{\beta z}; f_{\alpha v}$ = higher natural frequencies of a coupled pair of modes

F = force

H = overall dimension of a machine

I = moment or product of inertia

k = stiffness of an isolator

K_{dyn} = ratio between dynamic (k_{dyn}) and static (k_{st}) stiffnesses

M, m = mass

t = time

T = torque

v = velocity

W = weighting factor

γ = coupling coefficient; coordinate angle around Z axis

Δ = tolerated amplitude of relative vibrations; damping constant

δ = logarithmic decrement

Φ = criterion of isolation quality

η = ratio of two principal stiffnesses of an isolator

μ = transmissibility

ρ = radius of inertia

σ = dynamic interconnection factor

cent accuracy as:

$$\rho_i = \sqrt{1/20 \, (H_j{}^2 + H_k{}^2)}, \qquad (1)$$

where H is the overall dimension in corresponding direction, and i, j, k cyclically attain meanings of x, y, z. If heavy units are situated on the periphery of a machine, e.g., a roll of cloth in looms (a parrallelopiped-like mass distrubution), the $1/12$ should replace $1/20$ in (1).

C. It is known [1], that if a machine with one vertical plane of symmetry (a typical case) is installed on isolating mounts whose vertical and horizontal stiffnesses are proportional to the weight loads on the mounts, then oscillations along Z and around Z (angle coordinate γ) would be uncoupled and there would be two pairs of coupled coordinates: X and β (around axis Y); Y and α (around axis X) - rocking modes. In opposite case all coordinates would be coupled.

The effect of deviations from these conditions on coupling was investigated using an electric analogy simulating a plane (three degrees of freedom) vibroisolation system (Fig. 1(a)) [2]. For the parameters of a typical lathe, this analogy has been run with different degrees of asymmetry, $a_{x1}/a_{x2} = 1.0 - 2.5$ and a stiffness ratio of isolators $\eta_x = k_x/k_z = 0.5 - 10$. Results for excitation by horizontal force of unit amplitude acting through c.g. are shown in Figs. 1(b–d) (amplitude A versus frequency f) for $\eta_x = 1.0$ (with other η_x results are analogous). The following conclusions can be drawn from Fig. 1:

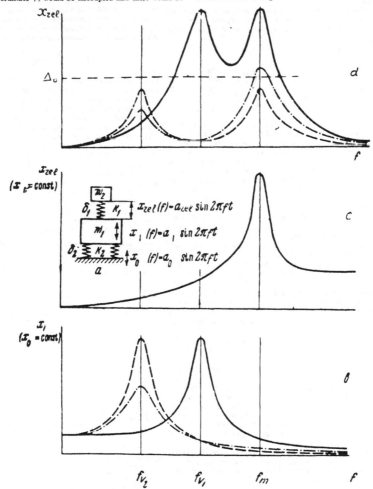

Fig. 2 Principle of isolation of vibration-sensitive equipment (a). b - vibration of the bed m_1, c - sensitivity of equipment (a) to vibrational displacements of the bed m_1; d - resulting amplitudes of displacements in the working area depending on isolators' parameters k_2, δ_2.

a. In the symmetrical case ($a_{z1} = a_{z2}$ or $k_{x1}a_{z1} = k_{x2}a_{z2}$) there are two rocking modes with frequencies $f_{x\beta}$ and $f_{\beta x}$; in asymmetrical cases, ($k_{x1}a_{z1} \neq k_{x2}a_{z2}$), there is a third resonance of rocking motion at f_z and undesirable vertical (z) motion of M at all three natural frequencies. The amplitude of the latter is 0.05; 0.12; 0.25 of maximum amplitude of X-motion at the first rocking-mode resonance ($f_{x\beta}$) when degree of asymmetry is, respectively, $(k_{x1}a_{z1})/(k_{x2}a_{z2}) = 1.2$; 1.7; 2.5. Similar coupling effects are found to occur in case of excitation by a vertical force or in case of foundation motion.

b. Response of the system at the higher rocking mode ($f_{\beta x}$) to the horizontal-force excitation is 1-2 orders of magnitude less than for lower rocking mode ($f_{x\beta}$). Opposite is true for torque excitation.

c. Natural frequencies of the plane system do not strongly depend on degree of asymmetry in the investigated range. This follows also from the analytical expression derived by method [3]:

$$\left(\frac{f_{x\beta}}{f_s}\right)_{\text{asym}} = \left(\frac{f_{x\beta}}{f_s}\right)_{\text{sym}} + \epsilon;$$

$$\epsilon = \left(\frac{a_{z2} - a_{z1}}{2a_s}\right)^2 \frac{\left[1 - \left(\frac{f_{x\beta}}{f_s}\right)^2_{\text{sym}} \eta_z\right]^2}{1 + \left[1 - \left(\frac{f_{x\beta}}{f_s}\right)^2_{\text{sym}} \eta_z\right]^2 \frac{\rho_y^2}{a_z^2}} \quad (2)$$

The conclusion *a* could be applied, for example, to precision machine tools, which typically are 2 to 3 times more sensitive to horizontal than to vertical floor vibrations, whereas the vertical floor vibrations usually are 1.5-2.0 times greater than horizontal. To be on the safe side, the resonance amplitude of horizontal vibrations of such machine tools due to vertical excitation thus must not exceed $1/(3)(2) \approx 0.16$ of the resonant amplitude of vertical vibrations. By interpolation between the previously given values, this implies a maximum permissible asymmetry value $(k_1a_{z1}) / (k_2 a_{z2}) \leq 2.0$.

For typical commercially available isolators, stiffnesses of adjacent isolators in a line differ by a factor of 1.6; i.e., the maximum error in stiffness in selecting an isolator can be $\pm \sqrt{1.6} = \pm 1.27$; tolerances on the stiffness value of an isolator with rubber resilient element are ± 5 units of durometer, or ± 17 percent. The accuracy of determining c.g. coordinates for real machines is about ± 10 percent, average variation of weight loads on some mounting points is about ± 30 percent or more due to motions of heavy parts, changing heavy blanks etc. Thus, the total variance in (ka) for a real machine installed on conventional isolators may be expected to be of the order of $1.27 \times 1.17 \times 1.1 \times 1.3 = 2.12$, which corresponds to a possible variations in asymmetry of $(k_1a_{z1}) / (k_2a_{z2}) = (2.12)^2 = 4.5$. This is substantially larger than the recommended value of 2.0. Thus, excessive coupling and inadequate isolation may necessitate the use of softer isolators (see Fig. 7 below).

This example shows that uncertainties in parameter values associated with conventional constant-stiffness isolators may lead to inadequate decoupling, even if the isolation system is selected on the basis of adequate analysis. The problem of decoupling can be overcome, however, by using "equi-frequential" isolators [2] whose stiffnesses are proportional to the weight they carry. For such an isolator, the stiffness adjusts automatically to the load acting on it, thus eliminating coupling as well as the need for cumbersome calculations.

D. The similarity of space mass distribution in real machines led to a semi-empirical formula for the natural frequency of the most important for isolation "lower" rocking mode, not containing the radius of inertia,

$$f_{x\beta} = f_s \sqrt{\frac{0.9}{\eta_x + \left(\frac{a_s}{a_z}\right)^2}}, \quad (3)$$

where $a_z = (a_{z1} + a_{z2})/2$. For the YZ plane y, α should replace x, β. The formula is accurate to within 5-10 percent for $0.25 \leq k_x/k_{z;y} \leq 4$; $0 \leq a_z/a_{z;y} \leq 1$. The formula is useful for practical calculations as well as for the synthesis of an isolation system. Approximation for the natural frequency of oscillations around axis Z is

$$f_\gamma = f_s \sqrt{\frac{5}{\eta}}; \eta = \frac{\eta_x + \eta_y}{2} \quad (4)$$

3 Isolation of Vibration-Sensitive Machinery

It was shown in [4] that maximum amplitudes of regular floor vibrations in metalworking shops averaged for a variety of plant sites are independent of frequency in a range of 3 - 35 Hz.

Assuming this and modelling a machine as a set of three single-degree-of-freedom oscillators (one for each axis), a criterion for selection isolation parameters for each axis is:

$$\Phi_1 = \frac{f_v}{\sqrt{\delta_v}} \leq \sqrt{\frac{\Delta_o f_1^2}{\pi a_o \mu_{f1}}}. \quad (5)$$

Here f_v, δ_v are natural frequency and logarithmic decrement of an isolation system for a considered mode; a_o - amplitude of floor vibrations in a corresponding direction; Δ_o - tolerated amplitude of relative vibrations in the working area of the machine; μ_{f1} - transmissibility at a frequency f_1 from the machine bed to the working area, determined experimentally or by computation.

The principle of isolation is illustrated in Fig. 2. Here m_1, m_2 are the effective masses of a bed and a tool head; k_2, δ_2—stiffness and logarithmic decrement of isolators; k_1, δ_1—effective structural stiffness and logarithmic decrement of the machine. Transmissibility of the isolation system is shown in Fig. 2(b), structural transmissibility (from the bed to the working area) in Fig. 2(c). The relative displacement vibration x_{rel} in the working area is shown in Fig. 2(d). The solid lines in Fig. 2(b, d) represent stiff mountings with moderate damping; the broken lines-compliant mountings with the same damping; the dotted lines represent the same stiffness as in the previous case but with

Table 1 Selection criterion K_{dyn}/δ for resilient materials versus vibration amplitudes

A. Rubbers

Type of rubber	No. of blend	Durometer	Amplitude of vibration, μ m		
			6	25	100
Natural	1	41	4.6	4.3	2.6
	2	56	5.4	4.2	3.0
	3	61	5.9	4.4	3.0
	4	75	5.0	3.8	2.2
Neoprene	1	42	4.6	3.85	3.0
	2	58	3.75	3.3	2.5
	3	74	3.1	2.0	1.6
	4	78	5.8	2.65	1.85
Nitrile (26%)	1	42	4.0	3.6	3.3
	2	56	3.1	2.9	2.5
	3	69	2.9	2.2	1.95
Nitrile (40%)	1	50	3.1	3.0	2.05
	2	80	2.6	2.3	1.85
Butil	1	52	2.8	2.6	2.2

B. Meshed materials

Felt "Unisorb"			15	7.0	4.0
Wire-mesh isolators	V439-0	F_o = 400N	30	9.5	1.5
"Vibrachoc" under		F_o = 1150N	45	4.8	-
different compressive loads F_o	W246-0	F_o = 870N	13	4.5	1.8
		F_o = 1150N	23	11.2	2.9
	W246-5	F_o = 2300N	27	4.1	1.55

increased damping. Increasing of isolator damping allows higher values of f_s (stiffer mountings) to be used with the same isolating effect.

A survey has shown (see [5]) that practical recommendations of natural frequencies of isolation systems by isolators manufacturers for the same precision equipment differ by 150 percent. Criterion (5) for these cases showed a scatter of only ±25 percent.

For practical resilient materials the greater the damping the larger is the ratio $K_{dyn} = k_{dyn}/k_{st}$ of effective stiffness during vibrations (k_{dyn}) to static stiffness k_{st}, and both δ and K_{dyn} depend on the amplitude of vibrations. In [5] a criterion of choice of resilient material was derived—a minimum of K_{dyn}/δ for a given amplitude range. For small amplitudes of floor vibrations, e.g., 3 - 10μm (precision machine tools), the best choice is a highly damped elastomer, for amplitudes 100-200 μm (protection of sensitive equipment in vehicles) wire mesh is better (See Table 1).

Using data on vibration sensitivity and structural dimensions of representative precision machine tools, advisable stiffness ratios $\eta_{x,y} = k_z/k_{x,y}$ of isolators were derived in [6] from (3) and (5). Typical values are $\eta_x = 0.5 - 1.0$; $\eta_y = 1.5–2.0$.

An important factor in vibration isolation of precision machinery is rigidity of the machine's bed as influenced by comparatively soft mounts. This influence and ways of enhancing the effective rigidity by proper arrangement of mounts were discussed in [4] and [7].

4 Isolation of Vibration-Producing Machines

Main cases in this group are: single-frequency or polyharmonic periodic; conservative or inertial pulse dynamic force.

A. For isolation of a machine with a *single-frequency* (f) dynamic force (e.g. created by rotating unbalanced parts) ratio $f/f_s = 4 - 5$ is usually recommended, where f_s is the vertical natural frequency of the isolation system. Thus, 15 to 25 times reduction of the dynamic force can be achieved, although with modern balancing technique this is often unnecessary. Consequently, f_s is very low: 6-7 Hz for rotors with a rotational speed of 1800 rpm, 3–4 Hz at 900 rpm, and therefore steel-spring isolators are being used. These isolators have low damping and poor performance in a high-frequency range, $f \geq 10f_s$, thus high-frequency excitations due to inaccuracies of bearings, clearances etc., are not attenuated and often even amplified. The frequency ratio can be reduced and specified according to the real circumstances, considering transmissibility in all modes of isolation systems (see Appendix 1).

The estimation of Appendix 1 was compared with experimental results on a vertical-axis hammer-crusher ($f = 12$Hz; $f_{x\beta} = f_{y\alpha} = 1.66$ Hz; $f_{\beta z} = 3.5$ Hz; $f_{\alpha y} = 4.0$ Hz, and $\mu_{F_z} = 0.02$; $\mu_{T_z} = 0.093$; $\mu_{T_y} = 0.125$); by (A1-3) with $W_F = 2W_F = 1.33$, $\mu = 0.042$. By measuring floor vibrations with and without isolators, $\mu = 0.0355$. From (A1-3) and (A1-4) with a given attenuation μ necessary natural frequencies could be easily determined. For $\mu = 0.20$ (five times attenuation, which is adequate in most cases), assuming typical values $\eta_{x,y} = 4.0$ and $f_{z\beta}$, $f_{y\alpha} = 0.45 f_z$; $f_{\alpha z}$; $f_{\beta y} = 1.2 f_z$; $f_\gamma = 1.15 f_z$, then for Z-axis rotors from (A1-3) $f/f_z \geq 2.0$; for X-axis rotors from (A1-4) $f/f_z \geq 2.6$.

B. Approach to isolation in case of *polyharmonic dynamic force* depends on a such factors as the spectral content of excitation, the absolute values of the lowest frequencies of dangerous components, the dynamic characteristics of structures and equipment to be protected, the mode of operation of the machinery to be isolated, etc. In some cases, especially with inertia blocks under the machine, when highly damped isolators (with internal damping which does not impair post-resonant transmissibility) are used, resonance (not of the most dangerous spectral component) could be a permissible working regime.

C. The principles of isolation when dynamic forces are of a "*conservative pulse*" type (e.g., stamping presses) were initially developed in [8] and updated using experimental data in [9]. The basic mode in this case is the lower rocking mode in the *X-Z* plane. All manufacturers of isolators, however, directly or indirectly recommend for installation of presses and similar equipment values of the vertical natural frequency f_z. A survey in [9] has shown that for a given capacity and weight of a press distance between its mounting holes in the *X*-direction could differ up to 2-2.5 times (see Fig. 3). Thus, real values of $f_{x\beta}$ could be very far from advisable and machine instability is the usual result. As a cure, standardization of minimal distances between mounting holes was proposed [10].

D. Machines generating *pulses of inertial nature* (e.g. forging hammers or molding machines) are the most hazardous industrial sources of vibration. Usually hammer is being installed on a massive foundation block, which is supported by a multitude of compliant steel springs. Powerful viscous damping units or large rubber cubes are being used, but damping is usually low ($\delta = 0.2$–0.3). Additionally, underanvil gaskets of oak beams or reinforced rubber belting are being used. Analytical expressions for transmissibility in this complicated system were derived using a modified "shock spectrum" technique [11]. The results of this in-

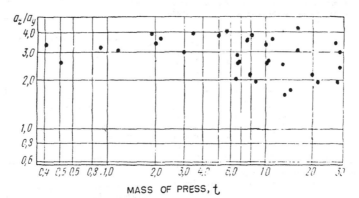

MASS OF PRESS, t

Fig. 3 Ratio of c.g. height a_z to average distance a_y between mounting holes and c.g. in Y direction for vertical stamping presses (each point represents a press model).

vestigation can be expressed in the criterion format,

$$\Phi_2 = f_v/\delta_v^{0.3} \qquad (6)$$

This expression shows that if for the lightly damped isolated foundations values of $f_v = 3 - 6$ Hz are conventional, for highly damped foundations those could be raised up to 4.5-9 Hz. It leads to less cumbersome and expensive installations. It was successfully confirmed in several cases by using rubber mats (see Section 7) instead of steel springs.

5 General Machinery

For this group, which is the largest and the most diversified one, selection of the isolation system parameters is influenced by such factors as dynamic stability of the production process and vibration level of the machine.

A. Dynamic Stability of the production, particularly cutting, process [12] could be disturbed by several mechanisms and all of them in effect introduce negative damping in a working area. Thus, when total damping balance in the structure becomes negative, chatter appears. Fastening the machine bed by means of stiff mounts to a rigid foundation improves the dynamic stiffness of the machine structure, as well as chatter resistance. On the other hand, compliant mounts (isolators) would have no such effect.

The influence of natural frequency and damping of the isolation system on the dynamic stability of the production process considered in Appendix 2 corresponds to the criterion (A2-4). Criterion $\Phi_3 = f_v{}^3\delta_v$ shows, that in this case a slight reduction of f_v (or stiffness of mounts) can be compensated only by substantially increasing the degree of damping.

An experimental verification of (A2-4) was performed on a medium-sized lathe. Chatter resistance, which was characterized by the maximum depth of cut without chatter (t_{lim}), was checked with different sets of mounting (Table 2, Fig. 4). Mountings A-1 and A-2 are the same all-metal jack mountings, and the scatter between dynamic data for two successive installations, 1 and 2, is typical for all-metal mounts. All other mountings (B-F) are equifrequential isolators with rubber resilient elements.

B. Vibration level of the machine. In some cases intensive vibrations with natural frequencies of the mounting system interfere with normal performance of the machine. A real machine has a multitude of vibration sources acting in different directions and with different frequencies; the dynamic characteristics of the building structure may also affect the vibration level. Though only an experimental selection of isolators is reliable, rough analytical criteria proposed proved to be useful in the development of optimal isolators.

a. Vibration spectra of an ordinary machine. The amplitude of the centrifugal force of an unbalanced rotating part is proportional to the second power of rotational frequency f. Tolerances on unbalance are stricter however, when f is higher. Thus the relation between f and the rated amplitudes F_{cf} of centrifugal forces for particular parts must be modified. By statistics for Russian industry, F_{cf} is proportional to $f^{0.5}$ for rotors of electric motors; to $f^{-0.5}$ for grinding wheels; to $f^{0.5} \cdot f^{-0.5} \approx f^0$ for unbalanced blanks for machining on lathes (forgings, castings and so on). As a first approximation, it can be assumed that rated amplitude of centrifugal force does not depend on frequency.

Same is often true for other sources. For example, vibration-displacement spectra of several types of machine tools, determined during idling and averaged over an ensemble can be considered as frequency independent. Thus, resonances in the system "machine on its mounts" are very probable.

b. Assuming amplitude F_o of exciting force independent on frequency,

$$(x_{m1})_{res} = \frac{F_o}{k_v}\frac{\pi}{\delta_v} = \frac{F_o}{4\pi f_v{}^2 m_1 \delta_v}, \qquad (7)$$

Table 2 Isolation systems data for experiments in lathe chatter-resistance

Mountings	f_z, Hz	δ_z	f_y, Hz	δ_y	$\bar{\Phi}_3$
A-1	45	0.39	13	0.09	63,000
A-2	75	0.9			2,160,000
B	30	0.57	10	0.67	69,000
C	24	0.5	7.8	0.5	6,900
D	20	0.9	6.3	0.88	7,200
E	17	0.42	5.4	0.44	2,140
F	12	0.38	3.2	0.30	660

Note: for cases A-1 and A-2, dynamic coupling between isolation system and a lathe dynamic system was considered for computation of $\bar{\Phi}_3$.

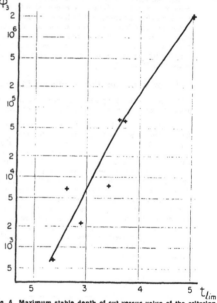

Fig. 4 Maximum stable depth of cut versus value of the criterion $\Phi_3 = f_v{}^3 \delta_v$.

where k_v, m_1 are effective stiffness of isolators and mass of this mode. Relative displacement in the working area of the machine then is:

$$(x_{rel})_{fv} = \frac{F_o}{4\pi f_v{}^2 m_1 \delta_v} \frac{1}{\dfrac{m_1}{m_1 + m_2}\dfrac{f_m{}^2}{f_v{}^2} - 1}. \qquad (7a)$$

If $f_v \ll f_m$, which is true in the case of installation on vibration isolators then

$$(x_{rel})_{fv} \approx \frac{m_1 + m_2}{m_1{}^2}\frac{F_o}{4\pi^2 f_m{}^2 \delta_v}, \qquad (7b)$$

that is, the maximum amplitudes of relative displacements in the working area, given the above assumptions, do not depend on the natural frequency of the isolation system, but only on its damping, which could be considered as a criterion,

$$\Phi_4 = \delta. \qquad (8)$$

In case of stiff mounts, when f_m and f_v are commensurable, the relative displacements would substantially increase. Fastening the machine to a massive foundation block is equivalent to increasing m_1 in (7a) with a corresponding reduction of relative vibrations.

Often the vibration-velocity level is considered as an index of the machine vibration status at frequencies higher than 5–8 Hz (e.g. [13]); an equally hazardous or annoying action of vibration on personnel also corresponds to an equal vibration-velocity level at frequencies higher than 2–8 Hz (International Standard ISO 2631). From (7), the maximum velocity amplitude

$$(v_o)_{f_v} = 2\pi f_v(x_{m1})_{f_v} = \frac{F_o}{2m_1 f_v \delta_v} \; ; \tag{9}$$

thus, a comparison of alternative mounting systems can be performed by the criterion

$$\Phi_b = f_v \delta_v \tag{10}$$

C. Experimental selection of isolators. Criterion (10) does not take into consideration such factors as dynamic data of the floor; specific features of machine vibrations in the different modes of the isolation system; dynamic data of the machine itself and isolators in a high-frequency range; real character of exciting forces; various regimes of machine performance, etc. Selection of isolators that consider these factors could give substantial advantages, e.g., a reduction of noise level. The best way of selection would be to compare several alternatives of machine installation, with specific reference to the given requirements, when each alternative is characterized by a single-rated parameter (selection criterion). Because each reinstallation of the machine is a time- and labor-consuming procedure, variable-stiffness isolators [14] can be useful.

In cases where several factors should be considered, analytical expressions for selection criterion sometimes may be used, but a common-sense judgment is often the best method. When the machine to be installed is used in different modes of operation, e.g., a machine tool with many spindle and feed speeds, obvious analytical expression for a selection criterion is

$$\Phi_\Sigma = \alpha_1 \beta_1 \Phi_1 + \ldots + \alpha_i \beta_i \Phi_i + \ldots \tag{11}$$

where Φ_i is the selection criterion for the ith mode of operation, α_i is the weighting factor attached to ith mode and β_i is a portion of full working time, during which the machine is used in the i-th mode of operation.

Common-sense judgment was used in selection of isolators for

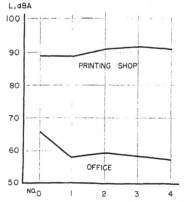

Fig. 5 Noise-level readings versus parameters of mountings for a knife-type folding machine.

a knife-type folding machine. In Fig. 5 dBA readings are shown for installation on all-metal jack mountings (0) and rubber mountings with same damping but different stiffness and f_s (10Hz - #1; 16 Hz - #2; 20 Hz - #4). Points of reference: 1) printing shop where the machine was installed; 2) office room under the printing shop. The #4 alternative was considered optimal. Note, that in this case optimal results were obtained with the isolators of maximum stiffness.

D. Rigidity of the machine bed (as it was mentioned for vibration-sensitive machinery) and *distances between mounting holes* should be considered for isolation of general machinery. Statistics for the latter factors show similar or even greater scatter as for presses (Fig. 3). Small distances between isolators lead to low values of natural frequencies for rocking modes (see (3)) which is inadvisable according to (10). The small distance in the plane of action of intensive exciters (e.g., the centrifugal force of an unbalanced machined part in the lathe) leads to large amplitudes of rocking motion and limits the performance of the machine. These distances can usually be increased to allow installation of the machine on isolators without any adverse effects.

6 Machines on Nonrigid Supports (on Upper Floors)

This problem is important for Russia, where about 25 percent of all new metal-working facilities are in multistoreyed buildings, and for some European countries. Even big machines are often installed on upper floors. Prestressed-beton structures of the floor are very strong (loads up to 10,000–20,000 N/m² are permitted), but their stiffness is low as is the damping ($\delta \approx 0.15$).

This problem was evaluated using the technique outlined in Appendix 2 [15]. A model is similar to that in Fig. 2(a), where m_2, k_2, δ_2 are in this case effective mass, stiffness and damping of the floor; k_1, c_1, m_1 are effective stiffness and damping of isolators and effective mass of the machine.

Some general considerations can be formulated:

The vibration level of the upper floor is usually 2–3 times higher than that of the ground and, accordingly, criterion $\Phi_1(\delta)$ should be 1.4-1.7 times less than it would be for the same device on the ground floor. Since a corresponding reduction of natural frequency would create instability, the proper way to comply with a reduced Φ_1 for installation of a vibration sensitive machine is to increase damping.

When a vibroactive or an ordinary machine is installed on the upper flooe, response of the latter to excitation is amplified because of small damping. Compensating it by using compliant isolators leads to instability. But, effective mass of the floor "attached" to the installed machine is substantially (usually 1.5-2.5 times) less than the mass of the machine. Thus, in a shop where several machines are installed, the dynamic characteristics, especially damping of the floor structure could be controlled by the proper choice of isolators. By using highly damped isolators it is possible to increase effective damping of the floor structure 2.5-4 times and correspondingingly reduce its sensitivity to external excitation. Thus, the main feature of an isolator to be used for an upper floor installation is high damping.

7 Vibration Isolators for Machinery

Formulation of vibration isolation requirements in the criteria format allowed to limit nomenclature of isolators.

A. Equifrequential isolators. The design of the rubber-metal equifrequential isolator [16,6], Fig. 6, is based on volumetric incompressibility of rubber and ensures a progressive reduction of the free surface area of the rubber element with increasing load. At low vertical load the free surface area is formed by the external cylindrical surface of element 1; external and internal cylindrical surfaces in a slot 3; internal cylindrical surface of the element 1; then stiffness is minimal. With an increased load, rubber is

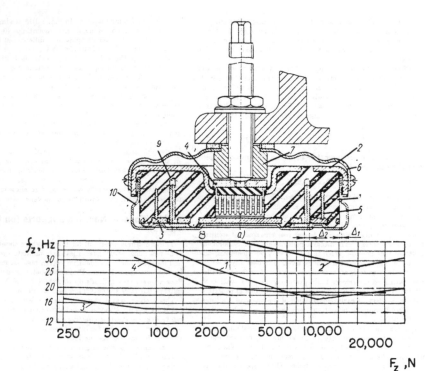

Fig. 6 Equifrequential rubber-metal vibroisolator (ERMI) by [16] and
[17]: *a*) design; *b*) load-frequency curves.

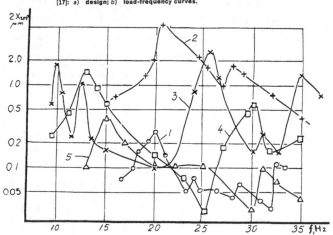

Fig. 7 Relative displacements between wheel and grinded surface
of a surface grinder when amplitude of floor vibrations is 5μm independent of frequency. Grinder is installed: 1 - on ERMI, f_z = 20 Hz;
2 - on cast-iron wedge-mounts, f_z = 27 Hz; 3 - on wire-mesh isolators
Vibrachoc (four V139-5 plus one V139-0), f_z = 25 Hz; 4 - Tico pads 1/4 in.
(0.63 cm) thickness, f_z = 30 Hz; 5 - rubber-metal isolators LM (four
LM5-20 plus one LM3-8), f_z = 15 Hz.

bulging at these surfaces. When the process of bulging is restricted by flange of upper cover 2 and by contacting internal and external surfaces in the slot 9, stiffness increases. By proper dimensioning, a proportionality between load and stiffness (equifrequential characteristic) could be achieved. Circular rib 5 ensures that transversal stiffness is also proportional to the vertical load; thus $\eta_{x,y} = 2.5 - 3.0$ in a whole load range. To comply with formulated requirements to optimal stiffnesses k_x and k_y an insert 10 could be used [17] which gives values $k_x/k_y = 1.9$, $k_z/k_x = 1.4$.

To simplify compliance of rubber with controversial requirements (low creep rate, damping, oil resistance, etc.), additional damping in the first version of the isolator has been introduced independently by a liquid damper 4. The load frequency curve 1 in Fig. 6b shows that this isolator is equifrequential in a load-range 3000–40,000N, but its working load range is 2000–40,000N, because small machines usually require higher natural frequencies.

Additional advantage of the equifrequential isolator is its low sensitivity to manufacturing tolerances. With changing durometer of rubber, the nominal frequency of the isolator does not change, but the load-frequency curve is shifting along the load axis. This shift is largely irrelevant due to the very broad load range.

This isolator became very popular in Russia for all groups of machinery. Fig. 7 shows amplitude-frequency curves for relative displacements in the working area for a surface grinder which was installed on different types of isolators (according to manufacturer's recommendations). The amplitude of vertical floor vibrations was kept in all cases equal to $5\mu m$. The superiority of the Equifrequential Rubber-Metal Isolator (ERMI) over the isolators even with lower f_s, can be explained by more perfect decoupling of vertical and horizontal vibrations. Performance of wire-mesh isolators is poor because of low amplitudes.

There also was developed a series of isolators on the same design principle with nominal frequencies 10 Hz (mainly for vibroactive equipment with harmonic dynamic force); 15 Hz (highly damped isolator, mainly for the upper floor installations-3 in Fig. 6b); 20 Hz (general purpose, mainly for precision machinery, 4 in Fig. 6); 35 Hz (general purpose, mainly for general machinery, 2 in Fig. 6(b).

B. *Rubber mats.* In addition to isolators with height adjusting devices, there is a need for a simple element whose size, stiffness, and load-carrying capacity could be easily adjusted. Thus a variety of vibration isolating mats made from rubber, cork, and other materials is used. Their common shortcoming is a very low stiffness in the horizontal plane, $\eta_{x,y} = k_x/k_{x,y} \geq 5$. Therefore, to avoid instability of an installed object, excessively high values of f_s are to be specified. With this in mind, two designs of rubber mats were developed [6]. Type 1, (Fig. 8) is isotropic, $\eta_{x,y} \approx 1.0$; its modification even has $\eta_{x,y} \approx 0.7$. It was achieved by the purposeful reduction of stiffness in the Z-direction (hollow cylindrical bosses on both sides: shear and bending of rubber between these bosses add to compression) and increasing stiffness in x,y directions (shear of the basic plate is practically eliminated or aggravated). Type 2, (Fig. 9) has higher load carrying capacity, but its main feature is a substantial anisotropy in x-y plane: $\eta_x \approx 1.3$; $\eta_y \approx 1.95$, which are close to the optimal values mentioned in Section 3.

A prototype of a rubber mat with equifrequential characteristic was also developed.

8 Conclusion

This paper described an attempt to solve the problem of vibration isolation and foundation-free installation of machinery by a systematic approach. Similarity of inertia-spatial characteristics of machinery permitted the derivation of a simplified technique for the evaluation and design of an isolation system. Such a technique was needed in view of the great variety of machines, with varied environments and modes of use.

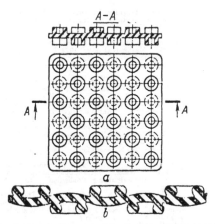

Fig. 8 Rubber mat, type 1. *a)* design; *b)* deformation under compressive load.

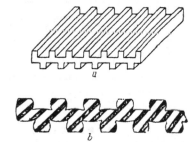

Fig. 9 Rubber mat, type 2. *a)* design; *b)* deformation under compressive load.

Generalized criteria were derived concerning main groups of machinery and thereby clarifying selection principles of isolators. These criteria show a possibility that in the majority of cases the natural frequencies of the isolation system (that is, stiffness of isolators) can be changed in the proper direction without creating adverse effects by increasing damping. The "change in the proper direction" corresponds to increasing frequencies for cases of sensitive and vibration producing machinery, and to reducing frequencies for general machinery. Such an approach has made it possible to construct a parametrical series of isolators for machinery [18]. The three major features of a good isolator were found to be: an equifrequential characteristic; high damping; a proper ratio between principal stiffnesses. Designs of the isolators developed on the basis of the achieved results possess these features.

A rational design of the bed for a machine, especially with regard to the location of mounting holes, can help in the optimization of an isolation system, particularly by increasing the effective stiffness of the machine structure and reducing vibrational displacements of the machine.

References

1 Himelblau, H., and Rubin, Sh., *Vibration of a Resiliently Supported Rigid Body*, Shock and Vibration Handbook, Vol. I, McGraw-Hill, N. Y., 1961.
2 Rivin, E. I., "Vibroisolation Systems with Equifrequential

Mountings," Izvestiya VUSov.," "Mashinostroyeniye," (Proceedings of Institutes for Higher Education. Machine-Building.), 1966, No. 3, (in Russian).

3 Banach, L. Ya., and Perminov, M. D., "Evaluation of Complex Dynamic Systems Considering Weak Coupling of Subsystems," "Mashinovedeniye," 1972, No. 4 (in Russian).

4 Kaminskaya, V. V., and Rivin, E. I., "Isolating Precision Machine Tools from Vibrations," *Machines and Tooling*, 1964, No. 11.

5 Rivin, E. I., "Review of Vibration Insulation Mountings for Machine Tools," *Machines and Tooling*, 1965, No. 8.

6 Rivin, E. I., "New Designs of Vibroisolating Mountings and Carpets," *Russian Engineering Journal*, 1967, No. 2.

7 Rivin, E. I., "Anti-Vibration Elements and Devices for Machine Tools," In "Detaly i Mekhanismy Metallorezhuschikh Stankov" ("Elements and Devices of Machine Tools"), ed. by Prof. D. Reshetov, Vol. 2, "Mashinostroyeniye" Publish. House, Moscow, 1972 (in Russian).

8 Crede, Ch. E., *Vibration ayd Shock Isolation*, J. Wiley & Sons, N. Y., 1959.

9 Rivin, E. I., "Principles of Vibroisolating Installation of Mechanical Presses," "*Kusnetchno-Shtampovotchnoye Proisvodstvo*" ("Metal-Stamping Production"), 1971, No. 10 (in Russian).

10 Verchenko, V. R., and Rivin, E. I., "The Unified System of Production Equipment Installation (on Vibroisolating Mountings)," "*Standarty i Katchestvo*" ("Standards and Quality") 1973, No. 10 (in Russian).

11 Rivin, E. I., "Design of Vibration Isolation Systems for Forging Hammers," *Sound and Vibration*, Vol. 12, No. 4, 1978.

12 Rivin, E. I., "Machine Tool Chatter Resistance as Influenced by the Installation Method," *Machines and Tooling*, 1974, No. 11.

13 Rathbone, T. A., "Proposal for Standard Vibration Limits," *Product Engineering*, 1963, Vol. 34, No. 5.

14 USSR Certificate of Authorship 312,995.

15 Rivin, E. I., "Dynamika Privoda Stankov (Dynamics of Machine Tools Drive)," Mashinostroyeniye, Publish. House, Moscow, 1966 (in Russian).

16 Rivin, E. I., Resilient Anti-Vibration Support, US Patent 3,442,475.

17 Rivin, E. I., Elastic Vibration-Proof Support, US Patent 3,460,786.

18 State Standard of the USSR "GOST" 17712-72: "Equifrequential Mountings for Machines. Parametrical Series. Technical Data."

APPENDIX 1

Isolation of a machine with unbalanced rotor. For horizontal (Y-axis) rotor, transmissibility of isolation systems is determined: for the Z-component of force (F_z) by natural frequency f_z; for F_x by natural frequency $f_{x\beta}$ of a lower rocking mode; for the Y-component of torque (T_y) by natural frequency $f_{\alpha y}$ of a higher rocking mode; for T_z by natural frequency f_γ. For the Z-axis rotor, accordingly, the correspondence is: $F_z - f_{z\beta}$; $F_y - f_{y\alpha}$ (lower rocking mode); $T_z - f_{\alpha y}$, $T_y - f_{\beta z}$. Modal coupling is not considered.

The effectiveness of isolation depends on attenuation of all components of force and torque; orthogonal components of both are phase-shifted by $\pi/2$. For a Z-axis rotor, if F_o is the centrifugal force amplitude, then

$$F_x = F_o \sin 2\pi f t; \quad F_y = F_o \sin\left(2\pi f t - \frac{\pi}{2}\right);$$

$$F'_x = F'_{x_o} \sin 2\pi f t; \quad F'_y = F'_{y_o} \sin\left(2\pi f t - \frac{\pi}{2}\right)$$

$$\mu_{F_x} = \frac{F'_{x_o}}{F_o}; \quad \mu_{F_y} = \frac{F'_{y_o}}{F_o};$$

$$\mu_F = \frac{\sqrt{(F'_{x_o}\sin 2\pi f t)^2 + (F'_{y_o}\cos 2\pi f t)^2}}{F_o}, \quad \text{(A1-1)}$$

where (') denotes components on the output side of the isolators; μ_{F_x}, μ_{F_y} are transmissibilities along respective axes; μ_F is overall force transmissibility. The amplitude of $(A\sin\phi)^2 + (B\cos\phi)^2$ is equal to max (A^2, B^2); therefore transmissibility

$$\mu_F = \frac{\max (F'_{x_o}, F'_{y_o})}{F} = \max (\mu_{F_x}, \mu_{F_y}). \quad \text{(A1-2)}$$

An analogous expression could be derived for the torque (μ_T). Thus, the effectiveness of isolation is determined by transmissibility in only two modes - the highest ones for force and for torque. For "effective transmissibility" a weighted average of both could be used. For the Z-axis and Y-axis rotors, respectively

$$\mu = \frac{W_F \mu_F + W_T \mu_T}{2}$$

$$= \frac{W_F \max(\mu_{F_z}, \mu_{F_y}) + W_T \max(\mu_{T_z}, \mu_{T_y})}{2} \quad \text{(A1-3)}$$

$$\mu = \frac{W_F \max(\mu_{F_y}, \mu_{F_z}) + W_T \max(\mu_{T_z}, \mu_{T_y})}{2}, \quad \text{(A1-4)}$$

where weighting factors $W_F + W_T = 2$. In general, transmission of torque is less dangerous since, with widely spaced isolators, forces creating output torque are small; with narrow spaced isolators action of the torque on the floor is local. Thus, in most cases $W_F = 2W_T$ could be assumed. Formulas (A1-3) and (A1-4) give conservative values since isolators reactions induced by forces and torques should be added up as vector quantities.

APPENDIX 2

Derivation of the criterion for dyanmic stability of a machine tool [12]. Influence of the vibration isolation system on the dynamic stability of the machine tool can be easily estimated considering coupling between dynamic systems of the machine itself (m_1), $m_2 - k_1 - m_1$ and the isolation system (ν), $m_1 - k_2$ in the two-mass model in Fig. 2(a). Damping constant $\Delta_2 = 2\pi f_2 \, (c/c_{cr})_2$ of the higher natural mode of the full system m_2-k_1-m_1-k_2 can be expressed in terms of damping constants in partial subsystems $\Delta_m = 2\pi f_m \, (c/c_{cr})_m$, $\Delta_\nu = 2\pi f_\nu \, (c/c_{cr})_\nu$ and dynamic interconnection factor [15]

$$\sigma = 2\gamma \frac{f_m f_\nu}{f_m^2 - f_\nu^2}, \quad \text{(A2-1)}$$

where γ is a coupling coefficient (in this case-inertia coupling coefficient $\gamma = \sqrt{m_2/m_1 + m_2}$; $f_m = 1/2\pi \sqrt{k_1 \, (m_1 + m_2/m_1 m_2)}$, $f_\nu = 1/2\pi \sqrt{k_2 \, (1/m_1)}$ are natural frequencies of respective partial subsystems. When $\sigma < 1$, then [15]

$$\Delta_2 = 1/2 \frac{(1 + \sqrt{1 + \sigma^2})\Delta_m - (1 - \sqrt{1 + \sigma^2})\Delta_\nu}{\sqrt{1 + \sigma^2}} \quad \text{(A2-2)}$$

Usually (e.g., for lathes and milling machines) $f_m \gg f_\nu$, then

$$\sigma^2 = 4\gamma^2 \frac{f_m^2 f_\nu^2}{f_m^2 - f_\nu^2} \approx 4\gamma^2 \frac{f_m^2 f_\nu^2}{f_m^4}$$

$$= 4 \frac{m_2}{m_1 + m_2} \frac{f_\nu^2}{f_m^2} < 1 \quad \text{(A2-1a)}$$

and

$$\Delta_2 \approx 1/2 \left[\left(1 + \frac{1}{1 + \frac{\sigma^2}{2}} \right) \Delta_m + \frac{\frac{\sigma^2}{2}}{1 + \frac{\sigma^2}{2}} \Delta_\nu \right] \quad \text{(A2-2b)}$$

At the stability limit, stablizing effect of structural damping is compensated by destabilizing effect of a cutting process so that total $\Delta_m \approx 0$. Then from (A3-1a) and (A3-2a), assuming $f_2 \approx f_m$ because of small σ [15] and introducing $(c/c_{cr})_\nu = \delta_\nu/2\pi$, where δ_ν is logarithmic decrement of the isolation system (or, in case of elastomeric isolators, of isolators themselves),

$$\left(\frac{c}{c_{cr}}\right)_2 \approx \frac{\sigma^2}{4}\left(\frac{c}{c_{cr}}\right)_v \frac{f_v}{f_m} = \frac{m_2}{m_1+m_2}\left(\frac{c}{c_{cr}}\right)_v \frac{f_v^3}{f_m^3}$$

$$= \frac{1}{2\pi}\frac{m_2}{m_1+m_2}\frac{f_v^3}{f_m^3}\delta_v. \qquad (A2\text{-}3)$$

It can be concluded from (A2-3) that the influence of mounting elements on chatter resistance is substantial only if the natural frequency of the machine structure (f_m) is not high enough or its design is poor (great value of $m_2/m_1 + m_2$). Of course, all relevant parameters ($m_2/m_1 + m_2$, f_m, $(c/c_{cr})_m$) could differ for different conditions of machining (e.g., position of support on the bed). From (A2-3) influence of mountings on chatter-resistance is determined by the criterion

$$\Phi_s = f_v^3\delta_v. \qquad (A2\text{-}4)$$

Design and Application Criteria for Connecting Couplings

E. I. Rivin

Professor,
Department of Mechanical Engineering,
Wayne State University,
Detroit, Mich. 48202
Mem. ASME

A classification of couplings as rigid, misalignment-compensating, torsionally flexible, and combination purpose is proposed. Selection criteria for two basic subclasses of misalignment-compensating couplings are derived, some standard designs are analyzed, and modifications of Oldham and gear couplings in which compensatory motion is accommodated by internal shear in thin elastomeric layers instead of reciprocal sliding are described. The new designs demonstrate very high efficiency and exert substantially reduced forces on the connected shafts. Factors determining the influence of torsionally flexible couplings on transmission dynamics are formulated as reduction of torsional stiffness, enhancement of damping, modification of nonlinearity, and inertia distribution. Compensation properties of combination purpose couplings are investigated analytically and a "design index" is introduced. A comparison of important characteristics of some commercially available types of combination purpose couplings is performed, to facilitate an intelligent comparison and selection of various coupling types. A line of approach for the improvement of torsionally flexible/combination purpose couplings by using highly nonlinear elastomeric elements is suggested.

1 Introduction

As stated in the Resolution of the First International Conference on Flexible Couplings [1], ". . . a flexible coupling, although it is relatively small and cheap compared to the machines it connects, is a critical aspect of any shaft system and a good deal of attention must be paid to its choice at the design stage." Everyone seems to agree with this statement. However, technical literature on connecting couplings is scarce and is dominated by trade publications or promotional literature originating with coupling manufacturers. Although couplings frequently are the critical components in a mechanical design, many textbooks on machine elements skip the issue. For example, Spotts [2] has just one short paragraph on couplings, referring the reader for further information to manufacturers' catalogs; Burr [3] mentions couplings in four locations but does not spend a single paragraph on the principles of their design; Shigley [4] does not even mention flexible couplings; the list can be extended.

On the other hand, there are probably more patents issued on various coupling designs than on any other machine element. This, together with a large variety of commercially available couplings and a lack of evaluation criteria, makes couplings one of the technically weakest links in mechanical systems.

Additionally, many coupling designs use elastomers in complex loading modes; have friction joints with limited travel distances in the joints, comparable to elastic deformations of coupling components; have severe limitations on

size and rotational inertia, etc. These factors make coupling design a very difficult task which can be helped by a more clear understanding of coupling functions.

The purpose of this paper is to formulate distinctly various couplings roles in machine transmissions, as well as to formulate criteria for comparative assessment of coupling designs and to show some ways toward coupling design optimization. To achieve these goals, a classification of connecting couplings is given and comparative analyses of some commercially available couplings are performed. Small and medium-size couplings are considered for the most part, although many conclusions are valid for both miniature and large couplings.

2 General Classification of Couplings

According to their role in transmissions, couplings can be divided in four classes:

1. Rigid Couplings. These couplings are used for rigid connection of precisely aligned shafts. Besides torque, they also transmit bending moment and shear force if any misalignment is presented, as well as axial force. The three latter factors cause substantial extra loading of the shaft bearings. The principal areas of application-long shafting; very tight space preventing use of misalignment-compensating or torsionally flexible couplings; inadequate durability and/or reliability of other types of couplings.

2. Misalignment-Compensating Couplings. Such couplings are required for connecting two members of a power-transmission or motion-transmission system that are not perfectly aligned. "Misalignment" means that components that are coaxial by design are not actually coaxial, due either

Contributed by the Power Transmission and Gearing Committee and presented at the Design Engineering Technical Conference, Cambridge, Mass., October 7–10, 1984 of THE AMERICAN SOCIETY OF MECHANICAL ENGINEERS. Manuscript received at ASME Headquarters, July 10, 1984. Paper No. 84-DET-97.

Reprinted from J. Mechanisms, Transmission, and Automation in Des., March 1986, Vol. 108, with permission of ASME.

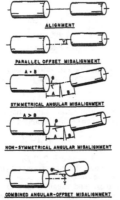

ALIGNMENT

PARALLEL OFFSET MISALIGNMENT

A = B

SYMMETRICAL ANGULAR MISALIGNMENT

A > B

NON-SYMMETRICAL ANGULAR MISALIGNMENT

COMBINED ANGULAR-OFFSET MISALIGNMENT

Fig. 1   Shaft misalignment conditions (from [6])

to assembly errors or to deformations of subunits and/or their foundations. The latter factor is of substantial importance for large turbine installations (thermal/creep deformations leading to drastic load redistribution between the bearings [5]) and for transmission systems on nonrigid foundations (such as ship propulsion systems).

Various types of misalignment as they are defined in AGMA Standard 510.02 [6] are shown in Fig. 1. If the misaligned shafts are rigidly connected, this leads to elastic deformations, and thus to dynamic loads on bearings, vibrations, increased friction losses in power transmission systems, and unwanted friction forces in motion transmission, especially control systems.

Misalignment-compensating couplings are used to reduce the effects of imperfect alignment by allowing nonrestricted or partially restricted motion between the connected shaft ends. Similar coupling designs are sometimes used to change bending natural frequencies/modes of long shafts.

When only misalignment compensation is required, rigidity in torsional direction is usually a positive factor, otherwise the dynamic characteristics of the transmission system might be distorted. To achieve high torsional rigidity together with high compliance in misalignment directions (radial or parallel offset, axial, angular), torsional and misalignment-compensating displacements in the coupling have to be separated by using an intermediate compensating member. Frequently, torsionally rigid "misalignment-compensating" couplings, such as gear couplings, are referred to in the trade literature as "flexible" couplings.

3. Torsionally Flexible Couplings. Such couplings are used to change the dynamic characteristics of a transmission system, such as natural frequency, damping, and character/degree of nonlinearity. The change is desirable or necessary when severe torsional vibrations are likely to develop in the transmission system, leading to dynamic overloads in power-transmission systems.

Torsionally flexible couplings usually demonstrate high torsional compliance to enhance their influence on transmission dynamics.

4. Combination Purpose Couplings are required to possess both compensating ability and torsional flexibility. The majority of the commercially available connecting couplings belong to this group.

This paper is dealing with couplings representing classes 2, 3, 4.

3   Misalignment-Compensating Couplings

As stated in Section 2, misalignment-compensating couplings have to reduce forces caused by an imperfect alignment of connected rotating members (shafts). Since components designed to transmit higher payloads usually can tolerate higher misalignment-caused loads, a ratio between the load generated in the basic misalignment direction (radial or angular) to the payload (rated torque or tangential force) seems to be a natural design criterion for purely misalignment-compensating couplings. For torsionally flexible and combination purpose couplings, torsional stiffness is usually an indicator of payload capacity. In such cases, the basic design criterion can be formulated as a ratio between the stiffness in the basic misalignment direction and the torsional stiffness. In this paper, only radial misalignment is considered.

Selection Criteria for Misalignment-Compensating Couplings. All known designs of misalignment-compensating (torsionally rigid) couplings belonging to Class 2 are characterized by the presence of an intermediate member located between the hubs attached to the shafts being connected. The compensating member has mobility relative to both hubs. The compensating member can be solid or composed of several links. There are two basic designs subclasses:

(*A*) Couplings in which the displacements between the hubs and the compensating member have a frictional character (examples: Oldham coupling, Cardan joint, gear coupling).

(*B*) Couplings in which the displacements are due to elastic deformations in special elastic connectors (e.g., Bendix Flexural Pivot [7], Alsthom Coupling [8], modified Oldham coupling described in the following).

For Subclass (*A*), the radial force F_{com}, acting from one hub to another and caused by misalignment, is a friction force equal to the product of friction coefficient μ and tangential force F_t at an effective radius R_{ef}, $F_t = T/R_{ef}$, where T is transmitted torque,

$$F_{com} = \frac{\mu T}{R_{ef}}. \qquad (1)$$

Since motions between the hubs and the compensating member are of a "stick-slip" character with very short displacements alternating with stoppages and reversals, μ might be assumed to be the static friction coefficient.

When the rated torque T_r is transmitted, then the selection criterion is

$$\frac{F_{com}}{T_r} = \frac{\mu}{R_{ef}}, \qquad (2)$$

or the ratio representing the selection criterion does not depend on the amount of misalignment; lower friction and/or larger effective radius would lead to lower forces.

For Subclass (*B*), assuming linearity of the elastic connectors,

$$F_{com} = k_{com} e, \qquad (3)$$

where e is misalignment value, k_{com} = combined stiffness of the elastic connectors. In this case,

$$\frac{F_{com}}{T_r} = \frac{k_{com}}{T_r} e. \qquad (4)$$

Unlike couplings from Subclass (*A*), Subclass (*B*) couplings develop the same radial force for a given misalignment regardless of transmitted torque; thus they are more effective for larger T_r. Of course, lower stiffness of the elastic connectors would lead to lower radial forces.

Designs of Misalignment-Compensating Couplings.

Misalignment-compensating couplings are used in cases where a significant torsional compliance can be an undesirable factor and/or a large allowable misalignment is required. More sophisticated Cardan joints (requiring long intermediate shafts) and linkage couplings are not frequently used, due to the specific characteristics of general-purpose machinery, such as limited space, limited amount of misalignment to compensate for, and cost considerations. The most frequently used couplings of this class are gear and Oldham couplings.

Conventional Oldham and Gear Couplings (Subclass A). Both Oldham and gear couplings compensate for misalignment of the connected shafts by means of sliding between the hub surfaces and their counterpart surfaces on the intermediate member. The sliding has a cyclical character, with double amplitude of displacement equal to radial misalignment e for an Oldham coupling and $D_p\theta$ for a gear coupling [9], where D_p is the pitch diameter and θ = angular misalignment. (If a radial misalignment e has to be compensated, then $\theta = e/L$, where L is distance between the two gears, or the sleeve length.) Such a motion pattern is not conducive to good lubrication since at the ends, where velocity is zero, a metal-to-metal contact is very probable. Stoppages are thus associated with increasing friction coefficients close to static friction values. This is the case for low-speed gear couplings [9] and Oldham couplings; for high-speed gear couplings the high lubricant pressure due to centrifugal forces alleviates the problem [9].

For the Oldham coupling, radial force from one side of the coupling (one hub-to-intermediate member connection) is a rotating vector with magnitude

$$F_1 = \mu \frac{T}{R_{ef}},\tag{5}$$

whose direction reverses abruptly twice during a revolution. The other side of the coupling generates another radial force of the same magnitude, shifted 90 deg. Accordingly, the magnitude of the resultant force is

$$F_r = \sqrt{2}\mu \frac{T}{R_{ef}} = \sqrt{2}\,\mu \frac{T}{R_{ef}};\tag{6}$$

its direction changes abruptly four times per revolution. Similar effects occur in gear couplings.

The frequent stoppages and jumps in direction of forces lead to the high noise levels generated by Oldham and gear couplings. As shown in [10], the gear coupling is the noisiest component of a large power-generation system; in our experiments, an Oldham coupling with $T_r = 150$ Nm, external diameter $D_{ex} = 0.12$ m, $e = 1$ mm, $n = 1450$ rpm generated $L_{eq} = 96$ dBA.

Due to inevitable backlashes and, associated with them, nonuniform contact loading, and also due to very poor lubrication conditions at stick-slip motion with high contact loading, friction coefficients in gear and Oldham couplings are very high, especially in the latter. Experimental data for gear couplings show $\mu = 0.3$ [11] or even $\mu = 0.4$ [12].

Assuming $\mu = 0.4$, for an Oldham coupling with rated torque $T_r = 150$ Nm, $D_{ex} = 0.12$ m and $D_{ef} = 0.8\,D_{ex}$ [14], the maximum radial force on the connected shafts according to (6) is 881 N, which is reasonably close to our experimental value 720 N.

Because of such high forces, deformations of the transmission system can be very substantial. If the deformations become equal to the existing misalignment, then no sliding will occur and the coupling behaves as a solid structure, being cemented by static friction forces [12]. It was concluded in [15] that this happens at misalignments below $e = 10^{-3}\,D_0$. This effect seems to be one of the reasons for the trend toward replacing misalignment-compensating couplings with rigid connectors in power generating systems.

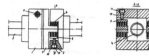

Fig. 2 Oldham coupling with thin elastomeric laminates. 1,2 = hubs; 3,4 = connected shafts; 5 = intermediate member; 6 = laminates; 9 = lips for laminate preloading; 11 = preloading screws.

Due to internal sliding with high friction, Oldham and gear couplings demonstrate substantial energy losses. Thus, the efficiency of an Oldham coupling for $e/D_0 \lesssim 0.04$ [13]

$$\eta \sim 1 - 10\,\frac{\mu}{\pi}\,\frac{e}{D_0},\tag{7}$$

or for $\mu = 0.4$ and $e = 0.01\,D_0$, $\eta = 0.987$. Similar (slightly better due to better lubrication) efficiency is characteristic for single gear connections.

A typical (and usually negative) feature of Subclass (A) couplings is backlash which is necessary to accommodate thermal distortions of the sliding pars and which increases as wear progresses.

Modified Oldham and Gear Couplings (Subclass B). The basic disadvantages of conventional Oldham and gear couplings (high radial forces, jumps in direction of the radial force, energy losses, backlash, nonperformance at small misalignments, noise) are all associated with reciprocal, short travel, poorly lubricated sliding motion in the hub-intermediate member connections. There are several known techniques of changing friction conditions.

Using rolling friction would greatly reduce friction (and radial) forces. However, roller bearings do not perform well in reciprocal motions of small amplitudes. There are several known designs of Oldham-type couplings with rolling friction for low rated torques (e.g., servo-control system applications). This concept can hardly be used for gear couplings.

Another possible option is using hydrostatic lubrication. This technique is widely used for rectilinear guideways, journal and thrust bearings, screw and worm mechanisms, etc. However, this technique seems to be impractical for rotating systems with high loading intensity (and thus high required oil pressures).

The most promising approach to design optimization of misalignment-compensating couplings seems to be the application of thin layered rubber-metal laminates [16], which demonstrate extremely high anisotropy: very high stiffness/load-carrying capacity in compression together with very low (2–3.5 decimal orders of magnitude lower) stiffness in shear. Such properties, investigated in detail in [16], are naturally suited for applications in misalignment-compensating couplings.

A design of an Oldham-type coupling using thin layered rubber-metal laminates is shown in Fig. 2 [17]. In this design, laminates 6 are installed between the intermediate member 5 and hubs 1, 2. The laminates are preloaded with bolts 11 to eliminate backlash, enhance uniformity of stress distribution and increase torsional stiffness.

In the coupling in Fig. 2, there is no actual sliding between the contacting surfaces; thus the expensive surface preparation necessary in conventional Oldham and gear couplings (heat treatment, high-finish machining, etc.) is not required.

To derive an expression for the efficiency of the Oldham coupling with laminated connections, let the shear stiffness of the connection between one hub and the intermediate member be denoted by k_{sh}, and the relative energy dissipation in the rubber for one cycle of shear deformation by ψ. Then,

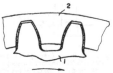

Fig. 3 Gear coupling with elastomeric coating on both profiles. 1 = gear; 2 = sleeve.

maximum potential energy in the connection (at maximum shear *e*) is equal to:

$$V_1 = k_{sh} \frac{e^2}{2} \qquad (8)$$

and energy dissipated per cycle of deformation is

$$\Delta V_1 = \psi \, k_{sh} \frac{e^2}{2} \qquad (9)$$

Each of two connections experiences two deformation cycles per revolution; thus total energy dissipated per revolution of the coupling is

$$\Delta V = 2 \times 2\Delta V_1 = 2\psi \, k_{sh} e^2 \qquad (10)$$

Total energy transmitted through the coupling per revolution is equal to

$$W = P_t \times \pi D = 2\pi T \qquad (11)$$

where P_t is tangential force reduced to the external diameter D and T is the transmitted torque. The efficiency of a coupling is therefore equal to:

$$\eta = 1 - \frac{\Delta V}{W} = 1 - \frac{\psi k_{sh} e^2}{\pi T} \qquad (12)$$

For the experimentally tested coupling ($D_{ex} = 0.12$ m), the parameters are:

$$\psi = 0.2; k_{sh} = 1.8 \times 10^5 \text{ N/m}; T = 150 \text{ Nm};$$

$$e = 1 \text{ mm} = 0.001 \text{ m};$$

thus

$$\eta = 1 - \frac{0.2 \times 1.8 \times 10^5 \times 10^{-6}}{\pi \times 150} = 1 - 0.75 \times 10^{-4} = 0.999925 \qquad (12a)$$

or losses at full torque are reduced 200 times compared to the conventional coupling.

Test results for conventional and modified Oldham couplings (both with $D_{ex} = 0.12$ m, a laminate with rubber layers 2 mm in thickness) showed that the maximum transmitted torque was the same but there was a 3.5 times reduction in radial force transmitted to the shaft bearings with the modified coupling. Actually, the coupling showed the lowest radial force for a given misalignment compared with any compensating coupling, including couplings with rubber elements. In addition to this, noise level at the coupling was reduced 13 dBA to $L_{eq} = 83$ dBA. Using ultrathin-layered laminates for the same coupling would further increase its rating by about one order of magnitude, and would therefore even require a redesign of the hardware to accommodate such a high transmitted load in a very small coupling.

A similar concept can be applied to gear couplings [18]; see Fig. 3. Again, the torque is transmitted via compression of a thin elastomeric layer; compensation is achieved via shear in this layer; direct contact between the intermediate member and the hubs is eliminated, as well as any need for lubrication.

Since allowable compression loading for the thin elastomeric layers (up to 100–300 MPa) is much higher than allowable contact loading in conventional gear couplings, the number of teeth and/or the coupling diameter can be reduced as compared with conventional couplings. The allowable amplitude of relative shear deformation of a rubber part is

0.5–1.0, depending on the rubber blend. For a gear coupling, relative displacement between the meshing teeth is

$$\pm (D_p)/(2) \, \theta,$$

where D_p is pitch diameter, θ = misalignment angle, rad. Accordingly, the thickness of the rubber coating has to be $(0.25{-}0.5) \, D_p \theta$. For a typical $\theta = 0.025$ rad, and $D_p = 100$ mm, thickness would be 0.63–1.25 mm. The efficiency of such coupling would be very close to 1.0, especially for large transmitted loads, similar to (12).

4 Torsionally Flexible Couplings and Combination Purpose Couplings

These two classes of couplings are usually represented by the same designs. However, in some cases only torsional properties are required, in other cases both torsional and compensation properties are important and, most frequently, these coupling designs are used as the cheapest available and users do not have any understanding of what is important for their applications. Accordingly, it is of interest to analyze what design parameters are important for various applications. The requirements for "torsionally flexible" and "combination purpose" couplings are considered separately, and then an analytical survey of some commercially available designs is performed and directions for an improved design are suggested.

Torsionally Flexible Couplings. Torsionally flexible couplings are used in transmission systems when there is a danger of developing resonance conditions and/or transient dynamic overloads [19, 20]. Their influence on transmission dynamics can be due to one or more of the following factors.

Reduction of Torsional Stiffness and, Consequently, Shift of Natural Frequencies. If a resonance condition occurs before installation (or change) of the coupling, then shifting of natural frequency can eliminate resonance; thus dynamic loads and torsional vibrations will be substantially reduced.

However, in many transmissions (e.g., in machine tools and reciprocating machine installations) the frequencies of the disturbances acting on the system and, sometimes, natural frequencies (in variable speed transmissions) may vary widely [19]. In such instances, a simple shift of the natural frequencies of the drive can lead to a resonance occurring at other working conditions, but the probability of its occurrence is not lessened. A reduction in the natural frequency of a drive, for example, is advisable for the drive of a milling machine only at the highest spindle speeds and may be harmful if introduced in the low-speed stages [19].

A shift of natural frequencies of the drive can prove to be beneficial in transmissions with narrow variations in working conditions. If, however, a drive is operated in the prerersonance region, an increase in torsional compliance will lead to increased amplitudes of torsional vibrations, and thus to nonuniform rotation. In some cases excessive torsional compliance may lead to a dynamic instability of the transmission and create intensive self-excited torsional vibrations.

An important feature of multispeed (or variable-speed) transmissions is fast changing of effective torsional compliances of their components with changing output speeds due to changing reduction coefficients (although the physical condition of the components does not change). As a result, the role of the coupling as a compliant member can change dramatically depending on the configuration of the drive. Figure 4 [19, 21] shows mathematical models of mechanical systems of the same milling machine drive at several kinematic configurations representing various output (spindle) speeds n_{sp}. Rotor of the driving induction motor is represented by the left disc in Fig. 4 ($I = 150 \times 10^{-3}$ kgm²). If compliance of the motor coupling (200×10^{-6} rad/Nm), second compliance from the left in the Fig. 4 models, is about

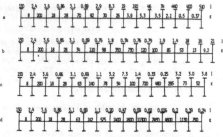

Fig. 4 Torsional models of some configurations of the spindle gear box of a milling machine (left disk represents inertia of the rotor of the driving induction motor, n = 1450 rpm, 14 kW; right disk represents inertia of the spindle, reduced to the motor shaft). $a - n_{sp}$ = 3000 rpm; $b - n_{sp}$ = 600 rpm; $c - n_{sp}$ = 235 rpm; $d - n_{sp}$ = 60 rpm. All moments of inertia (I) in 10^{-3} kgm^2, all torsional compliances (e) in 10^{-6} rad/Nm.

50 percent of total compliance at n_{sp} = 3000 rpm, and 10.2 percent at n_{sp} = 600 rpm, 8.5 percent at n_{sp} = 235 rpm, it is only a negligible fraction (less than 1 percent) at low n_{sp}, such as n_{sp} = 60 rpm. Thus, compliance of a coupling of any reasonable size installed in the high-speed part of the system (close to the driving motor, left in Fig. 4) would not have any noticeable effect at low output rpm. Compliance in a coupling installed in the low-speed part of the system (close to the spindle, right in Fig. 4) would be very effective, but the coupling size might be excessive due to high torques transmitted to the spindle at low rpm.

Increasing Effective Damping Capacity of a Transmission by Using Coupling Material With High Internal Damping or Special Dampers. When the damping of a system is increased without changing its torsional stiffness, the amplitude of torsional vibrations is reduced at resonance and in the near-resonance zone. Increased damping is especially advisable when there is a wide frequency-spectrum of disturbances acting on a drive; more specifically, for the drives of universal machines.

The effect of increased damping in a torsionally flexible coupling of a milling machine transmission, whose mathematical models are shown in Fig. 4, is illustrated in Fig. 5 for the configuration of n_{sp} = 600 rpm (natural frequencies f_{n_1} = 10 Hz, f_{n_2} = 20 Hz). Figure 5(a) shows the resonance of the milling cutter runout (10 rps) with f_{n_1} for an OEM coupling (flexible element made from neoprene rubber, log decrement δ = 0.5). After this element was made from butyl rubber (same compliance, but δ = 1.5), the peak torque amplitude was reduced ~1.8 times, the clearance opening (source of intensive noise) was eliminated, and oscillations with f_{n_2}, excited by the second harmonic, became visible (Fig. 5(b)). Similar tests for n_{sp} = 235 rpm demonstrated amplitude reduction of ~1.4 times, consistent with the lesser role of coupling compliance in this case.

A common misconception about using high-damping elastomers for coupling elements is their alleged high heat generation at resonance. This is easy to disprove. Maximum potential energy stored in a flexible element during a vibration cycle is

$$V = k \frac{A^2}{2},\qquad(13)$$

where k = generalized (torsional for couplings) stiffness, A = generalized (angular for couplings) amplitude. A fraction ψV of the energy is transformed into heat ($\psi = ~2\delta =$ relative energy dissipation). The most intensive heat

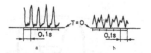

Fig. 5 Dynamic loads (tracings of oscillograms) in the milling machine drive with (a) manufacturer-supplied motor coupling (δ = 0.4) and (b) high damping (δ = 1.5) motor coupling

generation is at resonance when vibration amplitude is the highest,

$$A_{res} = A_{ex} \frac{\pi}{\delta},\qquad(14)$$

where A_{ex} = generalized amplitude of excitation, δ = log decrement. Accordingly, energy dissipation per cycle at resonance, responsible for heat generation, is

$$\Delta V_{res} = \psi k \frac{A_{res}^2}{2} = \psi k A_{ex}^2 \frac{\pi^2}{2\left(\frac{\psi}{2}\right)^2} = \frac{2\pi^2 k A_{ex}^2}{\psi},\qquad(15)$$

so that heat generation is *reduced* inversely proportionally to damping *increase*. The reason for this "paradox" is that a change in damping, first of all, changes the dynamic system and its characteristics, such as the amplitudes of its components; the absolute amount of the dissipated energy is a secondary effect.

The influence of a flexible element on the total energy dissipation in a transmission increases with an increase in its damping capacity, in the amplitude of the torque in the element, and in its compliance. For maximum efficiency, the flexible element of a coupling must therefore have as high internal energy dissipation as possible; it must also possess maximum permissible compliance, and must be located in the part of the system where the intensity of vibrations is the greatest.

Introducing Nonlinearity in the Transmission System. A nonlinear dynamic system becomes automatically detuned away from resonance at a fixed-frequency excitation, the more so the greater the relative change of the overall stiffness of the system on the torsional deflection equal to the vibration amplitude. For example, when damping is low, relative change of the stiffness by a factor of 1.3 reduces the resonance amplitude ~1.7 times; a relative change of stiffness by a factor of 2 reduces the resonance amplitude ~1.85 times [22].

Nonlinear torsionally flexible couplings can be very effective in transmissions where high-intensity torsional vibrations can exist and where the coupling compliance constitutes a major portion of the overall compliance.

Production machines usually have variable speed transmissions. To keep the coupling size small, it is conventionally installed close to the driving motor, where it rotates with a relatively high speed and transmits a relatively small torque. At the lower speeds of an output member, the installed power is not fully utilized and the absolute values of torque (and of amplitudes of torsional vibrations) on the high-speed shaft are small. Furthermore, the role of a coupling in the balance of torsional compliance is small at low speeds of the output shaft, as shown in the foregoing (see Fig. 4). Thus, only a very strong nonlinear characteristic of a coupling can manifest itself.

Another important advantage of couplings with nonlinear load-deflection characteristics is the feasibility of making a resonably small coupling with low torsional stiffness and high rated torque. An overwhelming majority of power-transmission systems are loaded 80–90 percent of the total "up" time with less than 0.5 T_r. A nonlinear coupling with a

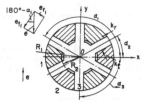

Fig. 6   Schematic of a spider coupling. 1,2 = hubs; 3 = rubber spider.

hardening load-deflection characteristic provides low torsional stiffness for most of the time, but since its stiffness at the rated torque is much higher, its size can be relatively small.

Introducing Additional Rotational Inertia in the Transmission System. This is a secondary effect since couplings are not conventionally used as flywheels. However, when a large coupling is used, this effect has to be considered. As shown in [23], it is better to install a flywheel in a transmission on its output shaft (work organ of the machine). Since couplings are usually installed close to input shafts, their inertia can have a negative effect on transmission dynamics, increase the nonuniformity of the work organ rotation, and also reduce the influence of the coupling compliance, damping, and nonlinearity on transmission dynamics. Accordingly, reduction in the coupling inertia would be a beneficial factor.

Analysis of Combination Purpose Couplings. Combination purpose couplings do not have a compensating member. As a result, compensation of misalignment is accomplished, at least partially, by the same mode(s) of deformation of the flexible element which are called forth by the transmitted payload.

To better understand the behavior of combination purpose couplings, an analysis of the compensating performance of a typical coupling with a spider-like flexible element is helpful.

The coupling in Fig. 6 consists of hubs 1 and 2 connected with a rubber spider 3 having an even number $Z = 2n$ of legs, with "n" legs ("n" might be odd) loaded in forward direction and the other n legs loaded during reverse rotation. Deformation of each leg is independent. Assuming that the radial misalignment e of the coupled shafts is in Y direction, then the following relationships exist for the ith leg:

$$e_{t_i} = e \cos \alpha_i; e_{r_i} = e \sin \alpha_i;$$

$$F_{t_i} = k_t e_{t_i} = k_t e \cos \alpha_i; F_{r_i} = k_r e_{r_i} = k_r e \sin \alpha_i;$$

$$F_{x_i} = -F_{t_i} \sin \alpha_i + F_{r_i} \cos \alpha_i = -k_t e \sin \alpha_i \cos \alpha_i$$
$$+ k_r e \sin \alpha_i \cos \alpha_i = e(-k_t + k_r) \sin \alpha_i \cos \alpha_i;$$

$$F_{y_i} = F_{t_i} \cos \alpha_i + F_{r_i} \sin \alpha_i = k_t e \cos^2 \alpha_i$$
$$+ k_r e \sin^2 \alpha_i = e(k_t \cos^2 \alpha_i + k_r \sin^2 \alpha_i), \quad (16)$$

where subscripts t, r denote tangential and radial components, respectively; k_t, k_r = stiffness of a leg in compression (tangential direction) and shear (radial direction); e_t, e_r = components of deformation of the ith leg produced by misalignment e; F with subscripts = corresponding components of compensating force from the ith leg. Overall components of the compensating force in x, y directions are sums of F_{x_i} and F_{y_i} for all loaded legs. For a four-leg spider (two loaded legs, $n = 2$) $\alpha_{i+1} = \alpha_i + 180$ deg and

$$\Sigma F_x = \frac{1}{2} e(k_r - k_t)[\sin 2\alpha + \sin (2\alpha + 360 \deg)]$$

$$= e(k_r - k_t) \sin 2\alpha = -k_t \left(1 - \frac{k_r}{k_t}\right) e \sin 2\alpha; \quad (17)$$

$$\Sigma F_y = e\{k_t[\cos^2 \alpha + \cos^2 (\alpha + 180 \deg)] + k_r[\sin^2 \alpha$$
$$+ \sin^2 (\alpha + 180 \deg)]\} = 2k_t e\left(\cos^2 \alpha + \frac{k_r}{k_t} \sin^2 \alpha\right);$$

$$(18)$$

$$F = \sqrt{(\Sigma F_x)^2 + (\Sigma F_y)^2} = 2k_t e \sqrt{\cos^2 \alpha + \frac{k_r^2}{k_t^2} \sin^2 \alpha}, \quad (19)$$

thus the total radial force F fluctuates both in magnitude and in direction. The compensation stiffness is

$$k_{com} = \frac{F}{e} = 2k_t \sqrt{\cos^2 \alpha + \left(\frac{k_r^2}{k_t^2}\right) \sin^2 \alpha} \quad (20)$$

When $n \geqq 3$

$$\Sigma F_x = \sum_{k=0}^{n-1} e(-k_r + k_t) \sin \left(\alpha + k \frac{360 \deg}{n}\right) \times$$

$$\cos \left(\alpha + k \frac{360 \deg}{n}\right)$$

$$= \frac{e}{2} (k_r - k_t) \sum_{k=0}^{n-1} \sin \left(2\alpha + 2k \frac{360 \deg}{n}\right) = 0; \quad (21)$$

$$\Sigma F_y = e \sum_{k=0}^{n-1} \left[k_r \cos^2 \left(\alpha + k \frac{360 \deg}{n}\right) + \right.$$

$$\left. k_t \sin^2 \left(\alpha + k \frac{360 \deg}{n}\right)\right]$$

$$= e \sum_{k=0}^{n-1} \left[k_r \frac{1 + \cos \left(2\alpha + 2k \frac{360 \deg}{n}\right)}{2} \right.$$

$$\left. + k_t \frac{1 - \cos \left(2\alpha + 2k \frac{360 \deg}{n}\right)}{2} \right] = \frac{n}{2} e(k_r + k_t),$$

$$(21a)$$

thus with $Z \geqq 6$ the total radial force F is constant and directed along the misalignment vector;

$$k_{com} = \frac{F}{e} = \frac{n}{2} (k_t + k_r) \quad (22)$$

Since maximum allowable radial misalignments (e) for the spider couplings do not exceed 0.007–0.01 of the outside coupling diameter D_{ex} and the spider leg width $b = (0.2-0.25)$ D_{ex}, the maximum shear of the leg does not exceed about 0.03. Accordingly, the value of the shear modulus G has to be modified as compared with the conventional $G(H)$ relationship (H = rubber durometer) as shown in [24] and, thus, ratio k_r/k_t varies with changing H. For typical spider proportions, $k_r/k_t = 0.26$–0.3 for medium durometer $H = 40$–50, and $k_r/k_t = \sim 0.4$ for hard rubber spiders, $H = 70$–75.

Assuming that the tangential load is evenly distributed along the leg surface and that $R_1 = \sim 2R_2$ (typical for commercially available spider couplings), the resultant force will be acting on the leg at the distance $R_{eff} = (R_1 + R_2)/2 = \sim 0.75 R_1 = 0.75 R_{ex}$ from the coupling center 0, where R_{ex} − external radius of the coupling. Torsional stiffness of the coupling

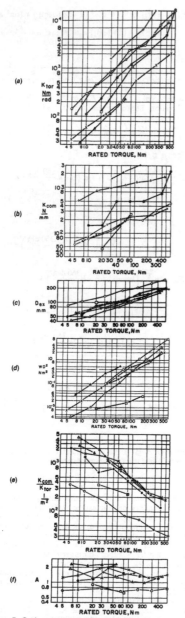

$$k_{tor} = \frac{T}{\phi} = \frac{F_t R_{eff}}{\dfrac{\Delta}{R_{eff}}} = \frac{F_t}{\Delta} R_{eff}^2 = n k_t R_{eff}^2, \qquad (23)$$

where T is transmitted torque; ϕ = angular deformation of the coupling; Δ = tangential deformation of one loaded spider leg. From (20) and (23), (22) and (23), respectively, we have:
for $Z = 4$, $n = 2$

$$\frac{k_{com}}{k_{tor}} = \frac{\sqrt{\cos^2\alpha + \left(\dfrac{k_r}{k_t}\right)^2 \sin^2\alpha}}{R_{eff}^2};$$

$$\left(\frac{k_{com}}{k_{tor}}\right) max = \frac{1}{R_{eff}^2} = \sim \frac{1.8}{R_{ex}^2}; \qquad (24)$$

for $Z \geqq 6$, $n \geqq 3$

$$\frac{k_{com}}{k_{tor}} = \frac{1 + \dfrac{k_r}{k_t}}{2R_{eff}^2}; k_{tor} = \sim \frac{1.15}{R_{ex}^2} \ (H = 40 = 50);$$

$$\frac{k_{com}}{k_{tor}} = \sim \frac{1.25}{R_{ex}^2} \ (H = 65-75); \qquad (25)$$

The values of k_{com}/k_{tor} from (24) and (25) are very close to actual values from the manufacturers catalogs, plotted in Fig. 7(e) and, indirectly, in Fig. 7(f).

Several conclusions can be reached from the preceding analysis:

Some combination purpose couplings are characterized by undesirable fluctuations of the force they exert on the connected shafts, both in magnitude and in direction, see (24);

For a given design and value of torsional stiffness, a coupling's stiffness in radial directions diminishes with increasing external radius;

The ratio of radial (compensating) stiffness and torsional stiffness of a combination purpose flexible coupling can be represented as

$$\frac{k_{com}}{k_{tor}} = \frac{A}{R_{ex}^2} \qquad (26)$$

where the "Coupling Design Index" A allows one to select a coupling design better suited to a specific application. If the main purpose is to reduce misalignment-caused loading of the connected shafts and their bearings, for a given value of torsional stiffness, then the least value of A is the best, together with large external radius. If the main purpose is to modify the dynamic characteristics of the transmission, then minimization of k_{tor} is important.

Comparison of Existing Flexible Coupling Designs. The bulk of designs that are used as torsionally flexible or combination purpose couplings are couplings with elastomeric (rubber) flexible elements. Couplings with metal springs possess the advantages of being more durable and of having characteristics less dependent on frequency and amplitude of torsional vibrations. However, they have a larger number of parts and higher cost, especially for smaller sizes. As a result, couplings with metal flexible elements have found their main applications in large transmissions, usually for rated torques 1000 Nm and up, which is beyond the scope of the present paper.

Couplings with elastomeric flexible elements form two subgroups:

Fig. 7 Basic characteristics of frequently used torsionally flexible/combination purpose couplings. (a) torsional stiffness; (b) radial stiffness; (c) external diameter; (d) flywheel moment; (e) ratio between radial and torsional stiffness; (f) coupling design index A. Legend: △ = rubber spider; ▲ = modified spider; + = finger sleeve; o = toroid shell; □ = rubber disk, ● = centaflex.

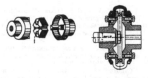

Fig. 8 Some coupling designs surveyed in Fig. 7: (a) modified spider coupling; 1 – lip, providing bulging space; (b) toroid shell coupling

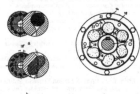

Fig. 9 Couplings, utilizing streamlined rubber elements. (a) and (b) Rolastic with rubber rollers, no load (a) and under load (b); (c) Elliott with rubber spheres.

(*a*) Couplings in which the flexible element contacts each hub along a continuous surface (tubular/sleeve types, with a toroidal shell, with a solid rubber disc/cone etc.); usually, torque transmission in these couplings is associated with the shear deformation of rubber;

(*b*) Couplings in which the flexible element consists of several independent or interconnected sections (disk-finger and finger sleeve types, spider couplings, couplings with rubber blocks, etc.); usually, torque transmission in these couplings is associated largely with the compression or "squeeze" of rubber.

Comparative evaluation of the commercially available couplings is not an easy task, since only a few manufacturers provide users with such necessary data as torsional stiffness, stiffness in the basic misalignment directions, the ratio between static stiffness and effective stiffness in vibratory conditions (dynamic stiffness), etc. This data (not always reliable) is available in the catalogs of large companies which can afford the testing program. Another factor is the wide diversity of coupling designs. For example, widely used toroidal shells can be reinforced with cord or made of plain rubber, use the outer or inner half of the torus, etc. Another popular flexible element design, the rubber spider, is also available in numerous forms: spider with plain straight rectangular cross-sectional legs, with legs that are barrel-shaped in cross or axial section; with lips or dots, etc. All these relatively minor variations have a substantial effect on stiffnesses (although a lesser effect on their ratios). Some coupling designs demonstrate pronounced effects of speed and transmitted torque on radial and axial loads [25].

There are very few publications describing data on various couplings measured on the same test rigs. The validity of the most comprehensive one [25] is reduced since data is not given on the basic parameters of the couplings tested.

In spite of these reservations, it seems that even bringing together the available manufacturer-supplied data on flexible couplings would be useful. This data is given in Fig. 7. Plots in Fig. 7(*a–d*) give data on such basic parameters as torsional stiffness k_{tor}, radial stiffness k_{rad}, external diameter D_{ex} and flywheel moment WD^2; plots in Fig. 7(*e,f*) give derivative information: ratio k_{rad}/k_{tor} (Fig. 7(*e*)) and design index A (Fig. 7(*f*)).

Some remarks about Fig. 7 seem to be warranted:

(*a*) The "modified spider"[1] coupling (Fig. 8(*a*)) is different from the conventional spider coupling shown schematically in Fig. 5 by four features: legs are tapered, instead of straight; legs are made thicker even in the smallest cross section, at the expense of reduced thickness of bosses on the hubs; lips 1 on the edges provide additional space for bulging of the rubber when legs are compressed; the spider is made of a very soft rubber. All these features lead to substantially reduced stiffnesses while retaining small size, which is characteristic for spider couplings.

(*b*) Data for "toroid shell" couplings are for the coupling as shown in Fig. 8(*b*).

(*c*) The "spider coupling" for $T_r = 7$ Nm has a four-legged spider ($z = 4$) while all larger sizes have $z = 6$ or 8. This explains the differences in A ($A = 1.96$, close to theoretical 1.8, for $z = 4$; $A = 0.98$–1.28, close to theoretical 1.15–1.25, for $z = 6,8$).

(*d*) Values of A are quite consistent for a given type of coupling. Variations can be explained by differences in design proportions and rubber blends between the sizes.

Using plots in Fig. 7, one can more easily select a coupling type whose stiffnesses, inertia, and diameter, are best suited for a particular application. These plots, however, do not address issues of damping and nonlinearity. Damping can be easily modified by proper selection of elastomer. As shown previously, high damping is very beneficial for transmission dynamics, and may even reduce thermal exposure of the coupling. More complex is the issue of nonlinear characteristics, which is specifically addressed below. Couplings represented in Fig. 7 are linear or only slightly nonlinear.

A Suggestion for Design Development of Torsionally Flexible/Combination Purpose Couplings. From the preceding discussion, it can be concluded that an optimal torsionally flexible coupling would have a strong nonlinear characteristic together with high damping. A nonlinear characteristic also allows one to obtain very high torsional compliance for the most frequently used fractional loading in a relatively small coupling. Accordingly, the misalignment-compensating properties of a highly nonlinear coupling would be superior at fractional loads.

Since damping can be modified by change of material, the design of the flexible element must first be selected to achieve a desired nonlinearity.

Nonlinearity in elastomeric couplings can naturally be achieved by utilizing the nonlinear load-deflection characteristic of cylinders, spheres, etc. under compression. Strangely enough, two commercially available couplings with such elements, the Rolastic [26] with cylindrical elements, Fig. 9(*a,b*) and the Elliott [27] with spherical elements, Fig. 9(*c*) (upper half-unloaded, lower half-loaded), use "squeezing" loading modes, associated with rather slight nonlinearity. Thus, the stiffness of the Rolastic coupling at rated torque (torsional deflection 10 deg) is only two times higher than its stiffness at no load.

In our experiments with a rubber cord $D = 3.2$ mm (1/8 in.), $L = 190$ mm, its compression stiffness was $k = 250$ N/mm in the deformation range 0–0.04D (0–0.13 mm), $k = 963$ N/mm for deformations (0.23–0.28)D, $k = 2,675$ N/mm for deformations (0.47–0.50)D. Similar data have been measured for a spherical sample $D = 15.8$ mm (5/8 in.). For deformations (0–0.01)D, $k = 4.5$ N/mm; for (0.16–0.24)D, $k = 32$ N/mm; for (0.4–0.48)D, $k = 110$ N/mm. Thus, stiffness change exceeding a decimal order of magnitude can be observed for deformations below relative compression 0.5.

---

[1] This generic name is used since the coupling is sold under at least, three trade names.

It is shown in [28] that spherical rubber parts have excellent fatigue endurance at cyclical compression $0-0.5D$ (relative compression 0.5 as compared with allowable relative compression 0.05–0.15 for rectangular rubber blocks). Similar behavior can be assumed for cylindrical samples.

Thus, the use of rubber flexible elements of this geometry seems to be a promising direction for a substantial improvement in coupling designs.

5 Conclusions

Classification of couplings according to application requirements and subsequent analysis of each class allows the development of evaluation criteria and design principles that can be useful for selection from existing coupling designs as well as for the development of new couplings. Comparison of various coupling designs, based on the unified criteria, may facilitate intelligent selection of a coupling type for a given application. Ways for making substantial improvements are demonstrated for misalignment-compensating couplings, and outlined for torsionally flexible and combination purpose couplings.

Acknowledgment

This work was partly supported by the grant MEA-8308751 from the National Science Foundation. The funding is gratefully appreciated.

References

1 *Proceedings of International Conference on Flexible Couplings for High Powers and Speeds*, University of Sussex, England, 1977.
2 Spotts, M. F., *Design of Machine Elements*, 5th Ed., Prentice-Hall, Englewood Cliffs, N.J., 1978, 684 pp.
3 Burr, A. H., *Mechanical Analysis and Design*, Elsevier, New York, 1982, 640 pp.
4 Shigley, J. E., *Mechanical Engineering Design*, McGraw-Hill, New York, 1977, 695 pp.

5 Ettles, C., Wells, D. E., Stokes, M., and Matthews, J. C., "Investigation of Bearing Misalignment Problems in a 500 MW Turbo-Generator Set," *Proc. of the Inst. of Mechanical Engineers*, Vol. 188, No. 35/74, 1974.
6 "AGMA Standard Nomenclature for Flexible Couplings," No. 510.02, AGMA, 1969.
7 The Bendix Corporation, Publication No. 00U-6-792C.
8 Duditza, F., Querbewegliche Kupplungen, *Antriebstechnik* (in German), Vol. 10, No. 11, 1971, pp. 409–419.
9 Calistrat, M. M., "Gear Couplings," *Wear Control Handbook*, Peterson, M. B., and Winer, W. O., eds., ASME, 1980, pp. 831–841.
10 Pleeck, G., "Noise Control in the Turbine Room of a Power Station," *Noise Control Engineering*, Vol. 8, No. 3, 1977, pp. 131–136.
11 Crease, A. B., "Forces Generated by Gear Couplings," in *Proceedings of International Conference on Flexible Couplings for High Powers and Speeds*, University of Sussex, England, 1977.
12 Yampolskii, M. D., Palchenko, V. I., Gordon, E. Ya., "Dynamics of Rotors Connected with a Gear Coupling," *Mashinovedenie* (in Russian), No. 5, 1976, pp. 29–34.
13 Reshetov, D. N., *Machine Elements* (in Russian), Mashinostroenie Publ. House, Moscow, 1974, 655 pp.
14 Reshetov, D. N., and Palochkin, S. V., "Damping in Compensating Couplings," *Izvestia VUSov. Mashinostroenie* (in Russian), No. 12, 1981, pp. 13–18.
15 Rivin, E. I., "Anti-Vibration Elements and Devices in Machine Tools," in *Components and Mechanisms of Machine Tools* (in Russian), Reshetov, D. N., ed., Mashinostroenie Publ. House, Moscow, Vol. 2, 1972, pp. 455–516.
16 Rivin, E. I., "Properties and Prospective Applications of Ultra Thin Layered Rubber-Metal Laminates for Limited Travel Bearings," *Tribology International*, Vol. 18, No. 1, 1983.
17 USSR Certificate of Invention, 252, 777.
18 Rivin, E. I., "Gears Having Resilient Coatings," U.S. Patent, 4,189,380.
19 Rivin, E. I., *Dynamics of Machine Tool Drives* (in Russian), Mashinostroenie Publ. House, Moscow, 1966, 204 pp.
20 Eshleman, R., and Schwerdlin, H., "Combating Vibration With Mechanical Couplings," *Machine Design*, September 25, 1980.
21 Rivin, E. I., "Compilation and Compression of Mathematical Model for a Machine Transmission," ASME Paper 80-DET-104.
22 Nestorides, E. J., *A Handbook of Torsional Vibrations*, Cambridge University Press, 1958.
23 Rivin, E. I., "Role of Induction Motor in Transmission Dynamics," ASME Paper 80-DET-96.
24 Rivin, E. I., "Horizontal Stiffness of Anti-Vibration Mountings," *Russian Engineering Journal*, No. 5, 1965, pp. 21–23.
25 Schwerdlin, H., "Reaction Forces in Elastomeric Couplings," *Machine Design*, July 12, 1979.
26 Eurodrive, Inc., Publication E-0080.
27 Eliott Company, Bulletin Y-50C; also Swiss Patent 630,708.
28 Schmitt, R. V., and Kerr, M. L., "A New Elastomeric Suspension Spring," SAE Paper 710058, SAE, 1958.

────── D I S C U S S I O N ──────

H. Schwerdlin. The author's concerns about the
dissemination of coupling design data is well founded. However, this void is about to be filled in the soon-to-be-published text *Couplings and Universal Joints: Design, Selection and Application*, [29] by Jon Mancuso of Zurn Industries and *The Standard Handbook of Machine Design* [30] (coupling chapter authored by Howard Schwerdlin of Lovejoy, Inc.). At the university level, machine and mechanisms courses only cover universal joints and gearing, while neglecting other types of couplings. I have personally used manufacturers' catalogs as handouts to my students to describe the different types of couplings available and their limitations, due to the absence of any substantial text on the subject.

The thin rubber laminate couplings discussed in the article seem too good to be true with an efficiency of greater than 99.99 percent at misalignments of 0.04 in. (1 mm) for both the Oldham and gear coupling types.

Concerning the stiffness equations (22–25) for jaw type couplings that we at Lovejoy patented 60 years ago, those look very good. However, most commercial couplings are manufactured with higher durometer typically 80–85 Shore A. I suspect that the K_r/K_t ratio would then be about 0.5 and $K_{com}/K_{tor} = 1.33/R_{ex}^2$ for a coupling design index of 1.33 in basic agreement with Fig. 7(f). However, the calculated stiffnesses (K_{com}) for the spider and toroid shell couplings are twice as stiff as actual measurements. The calculated torsional stiffness of the jaw coupling is also 50 percent high, while the torsional stiffnesses of the other types are correct as per published catalog data.

The basic data from my earlier work, cited by the author, concerning coupling compliance is as follows:

Type	Rating (lb.in.)	OD (in.)	K_{com} (lb.in.)	K_{tor} (lb.in./Rad)	A
Spider	1071	3 3/4	3300	18,000	0.65
Finger sleeve	908	5	1400	4,400	1.99
Toroidal shell	1135	6 1/2	500	23,700	0.22
Centaflex rubber precompressed (Radially restrained)	708	4 3/4	1500	10,600	0.80
Rubber precompressed (Axially restrained	1260	6 3/4	1300	21,600	0.69

While the author proposes the following values from a theoretical basis:

Type	Rating (lb.in.)	OD (in.)	K_{com} (lb.in.)	K_{tor} (lb.in./rad)	A
Spider	1070	4	6850	26,500	1.03
Finger sleeve	908	4 3/4	1313	4,400	1.69
Toroidal shell	1135	7.8	1142	23,700	0.73
Centaflex	708	4 3/4	2800	10,600	1.50

We can see that the calculated K_{com} values vary from the experimental results. Except for Centaflex the radial stiffness (K_{com}) is not published by any manufacturer. These differences in calculated and experimental values make the numerical value coupling design index difficult to determine. However, the concept of this index is valid and does show the way radial and torsional compliance are interrelated for elastomeric couplings. The coupling design index will have a very small value for linkage type couplings such as the Control-flex$_{Tm}$ coupling manufactured by Schmidt.

Couplings similar to the modified spider type (the author's Fig. 8a) are sold by Pirelli Rubber Co. under the name Guibomax.

References

29. Mancuso, J., *Couplings and Universal Joints: Design, Selection and Application*, Marcel Dekker, New York, 1985.

30. Shigley, J. E., and Mischke, C. R., *Standard Handbook of Machine Design*, McGraw-Hill, New York, 1985.

Author's Closure

It is very pleasant to see an acceptance of analytical concepts suggested in the subject paper by one of the largest coupling manufacturers. Two remarks by Mr. Schwerdlin are about Oldham gear couplings in which physical sliding is replaced by shear in thin elastomeric laminates and about discrepancy between coupling parameters in Fig. 7 and his experimental results.

On the former issue, the projections given in the paper are based both on test data for a prototype Oldham coupling and on an extensive experimental study of laminated elements described in [16].

As to the latter issue – all the data in Fig. 7a–d are taken from manufacturer catalogs. Thus, the substantial discrepancy with Mr. Schwerdlin's test results only emphasizes the pressing need to develop a reliable database on couplings for their users.

Properties and prospective applications of ultra thin layered rubber-metal laminates for limited travel bearings

E. I. Rivin*

Limited and/or oscillating motions represent the most severe operating conditions for conventional bearings, both sliding and antifriction. Thin-layered rubber-metal laminates seem to be ideal substitutes for conventional bearings for oscillating motions. This paper describes experimental investigations of the compression and shear properties of flat and spherical laminates. The very high compression stiffness and strength of the laminates are accompanied by low shear stiffness. Strong non-linearity of the hardening type in compression is accompanied by weak non-linearity of the softening type in shear. Substantial non-linearity in compression starts as early as at relative compression 0.001. Applications of thin-layered laminates for compensating couplings, U-joints, gears, vibration isolators and impact cushioning in joints of mechanisms are described. Expressions have been derived for efficiency of couplings and joints equipped with laminates

Keywords: *bearings, limited travel bearings, rubber-metal laminates*

Conventional machine design textbooks are based on the use of external friction to formulate important relationships in mechanisms, such as expressions for energy efficiency (eg cylindrical joints of limited displacement, gears, U-joints, couplings), conditions for self-locking in wedge-type mechanisms, etc. In many of these mechanisms only limited motion occurs in the joint, as is the case in U-joints, Oldham couplings, and gear meshes. Accordingly, conditions for full hydrodynamic lubrication do not develop in such joints. Near the points where the relative velocity of the contacting surfaces changes direction, the friction characteristics are similar to dry (Coulomb) friction. In other areas of the engagement cycle there is a mixture of boundary and elasto-hydrodynamic lubrication (ehl). Occurrences of dry and/or boundary lubrication conditions lead to reduced energy efficiency and to increased levels of vibration and noise in the mechanisms caused by the impulsive character of friction forces at the point of reversal of relative velocity in a joint or in a gear mesh. Intensive heat generation causes thermal expansion of components and the possibility of jamming; thus, initial clearances are required. An efficient lubrication system, together with very hard contacting surfaces, is required to reduce wear, etc.

Recently, it has been demonstrated that it is possible to replace external friction in some structural joints by the judicial application of elastomeric materials with restricted free surface area, mostly in thin-layer form[1-4]. Such an approach could have a great impact on design engineering. Some of the prospective advantages of thin-layered elastomers in mechanism applications are: substantial reduction of energy dissipation (ie improvement in efficiency); elimination of the lubrication system and the dirt-protection (sealing) system; reduction in vibration and noise generation

*Department of Mechanical Engineering, Wayne State University, Detroit, MI 48202, USA

and transmission; easing of material, heat treatment and machining specifications for contacting surfaces; elimination of clearances in joints.

However, these bright prospects have not yet materialized fully, largely because of a lack of information on the basic properties of thin-layered elastomeric materials, especially of the most promising ultrathin-layered materials where the thickness of a single elastomeric layer is in the range 0.01-0.50 mm.

This paper presents experimental data on properties of ultrathin rubber-metal laminates, together with discussions of some possible applications. The discussion in this paper is mostly confined to the static properties of the laminates. Many other issues, such as thermal and fatigue resistance, optimal material selection etc, certainly have to be addressed by future researchers. However, since energy losses in the laminated connections are, as will be shown below, several orders of magnitude less than in the conventional connections, then both the heat dissipation problem and the closely related (in elastomers) fatigue problem do not seem *a priori* to be critical.

Experimental arrangement to determine stiffness characteristics of ultrathin-layered rubber-metal laminates

Two types of laminates have been tested: (a) flat and (b) spherical (Fig 1). The laminates consisted of n layers of metal and $n - 1$ layers of rubber. Metal layers in the experimental laminates were made of 0.05 mm brass foil or 0.1 mm steel foil; rubber was bonded to the metal layers.

The issues considered when selecting the rubber were:

(a) A structural laminate should have the maximum possible compressive stiffness k_z to assure minimum dimensional variation under load. On the other hand, the shear stiff-

Reprinted from Tribology Int., 1983, Vol. 18, No. 1, with permission of Elsevier Science.

ness k_x should be as low as possible for better compensation of misalignment and/or lower resistance to the working displacement. Thus, a maximum value for the ratio k_z/k_x is desirable.

(b) Compressive stiffness k_z depends highly on the deviation of Poisson's ratio from $\nu = 0.5$ — the ultimate value for a volumetric incompressible material. The ratio ν is closest to 0.5 for soft rubber and the deviation increases with increasing rubber durometer H; eg for chlorophroe rubber $\nu = 0.4997$ for $H = 0.4990$ for $H = 75^1$.

(c) The compressive stiffness k_z of thin-layered laminates increases for harder rubbers more slowly than their modulus. This is because higher hydrostatic pressures in the harder rubber cause noticeable stretching of the metal interleaves, equivalent to limited slippage on the bonded surfaces.

(d) Shear stiffness k_x does not depend on factors (b) and (c) and depends only on shear modulus G.

On the basis of these factors, soft rubber ($H = 42^*$) was the main choice for this work. However, two control samples were fabricated and tested using rubber with $H = 58^*$ and $H = 75^*$. The parameters of the type (a) flat samples tested are listed in Table 1, and of the type (b) spherical samples in Table 2.

Flat samples were tested in compression to determine compression stiffness k_z versus specific compressive load p_z, and in shear to determine shear stiffness k_x versus specific shear load p_x. Spherical samples were tested in compression, in shear around the x axis (k_α in α-direction, Fig 1b), and in torsion around the z-axis (k_γ in γ-direction, Fig 1b).

Tests were performed on universal precision testing machines — an Instron TT-DM (0.1 MN maximum capacity) and TT-KM (0.25 MN maximum capacity). These machines have very high structural stiffness and sensitive extensometers. However, both parameters were found to be inadequate for testing ultrathin-layered laminates in compression. Ultra-sensitive displacement transducers (Fig 2(a)) were used to eliminate the influence of the testing machine structural stiffness on test results. The transducer was

Table 1 Parameters of the flat laminates

No	A cm^2	t_r mm	n	h_r mm	t_m mm	H	Metal
1	21.3	0.16	33	5.1	0.05	42	Brass
2	26.4	0.33	33	10.6	0.05	42	Brass
3	23.7	0.39	17	6.2	0.05	42	Brass
4	25.9	0.25	9	2.0	0.05	42	Brass
5	23.5	0.53	17	8.5	0.05	58	Brass
6	23.1	0.58	11	5.8	0.05	75	Brass
7	12.3	0.106	15	1.7	0.1	42	Steel
8	36	0.28	15	4.5	0.1	42	Steel
9	12.3	0.44	14	6.6	0.1	42	Steel

Table 2 Parameters of spherical laminates

No	D mm	d mm	R_s mm	t_r mm	t_m mm	n	h_r mm	A_{pr} cm^2
1	101	50	80	0.7	0.6	7	4.2	60.5
2	101	50	80	1.24	0.6	11	12.4	60.5
3	101	50	80	0.55	0.6	6	2.8	60.5
4	101	50	80	0.47	0.6	6	2.4	60.5
5	101	50	80	0.42	0.6	11	4.2	60.5
6	101	50	80	0.7	0.6	4	2.1	60.5

machined from a solid block of low-hysteresis (spring) steel, thus eliminating friction in the joints which could affect transducer sensitivity. Four strain gauges provided compensation for machining asymmetry and thermal effects. Using good strain-gauge amplifiers, these transducers can reliably measure displacements as small as 0.05–0.1 μm. Standard extensometer amplifiers were used with the testing machines to give resolutions of 0.25 μm actual displacement per mm on graph paper. Signals from two transducers (Fig 2(b)) were averaged to reduce the adverse influence of machining errors and of the asymmetry of motion of the left and right driving screws of the testing machines.

The shear stiffness of the flat samples were measured under variable compression. To alleviate friction effects, a standard set of rollers in a cage and hardened and ground end plates (Fig 3(a)) were used to apply compression force p_z. In measuring shear k_α and torsional k_γ stiffness of the spherical samples, two identical samples with intermediate solid double-convex lens-shaped block were used for a similar purpose (Fig 3(b)). Compression load on the spherical samples was applied through solid concave lens-shaped blocks. When the double-sample set was used, as in Fig 3, the measured value of compression stiffness was $k_z/2$, and of shear and torsional stiffness, $2k_\alpha$ and $2k_\gamma$, respectively.

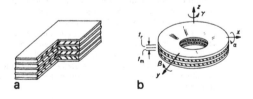

a   **b**

Fig 1 Test samples of (a) flat and (b) spherical laminates

Experimental results

Compression stiffness

Compression stiffness versus specific compression load $p_z = P_z/A$, where A is the surface area of the sample, is shown in Fig 4 in a double-logarithmic scale. Compression stiffness is expressed in terms of differential (local)

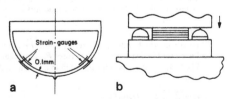

Strain-gauges
0.1mm

a   **b**

Fig 2 Displacement transducers: (a) design (b) arrangement

---

* Hardness was measured on standard cylindrical samples (2.54 cm diameter, 1.27 cm high) made from a given rubber. The actual hardness of a thin layer cannot be directly measured; it can be different from the hardness of standard samples made of the of the scme rubber blend. The actual hardness can be judged by the measured shear modulus shown in Fig 6.

compression modulus E referred to the total thickness of rubber in the sample h_r:

$$E = \frac{\Delta p_z h_r}{\Delta z} \qquad (1)$$

and was calculated from the load-deflection diagrams. Here Δz is the increment in compression deformation caused by an increment Δp_z of the specific compression force. With such a format of stiffness expression, data plotted in Fig 4 does not depend on n and A. As shown in Table 1, all the samples have different values of surface area. The comparison of the properties of such samples in a dimensionless format, as given in Fig 4, became possible after it was shown experimentally that, with the shape factor $S > 20$, compression stiffness is proportional to the loaded surface area of the sample (the shape factor S is the ratio of one loaded surface area to the force-free surface area of the sample). The nonlinearity starts from the smallest deformations which could be measured, in some cases from $\epsilon = \Delta z/h_r = 0.001$. Modulus E increases as much as one decimal order of magnitude when compression load increases about 1.5 decimal orders of magnitude. The effect is similar (but much more pronounced) to the nonlinearity for laminates with rubber layers 2–4 mm in thickness described in Ref 5.

Samples with metal layers made of brass failed at the specific compression load $p_z = 45$ MN/m² and those with steel metal layers at $p_z > 250$ MN/m². In all cases, failure was due to rupture of the metal layers; the rubber remained intact. This suggests that, by using stronger metals for the metal layers,

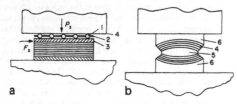

Fig 3 *Testing set-up for (a) flat and (b) spherical laminates. 1-rollers; 2-cage; 3-hardened steel plate; 4-spherical laminate; 5-double-convex adapter; 6-concave supporting plates*

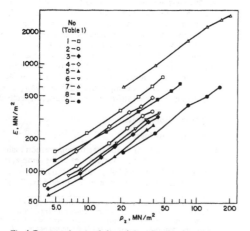

Fig 4 *Compression modulus of the ultrathin-layered rubber-metal laminates*

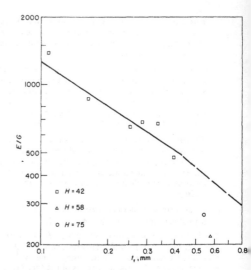

Fig 5 *Compression-to-shear stiffness ratio for flat ultrathin-layered rubber-metal laminates (E-value taken at $p_z = 30$ MN/m², G-value taken at $p_x = 0.1$ MN/m²)*

the compression strength of thin-layered laminates could be substantially increased. It is worth mentioning that the maximum (destructive) compression load did not depend in these experiments on rubber durometer.

From the data in Fig 4, some important conclusions can be reached about the properties of laminates:

● Thin-layered rubber-metal laminates in compression demonstrate a very substantial nonlinearity of the hardening type.

● Compression modulus E depends monotonously on thickness t_r of a rubber layer for a given durometer (the thinner the layer, the higher the modulus). The dependence is very steep (see Fig 5, data for $p_z = 30$ MN/m²).

● Compression modulus E increases with the increasing durometer of the rubber, but not as fast as the shear modulus G of the rubber. This is clear from Fig 5, where ratios E/G are plotted. This conclusion cannot be considered as final because it is based on limited data. The effect might be caused by a more pronounced contribution of metal stretching with increased rubber durometer, as well as by lower values of ν for the higher durometer rubbers.

● Absolute values of compression stiffness for thin-layered laminates are very high. Thus, for sample 7 at $p_z = 100$ MN/m², $E = 1800$ MN/m², which is equivalent to 1 μm deflection per 1.06 MN/m² compressive force. This is in the same range as the values of contact stiffness in a joint between two flat, ground-steel surfaces[6,7].

Shear stiffness

Shear stiffness was expressed in terms of shear modulus G and is plotted in Fig 6 versus shear force P_x. Analogously to Fig 4, shear modulus was calculated in a differential format:

$$G = \frac{\Delta p_x \cdot h_r}{\Delta x} \quad (\text{N/m}^2) \qquad (2)$$

where Δx is the increment of shear deformation caused by an increment Δp_x of the specific shear force. In shear, as opposed to the compression data shown in Fig 4, laminates demonstrate slight nonlinearity of the softening type.

Although the samples were loaded with high compression forces, tests showed no noticeable correlation between the value of the compression force and the shear modulus in the range of $p_z = 0.5 - 150$ MN/m². Fig 6 shows $G = G(p_x)$ at $p_z = 0.5 - 4.2$ MN/m².

As was shown in Ref 5, for thicker rubber layers ($t_r = 2-4$ mm) the apparent shear modulus increases as, approximately, $p_z^{1/2}$.

For pratical applications, a very improtant characteristic of laminates is the ratio between their compression and shear stiffnesses – the E/G ratio. The E/G ratios for the samples investigated for $p_z = 30$ MN/m² and $p_x = 0.1$ MN/m² are shown in Fig 5. It can be seen that the E/G ratio can attain very high values for thin layers of soft rubber, and that an increase in rubber durometer leads to a substantial drop in E/G.

Compression stiffness of spherical samples

All spherical samples were fabricated using nitrile rubber $H = 42$ (the same as that used for flat samples) and a mild carbon steel. In this case, the specific compression load p_z was calculated as the ratio of compression force p_z (in Newtons) to the surface area of axial projection A_{pr} of the sample. The latter is equal to:

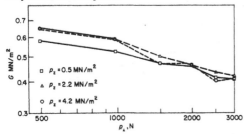

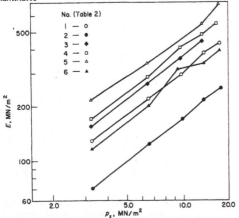

Fig 6 Shear modulus of flat ultrathin-layered rubber-metal laminates

Fig 7 Compression modulus of spherical ultrathin-layered rubber-metal laminates

$$A_{pr} = \frac{\pi}{2}(D^2 - d^2) \quad (m^2)$$

Compression modulus E calculated relative to the projection A_{pr} is shown in Fig 7. As for the flat samples, E increases with reducing t_r and demonstrates a strong nonlinearity of the hardening type.

Shear and torsional stiffness of spherical samples

In torsion (Fig 1(b)) the measured values were those of differential (local) stiffness k_γ

$$k_\gamma = \frac{\Delta T_\gamma}{\Delta \gamma} \quad (\text{Nm/arc min}) \qquad (3)$$

where $\Delta \gamma$ is an increment in angular deformation of the sample for an increment ΔT_γ of torque in the same direction. Plots of torsional stiffness k_γ are given in Fig 8; they demonstrate a slight nonlinearity of the softening type. As with the flat samples, k_γ (and k_α) do not depend on compression load.

Shear stiffness k_α was also evaluated in the differential manner:

$$k_\alpha = \frac{\Delta T_\alpha}{\Delta \alpha} \quad (\text{Nm/arc min}) \qquad (4)$$

Plots of shear stiffness k_α versus T_α are given in Fig 8(b); they demonstrate a substantial nonlinearity of the softening type. It was found that, with all given parameters t_r, h_r, t_m, D and d, the shear stiffness k_α increases with reducing sphere radius R_s ($k_\alpha = 1.2-1.6$ MN/arc min for $R_s = 80$ mm versus $k_\alpha = 0.9-1.0$ MN/arc min for $R_s = 120$ mm). Comparison of the plots in Fig 8(b) with the plots for the same samples in Fig 8(a) shows that k_γ values are about three times lower than k_α values. These differences are explained by the fact that all metal parts (shells) for the spherical samples tested were identical. However, for an ideal sphere each shell must have a distinct radius depending on its location in a sample – ie the smallest radius for the internal shell, the largest for the external shell. This effect is, of course, more pronounced (with the same t_r and t_m) for smaller values of R_s. On the other hand, deformation about the γ-axis is pure shear with any axisymmetrical shape of shell. The noted differences in k_α are indicative of the importance of paying due attention to the proper shape of the metal shells during fabrication of spherical laminates.

Design features and some prospective applications

The properties of ultrathin-layered rubber-metal laminates discussed above make them ideally suited for many machine-design applications. In the author's opinion, the slow development of these applications has been due to a lack of information about the properties of the laminates. The prospects for ultrathin-layered laminates can be seen from the fact that laminates with thicker rubber layers (4–10 mm) and inferior properties – compression strength of 2–4 mm rubber layers bonded to metal is about[5] 15–20 MN/m², compared to 45–250 MN/m² for the ultrathin-layered laminates described above – have found very successful applications. The best known of these applications are compensating supports for bridges and helicopter rotors[1].

Ultrathin-layered laminates have been successfully tested in such basic machine components as compensating couplings and U-joints, and compensating washers. They also look promising for applications such as gears, screw mechanisms and guideways for limited displacements. Some of these applications are reviewed in the remainder of this paper.

The Oldham coupling

This is used for compensation of radial misalignments between power-tranmission shafts. It consists of two hubs (1 and 2) and an intermediate member (3) (see Fig 9(a)). Torque is transmitted between the driving member (1) and the intermediate member (3), and between the intermediate member (3) and the driven member (2), by means of two orthogonal sliding connections. Because of the decomposition of a misalignment vector into two orthogonal components, this coupling theoretically assures ideal compensation, being at the same time both torsionally rigid and possessed of a high torque/weight ratio. However, this ingenious design now finds only very infrequent applications. The main reasons for this are as follows:

- Clearance is necessary for the normal operation of a sliding connection; thus contact stresses in the connections are distributed nonuniformly (Fig 9(b)), with pressure concentration at the edges. This, compounded with the two 'dead points' in each connection during a revolution – where relative sliding velocity becomes zero and friction becomes static – and with poor lubrication, leads to a rapid rate of wear. To reduce contact stressed, the coupling dimensions have to be increased.
- Maximum forces acting on the shaft bearings are equal to the maximum friction forces in the connections. Maximum friction forces occur at the reversals ('dead points') and are very high because the static friction coefficient is normally in the $f = -.1-0.2$ range. Thus, the Oldham coupling exerts very high loads on connected shafts.
- For the same reason, such couplings are not suitable for compensation of small misalignments (the limit value is usually about $0.5-1.0 \times 10^{-3}D$, where D is external diameter of the coupling). At these small misalignments all structural members of the coupling stay cemented by the static friction forces and sliding (ie compensation) does not occur.
- The intermediate member (3) must transmit torque and, at the same time, must have very hard and wear-resistant surfaces in the sliding connectors b and d. Thus, the only feasible material for fabrication of this member (3) is heat-treated steel. Since the centre of the intermediate member during rotation describes a circle with diameter e and rotates at twice the rotational speed of the connected shafts, then the centrifugal forces within the steel intermediate member can be of very substantial magnitude.
- Continuous sliding (and friction force) reversals in two sliding connections lead to intense noise generation, especially for high-speed applications.
- Two sliding connections with substantial friction lead to noticeable energy losses. For small misalignments, $e/D \leqslant 0.04$, efficiency[8]

$$\eta = 1 - 8\frac{f}{\pi}\frac{e}{cD} \qquad (5)$$

where the factor c, describing the effective diameter of tangential force application (see Fig 9(b)), depends on

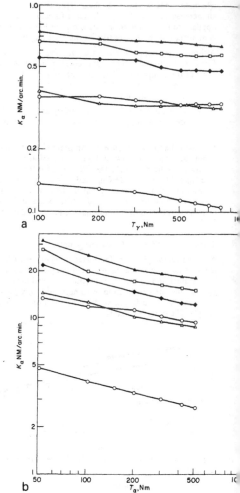

Fig 8 (a) Torsional and (b) shear stiffness of spherical ultrathin-layered rubber-metal laminates

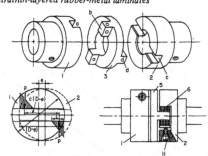

Fig 9 Oldham-type compensating coupling; (a) exploded ↗ of basic conventional design; (b) stress distribution on contact surfaces in conventional coupling design; (c) Oldh coupling with preloaded rubber-metal laminates

the amount of clearance in the coupling (typical value $c \cong 0.8$). For representative values of $e (= 0.03\,D)$ and $f (= 0.15)$ the efficiency is only 0.985.

- Angular misalignments of the coupling elements $(1, 2, 3)$ lead to a further reduction of effective load-carrying area in the connections and to increased peak pressures. Thus, angular misalignments are allowable only up to $0.5°$.

It seems natural to use thin-layered rubber-metal laminates to eliminate sliding friction in the Oldham coupling, and this has been proposed[9]. One of the embodiments of the proposed design is shown in Fig 9(c), where the laminates are installed between the hubs (1 and 2) and the intermediate member (5), and preloaded with bolts (11) to eliminate backlash and increase torsional stiffness. This design effectively eliminates all the disadvantages of the conventional Oldham coupling listed above. The questions of the centrifugal forces in the steel intermediate member and the energy losses deserve to be elaborated upon.

In the design shown in Fig 9(c) two main functions of the intermediate member — to transmit torque and to accommodate displacement between the hubs — are separated. Torque transmission is taken up by the body of the intermediate member and displacement accommodation by the laminated elements. Thus, only bulk strength is required from the intermediate member, and not contact durability, and it can be fabricated from a light and strong material without regard to its hardness. This can substantially reduce centrifugal forces in the transmission.

The efficiency of the coupling in Fig 9(c) is very different from the coupling in Fig 9(a). In the latter design, (frictional) resistance to the relative movement in the connections a-b and c-d is proportional to the transmitted torque; thus energy losses are less under light load than at higher loads and efficiency η is independent of transmitted torque, which is reflected in Eq (5). On the other hand, resistance to compensatory movement in the design in Fig 9(c) is produced by the shear deformation of rubber and, as shown earlier, does not depend on the compression force (ie, transmitted torque); energy losses at a given design and amount of misalignment are directly related to hysteretic losses in rubber. Thus, the absolute losses do not depend on transmitted power and, accordingly, efficiency increases with increased transmitted power.

To derive an expression for efficiency of the Oldham coupling with laminated connections, let the shear stiffness of the connection between one hub and the intermediate member be denoted by k_{sh}, and relative energy dissipation in the rubber for one cycle of shear deformation by ψ. Then, maximum potential energy in the connection (at maximum shear e) is equal to:

$$V_1 = k_{sh}\frac{e^2}{2} \tag{6}$$

and energy dissipated per cycle of deformation is equal to:

$$\Delta V_1 = \psi\,k_{sh}\frac{e^2}{2} \tag{7}$$

Each of two connections experiences two deformation cycles per revolution, thus total energy dissipated per revolution of the coupling is

$$\Delta V = 2 \times 2\Delta V_1 = 2\psi\,k_{sh}e^2 \tag{8}$$

Total energy transmitted through the coupling per revolution is equal to

$$W = P_t \times \pi D = 2\pi T \tag{9}$$

where P_t is tangential force reduced to the external diameter D and T is the transmitted torque. The efficiency of a coupling is therefore equal to:

$$\eta = 1 - \frac{\Delta V}{W} = 1 - \frac{\psi k_{sh}e^2}{\pi T} \tag{10}$$

For the experimentally tested coupling ($D = 0.12$ m), the parameters are:

$\psi = 0.2; k_{sh} = 1.8 \times 10^5\,\text{N/m}; T = 150\,\text{Nm};$
$e = 1\,\text{mm} = 0.001\,\text{m};$

thus

$$\eta = 1 - \frac{0.2 \times 1.8 \times 10^5 \times 10^{-6}}{\pi \times 150} = 1 - 0.75 \times 10^{-4}$$
$$= 0.999925$$

or losses at full torque are reduced 200 times compared to the conventional coupling.

Comparison of test results for conventional and modified Oldham couplings (both with $D = 12$ m) is given in Ref 10. These results (for a laminate with rubber layers 2 mm in thickness) showed that the maximum transmitted torque was the same but there was a 3.5 times reduction in radial force transmitted to the shaft bearings with the modified coupling. Actually, the coupling showed the lowest radial force for a given misalignment compared with any compensating coupling, including couplings with rubber elements. In addition to this, noise level at the coupling was reduced 13 dBA. Using ultrathin-layered laminates for the same coupling would further increase its rating by about one order of magnitude, and would therefore even require a redesign of the hardware to accommodate such a high transmitted load in a very small coupling.

The universal (Cardan) joint

This joint is widely used for power transmission between shafts permanently or variably inclined against one another (as in automobiles, construction machinery, etc). A typical joint has two yokes attached to the shafts to be connected and a spider with four trunnions, each pair of the trunnions rotationally engaged with its respective yoke and the axes of two pairs in one plane and orthogonal. Both sliding and rolling friction bearings are used in universal joints, and for both types their use in these joints is one of the most trying possible applications because of the oscillatory character of the motion.

Again, the application of thin-layered rubber-metal laminates for U-joint yoke bearings (Fig 10)[11] seems to be a logical solution of the problem. Detailed calculations have shown that using the types of laminates discussed in the first part of this paper, with one-layer thickness of 0.01–0.1 mm, only a small fraction of the load-supporting area of the trunnion is needed for transmission of the rated load for a given size of the joint. Reduction of this area greatly reduces the shear stiffness of the laminated bearings.

The greatest advantages of universal joints with rubber laminated bearings are: elimination of lubrication and sealing devices; elimination of wear and backlash in the connection; very substantial attenuation of radial forces and/or vibrational excitations transmitted through the connection. To clarify this last statement, it should be noted that in conventional joints, especially those with antifriction bearings, high transmitted (tangential) forces produce significant friction forces in the radial direction of the connected shafts. These friction forces effectively render the connection rigid in the radial direction. Accordingly, vibratory forces transmitted from one connected shaft to another and exciting bending vibrations are transmitted in full without an attenuation, unless they exceed the friction force. Transmission of high-frequency torsional vibrations is also substantially attenuated by the laminated bearings.

The expression for efficiency of universal joints with rubber-metal laminated bearings is similar to Eq (10). With the angular misalignment of the connected shafts equal to α, each of four elastic bearings is cyclically twisted during rotation for $\pm \alpha$ per revolution. If the angular stiffness of each bearing is k_{ang}, then the maximum deformation energy of one bearing is $k_{ang}\,\alpha^2/2$ and the energy dissipation per revolution is:

$$\Delta V_1 = \psi k_{ang}\,\frac{\alpha^2}{2} \tag{7'}$$

Total energy dissipation in four bearings is

$$\Delta V = 4\,\Delta V_1 = 2\psi k_{ang}\,\alpha^2 \tag{8'}$$

Work transmitted by the joint during one revolution is

$$W = 2\pi T \tag{9'}$$

where T is transmitted torque; thus the efficiency of a universal joint with rubber-metal laminated bearings is

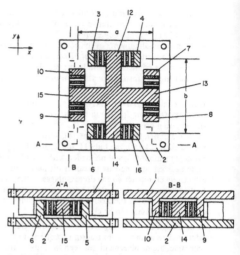

Fig 11 Vibration isolator with independently set stiffness values in two linear and one angular direction

$$\eta = 1 - \frac{\Delta V}{W} = 1 - \frac{\psi k_{ang}\alpha^2}{\pi T} \tag{10'}$$

as compared with the efficiency of conventional joints[8]

$$\eta = 1 - f\frac{d}{R}\,\frac{1}{\pi}\,(2\tan\frac{\alpha}{2} + \tan\alpha) \tag{11}$$

where d is the effective diameter of the trunnion bearing, and $2R$ the distance between the centres of the opposite trunnion bearings. Again, as with Oldham couplings, the efficiency of the joints with rubber-metal laminated bearings is not constant but increases with increasing load. Losses for the loads close to the rated load are 1.5–2.0 orders of magnitude less than for conventional joints.

Vibration isolator with high angular stiffness

The design principles used in the Oldham coupling and the universal joint can be applied to build a vibration isolator with low and independently adjustable linear stiffness in two directions, and with high and also independently adjustable angular stiffness. Conventional isolators have very low angular stiffness, to such an extent that they are neglected in the equations of motion (eg, see Refs 12 and 13). However, in many instances, such as the mounting of gyroscopes or of plant machinery with high internal dynamic forces, high angular stiffness could be of great advantage. The design of such an isolator utilizing thin-layered rubber-metal laminates is shown in Fig 11[14]. Linear stiffnesses in x and y directions can be tuned by assigning dimensions of rubber-metal laminates experiencing shear in the x and y directions, respectively. Angular stiffness in α direction can be adjusted by selection of the arm lengths a, b of intermediate member (12), (13), (14), (15).

The ball screw

The screw mechanism with rolling friction between male and female threads has wide application whenever low friction and backlash elimination is required. However, there are cases where only a limited relative displacement between the threaded parts is required. In such cases, the

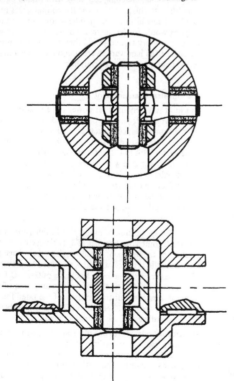

Fig 10 Universal joint with rubber-metal laminated bushings

use of expensive and heavy ball screws puts certain burdens on the design. As in every sliding joint with limited displacement, the use of rubber-metal laminated bearings could be a solution in this case (Fig 12)[15] . In this mechanism, a thin rubber layer is bonded to both male and female threaded surfaces. Due to drastic differences between the compression and shear stiffness of the rubber layer, the system acts as a very precise screw mechanism. There are cases where the elastic character (with spring-type restoring force) of this mechanism could be an additional beneficial feature.

Power-transmitting gears

The most frequently-used machine components where sliding on a limited sliding path occurs, are power transmitting gears. This sliding (together with rolling) takes place during engagement of each meshing pair of teeth. In conventional gears, friction accompanying sliding during gear engagement is responsible for the major shortcomings of gear tranmissions. These include wear and pitting of the profiles, energy losses, heat generation leading to thermal expansion of teeth (and consequently to a need for providing clearance) and noise generation initiated by the jump of the friction force at the point of reversal of the sliding velocity.

Total sliding path during engagement of a pair of teeth is equal to[16]:

$$4 \frac{25.4 \, \pi^2}{p_d \cos \alpha} \left(\frac{1}{Z_1} + \frac{1}{Z_2} \right) \quad \text{(mm)} \tag{12}$$

where p_d is diametral pitch (1/in), α is pressure angle, and Z_1, Z_2 represent the number of teeth on engaging gears. For conventional involute gears the length of the sliding path is of the order of magnitude of $m/10$, where $m = 25.4/p_d$, the module of the gearing in millimetres. The sliding path for a medium size gear ($m = 4$, diametral pitch $\simeq 6$) is about 0.4 mm, and can be accommodated by thin rubber coating(s) on the engaging teeth. This concept of eliminating sliding friction and taking up the inevitable geometrical sliding by shear deformation of an elastomeric (rubber) coating has been proposed in Ref 4 (Fig 13); it seems to be especially suitable for conformal (Wildhaber-Novicov) gears and gear couplings.

Fig 13 Gear design in which geometrical sliding is accommodated by internal shear in elastometric coating

In addition to the single-layer coating shown in Fig 13(a), modifications could be made to accommodate larger amounts of sliding. One modification, shown in Fig 13(b), uses-a multi-layered laminate which has the same allowable specific load in compression but allows greatly increased shear deformation. Another modification, shown in Fig 13(c), solves this problem in a different way. In this case, the coating is divided into several sections by narrow notches. Thus, when the engagement of two teeth commences, only the first segment experiences shear. After the contact point progresses along the tooth and reaches the end of the segment, shearing of the adjacent starts and the first segment becomes free and can return to its initial (unstressed) condition. With such a design, the allowable shear to absorb the sliding path is equal to the shear allowable for the given thickness of the coating, multiplied by the number of segments.

Impact cushioning

High compression strength, together with steep nonlinearity on a very short travel, makes laminates suitable for impact cushioning in the confined spaces of real mechanism joints. Two applications of such a concept have been proposed in Ref 18.

One of these applications is for cushioning the impacts due to clearances in the ball joint between the connecting rod and the slide of a stamping press. Treatment consists of inserting a laminated spacer between the two halves of the bronze joint bearing. Three additional pieces between the upper half of the bearing and its housing serve to cushion a random lateral motion. The main spacer is preloaded by the bolts joining the two halves, and cushions tension impacts at the bottom dead point. This treatment has reduced the equivalent noise level of the press at the idling condition from 99.9 to 96.9 dBA. The case is typical for many link-type mechanisms.

Another application is for cushioning impacts in a blanking die between the stripper plate and its keepers. In many cases, these impacts contribute more to stamping press noise then the stamping (blanking) operation itself. Cushioning with conventional rubber or polyurethane pads 3–5 mm thick was not successful. This was because the contact area was very narrow (about 5 mm), and the contact pressures (4–4.5 MPa) far exceeded permissible loads for rubber and polyurethane. However, these loads are very low for thin-layered laminates, which do not show any deterioration after tests. Noise reduction for the relatively 'quiet' die tested was from 96.9 to 94.3 dBA.

Conclusions

1. Ultrathin-layered rubber-metal laminates demonstrate extremely high compressive strength and stiffness, together with very steep nonlinearity of the hardening type in compression, while retaining the properties of a conventional rubber in shear. Thus, the ratio between

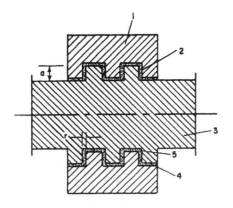

Fig 12 Frictionless threaded connection for limited displacements

the effective compression and shear moduli could be in the range of thousands.

2. These unique properties make the laminates ideal for design applications where limited relative displacements have to be executed, eg couplings, joints, gear meshes.

References

1. **Freakley P.K. and Payne A.R.** Theory and Practice of Engineering with Rubber. *Applied Science Publishers, London, 1978*

2. **Hinks W.L.** Static Load Bearings. *US Patent 2,900,182*

3. **Rivin E.I.** Antivibration Elements and Devices for Machinery (in Russian) *In "Detaly i Mekhanismy Metallorezhuschikh Stankov", Moscow, 1972*

4. **Rivin E.I.** Gears Having Resilient Coatings, *US Patent 4,184,380*

5. **Rivin E.I. and Aronshtam L.I.** Investigation of Compression and Shear of Thin-Layered Rubber-Metal Elements, *Kautchuk i Resina, 1967, No 7* (in Russian)

6. **Levina Z.M. and Reshetov D.N.** Contact Stiffness of Machines (in Russian) *Moscow, 1971*

7. **Rivin E.** Compilation and Compression of Mathematical Model for a Machine Transmission, *ASME Paper 80-DET-104*

8. **Reshetov D.N.** Machine Elements. *Moscow, 1974* (in Russian)

9. **Rivin E.I.** Cross-Sliding Coupling, *USSR Certificate of Invention (CI) 190,731; 228,410; 242,777; 268,805*

10. **Rivin E.I.** Comparison and Selection of Couplings. *Proceedings of 5th National Conference on Power Transmission, Illinois Institute of Technology, Chicago, 1978*

11. **Rivin E.I.** *USSR CI 217,153*

12. **Himelblau M. and Rubin Sh.** Vibration of a Resiliently Supported Rigid Body. *In "Shock and Vibration Handbook", McGraw-Hill, NY, 1976*

13. **Rivin E.I.** Principles and Criteria of Vibration Isolation of Machinery. *Trans. ASME, Journal of Mechanical Design, Vol 101, p. 682, 1979*

14. **Rivin E.I.** Vibration Isolator. *USSR CI 261,834*

15. **Rivin E.I.** Screw Mechanism. *USSR CI 396,496*

16. **Tuplin W.A.** Involute Gear Geometry. *Ungar Publishing Co, NY, 1963*

17. **Rivin E.I.** New Design of Power Transmission Gearing and Gear Couplings. *Proc. 9th National Conf. on Power Transmission, 119 Chicago, 1982*

18. **Rivin E.I. and Shmuter S.** Metal Stamping Presses Noise Investigation and Abatement. *SAE Paper 800495, February, 1980*

Nomenclature

A	Loaded surface area	p	Specific load
A_{pr}	Projection of loaded surface area for a spherical laminate	P_d	Diametral pitch
		R_s	Spherical radius of a spherical laminate
D	External diameter	r	Pitch radius
d	Internal diameter	S	Shape factor
E	Compression modulus	T	Torque
e	Excentricity	t_m	Thickness of a metal layer
f	Friction coefficient	t_r	Thickness of a rubber layer
G	Shear modulus	V	Potential energy
H	Rubber durometer	W	Transmitted energy
h_r	Total thickness of rubber in a laminate	Z	Number of teeth
k	Stiffness	Δh_r	Deformation of rubber
n	Number of metal layers in a laminate	ϵ	Relative deformation
P	Load	ν	Poisson's ratio
		η	Efficiency
		ψ	Relative energy dissipation per cycle

Improvement of Machining Conditions for Turning of Slender Bars by Application of Tensile Force

E. Rivin, P. Karlic, and Y. Kim
Department of Mechanical Engineering
Wayne State University
Detroit, Michigan

ABSTRACT

Low stiffness of part or tool in a machining system leads to geometric distortions of the machined part, to inferior surface finish, to chatter vibrations, etc. While special designs and materials can be used for tooling, such approaches are not feasible for parts. Use of steady rests requires the use of many tool holding surfaces in CNC turning centers, thus limiting their flexibility; steady rests are not suitable for parts with abrupt changes in diameters.

The proposed paper describes application of a tensile force to the part being machined, which leads to increased bending stiffness ("reverse buckling" or "guitar string" effect). The application of the tensile force to slender parts was studied using a specially designed tail stock modification. Static, dynamic, and cutting tests have been performed. Static stiffness increase of 2–3 times has been

Reprinted from "Fundamental Issues in Machining," ASME PED-Vol. 43, 1990, with permission of ASME.

observed, with corresponding increases in natural frequencies of the system. Cutting tests on steel bars demonstrated dramatic improvements in surface finish (from R_a = 300–400 μin down to 40–50 μin for 0.5 in diameter, 17 in long part) and in cylindricity (from 0.004–0.005 in down to 0.0001–0.0002 in).

A secondary effect of improvement of surface finish and cylindricity even on relatively short parts when the tensile force was applied has been observed.

1. INTRODUCTION

Stiffness is one of the most important characteristic parameters of a machining system. Low stiffness of a part (e.g., a slender workpiece being machined on a lathe) or of a tool (e.g., a cantilever tool with a large L/D ratio) leads to geometric distortions of the machined part, to inferior surface finish, to chatter vibrations resulting in reduced productivity and shortened tool life, etc. While special designs, materials with higher Young's modulus and enhanced damping, etc. can be used for tooling, such approaches are not feasible for parts, although in some cases a proper "tuning" of the tool (assignment of its stiffness and damping) could partially compensate for part compliance (Rivin and Kang, 1989a). Machining of slender parts is typically done on lathes, and steady rests are used to provide an intermediate support and thus enhance effective part stiffness. Since stationary steady rests interfere with the machining, travelling steady rests are frequently employed. Problems with travelling steady rests include difficulties in servicing parts with abrupt changes in diameter and limitation of the flexibility of automated machines. For example, installation of a travelling steady rest on a CNC turning center requires up to three faces of its turret, thus significantly reducing availability of cutting tools.

The issue of turning a slender part is one case of an important generic problem of improving dynamic quality of machining systems with low dynamic stiffness. The problem also includes low stiffness (usually, cantilever) tooling structures. Some aspects of this problem, especially related to cantilever boring bars, were addressed, e.g., by Hahn (1951), Thomas, et al. (1970), Peters and Vanherk (1981), Rivin and Kang (1989b). The issue of low stiffness parts has received much less attention with the exception of some stiffening techniques involving the use of low melting temperature material (ice, Wood alloy, etc.) for filling up thin-walled parts. Besides Rivin and Kang (1989a) and Masuda and Watanabe (1982), we could not find publications on passive techniques for enhancement of dynamic quality for slender parts being machined. In the semi-active method described by Masuda and Watanabe (1982), the part is measured after the first cut, then radial tool positions along the part length which are necessary to compensate part geometry are calculated, and during the cut the tool is programmed to attain these precalculated positions using a micromanipulator. Active techniques (with on-line measurement of diameter deviations) are still in their infancy. Even if sensors for on-line measurements were available, enhancement of chatter resistance would be a difficult task.

This paper describes a concept which results in a significant enhancement of the part stiffness (and, it seems, also of the machine stiffness) during machining, which allows to achieve a stable cutting process and high geometric accuracy and surface finish without resorting to steady rests. Such an effect is achieved by applying tensile force to the part being machined during the machining process.

2. THE CONCEPT

It is well known that application of a compressive force to a relative slender beam would cause a collapse (buckling) of the beam after the force reaches its critical value, usually referred to as the Euler force P_{cr}. However, only very infrequently is attention devoted to the development of the buckling phenomenon. If the beam which is compressed by force P is also acted upon by a bending moment M^1, then the effective moment M would have a value (Blake, 1985).

$$M = \frac{M^1}{1 - P/P_{cr}} \tag{1}$$

Since the effective bending moment determines the deformation (or, in other words, stiffness) of the beam, the buckling process can be described alternatively as a gradual reduction of the bending stiffness of the beam, the stiffness reducing to zero when $P = P_{cr}$.

If the beam is loaded with a tensile force, the equations leading to determining the Euler force or the derivation of expression (1) do not change, the only difference being the sign change for the force P. Thus, the effective moment acting on a beam loaded with an axial tensile force P would be

$$M = \frac{M^1}{1 + P/P_{cr}} \tag{1'}$$

and its stiffness accordingly increases with increasing tensile force. As a result, the natural frequencies of the beam also increase,

$$f_n = f_n^1 \sqrt{1 + \frac{1}{n^2} \frac{P}{P_{cr}}} \tag{2}$$

where f_n^1, f_n are the nth natural frequency values of the beam without and with tensile force respectively (Rivin, 1988). This process is responsible for a changing pitch of a guitar string when its stretch is adjusted.

Turning of long parts usually is performed while the part is installed between two centers, or clamped in the chuck and supported by the tailstock center. In both cases, a significant compressive force is applied to the part which, as shown above, tends to reduce its effective stiffness below it's low original value.

2.1. Effects of Static Stiffness

In turning long slender workpieces between headstock and tailstock centers, the turning machine, the workpiece, and the tool system comprise a flexible system which is subjected to deflection due to the cutting force. The inaccuracy in the diameter is mainly due to deflections of the system components: headstock, carriage, tailstock, workpiece and the tool. The diameter inaccuracy is mainly affected by the transverse cutting force P_y as shown in Fig. 1. The diameter in-

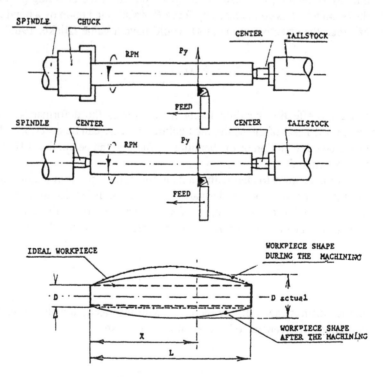

Fig. 1. Generation of "barrel shape" of a slender part machined with a rigid tool; solid line–actual shape, broken line–ideal cylindrical shape with rigid both part and tool; chain line–deformed part during cutting slender part with the rigid tool.

crease of the workpiece y_x at a distance x from the headstock is a result of superposition of deflections of the system components (headstock and tailstock) and deflection of the workpiece with simple supports (two centers). This is expressed according to Koenigsberger and Tlusty (1970) as

$$y_x = P_y\left[\frac{1}{K_t} + \left(\frac{L-x}{L}\right)^2\frac{1}{K_H} + \left(\frac{x}{L}\right)^2\frac{1}{K_T} + \frac{1}{K_w}\right] \qquad (3)$$

Here L is workpiece length; x is position along the workpiece; K_t is the equivalent stiffness of cutting tool; K_H is the stiffness of tailstock. K_w is the stiffness of the workpiece and varies along its length; for a constant diameter workpiece, $K_w = [(L - x)^2x^2]/3EIL$, where E is Young's modulus, and I is the cross-sectional moment of inertia of the workpiece.

For relatively slender parts, the last term in (3)—deflection of the part itself—is predominant. For the lathe which was used in the experimental studies described below (a relatively old machine), $K_H = 6.9 \times 10^5$ lb/in (123×10^6 N/m), $K_T = 0.625 \times 10^5$ lb/in (11.2×10^6 N/m), $K_t = 1.63 \times 10^5$ lb/in (29.1×10^6 N/m) (Rivin and Kang, 1989a), and the stiffness of a double-supported cylindrical beam (part) is shown in Fig. 2 as a function of L/D and diameter D. The part stiffness at $L/D \geq 5$ and $D \leq 0.7$ in is much lower than the machine and tool stiffnesses. This effect is demonstrated by Fig. 3 (Rivin and Kang, 1989a), which shows P_y/y_x along a cylindrical bar 0.7 in (17.8 mm) diameter and 15 in (380 mm) long for two values of the tool stiffness. At the maximum deflection point (midspan), a 15-fold reduction in tool stiffness resulted in only about 50% reduction in overall stiffness. Accordingly, enhancement of the part (workpiece) stiffness is important for improving part geometry.

2.2. Effect of Tensile Force

As was stated above, the enhancement of the effective bending stiffness of slender parts having low Euler forces P_{cr} can be achieved by application of tensile forces. The most important case in turning is when the workpiece is supported by two centers (dead center at the headstock and live center at the tailstock). Such a case can be modelled as a doubly supported beam. Since the largest deflection of such a beam occurs when the cutting force is applied at the midspan, the effect of the tensile force on deflection in the middle of the beam under the bending load (cutting force) applied in the same point should be analyzed. This case is analyzed in (Rivin, 1988), and it was shown that the maximum deflection with the tensile force y_{max} is related to the maximum deflection y^1_{max} without the tensile force as

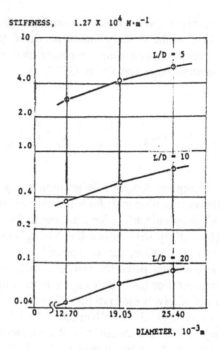

Fig. 2.   Stiffness of a double supported cylindrical steel rod at its midspan as a function of L/D and D; stiffness units 1.27×10^4 N/m.

$$y_{max} = y_{max}^1 \frac{(\alpha L/2) - \tanh(\alpha L/2)}{1/3(\alpha L/2)^3} \tag{4}$$

where $\alpha = \sqrt{P/EI}$. This expression is illustrated in Fig. 4.

The product αL can be transformed in the case of a constant diameter cylindrical beam in order to better visualize the stiffening effect. For such a beam, the cross-sectional area $A = \pi D^2/4$, $I = \pi D^4/64$, $I/A = D^2/16$, and according to Rivin (1988)

$$\alpha L = L\sqrt{\frac{P}{EI}} = L\sqrt{\frac{P1A}{AEI}} = L\sqrt{\frac{\sigma_r}{E}\frac{16}{D^2}} = 4\frac{L}{D}\sqrt{\frac{\sigma_r}{E}} = 4\frac{L}{D}\sqrt{\epsilon_T} \tag{5}$$

where σ_r is the tensile stress caused by the tensile force P, and ϵ_T is the respective relative elongation from the tensile force.

Thus, the effect of the tensile force on the stiffness of a slender workpiece quickly increases with increasing "slenderness ratio" L/D. It explains the very high range of stiffness (or pitch) change during stretching of a guitar string, whose

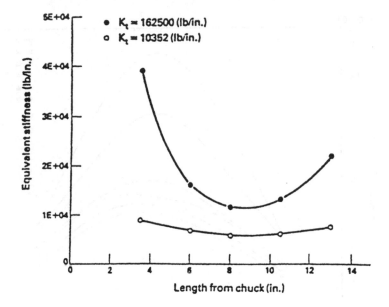

Fig. 3. Equivalent static stiffness between workpiece and tool at various points along the workpiece; K_t–stiffness of tool.

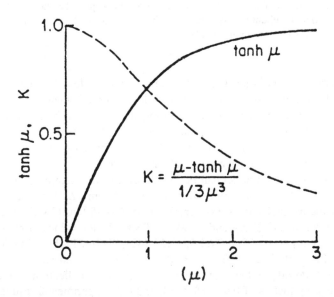

Fig. 4. Stiffening factor ($K = y_{max}/y_{max}^1$) for a double supported bar with applied tensile force; $\mu = \alpha L/2$.

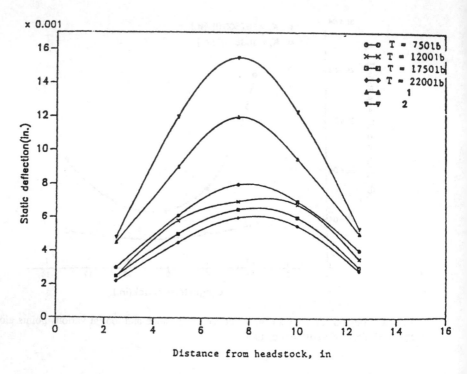

Fig. 5. Static deflection diagrams under force 90 N applied at the midpoint of cylindrical part with and without tensile force; 1—chuck and tailstock dead center; 2—between two dead centers (both without tensile force).

L/D ratio is very large. The maximum stiffening effect depends on the allowable tensile strength. While machining of high strength materials generates higher cutting forces and deflections, high tensile forces can be applied due to higher allowable tensile stresses of such materials. The stiffness effect is more pronounced for lower modulus materials (due to higher $\sigma_r/E = \epsilon_T$). It is important since deflections for such materials are larger.

The stiffening effect can be illustrated for the example of a steel workpiece with D = 0.5 in (12.7 mm), L = 15 in (387 mm), L/D = 30, which was used for experimental studies. For tensile loads P = 750 lbs (3300 N), 1200 lbs (5330 N), 1750 lbs (7770 N), and 2200 lbs (9800 N), the values of σ_r, ϵ_T and αL are, respectively, σ_r = 3820 (26.25), 6120 (42), 8900 (61), 11220 (77) psi (MPa); ϵ_T = 1.27 × 10⁻⁴; 2 × 10⁻⁴; 3 × 10⁻⁴; 3.7 × 10⁻⁴ and αL = 1.35; 1.70; 2.08; 2.30. Accordingly, the reduction in the workpiece deflection would be, approximately, by factors of 0.65, 0.5, 0.38, and 0.3. Experimental results are shown in Fig. 5 and discussed later.

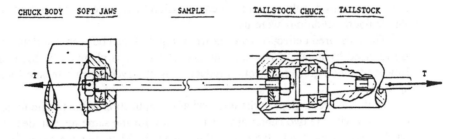

CHUCK BODY   SOFT JAWS   SAMPLE   TAILSTOCK CHUCK   TAILSTOCK

Fig. 6. Set-up for application of tensile force to part.

3. EXPERIMENTAL STUDY

The effects of the application of tensile force to the machining system have been studied experimentally under static conditions, under dynamic excitation, and during actual machining. The test program was performed on an engine type lathe. Static stiffness values of the spindle (headstock), of the tailstock, and of the tool holder of the lathe had been measured for the machine.

Two specimens were used in the experiments: a steel bar 0.5 in (12.7 mm) dia, 8 in (203 mm) long between the supports, and another steel bar of the same diameter and 17 in (432 mm) long between the supports. Both specimens were made of mild steel 1141.

To apply a tensile force to the specimen, one end was clamped by a three jaw chuck. The other end is threaded. To apply the tensile force, a nut is screwed on the threaded end and then locked in cavity A of the modified tailstock quill, Fig. 6. The applied tensile force is measured by strain gages attached to the quill. The quill was calibrated for measuring tensile force using Instron Model 1350 servohydraulic test sytem.

Special workpiece specimens with lengths between supports of 8 in and 17 in were made for the tests in which the workpiece is supported between two centers.

3.1. Static Tests

Static deformations of slender cylindrical bars with various support conditions without and with the tensile force were measured using shortened specimens (0.5 in dia and 15 in length between supports). The static deformations were measured under 20 lbs (90 N) force which represents a reasonable cutting force magnitude for such a workpiece. In cases when one center (chuck–tailstock center) or two centers were used, only a minimum compression force necessary to provide reliable support conditions were used.

The measured deflections are shown in Fig. 5. It can be seen that deflections are reduced by about 25% when one end of the bar is clamped by the chuck. However, much greater deflection reduction is observed when a tensile force is applied.

Deflection reductions at the midspan after application of the tensile force (as compared with the two center arrangement) are, for the same magnitudes of the tensile force, respectively, 0.52, 0.45, 0.42, and 0.38. There is a general agreement between the values predicted by equation (4) (0.65, 0.5, 0.38, 0.3) and measured values. The discrepancy is due to the difference between the supporting conditions (the tailstock center was replaced by the modified tailstock as shown in Fig. 6), and due to the fact that the tailstock deflections were not considered. The latter becomes important when the workpiece stiffness is high. It can be seen both from computed and especially from measured results that increasing the tensile force beyond a certain limit results in a somewhat diminished return.

3.2. Dynamic Tests

Frequency response functions were measured on the specimens with various support conditions, and without and with tensile force. Various excitation and measuring points along the specimen have been used, but most of the tests were performed with impact excitation at $0.25L$ from the headstock and accelerometer location at $0.125L$ from the headstock, where L is the specimen length between the supports. Fig. 7 shows frequency response (inertance or accelerance) plots for the long (17 in) specimens with various support conditions. When the specimen was supported by the tailstock center (plots 2, 3), compression force $P = -500$ lb (2200 N) has been applied, which can be considered as a realistic value. The plots show dramatic differences in dynamic characteristics depending on support conditions.

For the least rigid arrangement, curve 3 (the specimen between two centers), there are four distinct peaks (natural frequencies) on the frequency response plot (see Table 1). When the specimen is clamped in the chuck and supported with the tailstock center (curve 2), three prominent peaks are observed, Table 1. When the tensile force of 2000 lbs (8800 N) is applied (curve 1), there are also three natural frequencies in the 0–2000 Hz range, but they are much higher, see Table 1.

Damping at the two lowest resonant peaks is also slightly higher for the case with tensile force. While damping at the third resonant frequency f_3 is the highest for the two center support case, the overall height of the response plot in the 500 Hz–1400 Hz range is the lowest for the case with the applied tensile force.

Fig. 8 and Table 1 demonstrate the effect of a changing tensile force. The effect is more significant at the fundamental frequency f_1 (15% increase in fre-

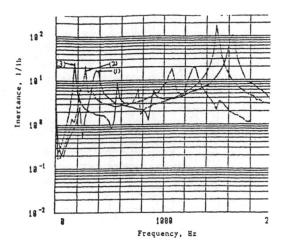

Fig. 7. Frequency–response characteristics of part (L = 17, in D = 0.5 in) under impact excitation (impact position at 0.25 L from headstock, measuring position at 0.125 L from headstock); 1–tensile force T = 2,000 lb (9,000 N), fundamental natural frequency center), f_1 = 366 Hz; 2–compression force 500 lb (2,250 N) (chuck and dead center), f_1 = 293 Hz; 3–compression force 500 lb (2,250 N) (two dead centers), f_1 = 171) Hz.

quency with the tensile force change from 250 lb to 2000 lb) than at the higher frequencies (8% at f_2 and 4% at f_3).

The noted behavior of the resonant frequency values can be explained as follows: the tensile force, besides stiffening the workpiece, also results in closing and tightening joints in the machine tool structure. Thus the effect of the tensile force goes beyond just stiffening of the workpiece. While the compression force, which is present when the traditional support conditions are used, would also

TABLE 1 Resonant Frequencies Measured on 17 in Bar with Various Support Conditions

Support conditions	Resonance frequency, Hz			
	f_1	f_2	f_3	f_4
Chuck–center 500 lb compression	293	675	1490	
Center–center 500 lb compression	171	560	1060	1280
250 lb tension	317	820	1565	
500 lb tension	317	845	1580	
1,000 lb tension	342	870	1605	
2,000 lb tension	366	900	1630	

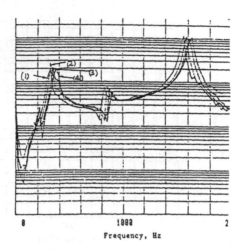

Fig. 8. Frequency–response characteristics of part (L = 17 in, D = 0.5 in) under impact excitation (impact and measuring positions as in Fig. 7) and subjected to tensile force of various magnitudes; 1—T = 250 lb (1125 N), f_1 = 317 Hz; 2—T = 500 lb (2250 N), f_1 = 317 Hz; 3—T = 1500 lb (6750 N), f_1 = 342 Hz; 4—T = 2000 lb (9000 N), f_1 = 366 Hz.

tighten the joints, stiffness reduction of the workpiece leads to deterioration of the overall picture. Also, the compression force may have a different effect from the tensile force due to design specifics of the lathe.

This hypothesis is reinforced by the dynamic test of the short (8 in long) specimen. The inertance frequency response plot in Fig. 9 compares the cases of 500 lb compression between two centers (curve 3, f_1 = 635 Hz), 500 lb compression using the modified tailstock and the specimen designed for application of the tensile force (curve 2, f_1 = 953 Hz), and 2,000 lb tensile force using the same arrangement (curve 1, f_1 = 1245 Hz). Such a pronounced effect (about 30% frequency change between curves 2 and 1, corresponding to a 70% change in the effective stiffness) cannot be explained just by changing the bending stiffness of the specimen when the axial force changes from −500 to +2,000 lb. It reinforces the hypothesis about structural changes in the machine under the influence of the tensile force.

3.3. Cutting Tests

The cutting tests were performed using inserts TPT-321 made of uncoated carbide (grade C-8) with nose radius 1/64 in. Some results of the tests are shown in Figs. 10a,b. Fig. 10a shows surface finish and deviations from cylindricity for a long (17 in) specimen machined at 99.5 sfm and feed rate 0.002 in/rev. Machining

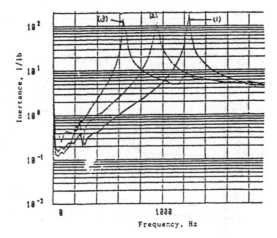

Fig. 9. Frequency–response characteristics of a shorter part (L = 8 in, D = 0.5) under impact excitation (impact and measuring positions as in Fig. 7); 1—tensile force, T = 2000 lb, f_1 = 1245 Hz; 2—compression force 1500 lb (6750 N) (same tailstock as for tensile force application), f_1 = 952 Hz; 3—compression force 500 lb (2250 N) (part between two dead centers), f_1 = 635 Hz.

without tensile force results in unacceptable surface roughness and larger deviations from cylindricity. It is interesting to note that while the surface roughness in the midspan is the worst, even at the headstock and tailstock where stiffness of the part is reasonably high (e.g., see Fig. 5), surface finish is very poor even at a small depth of cut, only t = 0.005 in. Fine cuts (t = 0.0025 − 0.005 in) with a 1,000 lb tensile force result in a very good surface finish R_a = 50 − 60 μin (1.25 − 1.5 μm), and very good cylindricity (within 0.0001 in). However, surface finish and especially cylindricity deteriorate with increasing depth of cut, t = 0.00 in. With further increase in depth of cut to t = 0.010 in, the surface finish becomes very bad even near the supports.

Similar effects are observed with an increasing feed rate to f = 0.003 in/ rev, Fig. 11a,b. In this case, deterioration of the surface finish (but not cylindricity) begins at a lower depth of cut. Similar qualitative effects are developing with increasing cutting speed, Fig. 12a,b. At f = 0.002 in./rev, t = 0.005 in, and P = 1,500 lbs, increase of cutting speed from V = 99.5 to 161 sfm resulted in deterioration of the surface finish especially in the midspan area. One explanation for this can be increased deformations due to centrifugal forces caused by eccentricity of the part. All these effects indicate that the magnitude of the cutting force is responsible for deterioration of the part surface at higher depths of cut.

Substantial improvements in both surface finish and cylindricity were observed also on the short (8 in long) specimens, Fig. 13a,b. However, when the

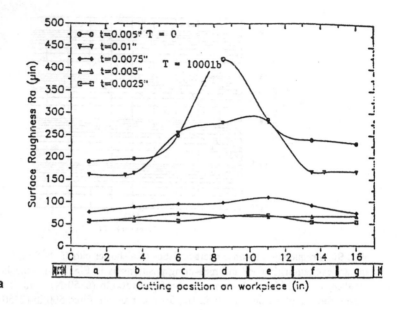

a

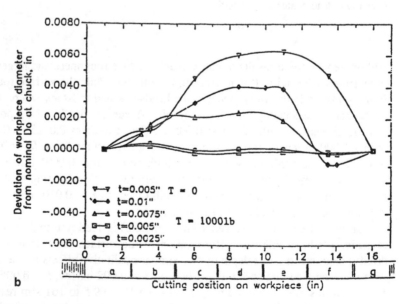

b

Fig. 10. Surface roughness (a) and cylindricity (b) of machined part (L = 17 in, D = 0.5 in, f = 0.002 in/rev, v = 99.5 sfm) without tensile force and with 1,000 lb [(4,500 N) tensile force and variable depth of cut.

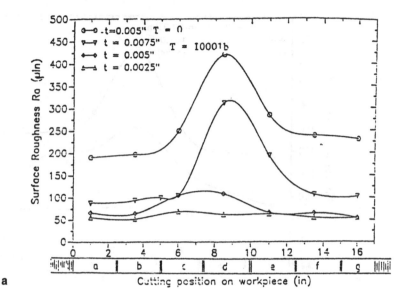

a

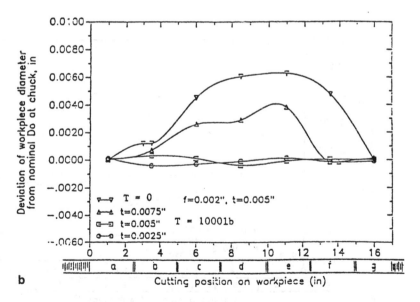

b

Fig. 11. Surface roughness (a) and cylindricity (b) of machined part (L = 17 in, D = 0.5 in, f = 0.003 in/rev, v = 99.5 sfm) without tensile force and with 1,000 lb (4,500 N) tensile force and variable depth of cut.

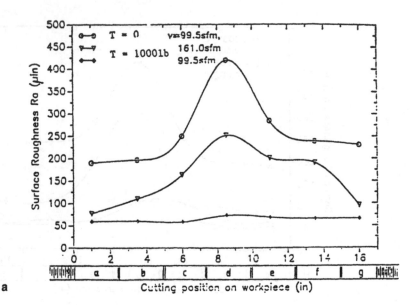

a

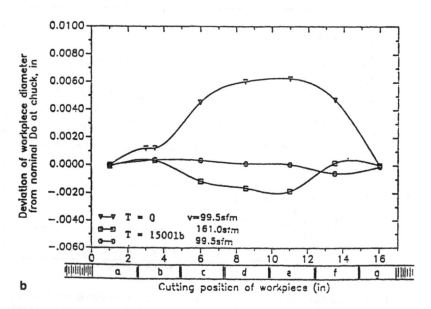

b

Fig. 12. Surface roughness (a) and cylindricity (b) of machined part (L = 17 in, D = 0.5 in, f = 0.002 in/rev, t = 0.005 in) without tensile force and with 1500 lb (6750 N) tensile force, and variable cutting speed.

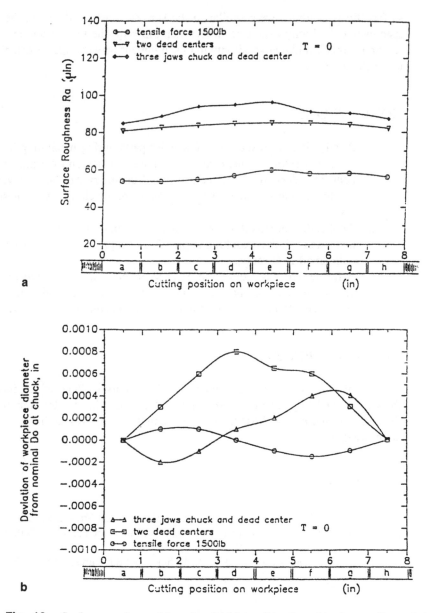

Fig. 13. Surface roughness (a) and cylindricity (b) of machined part (L = 8 in, D = 0.5 in, f = 0.002 in/rev, t = 0.005 in, v = 99.5 sfm) with different support conditions.

surface finish readings were similar to the surface finish readings on the long specimen, cylindricity results are worse than those for the long specimens. One reason for this is a higher sensitivity to misalignment between the headstock and tailstock supports.

CONCLUSIONS

1. Application of tensile force to slender parts during machining leads to improved static stiffness, dynamic characteristics and cutting process behavior. Very good surface finish and geometry can be achieved for machining parts which otherwise cannot be machined without steady rests.
2. Application of tensile force is beneficial to both the workpiece and the machine tool, thus resulting in significant improvements even for parts with small L/D ratios.
3. A further work is required in order to better understand interactions of various system parameters during cutting of the stretched workpieces, and to develop practical techniques for application of the tensile force.

ACKNOWLEDGMENT

Support of this project by the Wayne State University College of Engineering through the DeVlieg Award for Research in Manufacturing for the first and the third authors and by the WSU Institute for Manufacturing Research is gratefully acknowledged.

REFERENCES

Blake, A. (Ed), 1985, "Handbook of Mechanics, Materials and Structures," John Wiley.
Hahn, R. S., 1951, "Design of Lanchester Damper for Elimination of Metal-Cutting Chatter," Trans. ASME, Vol. 73, No. 3.
Koenigsberger, F., and Tlusty, J., 1970, "Machine Tool Structures," Vol. 1, Pergamon Press.

ENHANCEMENT OF DYNAMIC STABILITY OF CANTILEVER TOOLING STRUCTURES

Eugene I. Rivin† and Hongling Kang‡

(Received 15 March 1991; in final form 17 June 1991)

Abstract—The least rigid components of machining systems are cantilever tools and cantilever structural units of machine tools (rams, spindle sleeves, etc.). These components limit machining regimes due to the development of chatter vibrations, limit tool life due to extensive wear of cutting inserts, and limit geometric accuracy due to large deflections under cutting forces. Use of high Young's modulus materials (such as sintered carbides) to enhance the dynamic quality of cantilever components has only a limited effect and is very expensive. This paper describes a systems approach to the development of cantilever tooling structures (using the example of boring bars) which combine exceptionally high dynamic stability and performance characteristics with cost effectiveness. Resultant success was due to: (1) a thorough survey of the state of the art; (2) creating a "combination structure" concept with rigid (e.g. sintered carbide) root segments combined with light (e.g. aluminum) overhang segments, thus retaining high stiffness and at the same time achieving low effective mass (thus, high mass ratios for dynamic vibration absorbers, or DVAs) and high natural frequencies; (3) using the concept of "saturation of contact deformations" for efficient joining of constituent parts with minimum processing requirements; (4) suggesting optimized tuning of DVAs for machining process requirements; (5) development of DVAs with the possibility of broad-range tuning; (6) structural optimization of the system; and (7) using a novel concept of a "Torsional Compliant Head", or TCH, which enhances dynamic stability at high cutting speeds and is suitable for high rev/min applications since it does not disturb balancing conditions. The optimal performance and interaction of these concepts were determined analytically, and then the analytical results were validated by extensive cutting tests with both stationary and rotating boring bars, machining steel and aluminum parts. Stable performance with length-to-diameter ratios up to $L/D = 15$ was demonstrated, with surface finish 20–30 μin with both steel and aluminum at $L/D = 7$–11. Comparative tests with commercially available bars demonstrated the advantages of our system.

INTRODUCTION

THE PRINCIPAL parameters of any machining operation, productivity and accuracy/surface finish, are determined, in large part, by the static and/or dynamic stiffness (rigidity) of the machine-fixture-tool-part system. Since the overall stiffness is as high (or as low) as the stiffness of the weakest component, the latter obviously deserves serious investigation. The least rigid components of a typical machine tool are cantilever structural units, such as tail-stock spindles in lathe-type machines, and rams and sliding spindle sleeves in horizontal and vertical boring mills, NC machining centers, etc. Similarly, the least rigid tooling components are cantilever tool-holders and tools, such as cantilever (stationary and rotating) boring bars, internal grinding mandrels (quills), end mills, drills, standard tooling for NC machines with tool changers, measuring probes, etc. If the static/dynamic rigidity of these cantilever elements is inadequate, they:

directly limit the attainable accuracy, due to their easy deflection, even under low-magnitude cutting forces;

indirectly limit accuracy, since their high-frequency micro-vibrations lead to noticeable (in the 0.0001 in range) wear of cutting inserts during each cutting cycle, resulting in tapered surfaces instead of the required cylindrical ones; and

limit machining regimes through the generation of self-excited vibrations ("chatter") at relatively low cutting regimes when the length-to-diameter (L/D) ratio of the overhang segment exceeds 4 : 1–5 : 1

While the L/D ratio is of critical importance for the performance of cantilever tooling, it must be remembered that in order to machine or inspect a hole, the diameter, D,

---

†Department of Mechanical Engineering, Wayne State University, Detroit, MI 48202, U.S.A.
‡ACME Engng & Mfg Corp., Muskegee, OK 74402, U.S.A.

Reprinted from Int. J. Mach. Tools Manufact., 1992, Vol. 32, No. 4, with permission of Pergamon Press.

of the tool should be smaller than the hole diameter, to accommodate a cutting or measuring head, but that its length, L, should be longer than the hole depth to eliminate interference between the tool-holding spindle and the part being machined. In automated tool-handling systems (machining centers, flexible manufacturing systems, etc.) the useful length of tools is further reduced by the need for special design features such as gripping surfaces for tool changers, surfaces for identification codes, etc.

Even with all these limitations, cantilever cutting and measuring tools are widely used in manufacturing. They are very versatile and can be used in blind cavities, holes with varying diameters, etc. They do not require support structures, and thus are uniquely suitable for applications in automated systems, such as machining centers and flexible manufacturing cells.

Substantial research efforts have been aimed at developing more stable cantilever tools working with longer overhangs. More stable (chatter-resistant) tools make possible deeper cuts, and thus fewer cuts, resulting in higher productivity. Chatter-resistant tools also combine better surface finish with more intensive machining regimes. They demonstrate slower cutter wear due to reduced high-frequency vibration amplitudes, which may result in better cylindricity of the machined surfaces, especially in machining cast iron. However, the most important contribution made by stable cantilever tools with long overhang would be to mechanical design. Tools capable of machining long blind holes, especially in modern high-strength, hard-to-machine alloys, would allow much simpler designs which do not require the changes forced upon the designer by an imperfect manufacturing technology. These changes complicate the design in order to make feasible the machining of needed internal cavities, and tend to add extra assembly operations.

Typical representatives of cantilever tooling structures are cantilever boring bars, which perform critical finishing operations on internal cylindrical cavities (holes), grooves, etc. This paper describes a systems approach to the problem of enhancing chatter resistance and general performance of cantilever structures, using the example of cantilever boring bars.

There are two basic types of boring bars: stationary bars for performing boring operations on rotating parts (e.g. on lathes), and rotating bars used on machining centers, boring mills, jig borers, etc. Effort was mainly concentrated on studies of stationary bars, which require much simpler facilities for cutting tests, but the principal solutions are applicable to rotating bars as well.

STATE OF THE ART

Several techniques are known for enhancing the dynamic stiffness and stability (chatter-resistance) of cantilever tools and, thus, for increasing allowable overhang. The four most widely-used and most universal approaches are: (1) use of anisotropic mandrels (bars) with specifically assigned orientations of the stiffness axes; (2) use of high Young's modulus and/or high damping materials; (3) use of passive dynamic vibration absorbers (DVAs); and (4) use of active vibration control means.

The use of *anisotropic mandrels* (surveyed by Thomas *et al.* [1]) is based on a theory explaining the development of chatter vibrations during cutting by an intermodal coupling in the two-degrees-of-freedom system referred to the plane orthogonal to the mandrel axis and passing through the cutting zone (e.g. see Tlusty [2]). According to this theory (and to practical experience) there is a specific orientation of stiffness axes relative to the cutting forces, resulting in a significant increase in dynamic stability (chatter resistance). Since precise relative orientation of the force vector and stiffness axes is hardly possible, and creation of the anisotropic structure requires a deliberate weakening of the mandrel in one direction, the overall effect could be marginal or even negative for a mandrel designed for universal applications. This conclusion is confirmed by experiments by Mescheriakov *et al.* [3].

The most frequently-used *high Young's modulus materials* are sintered tungsten carbide, $E = \sim 5.5 \times 10^5$ MPa (80×10^6 psi), and machinable sintered tungsten alloy with an added 2–4% of copper and nickel, $E = \sim 3.5 \times 10^5$ MPa (50×10^6 psi) (Kennertium, Malory No-Chat, etc.). Both materials are expensive. For sintered carbide, manufacturing (machining of the surfaces to be joined with other parts of the mandrel, and the joining operation itself) is even more expensive. Sintered tungsten alloy is relatively easy to machine, and is characterized by relatively high damping, which compensates for the significantly lower Young's modulus of this material. Frequently stability of the cutting process is determined by the criterion $K\delta$, where K is the stiffness of the critical structure, and δ is its log decrement; see, e.g. Tlusty [2]. Thus, solid mandrels made of both these materials allow stable cutting with ratios $L/D \leq 7$.

At present, the commonest approach to enhancing the dynamic stability of cantilever structures is the application of *dynamic vibration absorbers (DVAs)* with an inertia mass. DVAs are successfully used; not only for manufacturing tools, but also for other cantilever structures, such as high-rise buildings in which unwanted and dangerous vibrations are excited by aerodynamic forces. Successful applications include TV transmission towers and many ultra-high office buildings, such as those described in Ref. [4] and analysed by Jenniges and Frohrib [5].

The effectiveness of a DVA for a given mass depends on the vibration amplitude at its attachment point. Accordingly, absorbers are usually installed at the furthest available positions along the cantilever. Another factor determining their effectiveness is the mass value of the inertia weight of the DVA. While in high-rise buildings the inertia mass can be located on the outside walls and on designated floors, thus allowing them to be relatively large (66 tons in Ref. [4]), in hole-machining tools, such as boring bars, the DVA has to be placed inside the structure. This limits both the position of the DVA along the cantilever axis (since the end of the cantilever is occupied by the tool head) and the size of the inertia weight, which has to be positioned in an internal cavity whose diameter must be much smaller than the bar diameter. To achieve a reasonable degree of DVA effectiveness, materials with very high specific density must be used for inertia weights. The most frequently-used material is sintered tungsten alloy (specific density $\gamma = \sim 16$).

Various types of DVA are used for cantilever tooling, with the majority of applications being for boring bars. These include impact dampers (e.g. Thomas *et al.* [1], Iwata and Moriwaki [6]), granular inertia mass dampers (e.g. Briskin [7], Popplewell *et al.* [8], and Araki *et al.* [9]), and tuned DVAs (e.g. Donies and Van den Noortgate [10]), etc.

The most widely-used DVAs for cantilever tools are Hahn dampers [11], consisting of a cylindrical slug with a calibrated clearance of $\sim$1–5 $\times 10^{-3}$ d into a coaxial cavity in the mandrel; d is the slug diameter. The Hahn damper is analogous to the Lanchester damper [12], with some modifications as discussed in Ref. [6]. Lanchester-like dampers are far less effective than tuned DVAs, as shown in Den Hartog [12]. However, they have the advantage of low sensitivity to system parameters. This feature is especially important for cantilever tooling, which can be used with various overhangs and thus is characterized by the variable natural frequency and effective mass of the system. Steel boring bars with Hahn dampers are stable at $L/D = 7$–8; sometimes very expensive solid carbide bars with absorbers are used, for which $L/D = \sim 10$.

The best designs of tuned DVAs, which have both optimal tuning and optimal damping, allow performance of boring operations with steel boring bars at L/D up to 9.5 [10]. Although there are certain inconveniences in using tuned DVAs, as noted before, this value is certainly the highest reported L/D value for a boring operation with a cantilever boring bar made of steel.

Active vibration dampers are effective, but require vibration sensors and actuators generating forces opposing the deflections of cantilever structures during the vibratory (chatter) process. The most frequently used are active systems with cavities in the mandrel body which are filled with pressurized oil. Pressure values in the cavities vary according to the output of the control system, thus generating dynamic deformations to cancel the chatter vibrations. The development and testing of boring bars with active vibration control systems are described, e.g. in Refs [13] and [14]. Present designs of active dampers are not very reliable in the shop environment, and may require frequent adjustment. On the other hand, they can also be used for compensation of static deflections of the cantilever system, with the appropriate sensors. The ultimate L/D ratio for a boring bar with an active damper, as reported by Dornhöfer and Kemmerling [14], is 10–12.

Periodic speed variation (e.g. Ref. [15]) is a universal chatter reducing technique being used for turning, grinding, and other operations. Its effectiveness depends on cutting paramters as well as on structural dynamic characteristics of the machining system. In some cases it is only marginal. Also, this technique requires a major retrofitting effort for older machine tools. On newer machine tools, equipped with variable speed electric drives, it seems to be promising.

THE SCOPE OF THE STUDY

Since the state of the art survey shows that the most effective means of performance enhancement for cantilever boring bars are the use of high Young's modulus materials and the use of dampers and dynamic vibration absorbers, the natural first step is to optimize these approaches, to make them cost-effective, and to develop simple analytical techniques for designing optimized structures. Then, other techniques were developed to supplement known ones and to enhance their effectiveness further. In this study, efforts were concentrated on structural design issues. Deliberately, there was no attention paid to the geometry and material of cutting inserts. All cutting tests were planned to be of a comparative nature and using the same, not necessarily optimum, cutting inserts, for the design embodiments being tested. Thus, the achieved results can be further improved in the future if more advanced insert geometries and materials are used.

DESIGN OF CANTILEVER BORING BARS WITH DVA

Combination structure

The "combination structure" proposed by Rivin and Lapin [16] and described by Rivin [17, 18], was a major breakthrough in the technology for boring bars, and for cantilever structures in general. In the combination structure, Fig. 1(a), the root segment of the cantilever is made of a material with a relatively high Young's modulus, while no special attention is paid to its specific density, which can be quite high. However, even the high density of the root segment does not lead to a substantial increase in the effective mass of the structure in its fundamental mode. The overhanging (free end) segment of the combination structure, on the other hand, is designed to be light, while no special attention is paid to its Young's modulus, since the effective structural stiffness does not depend noticeably on the stiffness of the free end segment. These statements can be illustrated by the following numbers: if a solid bar is deflected by an end loading, $\frac{7}{8}$ of the potential deformation energy is contained in the root half of the bar, with $\frac{1}{8}$ in the overhang half. Also, if the bar is vibrating in the first (fundamental) mode, $\frac{3}{4}$ of the kinetic energy is contained in the overhang half, and only $\frac{1}{4}$ in the root half. Figures 1(b) and (c) show the effective stiffness and natural frequency of a combination sintered tungsten carbide–aluminum structure as a function of the ratio between the "rigid" and "light" segments. It can be seen that for the ratio corresponding to the highest natural frequency (about $L_1/L_2 = 0.5$), effective stiffness reduced to the end is only about 15% less than the stiffness of the solid structure made of the "rigid" material, while its effective mass is 3.5 times less than for the solid carbide bar. The natural frequency is 1.8 times higher than for the solid carbide bar. The most significant effect of such a combination structure is a reduction in the effective mass, which allows for an enhanced mass ratio of the absorber for a given size of inertia weight, and thus greatly improves its effectiveness.

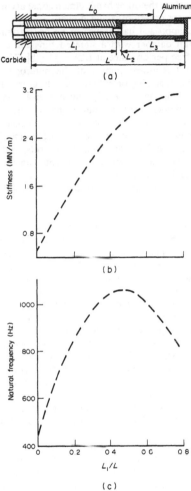

FIG. 1. Effective stiffness (b) and natural frequency (c) for combination boring bar (a).

Computational optimization of the combination structure can be performed using various criteria. High stiffness is, of course, important for maintaining the geometric accuracy of the machined hole, while the dynamic quality (chatter resistance) of the system is a function of the criterion $K\delta$, where K is effective stiffness of the machining system, and δ is its damping characteristic (in this case, log decrement), as mentioned above. While K increases with the longer "rigid" segment, δ can be enhanced by using a dynamic vibration absorber, the more so the smaller the effective mass of the structure (or the longer the "light" segment). Optimization by this criterion requires a more sophisticated model and is described later. The relative importance of these optimization parameters depends, to a certain degree, on the prevalent application of the tool. If chatter resistance is of higher importance (e.g. for rough machining with intensive chip removal), then the mass ratio of the dynamic vibration absorber would be a more important criterion. If on the other hand, dimensional accuracy is of critical importance, then the static stiffness criterion should prevail.

Such a structural approach has proven to be an extremely effective one for boring bars. Use of a Hahn damper with a sintered tungsten carbide-aluminum boring bar resulted in chatter-free performance at $L/D = 11$ [18]. This latter design also incorporated other important features: joining of the heterogeneous segments without any positive means, such as brazing or adhesive bonding, but using the saturation property of non-linear joint compliance [19, 20], and use of a very simply-shaped (and simply machined) carbide element with a reduced mass. These features combine superior performance with very reasonable cost, and with savings of a strategic material.

The same principles of structural design can, of course, be applied to other types of cantilever tooling, such as internal grinding mandrels, end mills, drills, NC machining center tooling, etc. The same structural concepts can be used for machine tool and other production machinery frames. For example, the problem of the dynamic stiffness/stability of sliding overhang rams is quite important (see, e.g. Seto [21]).

Mathematical model

To advance the state of the art of combination structures applied to boring bars by means of dynamic vibration absorbers and other vibration control devices, it is important to have an adequate mathematical model. This will allow designers to perform a reliable simulation of a boring bar, together with the cutting process, and on this basis to optimize a design with a smaller number of time-consuming cutting tests.

The mathematical model of a machining system should include structural parameters of the machine tool and of the tool holder, and also the dynamic characteristics of the cutting process. Since boring bars of $L/D > 5$-7 have relatively low stiffness, the influence of a usually rigid machine tool structure is very small and can be neglected. The boring bar structure is a beam-like distributed parameter structure. However, it has definitely been established by numerous studies that the chatter vibrations of cantilever boring bars always develop at their lowest (fundamental) natural frequency. Accordingly, the dynamic model of the boring bar was assumed to be a single-degree-of-freedom lumped parameter system whose stiffness and mass are the effective stiffness and mass of the bar reduced to the point of attachment of the cutting insert (or, for analysis of the bar equipped with a vibration control device, to the center of such a device). See Rivin [17].

The dynamic characteristic of the cutting process is an expression for the variation of the cutting force ("dynamic cutting force") as a function of cutting parameters. Several such expressions have been proposed (e.g. in Tobias [25] and Tlusty [23]). The former expression for variation of the cutting force component F_x perpendicular to the machined surface (the most important component for generation of chatter vibration) is:

$$dF_x = k_1 \left[x(t) - \mu_x x(t-T) \right] + k_2 \frac{dx}{dt} \qquad (1)$$

where x = chip thickness, T = the time for one revolution of the tool or the workpiece, k_1, k_2 are dynamic coefficients, and $\mu_x = 0$-1—an overlap factor indicating the degree of influence of vibration marks on the machined surface from the previous pass on the chip thickness variation as perceived by the cutting tool. Since analysis of the chatter resistance deals with the onset of self-excited vibrations when the non-linearity of the vibratory process is not yet strongly pronounced, the process (chip thickness variation) can be assumed [22], to be harmonic and:

$$x(t) = Ae^{i\omega t} \qquad (2)$$

where A is an indefinite amplitude constant, and ω is the chatter frequency. Substituting equation (2) into equation (1), the latter can be rearranged as:

$$dF_x = K_{cx} x + C_{cx} \frac{dx}{dt} \tag{3}$$

where

$$K_{cx} = K_1 \left(1 - \mu_x \cos 2\pi \frac{\omega}{\Omega}\right); \quad C_{cx} = K_1 \frac{\mu}{\Omega} \sin 2\pi \frac{\omega}{\Omega} + K_2. \tag{4}$$

Here, $\Omega = 2\pi/T$ and K_{cx}, C_{cx} can be defined as "effective cutting stiffness" and "effective cutting damping", respectively. It can be seen that the dynamic cutting force can be modeled as a spring with "effective cutting stiffness" as the spring constant and a damper representing "effective cutting damping". Expression (3) would be the same if it described dynamic forces in a parallel connection spring K_{cx}–damper C_{cx}. Thus, the cutting process can be presented as a damped spring connection. The effective cutting stiffness and cutting damping are functions not only of cutting conditions but also of system (workpiece and/or tool) parameters (stiffness and mass), which enter equation (4) via frequency ω. The dynamic cutting force depends not only on displacement $x(t)$ but also on velocity dx/dt, and it is the velocity dependent term which may cause system instability.

For the general case of a three-dimensional cutting process, the dynamic cutting force can generally be written as:

$$dF = K_c x + C_c \frac{dx}{dt} + C_t \frac{d\phi}{dt} \tag{5}$$

where ϕ is (torsional) vibration in the cutting speed direction.

This representation of the cutting process as a damped linear spring can be illustrated by Fig. 2, where M is effective mass of the tool, K,C are its structural stiffness and damping, and K_c, C_c are effective stiffness and damping of the cutting process. This model allows easy determination of cutting stiffness from simple tests. When the system does not perform cutting, its (structural) natural frequency is:

$$f_{n1} = \frac{1}{2\pi} \sqrt{\frac{K}{M}} \tag{6}$$

since K_c, C_c are not present. During cutting, the natural frequency becomes:

$$f_{n2} = \frac{1}{2\pi} \sqrt{\frac{K + K_c}{M}}. \tag{7}$$

Thus the effective cutting stiffness is:

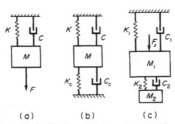

(a)   (b)   (c)

FIG. 2. Equivalent dynamic system of a boring bar with dynamic cutting force (a), its representation with cutting stiffness and damping (b), and boring bar with DVA (c).

$$K_c = \left(\frac{f_{n2}^2}{f_{n1}^2} - 1\right) K. \tag{8}$$

The cutting stiffness was determined for boring steel using the set-up in Fig. 3, in which a combination (carbide--aluminum) boring bar 1.25 in. in diameter and 12.75 in long was used, and $f_{n1} = 321$ Hz, $K = .3263$ lb/in. Vibrations were measured using LVDT transducers. The values of f_{n2} and of K_c determined by equation (8) are shown in Fig. 4.

Instability of the system in Fig. 2 (development of chatter vibration) would occur if the effective damping of the cutting process C_c is negative, and its effect overcomes the effect of the always-positive structural damping C.

Presenting the cutting process as a spring and damper simplifies and makes more transparent the comparative analysis of the stability of various designs of boring bars.

Dynamic vibration absorbers

Enhancement of chatter resistance of cantilever boring bars by dampers or dynamic vibration absorbers based on the use of inertia masses (whose basic concepts are described, e.g. by Den Hartog [12]) is hampered by the very small space available for an inertia mass in boring bar designs. The inertia mass must be located inside the relatively thin structure. To make the inertia mass more effective, heavy machinable tungsten alloy is utilized (its specific density about 16). However, even with this specific density, the mass ratio of dampers/absorbers does not exceed 0.75–1.0 for steel bars, and 0.4–0.55 for solid tungsten carbide bars. The use of combination structures can increase their ratio up to 1.4–2.0, with effective stiffness only slightly less than that of a solid carbide bar.

Due to enhanced mass ratios for combination boring bars, much higher stable slenderness ratios can be obtained. It was shown [18] that stable cutting in a narrow range of cutting regimes was possible at $L/D = 15$ for combination boring bars with a Hahn damper supplemented with a cavity filled with particles of machinable tungsten alloy.

Although it is well known that tuned DVAs are much more effective than Lanchester dampers (Hahn dampers in boring bars), they are highly sensitive to frequency and damping tuning. When a DVA is used in a boring bar, [10], it is tuned according to the classic scheme proposed by Den Hartog [12] for harmonic excitation of the main mass. However, there is no harmonic excitation during the cutting process. Accordingly, two cases were explored: a boring bar with a dynamic vibration absorber attached to it and acted upon by the dynamic cutting force as described by equation (3); and a boring bar with a DVA with the boring bar acted upon by random white noise-like vibrations, after the observation of Wu [24].

A boring bar (equivalent mass M_1, stiffness K_1, damping C_1) with a dynamic vibration absorber (mass M_2, stiffness K_2, damping C_2) is schematically shown in Fig. 2(c). For the first case (cutting force excitation), equations of motion for this system will be:

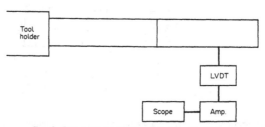

FIG. 3. Set-up for dynamic testing of boring bar.

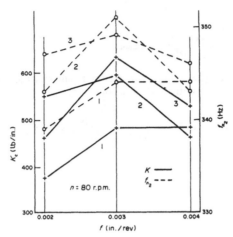

FIG. 4. Vibration frequency during cutting and effective cutting stiffness for various regimes: $L = 12.75$ in; $D = 1.25$ in; $F_n = 321$ Hz; $K_w = 3263$ lb/in; $1 - t = 0.01$ in; $2 - t = 0.02$ in; $3 - t = 0.03$ in.

$$M_1 \ddot{X}_1 + (C_1 + C_2)\dot{X}_1 + K_1 X_1 - C_2 \dot{X}_2 = F_x$$
$$M_2 \ddot{X}_2 + C_2 \dot{X}_2 + K_2 X_2 - C_2 \dot{X}_1 - K_2 X_1 = 0 \qquad (9)$$

or, after transformation,

$$\ddot{X}_1 + 2(\xi_1 \omega_1 + \xi_2 \omega_2 \mu)\dot{X}_1 + (\omega_1^2 + \omega_2^2 \mu) X_1$$
$$- 2\xi_2 \omega_2 \mu \dot{X}_2 - \omega_2^2 \mu X_2 = \frac{F_x}{M_1}$$
$$\ddot{X}_2 + 2\xi_2 \omega_2 \dot{X}_2 + \omega_2^2 X_2 - 2\xi_2 \omega_2 \dot{X}_1 - \omega_2^2 X_1 = 0 \qquad (9')$$

where

$$\frac{K_1}{M_1} = \omega_1^2; \frac{K_2}{M_2} = \omega_2^2; \frac{M_2}{M_1} = \mu; \frac{C_1}{\sqrt{K_1 M_1}} = 2\xi_1; \frac{C_2}{\sqrt{K_2 M_2}} = 2\xi_2,$$

where ω_1, ω_2 are the partial natural frequencies of the boring bar and of the absorber subsystems, μ is the mass ratio of the absorber, and ξ_1, ξ_2 are the damping ratios of the boring bar and absorber subsystems respectively.

Since only the variable part dF_x of the cutting force should be considered for the dynamic analysis, F_x in equation (9') can be replaced by expression for dF_x from equation (3). Stability of the resulting equation can be analysed using the Routh–Hurwitz criterion for the characteristic equation for the system [equation (9')] which is:

$$S^4 + A_3 S^3 + A_2 S^2 + A_1 S + A_0.$$

Here:

$$A_3 = 2(\xi_0 \omega + \xi_2 \omega_2 (1 + \mu)); A_2 = \omega^2 + \omega_2^2 (1 + \mu) + 4\xi_0 \xi_2 \omega \omega_2$$
$$A_1 = 2(\xi_0 \omega \omega_2^2 + \xi_2 \omega_2 \omega^2); A_0 = \omega^2 \omega_2^2$$

where

$$\frac{K_1 + K_{cx}}{M_1} = \omega^2; \frac{C_1 + C_{cx}}{M_1} = 2\xi_0\omega.$$

The Routh–Hurwitz criterion for stability requires that all coefficients of the characteristic equation be positive, which amounts to:

$$\xi_0 > -r\xi_2(1+\mu); \xi_0 > -\left(\frac{1}{4\xi_2 r} + \frac{r(1+\mu)}{4\xi_2}\right); \xi_0 > -\frac{\xi_2}{r} \tag{10'}$$

where:

$$r = \frac{\omega_2}{\omega}$$

and also that:

$$A_1 A_2 A_3 > A_1^2 + A_3^2 A_0. \tag{10''}$$

The critical value of ξ_0 can be obtained by replacing inequalities in equation (10') with equalities from which the extremal value can be determined and then checked with equation (10''). If the latter is not satisfied, the critical value of ξ_0 can be determined by iterations.

Figure 5 gives the critical values of ξ_0 at various frequencies, mass, and damping ratios of the absorber. It can be seen that there exists an optimal tuning frequency ratio ω_2/ω for given mass and damping ratios of the absorber at which the critical value of ξ_0 is maximum negative (maximum effectiveness of the absorber). A higher mass ratio gives a better system stability.

The influence of the absorber damping on system stability under optimal tuning conditions can be seen in Fig. 6. There exists also an optimal damping ratio of the absorber which results in the boring bar remaining stable at a higher negative damping of the cutting process for a given mass ratio.

For the case of random excitation acting on the boring bar (mass M_1 in Fig. 2(c)), optimal tuning of DVA would obviously depend on the excitation parameters. The

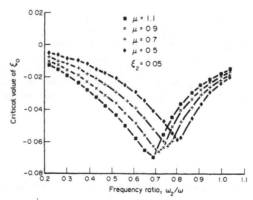

FIG. 5. Stability boundaries at various mass ratios of DVA (stable regions above the curves).

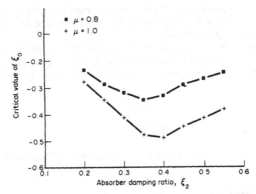

FIG. 6. Stability boundaries at optimal frequency ratios of DVA.

tuning was studied with the assumption of a white noise excitation in accordance with Wu [24].

For the main mass M_1 in Fig. 2(c), frequency response function with equations of motion (9') is:

$$H_1(\omega) = \frac{X_1}{\left(\dfrac{F_x}{M_1}\right)} = \frac{\omega_2^2 + 2\xi_2\omega_2(i\omega) + (i\omega)^2}{A_0 + A_1(i\omega) + A_2(i\omega)^2 + A_3(i\omega)^3 + (i\omega)^4} \qquad (11')$$

where:

$$A_3 = 2(\xi_1\omega_1 + \xi_2\omega_2(1 + \mu)); \quad A_2 = \omega_1^2 + \omega_2^2(1 + \mu) + 4\xi_1\xi_2\omega_1\omega_2$$
$$A_1 = 2(\xi_1\omega_1\omega_2^2 + \xi_2\omega_2\omega_1^2).$$

Without the DVA, the frequency response function is:

$$H_{10}(\omega) = \frac{1}{(i\omega)^2 + 2\xi_1\omega_1(i\omega) + \omega_1^2}. \qquad (11'')$$

The mean square response of the mass M_1, which represents the total energy of the main system under white noise excitation, is then given as:

$$E(X_1^2) = \int_{\infty}^{\infty} |H_1(\omega)|^2 S_0 d\omega. \qquad (12)$$

The ratio of mean square response of M_1 with DVA to mean square response of M_1 without DVA, which reflects the effect of the absorber on the main mass, is:

$$\frac{E(X_1^2)}{E(X_1^2)_0} = 2\xi_1\omega_1^3 \frac{A}{B} \qquad (13)$$

where

$$A = -A_0A_1 - A_0A_3(4\xi_2^2\omega_2^2 - 2\omega_2^2) + \omega_2^4(A_1 - A_2A_3)$$
$$B = A_0(A_0A_3^2 + A_1^2 - A_1A_2A_3).$$

This ratio is given in Fig. 7 for a main system damping $\xi_1 = 0.02$. The smaller the ratio, the smaller is the response of the main mass and the higher the effect of DVA on the main system behavior.

The classic case of harmonic F_x in equation (9') was analyzed in a close form by Den Hartog [12] for zero damping in the main system (a good approximation for the boring bar system), giving an optimal combination of tuning fequency and damping ratios for DVA as

$$\frac{\omega_1}{\omega_2} = \frac{1}{1+\mu}; \qquad \xi = \sqrt{\frac{3\mu}{8(1+\mu)}}$$

compliance with which results in the vibration amplitude of the main mass:

$$X_1 = X_{st} \sqrt{1 + \frac{2}{\mu}}$$

where X_{st} is the static displacement of the main mass.

Comparison of optimal frequency and damping as a function of mass ratio μ is shown in Fig. 8 for these three cases of excitation. The differences are quite pronounced.

Design of tunable DVA

In order to realize the needed tuning, the absorber design should have means for frequency tuning of the DVA. It is also important for restoring optimal tuning if a boring bar with DVA is used with various overhangs. The developed design of the absorber is shown in Fig. 9(a). It has inertia weight suspended on two rubber O-rings which have non-linear load-deflection characteristics in the radial direction. Turning the adjustment screw causes closing or separation of tapered bushings engaged with split bushings supporting the O-rings, thus changing their radial preload and, as a result, the suspension stiffness and natural frequency of the absorber subsystem. Damping of the absorber system can be varied by selecting the right rubber blend. Three rubber blends used in this project had damping ratios $\xi = 0.07$, 0.2 and 0.45 (log decrement $\delta = 0.45$, 1.25 and 2.8). Figure 9(b) shows the tuning capabilities for the absorber design in Fig. 9(a). With the high damping rubber elements ($\xi = 0.45$), the absorber frequency can be changed in the range 3.5 : 1, for medium and low damping elements, 2.2 : 1. Such broad ranges make the boring bar an extremely versatile tool.

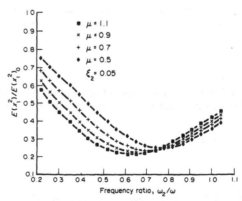

FIG. 7. Normalized mean square response at various mass ratios of DVA.

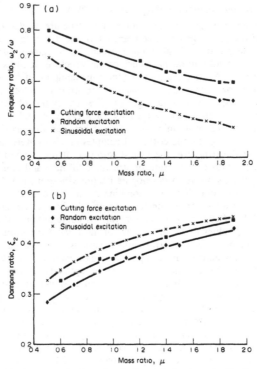

FIG. 8. Optimal tuning parameters (a: frequency; b: damping) for dynamic vibration absorber for various dynamic conditions of the main mass.

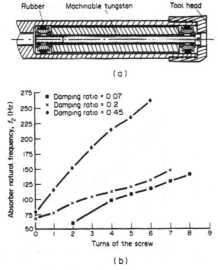

FIG. 9. Dynamic vibration absorber for boring bar (a) and its tuning range (b).

Structural optimization of boring bar with DVA

The general concept of the combination structure is described above. The high stiffness of the root segment is combined with the low specific weight of the overhang segment. The position of the joint between these segments determines the effective stiffness, mass, and natural frequency of the structure. If the structure is fitted with a damper or a DVA, its effectiveness (and thus damping of the structure) will increase with a reducttion in the effective mass of the structure. Optimization of the structure of a given design, such as in Fig. 1(a) (i.e. composed of certain materials, having certain design features such as a cavity for a DVA, a tool head of a certain weight, etc.), relates to the positioning of the joint between the segments. Optimization can be performed using various criteria. Examples of such criteria are: the highest natural frequency of the structure; the lowest response of the structure to external harmonic excitation; the best combination of structural stiffness and damping; etc.

For boring bars prone to developing chatter vibration, the third criterion would be the most suitable one. The best combination in this case is a maximum value of the product $K\delta$. Another approach to realizing "the best combination" is analysis, using equation (9), of system stability for various combinations of effective stiffness and corresponding values of the mass ratio of the DVA for various ratios of L_1/L in Fig. 1(a). This analysis can be performed for various combinations of structural materials and other design features. By computing stiffness values K_t at the bar end and (negative) value of ξ_{cr} of the combination bar under optimal frequency tuning and damping conditions, the "performance index" $K_t\xi_{cr}$ can be obtained as shown in Fig. 10. Figure 10 presents optimization results for a boring bar with outside diameter $D = 1.25$, and overall length 18 in. The root segment is made of sintered tungsten carbide with Young's modulus $E_1 = 80 \times 10^6$ psi and specific gravity $\rho_1 = 0.516$ lb/in³; the overhang segments are made of aluminum ($E_2 = 10 \times 10^6$ psi, $\rho_2 = 0.0938$ lb/in³); the tool head mass is $M_t - 4.86 \times 10^{-4}$ lb s²/in. The inertia mass of the DVA was made from machinable tungsten alloy (specific gravity $\rho_a = 0.65$ lb/in³). Three lengths of absorber cavity L_3 = 4 in., 5 in., and 6 in. were analyzed. It can be seen that the optimal ratio L_1/L is close to 0.5 for all studied values of L_3, although a larger L_3 results in better (more negative) values of the performance index. The important fact is robustness of the optimal ratio within a rather broad range, $L_1/L = 0.35–0.6$.

CUTTING TESTS

Cutting tests were performed using a boring bar with $D = 1.25$ in, $L = 18.75$ in, $L/D = 15$, $L_1/L = 0.6$, and $L_3 = 6$ in. The mass ratio of the DVA was $\mu = 1.07$. The

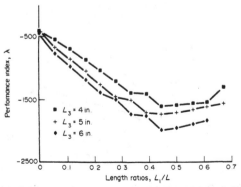

FIG. 10. Performance index vs length ratio for combination boring bar in Fig. 1 with DVA.

tests were performed on a lathe and on a machining center (rotating bar, stationary workpiece).

Stationary bar

Figure 11 shows relative vibrations between the tool and the workpiece in the radial directions for various tunings of the DVA. It can be seen that, while the bar with the DVA tuned in accordance with Den Hartog [12] results in a three to five times reduction in vibrations, tuning for the "dynamic cutting force" case results in an additional 50% reduction in vibration amplitudes.

Table 1 gives a sample of the surface finish of holes machined during these tests at various cutting speed (V) and depth (t) settings. The numbers are averaged from several cuts. The insert material is C5 sintered carbide with rake and side angles of 5°. St. 1045 is not an easy steel to machine, thus the R_a values can be considered satisfactory.

Machining of aluminum 6061–T6 at $L/D = 15$ (the same bar as above) was performed using a C-2 sintered carbide insert with a nose radius of 1/64 in (this was found to be the best for finish boring). A 2 in hole was machined at various cutting regimes (Table 2).

It is interesting to note that the best results were achieved at the largest depth of cut, $t = 0.04$ in and can be considered as excellent.

Several comparative tests were performed on shorter boring bars ($L/D = 7$), since longer bars were not commercially available.

Figure 12 shows the results of dynamic testing performed at and by the Manufacturing Development Center of Ford Motor Company on three boring bars: a Kennametal steel bar (520-KTFPRS); a Kennametal solid carbide boring bar (C-6420); and a combination carbide–aluminum bar with optimally tuned DVA. All bars were 1.25 in diameter. The first two bars had $L/D = 7$; two combination bars were tested with $L/D = 7$ and 10.

For $L/D = 7$, the minimum dynamic stiffness of the combination bar (3403 lb/in at 445 Hz) is much higher than for the solid carbide bar (1474 lb/in at 269 Hz) and, of

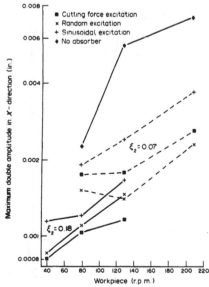

Fig. 11. Maximum double amplitudes of boring bar $L/D = 15$ vibration in radial (x) direction at various tuning conditions of DVA ($\xi = 0.18$). Diameter of machined hole $d = 2.3$ in; $s = 0.003$ in/rev; $t = 0.001$ in.

TABLE 1. SURFACE FINISH R_u μIN OF STEEL 1045 WORKPIECE (FEED RATE $f = 0.002$ IN); $L/D = 15$; $n = 80-400$ REV/MIN

t (in)	V (sfm)						
	48	80	109	129	177	208	245
0.01	126	90	97	94	98	89	96
0.02	167	77	81	97	88	63	109
0.03	223	170	95	88	85	49	87

TABLE 2. SURFACE FINISH R_u μIN. OF 6061–T6 ALUMINUM WORKPIECE (FEED RATE 0.002 IN/REV); $L/D = 15$; $n = 340-890$ REV/MIN

t (in)	V (sfm)				
	162	190	261	309	423
0.02	—	139	136	122	121
0.03	—	137	104	83	80
0.04	20	91	89	54	46

course, for the steel bar (285 lb/in at 263 Hz). Although the natural frequency for the combination bar with $L/D = 10$ is much lower (145 Hz), its dynamic stiffness is still much higher than for the much shorter conventional bars (2095 lb/in).

Boring of 6061-T652 aluminum at $L/D = 7$ (D = 1.25 in) was performed on a lathe using an optimized combination boring bar and commercially available Criterion Cridex Damping Bar (solid steel bar with a proprietary absorber/damper). The test results are given in Table 3.

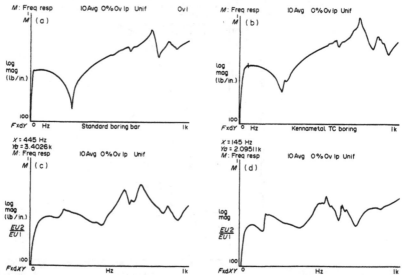

FIG. 12. Dynamic stiffness of boring bars $D = 1.25$ in, $L/D = 7$; (a) Kennametal S20-KTFPR3 solid steel bar; (b) Kennametal C-6420 solid carbide bar; (c) combination bar with optimally tuned DVA; (d) combination bar with DVA, $L/D = 10$.

TABLE 3. SURFACE FINISH R_u μIN OF 6061–T652 ALUMINUM WORKPIECE AT L/D = 7, FEED RATE 0.002 IN/REV (UPPER LEVEL: COMBINATION BAR; LOWER LEVEL: CRIDEX BAR)

	V (sfm)						
t (in)	288	340	400	470	550	760	890
0.01	19	19	—	22	26	23	28
	—	—		—	—	—	—
	-	-		-	-	-	-
0.02	24	21	-	33	39	32	33
	—	—		—	—	—	—
	39	67	92	-	128	-	-
0.03	23	26	—	39	41	41	31
	—	—		—	—	—	—
	-	-		-	-	-	-
0.04	-	-	-	-	-	-	-
	—	—	—	—	—	—	—
	50	82	109	-	108	—	-

Rotating bar

Cutting tests were peformed on the Leblond–Makino Vertical Machining Center at the General Motors Advanced Manufacturing Laboratory. Table 4 shows surface finish for boring aluminum 6061-T652 specimens (hole diameter 1.85 in) with the combination bar ($D = 1.25$ in, $L/D = 7.2$) and Cridex Damping Bar ($D = 1.25$ in, $L/D = 7$) using carbide insert C5-TPG221.

For very long overhangs, balancing of the bar becomes an important consideration. There are two main sources of unbalance: residual structural unbalance; and shift of the DVA inertia weight under centrifugal forces. The latter effect is greatly alleviated due to the higher tuning frequency (and thus stiffness of the suspension elements) in the proposed tuning arrangements. The former effect requires careful balancing, which could not be accomplished in the tests. Accordingly, results for $L/D = 12$ (Table 5) and $L/D = 15$ (Table 6) are inferior compared to the results for $L/D = 7$. In future, thorough balancing of the bar will be attempted.

TORSIONAL COMPLIANT HEAD (TCH)

While dynamic vibration absorbers, especially when applied to combination structures and with optimized tuning, are very effective in enhancing the dynamic qualities of cantilever boring bars, they have some shortcomings:

TABLE 4. SURFACE FINISH R_u μIN FOR MACHINING OF 6061–T652 ALUMINUM PARTS WITH ROTATING BORING BARS; UPPER LEVEL: COMBINATION BAR; LOWER LEVEL: CRIDEX BAR

		n (rev/min)				
t (in)	f (in/rev)	650	980	1200	1500	2000
0.023	0.004	—	25	28	27	26
			—	—	—	—
0.039	0.004		29	28	30	28
		106	103	88	—	-

TABLE 5. SURFACE FINISH R_a μIN FOR MACHINING STEEL AND ALUMINUM PARTS (1.85 DIA) WITH ROTATING COMBINATION BORING BAR $L/D = 12$, INSERT C5-TPG221

		n (rev/min)					
t (in)	f (in/rev)	180	340	480	650	980	1200
Steel 0.023	0.002	—	—	—	95	—	—
1018	0.004	—	111	126	82	—	—
0.039	0.004	95	178	159	58	—	—
Aluminum 0.023	0.002	—	—	—	—	53	—
6061–T652	0.004	—	—	43	50	68	—
0.039	0.004	—	—	30	48	40	35

TABLE 6. SURFACE FINISH R_a μIN FOR MACHINING 6061–T652 ALUMINUM PARTS (1.85 DIA) WITH ROTATING COMBINATION BORING BAR $D = 1.25$ IN. $L/D = 15$, INSERT C5-TPG221

		n (rev/min)			
t (in)	f (n/rev)	180	340	480	650
0.023	0.002				272
	0.004		130	194	221
0.039	0.004	162	234	169	193

(i) boring bars with DVA are not very effective at high cutting speeds; and

(ii) the intertia weight of a DVA has some mobility thus creating imbalance at high rev/min, although the mobility and ensuing imbalance are alleviated when an optimally tuned DVA is used.

These shortcomings led to a search for other means for enhancing dynamic quality. It was demonstrated [25] that the stability of the machining process can be enhanced by intentional reduction of the stiffness of the machining system in the direction tangential to the machined surface. Such an approach was analytically studied in boring bars using the above-formulated representation of the cutting force [equation (3)]. Since tangential motion of the tool tip is considered, it adds a degree of freedom to the boring bar model (Fig. 13) and equations of motion become:

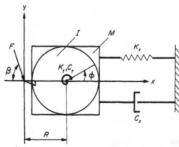

FIG. 13. Model of boring bar with torsionally compliant head.

$$MX + C_x X + K_x X = F \cos\beta$$
$$I\Phi + C_t\Phi + K_t\Phi = F R \sin\beta. \tag{10}$$

Here, X is linear vibratory displacement of the bar at the tool tip; Φ is torsional vibratory displacement of the tool head; M is equivalent mass of the boring bar (including the tool head) reduced to the tool tip position; I is moment of inertia of the tool head; C_x is translational damping coefficient of the bar; C_t is torsional damping coefficient of the tool head subsystem; K_x is bending stiffness of the bar; K_t is torsional stiffness of the tool head attachment; F is dynamic cutting force; R is distance from the tool tip to the axis of the boring bar; and β is angle between the resultant cutting force direction and X-direction with a mean value about 60 degrees [26].

The radial and tangential motions in equation (10) are dynamically coupled only through the cutting force F. Substituting equation (3) into equation (10), a set of coupled homogeneous equations are obtained:

$$X + 2 (\xi_x + \xi_{cx} \cos\beta) \, \omega_{nx} X + (\omega_{nx}^2 + \omega_c^2 \cos\beta) X + \frac{2\xi_{c\phi}\omega_{n\phi}\cos\beta}{\mu}\Phi = 0$$

$$\Phi + 2(\xi_\phi + \xi_{c\phi} \sin\beta) \, \omega_{n\phi} \, \Phi + \omega_{n\phi}^2\Phi + 2\xi_{cx}\omega_{nx} \, \mu\sin\beta X + \omega_c^2 \, \mu\sin\beta x = 0 \tag{14}$$

where:

$$\mu = \frac{M}{I} R^2; \quad \frac{C_x}{M} = 2\xi_x\omega_{nx}; \quad \frac{K_x}{M} = \omega_{nx}^2; \quad \frac{C_t}{I} = 2\xi_\phi\omega_{n\phi};$$

$$\frac{K_t}{I} = \omega_{n\phi}^2; \quad \frac{C_c}{M} = 2\xi_{cx}\omega_{nx}; \quad \frac{K_c}{M} = \omega_c^2; \quad \frac{C}{I} = 2\xi_{c\phi}\omega_{n\phi};$$

μ is mass ratio of the torsional head; ω_{nx}, $\omega_{n\phi}$ are partial angular natural frequencies of the boring bar and torsional head subsystems in X and torsional directions respectively, ξ_x, ξ_ϕ, are the respective damping ratios of these subsytems, ω_c is the angular cutting frequency, and ξ_{cx}, $\xi_{c\phi}$ are cutting damping ratios in X and the torsional directions. The stability of the system equation (11) was analyzed using Routh–Hurwitz criteria [27, 28] with the assumpion that the tangential loop of the machining system is always stable ($\xi_{c\phi} > 0$). The effectiveness of introducing compliance into the tangential loop ("compliant torsional head") can be characterized by the critical value of ξ_{cx}, which corresponds to the stability boundary of the system. The compliant torsional head is more effective, the larger the magnitude of the negative value of ξ_{cx} for which the torsional head can compensate. Thus smaller (more negative) values of ξ_{cx} correspond to better stability margins.

The analysis of equations (11) has shown that the critical value of ξ_{cx} does not depend on the mass ratio μ. Its magnitude (and the effectiveness of the TCH) increases with the increasing radius of the tool tip; the optimum TCH tuning corresponds to a frequency ratio $f_{n\phi}/f_{nx} = 0.9$–1.0; the effectiveness of the TCH increases with the reduction of damping of the torsional system (see Fig. 14).

The design of TCH should provide the required torsional stiffness and natural frequency and, at the same time, retain high stiffness in the radial (x) direction. These features were combined in a proprietary design [29], in which torsional stiffness is adjusted by varying the preload of non-linear rubber elements, and torsional guidance, together with high radial stiffness, are achieved by using rubber–metal laminated elements [30].

Vibration and cutting tests were performed on a tungsten carbide–aluminum combination bar 1.25 in in diameter.

Validation of tuning conditions was performed on a bar $L = 9$ in (the aluminum adapter was 2.5 in long). Both the bending natural frequency of the bar and the

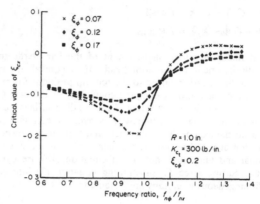

FIG. 14. Stability boundaries of a boring bar with TCH at various frequency and damping tunings.

torsional natural frequency of the TCH were measured using hammer excitation. Cutting of a St. 1045 specimen (hole diameter 3.5–4.0 in) was performed using Kennametal TPG 221KC850 carbide cutting inserts. Figure 15 shows that the relative vibration amplitude between the tool and the workpiece vs the frequency ratio has a minimum at $f_{n\phi}/f_{nx} \approx$ 1, in agreement with the analytical prediction.

Another test was performed with $D = 1.25$ in, $L = 14$ in, $L/D = 11.2$, machining St. 1045 with internal diameter 3.5 in. The elastic elements of the TCH were made of two different rubber blends: high damping ($\xi_\phi = 0.4$); and low damping ($\xi_\phi = 0.07$). Surface finish R_a μin vs rev/min is shown in Fig. 16 (the cutting speed in sfm is about 10% less than the numerical value of the rev/min). The results validate the analytical conclusion of detrimental effects of high damping in the TCH, as shown in Fig. 14. It can be seen that the effectiveness of the TCH is improved at higher cutting speeds.

The same boring bar ($L/D = 11.2$) was used for machining 6061 aluminum (2.45 in hole diameter; $t = 0.01$ in; $s = 0.002$ in/rev). The surface finish R_a was 20 μin at 890 rev/min (700 sfm), 19 μin at 1230 rev/min (965 sfm), and 23 μin at 1430 rev/min (1120 sfm). These excellent results were achieved on an old lathe incapable of machining with a better surface finish using any kind of tooling.

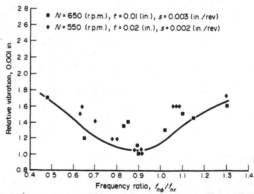

FIG. 15. Relative vibrations between tool and workpiece for various TCH tunings.

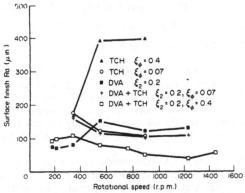

FIG. 16. Surface finish vs cutting speed for boring bar with various vibration control devices; $s = 0.002$ in/rev; $t = 0.01$ in; $L/D = 12$.

Figure 17 shows the torsional vibrations of the TCH relative to the boring bar structure while machining steel at $L/D = 7.2$. The cutting speed variation, depending on the regime, is within 5–50%. This effect, which is known to be beneficial for the enhancement of chatter resistance (e.g. Ref. [15]), seems to complement the dynamic effect derived from equations (14). Surface finish improves with increasing speed. The torsional vibrations have resulted also in a chip-breaking effect (2–4 in long chips vs continuous chip with the TCH deactivated). A significant advantage of bars with a TCH for high-speed applications is the absence of components with radial mobility.

COMBINATION OF TCH WITH DVA

A boring bar can be equipped with both a TCH and a DVA. Vibration absorber dynamics are described by the following set of three differential equations whose stability should be analyzed in order to derive tuning criteria for both TCH and DVA:

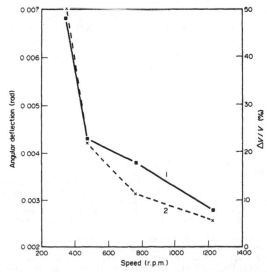

FIG. 17. Torsional vibrations of TCH (1 = displacement, 2 = velocity) and surface finish (3) for cutting steel 1045 with bar $L = 9$ in; $L/D = 7.2$; cutting edge radius $R_c = 1.23$ in; $s = 0.004$ in; $t = 0.01$ in.

$$M_1\ddot{X}_1 + (C_1 + C_2)\,\dot{X}_1 + (K_1 + K_2)\,X_1 - C_2\dot{X}_2 - K_2X_2 - F\cos\beta$$
$$M_2\ddot{X}_2 + C_2\dot{X}_2 + K_2X_2 - C_2\dot{X}_1 - K_2X_1 = 0$$
$$I\ddot{\phi} + C_t\dot{\phi} + K_t\phi = FR\sin\beta. \tag{15}$$

The analysis was performed by Kang [28], using also the Routh–Hurwitz method. It was shown that the combination has a synergistic effect, not just a summation of the partial effects, and that tuning of both the TCH and the DVA should be different than for cases where only one device was used. The optimal frequency ratio for the DVA is slightly lower than for the case when only the absorber is used, and the optimal frequency ratio for the TCH is 50–100% higher than for the case when only the TCH is used.

The effect of the DVA–TCH combination is shown in Fig. 16. It can be seen that the DVA is more effective at low cutting speeds, while the TCH (with optimal low damping) at high cutting speeds. An optimal combination of a properly-tuned DVA and TCH with low damping shows dramatic improvements in surface finish over a broad range of cutting speeds.

CONCLUSIONS

1. Application of a novel "combination structure" approach, together with dynamic analysis, optimization, and development of suitable vibration control means, results in significant enhancement of productivity and range of use of cantilever boring bars.

2. The use of the combination structure concept not only allows us to improve the dynamic characteristics of cantilever tools, but also improves the economics of advanced tooling, since it reduces by about 50% the need for expensive high Young's modulus materials. The suggested joining technique using high-pressure contact of flat surfaces reduces even more the need for complex machining of these hard-to-machine materials.

3. The results achieved can be applied to other cantilever tooling, and to cantilever structural components of machine tools, robots, etc.

4. It was demonstrated that while dynamic vibration absorbers in conjunction with the combination structure are more effective in the low/medium cutting speed range, they must be complemented with torsionally compliant heads to achieve stable performance over a broad range of cutting speeds.

5. Torsionally compliant heads are promising for high-speed machining with rotating tools, since they are more effective at high cutting speeds and do not have loose parts which may cause imbalance, and provide some chip breaking effect.

6. Use of the developed concepts allows simplifying the design of parts in which long precision holes have to be machined. In some cases, more integrated units can be adopted, resulting in substantial savings.

Acknowledgements—Financial support from the National Science Foundation (Grants DMC-8718911 and DDM-9005654) and from the Wayne State University Institute for Manufacturing Research is gratefully acknowledged.

REFERENCES

[1] M. D. THOMAS, W. A. KNIGHT, M. M. SADEK, Comparative dynamic performance of boring bars, *Proc. 11th Int. MTDR Conf.* Pergamon Press, Oxford (1970).
[2] J. TLUSTY, Criteria for static and dynamic stiffness of structures in, *Report of The Machine Tool Task Force*, Vol. 3. Machine Tool Mechanics, Lawrence Livermore Lab. (1981).
[3] G. N. MESHCHERIAKOV, L. P. TUSCHAROVA, N. G. MESHCHERIAKOV and A.N. SIVAKOV, Machine tool vibration stability depending on adjustment of dominant stiffness axes, *Ann. CIRP* **33**, 271–272 (1984).
[4] ANONYMOUS, Tuned mass dampers "calm" tall buildings during high winds, *Machine Design*, p. 12 (March 1978).
[5] R. L. JENNIGES and D. A. FROHRIB, Alternative tuned absorbers for steady state vibration control of tall structures, *ASME J. Mech. Des.* **100**, 279–285 (1978).
[6] K. IWATA and T. MORIWAKI, Analysis of dynamic characteristics of boring bar with impact damper, Memoirs of the Faculty of Engineering, Kobe University, pp. 85–94 (27/1981).
[7] E. S. BRISKIN, Damping of mechanical vibrations by dynamics absorbers with cavities partially filled with granular media, *Izvestia Vusov Machinostroen e.* No. 2, pp. 26–30 (1980) (in Russian).

[8] N. POPPLEWELL, K. MCLACHLAN, F. ARNOLD, C. S. CHANG and C. N. BAPAT, Quiet and effective vibroimpact attenuation of boring bar vibrations, *Proc. 11th Int. Congr. Acoust.*, pp. 125–128, Paris (1983).

[9] Y. ARAKI, Y. YUHKI, I. YOKOMICHI and Y. JINNOUCHI, Impact damper with granular materials, *Bull. JSME* 28, 1466–72 (1985).

[10] J. DONIES and L. VAN DEN NOORTGATE, *Machining of Deep Holes without Chatter*. Note Technique 10, Crif, Belgium (June 1974) (in Dutch).

[11] R. S. HAHN, Design of Lanchester damper for elimination of metal cutting chatter, *Trans. ASME* (1951).

[12] J. P. DEN HARTOG, *Mechanical Vibrations*. McGraw-Hill, New York (1956).

[13] D. J. GLASER and C. L. NACHTIGAL, Development of a hydraulic chambered, actively controlled boring bar, *Trans. ASME, J. Engng Ind.* 101, 362–268 (1979).

[14] R. DORNHÖFER and K. KEMMERLING, Boring with long bars, *VDI Z.* 128, 259–264 (1986) (in German).

[15] H. GRAB, Avoidance of chatter vibrations by periodic variation of rotational speed, Dissertation, TH Darmstadt (1976).

[16] E. I. RIVIN and YU. E. LAPIN, Cantilever tool mandrel, U.S. Patent 3,820,422 (1974).

[17] E. I. RIVIN, Chatter-resistant cantilever boring bar, *Proc. 11th North American Manufacturing Research Conf.*, SME, pp. 403–407 (1983).

[18] E. I. RIVIN, An extra-long cantilever boring bar with enhanced chatter resistance, *Proc. 15th North American Manufacturing Research Conf.* SME, pp. 447–452 (1987).

[19] Z. M. LEVINA and D. N. RESHETOV, *Contact Stiffness of Machines*. Machinostroenie, Moscow (1971) (in Russian).

[20] E. I. RIVIN, *Mechanical Design of Robots*. McGraw-Hill (1988).

[21] K. SETO, Effect of a variable stiffness-type dynamic absorber on improving the performance of machine tools with a long overhung ram, *Ann. CIRP* 27, 327–332 (1978).

[22] S. A. TOBIAS, *Machine Tool Vibration*. John Wiley, New York (1965).

[23] J. TLUSTY, Analysis of the state of research in cutting dynamics, *Ann. CIRP* 27 (1978).

[24] S. M. WU, Modeling machine tool chatter by time series, *Trans. ASME, J. Engng Ind.* 97, 211–215 (1975).

[25] M. E. ELYASBERG, A method for the structural improvement of machine tool vibration stability during cutting, *Sov. Engng Res.* 3, 59–63 (1983).

[26] S. G. KAPOOR, G. M. ZHANG and L. L. BAHNEY, Stability analysis of the boring process system, *Proc. 14th North American Manufacturing Research Conf.*, pp. 454–459S (1986).

[27] E. I. RIVIN and H.-L. KANG, Improving cutting performance by using boring bar with torsionally compliant head, *Trans. 18th North American Manufacturing Research Conf.*, SME, pp. 230–236 (1990).

[28] H.-L. KANG, Enhancement of dynamic stability and productivity for machining systems with low stiffness components, Ph.D. Thesis, Wayne State University (1990).

[29] E. I. RIVIN and H.-L. KANG, Torsional compliant head, Patent Disclosure, Wayne State University (1988).

[30] E. I. RIVIN, Ultra-thin-layered rubber-metal laminates for limited travel bearings, *Tribology Int.* 16, 17–25 (1983).

Design of Passive Damping Systems

C. D. Johnson
President,
CSA Engineering, Inc.,
2850 West Bayshore Road,
Palo Alto, CA 94303-3843

*This paper presents a brief review of techniques for designed-in passive damping f
vibration control. Designed-in passive damping for structures is usually based on one
four damping technologies: viscoelastic materials, viscous fluids, magnetics, or passi
piezoelectrics. These methods are discussed and compared. The technology of usi
viscoelastic materials for passive damping is discussed in more detail than the oth
methods since it is presently the most widely used type of damping technology. Testi
and characterization of viscoelastic materials and design methods for passive dampi
are discussed. An example showing the benefits of a passive damping treatment appli
to a stiffened panel under an acoustic load is presented.*

1 Introduction

Vibration and noise suppression are becoming more important in our society. Noise suppression of office machines, home appliances, aircraft, and automobiles makes for a more pleasant environment. Vibration suppression allows more precise medical instruments, faster and more compact disk drives, more precise images in ground- and space-based telescopes, safer buildings in the event of an earthquake, and lower stresses in products, generally leading to longer life and lighter weight. Passive damping is now the major means of supressing unwanted vibrations. The primary effect of increased damping in a structure is a reduction of vibration amplitudes at resonances, with corresponding decreases in stresses, displacements, fatigue, and sound radiation. However, damping is one of the more difficult issues to deal with in structural dynamics.

Passive damping may be broken into two classes: inherent and designed-in. Inherent damping is damping that exists in a structure due to friction in joints, material damping, rubbing of cables, etc. The level of inherent damping in a structure is usually less than 2 percent structural.[1] Designed-in damping refers to passive damping that is added to a structure by design. This damping supplements inherent damping, and it can increase the passive damping of a structure by substantial amounts.

To achieve a substantial increase in passive damping, a structural dynamist must have a good working knowledge of many factors: passive damping technologies, materials, concepts and implementations, in addition to design, analysis and predictions methods for passive damping systems. This paper will discuss each of these factors as related to the design of passive damping systems.

Much work has been done in the area of passive damping and this paper will not attempt to cite the many contributions made by a host of qualified individuals and companie: However, it is hoped that this paper will give the dynamist basic understanding of passive damping technology and er courage him or her to use passive damping as one mor design option in seeking solutions to structural dynamic prob lems.

In the last ten years, there have been many papers pub lished in the area of passive damping. The heaviest concer tration has been in the "Passive Damping" conferences be ginning in 1984. Since 1994, these conferences have become conference within the annual *North American Smart Struc tures and Materials Conference.* The reader is directed to th proceedings of these conferences (Air Force, 1984, 198(1989, 1991, 1993; Johnson, 1994) for many examples of pas sive damping applications.

2 Passive Damping Mechanisms

Designed-in passive damping for structures is usually base on one of four damping mechanisms: viscoelastic material: viscous fluids, magnetics, or passive piezoelectrics. Each (these damping mechanisms must be understood in order t select the most appropriate type of damping treatment. Th sections below describe each mechanism, and Table 1 pre sents a comparison.

The author believes that approximately 85 percent of th passive damping treatments in actual applications are base on viscoelastic materials, with viscous devices being the sec ond most actively used (the use of viscous devices is greate for isolation and shock). This paper will therefore concen trate on passive damping design using viscoelastic material: Damping using viscous and magnetic technology usually re quires either that the devices be purchased for the applica tion or a large effort be spent in the mechanical design. Thi paper will therefore discuss only the applications of suc devices, not their internal design. Passive piezoelectrics wi be discussed in broad terms since this technology has onl found limited applications to date.

---

[1] Structural damping g is equal to twice the viscous damping ratio ζ or the inverse of the quality factor, Q, in the case of harmonic excitation.

Contributed by the Design Engineering Division for publication in the *Special 50th Anniversary Design Issue.* Manuscript received Nov. 1994. Technical Editor: D. J. Inman.

Reprinted from Trans. ASME, 50th Anniversary of the Design Engineering Division, June 1995, Vol. 117(B), with permission of ASME.

Table 1 Primary passive damping mechanisms and related information

	TYPE OF DAMPING MECHANISM			
	Viscoelastic Materials	Viscous Devices	Magnetic Devices	Passive Piezoelectrics
Types of Treatments	All	Struts & TMDs	Struts & TMDs	Strut Dampers
Temperature Sensitivity	High	Moderate	Low	Low
Temperature Range	Moderate	Moderate	Wide	Wide
Loss Factor	Moderate	High	Low	Low
Frequency Range	Wide	Moderate	Moderate	Moderate
Weight	Low	Moderate	High	Moderate

2.1 Viscoelastic Materials. Viscoelastic materials (VEMs) are widely used for passive damping in both commercial and aerospace applications. Viscoelastic materials are elastomeric materials whose long-chain molecules cause them to convert mechanical energy into heat when they are deformed. For a detailed discussion of viscoelastic materials, see Aklonis and MacKnight (1983) or Ferry (1980). VEM properties are commonly described in terms of a frequency- and temperature-dependent complex modulus (G^*). Complex arithmetic provides a convenient means for keeping track of the phase angle by which an imposed cyclic stress leads the resulting cyclic strain. The complex shear modulus is usually expressed in the form

$$G^*(\omega, T) = G_0(\omega, T)[1 + i\eta(\omega, T)] \qquad (1)$$

The real and imaginary parts of the modulus, which are commonly called the storage or shear modulus and loss modulus, are given by $G_0(\omega, T) = G_R$ and $G_0(\omega, T)\eta(\omega, T) = G_I$, respectively. The loss factor (η) is a measure of the energy dissipation capacity of the material, and the storage modulus is a measure of the stiffness of the material. The shear modulus is important in determining how much energy gets into the viscoelastic material in a design, and the loss factor determines how much energy is dissipated. Although both are temperature and frequency dependent, temperature has a greater effect on damping performance in typical applications. Figure 1 shows the typical variation of material properties as a function of temperature and frequency. This figure also shows the optimum regions of VEMs for various types of damping devices. Region A is optimum for free-layer treatments, having a high modulus and high loss factor. Region B is optimum for constrained-layer treatments, having a low modulus and high loss factor. Region C is optimum for tuned-mass dampers, having a low modulus and low loss factor, while regions A and B are optimum for most other types of discrete dampers.

In order to design accurately passive damping systems using VEMs, one must know their material properties accurately. Since viscoelastic materials are temperature and frequency dependent, they must be tested over both temperature and frequency ranges to characterize the material. VEM test methods fall into two broad classes: resonant and nonresonant.

Resonant tests infer VEM properties from measured normal mode properties of some simple structure that includes the viscoelastic material, such as a sandwich beam. Resonant tests have the advantage of being relatively insensitive to both gain and phase errors in the transducing systems. However, a major disadvantage is that the measurement is indirect; material properties are inferred from modal properties by working backwards through some theoretical solution. Also, material properties are obtained only at discrete frequencies.

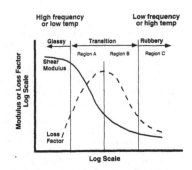

Fig. 1 Temperature- and frequency-dependence of VEMs

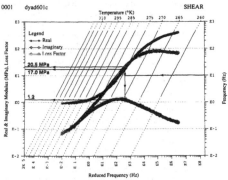

Fig. 2 Reduced-temperature nomogram

Nonresonant tests, often called complex stiffness tests, utilize a VEM sample connected to a rigid fixture and loaded dynamically, usually in shear. The force transmitted through the specimen and the resulting deformation across it are transduced directly. Damping is determined from the phase angle by which the displacement lags the force. Stiffness, or storage modulus, is determined from the ratio of in-phase force to displacement. Stiffness and loss factor are obtained as almost continuous functions of frequency and at discrete temperatures.

It was determined by Williams et al. (1955) and Ferry et al. (1952) that viscoelastic material test data could be shifted in temperature and frequency such that a relationship could be developed that characterizes the material at all combinations of temperature and frequency. This process is referred to as characterization and much work has been performed in this area (see, for example, Fowler, 1989; and Rogers, 1989). A layman's view is to determine a functional relationship (the temperature shift function α_T) between temperature and frequency such that both the storage modulus and loss factor at any temperature and frequency can be determined. Incorrect characterization processes can lead to major errors in property data. The end result of characterization is a viscoelastic material nomogram (also called the international plot, Jones, 1977). Figure 2 gives an example and its use for Soundcoat Dyad 601. These types of plots are the preferred method of presenting VEM data, and most damping material manufacturers present their material data in this form. It is therefore important that designers know how to use these plots.

To get modulus and loss factor values corresponding to 10 Hz and 273°K, for example, read the 10 Hz frequency on the right-hand scale and proceed horizontally to the 273°K temperature line. Then proceed vertically to intersect the curves along a line of reduced frequency. Finally, proceed horizontally from these intersections to the left-hand scale to read the values of 20.5 MPa for the real (shear) modulus, 17 MPa for the imaginary modulus, and 1.20 for the loss factor.

Once a material has been characterized accurately, its parameters may be placed into a database. The designer may then perform searches for materials that meet specific engineering criteria in a method exactly analogous to reading the international plot. Since these searches are based solely on the characterization parameters, the importance of quality data and characterization methods should not be underestimated.

2.2 Viscous Devices. These devices dissipate energy via a true velocity dependent mechanism, typically by forcing a fluid through a precision orifice. Although the actual viscous damping coefficient is usually not frequency dependent, the viscous damping force under periodic loading ($c\omega$) is obviously frequency dependent. Viscous dampers are most effective for axial deformations. The levels of loss obtainable by a viscous device are higher than those obtainable with VEM-based struts, but a price is paid in the "bandwidth" of effectiveness. That is, a viscous damper is usually effective at damping only modes in a relatively narrow frequency range because the damper is usually "tuned" to a frequency range. As with VEM damping treatments, the effectiveness of viscous dampers is affected by changes in temperature, but to a lesser degree. This change is due to the viscosity of the fluid changing.

Viscous damping mechanisms have been adapted to address bending deformations, but this is not the most direct or efficient use of the technology. This approach is thus not attractive for situations dominated by panel bending, such as acoustic-driven problems.

2.3 Magnetic Devices. With advancements in the production of powerful magnets, magnetic (eddy current) damping is proving to be a viable solution to problems where temperature extremes are a factor. The power and effectiveness of the magnetics are relatively unaffected by changes in temperature. As with fluid-based systems, this technology produces a true, velocity-dependent viscous damping force. However, the damping coefficients of magnetic devices are usually less than viscous devices per unit weight.

This is another technology that is not well suited for most bending problems. However, magnetic tuned-mass dampers (TMDs) have been shown to be effective in harsh environments where neither viscoelastic or viscous damping mechanisms are possible.

2.4 Passive Piezoelectrics. Piezoelectric ceramic materials have the unique ability to produce a strain when subjected to an electrical charge, and, conversely, they produce a charge when strained mechanically. This property has made them popular as actuators and sensors in active vibration control systems. This dual transformation ability also makes them useful as passive structural dampers (Forward, 1979). In passive energy dissipation applications, the electrodes of the piezoelectric are shunted with a passive electric circuit. The electrical circuit is designed to dissipate the electrical energy that has been converted from mechanical energy by the piezoelectric. Two major types of shunted circuits exist: a resistor alone and a resistor in series with an inductor. Other

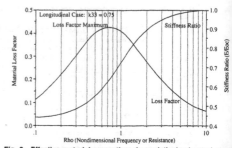

Fig. 3 Effective material properties of a resistively shunted piezo-electric assuming strain in the polarization (longitudinal) direction

circuits can be visualized and have been reported elsewhere (Hagood and von Flotow, 1991).

A resistor shunt provides a means of energy dissipation on the electrical side and thus increases the total piezoelectric loss factor above that of the unshunted piezoelectric. With a shunted resistor, the ceramic behaves like a standard first order viscoelastic material. The material properties of the resistive shunted piezoelectric can be represented as a complex modulus as is typically done for viscoelastic material. $\bar{E}^{eff}(\omega) = \bar{E}(\omega)(1 + i\eta(\omega))$, where $\bar{E}$ is the ratio of shunted stiffness to open circuit stiffness of the piezoelectric and η the material loss factor. The nondimensional expressions for η and E are

$$\eta(\omega) = \frac{\rho k^2}{(1 - k^2) + \rho^2}, \quad \bar{E} = 1 - \frac{k^2}{1 + \rho^2}, \quad \rho = RC^s\omega$$

where ρ is the nondimensional frequency ratio, k is the electromechanical coupling coefficient, R is the shunting resistance, and C^s is the clamped piezoelectric capacitance. These relations have been plotted versus ρ, the nondimensional frequency (or the nondimensional resistance) in Fig. for a typical value of k for longitudinal strain. As illustrated in the figure, for a given resistance the modulus of the piezoelectric changes from its short circuit value at low frequencies (about that of aluminum) to its open-circuit value at high frequencies. The transition occurs at the frequency RC^s. The material loss factor peaks at this transition frequency at a value of 44 percent for longitudinal or shear strain and 8 percent for transverse use. The point of maximum loss factor can be assigned to the desired frequency by the appropriate choice of resistor.

While the loss factor levels are not as high as those for viscoelastic materials, the high stiffness of the shunted piezoelectric materials (typically a ceramic) allows them to store many times the strain energy of a viscoelastic for a given strain. The piezoelectric material properties are also relatively temperature independent. The coupling coefficient for several common ceramic compositions vary by only $\pm$ percent over a temperature range from $-200°C$ to $+200°C$. The piezoelectric material density (~ 7500 kg/m^3) is much higher than that of viscoelastic materials, however.

Shunting with a resistor and inductor, along with the inherent capacitance of the piezoceramic, creates a resonant LRC circuit that is analogous to a mechanical tuned-mass damper, except that it counters vibrational strain energy instead of kinetic energy. High loss factors are possible. However, heavy shunt inductors are required for typical sized piezoceramics. More compact active inductors have been built, but this defeats the passivity of the system.

3 Passive Damping Concepts

Although passive damping is often attributed to friction or other such "accidental" mechanisms, designed-in damping using high-loss materials and techniques can yield energy dissipation that is orders of magnitude higher and much more predictable. All passive damping treatments share a common goal: to absorb significant amounts of strain energy in the modes of interest and dissipate this energy through some energy-dissipation mechanism. The effectiveness of all passive damping methods varies with frequency and temperature, though some more than others. For each of the basic passive damping *mechanisms*, there are several choices for implementation which can be divided into two major categories: discrete and distributed. Table 2 summarizes the primary passive damping concepts along with their typical uses. For a description of many types of damping devices, see Nashif, Jones, and Henderson (1985).

Discrete dampers can be very effective and may be easy to design and implement. Damped struts or links are commonly used in truss structures, although they can also be used to damp structures where two or more parts of the structure are moving relative to each other. Depending on the design constraints, the damping material may be in series or parallel with other structural members. Because of creep problems, one should not require damping materials to carry high static loads, but should provide alternate static load paths. A tuned-mass damper (TMD) is a discrete damping device attached to the structure at or near an antinode or a troublesome mode of vibration. These devices transfer energy at a particular resonance to two new system resonances, each highly damped. TMDs are in general the most weight-efficient damping devices for single mode damping.

One of the simplest passive damping methods, but the least weight effective, is the unconstrained or free-layer damping treatment. In this treatment, a high-modulus, high-loss-factor material is applied to a surface of the vibrating structure. Free layer treatments must be fairly thick in order to absorb sufficient amounts of strain energy, and are therefore not weight efficient. Constrained layer treatments are surface treatments where the damping material is sandwiched between the base structure and a constraining layer. The constraining layer causes shear in the damping material as the structure deforms. This type of damping treatment is most commonly used to damp bending modes of surfaces (shell-type modes). A small increase in damping may be achieved by placing damping materials in joints. The advantage of this type of damping is that it requires very little added weight. Embedded dampers have damping material embedded into a structural member (mainly composites) during manufacture.

4 Passive Damping Design Methods

The frequency and temperature dependencies of passive damping mechanisms must be taken into account during the design. Damping design is not just the selection of a high loss

mechanism (material, device) for the temperature range of interest; it is an integrated structural and materials design process. To achieve damping, two conditions must be met: significant strain energy must be directed into the high loss mechanism for all modes of interest, and the energy in the mechanism must be dissipated. The first condition requires most of the design effort and is dependent on structural properties, location, mode shapes, stiffness, wave lengths, thickness of material, etc. The second condition is met by selecting the mechanism with the proper loss factor that matches the designed stiffness.

Before the design of the passive damping treatment can begin, it is imperative that the true nature of the vibration problem be understood thoroughly. The designer must have in mind some figure of merit, which could be as simple as the response of a fundamental mode of a panel or as complicated as the RMS beam jitter of multiple optics in an optical system due to acoustic excitation. In any case, the engineer must determine whether the problem is a single mode or many modes over a broad frequency band. In the later case, the precise modes that are driving the figure of merit must be identified. Knowing all of this, the proper damping mechanism, analysis technique, and hardware can be chosen.

Passive damping treatments for complex structures are usually designed using finite element techniques. Methods for finite element analysis of damped structures can be classified as response-based or mode-based. Response-based methods use the bottomline dynamic response (e.g., RMS acceleration) to guide the design. Mode-based methods use a substitute metric which is easier to compute but which is known to influence the bottom line response significantly. Designs produced by mode-based methods should normally be verified by a final dynamic response computation.

The best known mode-based methods are modal strain energy (MSE) (Johnson and Kienholz, 1982) and complex eigenvalue analysis. Using the MSE method, the modal damping of a structure may be approximated by the sum of the products of the loss factor of each material and the fraction of strain energy in that material for each mode. In the case of a multimaterial system, the system loss is given by

$$\eta^{(r)} = \sum_{j=1}^{M} \eta_j \frac{SE_j^{(r)}}{SE^{(r)}}, \qquad (3)$$

where η_j = material loss factor for material j, $SE_j^{(r)}$ = strain energy in material j when the structure deforms in natural vibration mode r, and $SE^{(r)}$ = total strain energy in natural vibration mode r. In the MSE method, the material properties are real and a real eigenvalue analysis is performed. The underlying assumption of the MSE method is that the real eigenvalues are a good approximation to the complex eigenvalues. For high damping, this is not a good approximation. The more time-consuming complex eigenvalue method provides direct computations of $\eta^{(r)}$. The computational advantage of the MSE method is not as important as it once was, because computers are so much faster. However, MSE distri-

Table 2   Primary implementations of passive damping and associated design methods for structures

	TYPE OF TREATMENT					
	Strut / Link Dampers	Constrained Layers	Tuned-Mass Dampers	Joint/Interface Dampers	Embedded Dampers	Unconstrained or Free Layers
Target Modes	Global	Member Bending and Extension	Narrow Frequency Range, Any Mode Shape	Local or Global	Member Bending and Extension	Member Extension and Bending
Primary Design Method	Modal Strain Energy	Modal Strain Energy	Complex Eigenvalues	Joint Test or Modal Strain Energy	Modal Strain Energy	Hand Calculations Modal Strain Energy
Special Features	Removable, Lightweight	Flexible, Wide Bandwidth	Low Cost, Low Weight	Low Weight, Low Volume	Embedded, Low Outgassing	Low Cost, Easy Design

butions are also valuable for deciding where to place damping materials or devices, determining optimum design parameters, or for general understanding of the character of a structure's modes. In both the MSE and complex eigenvalue method, the analyses are performed with constant material properties.

Response prediction can be carried out in either the time or frequency domain. Modal superposition may be used to advantage in either case, but in frequency response, one pays a price in addition to the approximation due to modal truncation: there is no way to account for frequency-dependent materials. This is because modes must be computed using properties corresponding to a single selected frequency. A direct frequency response formulation (no modal superposition) can account for this dependency but at substantial computational cost.

For each of the analysis and prediction methods discussed above, the damping device or material must be represented in the analytical model. This may be as simple as representing a damped strut with a spring element, or as complex as modeling an embedded treatment in very precise detail with many finite elements. For a viscoelastic material damping treatment, the shear deformation of the VEM must be accurately captured and this is best done by using solid elements to represent the VEM (see, for example, Johnson et al., 1985). Large aspect ratios of the elements modeling the VEM can be accommodated in many finite element codes. Many finite element codes allow grid point off-sets, so that grid points at the corners of solid elements that model the VEM can also be used for the constraining and base layer plates, thereby saving many degrees of freedom in a model. For modeling purposes, Poisson's Ratio of VEMs is typically set to 0.4999. The amount of detail that is required in the finite element model is problem dependent.

The damping treatment design cycle for a VEM design using finite element techniques may be summarized as follows:

- Define the problem
  - Determine specifications, requirements, and constraints
  - Determine dynamics that cause high responses
- Perform preliminary design
  - Develop damping concepts
  - Perform analysis using reduced order modeling and appropriate analysis methods
  - Vary design variables to select good damping candidates
  - Perform response prediction analysis
- Perform final design
  - Develop detailed finite element model of damped structure
  - Vary design variables for best design
  - Select viscoelastic material based on analysis results
  - Perform response analysis using selected VEM properties
  - Determine if all specifications and constraints have been met

Current work is being performed by several researchers in the area of optimization of viscoelastic damping treatments. All of this work is based on performing sensitivity analysis of strain energy (see, for example, Gibson and Johnson, 1987).

5 Passive Damping Design Example Using VEM

For illustration purposes, consider the application of passive damping to a stiffened panel supporting a simulated component that is sensitive to its vibration environment. To simulate excitation by an acoustic field, a random pressure loading is applied over the surface of the panel. The chosen figure of merit is the RMS (10–1,000 Hz) values of normal displacements and their slopes. The analysis is performed with a finite element model using NASTRAN in which the

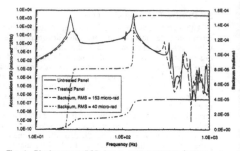

Fig. 4 Displacement PSD for θ_x with backsum (undamped and damped)

panel and ribs are modeled with plate elements, and the component is a lumped mass.

The first step is to determine which modes contribute the most to the figure of merit. For the displacement normal to the panel, it is easy to visualize that the fundamental bending mode dominates this response. However, for the rotations, the fundamental mode along with either the second (for θ_x) or third (for θ_y) modes are of equal importance. The displacement PSD (solid line) for the x rotation, along with its backsum (153 μrad), are shown in Fig. 4.

If only the normal displacement were important, this problem would be a good candidate for a tuned-mass damper, since this displacement PSD is strongly dominated by just the fundamental mode. However, the importance of the second and third modes makes a constrained-layer treatment more appropriate for this case.

There are five basic design parameters for a constrained-layer treatment: thickness of the constraining layer, modulus of the constraining layer, thickness of the VEM, modulus of the VEM, and placement of the treatment. In most practical situations, some of these parameters are determined by outside factors such as constraints on weight, thermal expansion, clearance, etc. Where weight is a factor, it is usually advantageous to make the constraining layer from advanced materials, such as metal matrix or graphite-epoxy. For this example, the constraining layer is made from the same material as the base panel: aluminum. The entire top surface of the panel, including under the component, is covered by the constrained-layer damping treatment.

A brief trade study with the remaining parameters showed that the VEM should be approximately 0.152 mm thick and have a shear modulus near 1.73 MPa. This trade study is documented in Table 3. One material that fits this closely for the three modes of interest is 3M's Y-966. The shear modulus and loss factor for this VEM at 25 and 130 Hz are approximately 0.65 MPa and 1.93 MPa, respectively. Two additional runs were then made with these values to get a better approximation for the MSE in the modes of interest. They are also reported in Table 3. These values of modal strain energy were subsequently multiplied by the loss factors for their respective frequencies and used to predict the responses of the structure with the added damping treatment. The RMS values are given in Table 4 and the effects are shown clearly by the PSD in Fig. 4 (dotted curve).

There are three variations on this concept that bear mentioning:

(*1*) The damping treatment could be shrunk so that it covered a smaller portion of the base panel.
(*2*) The panels themselves could be constructed from a sandwich of metal and VEM.
(*3*) Instead of a constraining layer, the VEM could be sand-

Table 3 Summary of trade study for add-on constrained-layer damping treatment

VEMT	VEMG	CL	% MSE in VEM		
(mm)	(MPa)	(mm)	Mode 1	Mode 2	Mode 3
0.254	1.73	1.27	5.41	6.57	5.43
0.254	1.73	2.54	10.0	11.43	8.93
0.254	1.73	3.81	13.65	15.02	11.49
use CLT = 2.54 mm as baseline, now vary VEMG					
0.254	0.35	2.54	5.61	4.93	3.15
0.254	13.79	2.54	7.18	9.44	10.05
0.254	6.89	2.54	9.09	11.82	11.42
use VEMG = 1.73 MPa for baseline, now vary VEMT					
0.051	1.73	2.54	8.05	10.53	10.44
0.127	1.73	2.54	9.77	12.14	10.53
0.102	1.73	2.54	9.49	12.01	10.77
0.152	1.73	2.54	9.93	12.11	10.22
Runs for final predictions of MSE in modes 1–3					
0.152	0.65*	2.54	8.88	n/a	n/a
0.152	1.93†	2.54	n/a	12.21	10.48

*3M Y-966 shear modulus at ~ 25 Hz
†3M Y-966 shear modulus at ~ 130 Hz

Table 4 Reductions in RMS values resulting from passive damping

	RMS		
	z (μm)	θ_x (μrad)	θ_y (μrad)
Untreated	12.3	153	191
With Damping	2.1	40	43

wiched between the base structure and built-up sections (I-beams, C-channels, hat sections, etc.).

Each of these alternatives is likely to save weight, though they are not discussed in this paper.

6 Conclusions

Passive damping can provide substantial performance benefits in many kinds of structures and machines, often without significant weight or cost penalties. Full realization of these benefits depends on (1) properly characterized materials, (2) knowledge of the strengths and weaknesses of the various materials and mechanisms, and (3) appropriate design/analysis methods and software.

References

1 Aklonis, J. J., and MacKnight, W. J., 1983, *Introduction to Polymer Viscoelasticity*, 2nd Edition, John Wiley and Sons.

2 Ferry, J. D., Fitzgerald, E. R., and Grandiene, L. D., and Williams, M. L., 1952, "Temperature-Dependence of Dynamic Properties of Elastomers: Relaxation Distributions," *Ind. Eng. Chem.*, Vol. 44, No. 4, Apr., pp. 703–706.

3 Ferry, John D., 1980, *Viscoelastic Properties of Polymers*, 3rd Edition, John Wiley and Sons.

4 Forward, R. L., 1979, "Electronic Damping of Vibrations in Optical Structures," *Applied Optics*, Vol. 18, No. 5, March.

5 Fowler, B. L., 1989, "Interactive Processing of Complex Modulus Data," *Dynamic Elastic Modulus Measurements in Materials*, ASTM STP 1045.

6 Gibson, W. C., and Johnson, C. D., 1987, "Optimization Methods for Design of Viscoelastic Damping Treatments," *Proceedings of the ASME Conference on Vibration and Noise*.

7 Hagood, N. W., and von Flotow, A., 1991, "Damping of Structural Vibrations with Piezoelectric Materials and Passive Electrical Networks," *Journal of Sound and Vibration*, Vol. 146, No. 1, April.

8 Johnson, C. D., and Kienholz, D. A., 1982, "Finite Element Prediction of Damping in Structures with Constrained Viscoelastic Layers," *AIAA Journal*, Vol. 20, No. 9, September.

9 Johnson, C. D., Kienholz, D. A., Austin, E. M., and Schneider, M. E., 1985, "Design and Analysis of Damped Structures Using Finite Element Techniques," *Proceedings of the ASME Conference on Mechanical Vibration and Noise*, Paper No. 85-DET-131.

10 Johnson, C. D., editor, 1994, *Proceedings, Passive Damping, Smart Structures and Materials 1994*, SPIE Volume 2193, February.

11 Jones, D. I. G., 1977, "A Reduced-Temperature Nomogram for Characterization of Damping Material Behavior," *48th Shock and Vibration Symposium*, October.

12 Nashif, A. D., Jones, D. I. G., and Henderson, J. P., 1985, *Vibration Damping*, John Wiley and Sons.

13 Rogers, L. C., 1989, "An Accurate Temperature Shift Function and A New Approach to Modeling Complex Modulus," *Proceedings of 60th Shock and Vibration Symposium*, November.

14 Williams, M. L., Landel, R. F., and Ferry, J. D., 1955, "The Temperature Dependence of Relaxation Mechanisms in Amorphous Polymers and Other Glass-Forming Liquids," *J. Amer. Chem. Soc.*, Vol. 77, p. 3701.

15 *Vibration Damping 1984 Workshop Proceedings*, Air Force Wright Aeronautical Laboratories, AFWAL-TR-84-3064, November.

16 *Damping 1986 Proceedings*, Air Force Wright Aeronautical Laboratories, AFWAL-TR-86-3059, May.

17 *Proceedings of Damping '89*, Air Force Wright Aeronautical Laboratories, AFWAL-TR-89-3116, November.

18 *Proceedings of Damping '91*, Air Force Wright Aeronautical Laboratories, AFWAL-TR-91-3078, August.

19 *Proceedings of Damping '93*, Air Force Wright Aeronautical Laboratories, AFWAL-TR-93-3103, June.

Trends in Tooling for CNC Machine Tools: Machine System Stiffness

Eugene I. Rivin
Department of Mechanical Engineering, Wayne State University, Detroit, MI 48202

Significant technological advances have been made in recent years in new cutting materials and machine tool design. Now, in many cases, the weakest link in a machining system is its tooling structure, which serves as an interface between the cutting insert and the machine tool. In order to develop advanced concepts in tooling system designs, the state of the art should be assessed, keeping in mind that the majority of publications on the subject are not in English. This article is a worldwide survey on the subject of how the stiffness of the overall machining system affects tool life and process stability. A companion paper, Trends in Tooling for CNC Machine Tools: Tool–Spindle Interfaces, *which follows in this issue on page 264 examines a related problem: the stiffness and accuracy of tool–machine interfaces, giving special attention to tooling for high-speed machine tools.*

INTRODUCTION

Modern machine tools are characterized by features such as high stiffness, high installed power, and high spindle rpm, which have all been incorporated into machine tools to realize better the beneficial properties of state-of-the-art cutting materials including coated carbides and high-speed steel (HSS), cubic boron nitride (CBN), and polycrystalline diamonds (PCD). Although significant progress has been achieved in both new cutting materials and machine tool design, in many cases the weakest link in the machining system is now the tooling structure that serves as an interface between the cutting insert and the machine tool.

The tooling structure is composed of attachment devices for the cutting inserts, the tooling itself, which can be a solid structure or a modular system consisting of several joined elements, and the tool–machine interface, which can be tapered, cylindrical, toothed, etc. With automated machine tools, tools are changed in accordance with a programmed sequence, and the majority of tools are of cantilever design with their external dimensions determined by the machined part design and by process limitations. This, together with inevitable compliances in the numerous joints and very heavy cutting regimes, which are typical for the state-of-the-art cutting materials, leads to inaccuracies; microvibrations,

resulting in poor surface finish and shorter insert life; and chatter vibrations, the onset of which during an automatic machining process could lead to serious damage both to the tool and the machined part.

The latest trend in machine tool design is to increase spindle rpm in order to utilize more fully the enhanced capabilities of advanced cutting materials. This trend complicates tooling design however, because very thorough balancing is required to maintain accuracy and surface finish, and to reduce dynamic loads on the spindle bearings. Balancing solid tools requires better designs and upgrading tolerances for all the joints. Second, balancing adjustable tools (such as boring bars) is hardly possible using presently available technology, thus limiting tooling capabilities and, as a consequence, the entire automated machining system.

These and other issues indicate a need for the development of tooling designs that do not degrade the performance of advanced machine tools and cutting materials. In fact, this area requires substantial research attention, and calls for the development of novel concepts in tooling design to address these issues.

Another important factor contributing to the case for advancements in machine tool research is that investment in tooling structures and cutting tools amounts to about ten percent or even more of the total cost for a CNC machine tool [1]. Thus, it is important to utilize tools that perform a

Reprinted from ASME Manufacturing Review, 1991, Vol. 4, with permission of ASME.

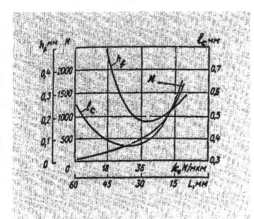

FIG. 1. Tool life indicators h_f, l_c, and N vs. machining system stiffness k_t (cutting insert overhang L of the face milling cutter)

maximum efficiency because they represent a large portion of overall investment. For example, Murata states that the effectiveness and reliability of flexible manufacturing cells and systems are determined to a large extent by the technology level of tooling and auxiliary systems [2]. A similar opinion is expressed in the report of the National Science Foundation (NSF) Workshop on Research Priorities for the NSF Strategic Manufacturing Research Initiative, which lists Tooling Flexibility as a research priority [3].

To assist the development of competitive, high productivity manufacturing systems, the state of the art in tooling R&D must be assessed. However, this is difficult to do because most of the publications on the relevant subjects are only available in languages other than English (especially German, Russian, Japanese, and Polish). The goal of this survey is to review these foreign language sources and shed some light on the R&D efforts in the area of tooling for CNC machine tools during the last five to 15 years. It is expected that such an assessment of R&D activities will help U.S. manufacturing companies and research institutions in their struggle to remain competitive.

THE EFFECT OF MACHINING SYSTEM STIFFNESS ON TOOL LIFE AND PROCESS STABILITY

It is a universally accepted notion that a high degree of stiffness in a machining system is a necessary condition for successful cutting process performance. However, recently a better understanding of this issue has begun to develop, one which has serious implications for the development of tooling and machining systems in general.

Machining system stiffness directly influences several aspects of a machining operation. The most important aspects effected are:

❑ Accuracy of the machined surface, as influenced by deformations in the machining system due to cutting forces

❑ Tool life (wear and cracking)

❑ Dynamic stability (chatter resistance) of the cutting process

Inaccuracies due to static or quasi-static deformations in the machining system are especially important for finishing passes, when cutting forces are relatively small. During rough regimes, when forces and deformations are large, accuracy is frequently a second order concern when compared to the rate of metal removal (productivity). Deformations can be reduced, of course, by stiffness enhancement, a reduction in cutting forces (using proper cutting angles, cutting insert materials and coatings and cutting fluids), and balancing cutting forces through the use of multi-edge tooling heads. However, for boring operations, multi-edge tooling heads with adjustable cutters are associated with stiffness problems for tool adjustment mechanisms. Kocherovskii, et al. demonstrated that deformations in tool adjustment mechanisms can be as high as 25 μ (0.001 in.) [4]. However, components of this deformations, caused by contact and bending deformations of design elements, can be adequately computed and their effect on machining accuracy reduced.

The Influence of Stiffness and Damping on Tool Life

The influence of stiffness on the life of cutting inserts is not straightforward. The most comprehensive studies of the correlation between stiffness and tool life are described by Chryssolouris and Fadeev, et al. Wear of superhard tool inserts was studied by Chryssolouris for two machining conditions: turning, using CBN inserts, and milling, using PCD inserts [5, 6]. In the first setup (turning), three cases of machining alloyed steel were studied: stiff workpiece/stiff toolholder [workpiece stiffness $k_w \sim 3$ N/μ (16,800 lb./in.), tool stiffness $k_t \sim 65$ N/μ (365,000 lb./in.)]; compliant workpiece/stiff toolholder [$k_w \sim 0.35$–2 N/μ (2,000–11,000 lb./in.)]; and compliant workpiece/compliant toolholder [$k_t \sim 3.5$ N/μ (20,000 lb./in.)]. Although crater wear increased with reduced stiffness, flank wear was minimized for the second case (about one-half the amount of flank wear when compared to the first and third cases). Interestingly, the static component of cutting force (for the same cutting regimes) was 870 N (195 lbs.) for the first case; 370 N (83 lbs.) for the second case; and only 270 N (60.5 lbs.) for the third case. In the second setup, (milling with PCD) cutter stiffness was $k_c = 9$–24 N/μ (50,000–135,000 lb./in.) for the first case, $k_c =$

7–10 N/μ (40,000–56,000 lb./in.) for the second case, and k_c = 3.5–4 N/μ (20,000–22,500 lb./in.) for the third case. The least amount of flank wear was observed for the first case (about one-half of the flank wear when compared to the second and third cases). The static cutting force component was lowest in the second case.

The influence of cutting insert clamping stiffness on wear in carbide inserts was studied by Fadeev, et al. [6] Milling with a two-tooth milling cutter was performed with different overhangs of the inserts, changing their stiffness within the range k_t = 18–54 N/μ (100,000–300,000 lb./in.). Figure 1 illustrates the correlation between k_t and flank wear h_f; number of loading cycles N before a microcrack occurs, and length l_c of the microcrack. It can be seen that there is an optimal stiffness [in this case about 36 N/μ (200,000 lb./in.)] associated with the minimum amount of flank wear. This effect is not the same for all grades of carbide inserts: for some, an increase in stiffness always correlates with a reduction in flank wear. It was observed that the fracture mechanism is different at the maximum stiffness (ductile intragrain fracture) and at the minimum stiffness (brittle intragrain fracture).

Golovin, et al. established that high frequency vibrations cause the development of microcracks in cutting inserts because they increase the concentrations of vacancies in the crystallic structure, increase energy dissipation on its defects, and impede heat transfer from the cutting edge [7]. From this theory it can be assumed that these effects depend selectively on vibration frequency, thus explaining the non-monotonous dependence of tool wear on stiffness.

While similar effects observed by Chryssolouris can be explained to a certain extent by damping variation for the studied cases, setups in seem to be characterized by the same damping [5, 6]. Damping positively influences tool life, however details of its effects on high frequency microvibrations of cutting inserts are not clear. For lower frequency ranges typical of chatter vibrations it is known that the effect of damping is similar to the effect of stiffness, thus chatter resistance can be characterized by a criterion $k\delta$, where k is the effective stiffness of the system and δ is a damping parameter (in this case, log decrement), for example [8].

Somewhat related to this issue is a study by Shustikov, et al., where the dynamics of the cutting process were studied for different frequencies of cutting insert vibrations in the 1–3 kHz range, substantially above chatter frequencies [9]. It was found that with increasing frequency, the chip shrinkage coefficient (i.e. the degree of metal deformation in the cutting zone) decreased and the depth of the deformed layer on the machined surface also decreased. This may explain a known positive influence for the application of ultrasonic vibration to the cutting zone.

Many publications describe the effects of the stiffness of different toolholder materials on tool life. In a study of turning tools with mechanically attached carbide inserts, Solnzev, et al. compared the tool life of inserts as a function of toolholder material [10]. Figure 2 shows that the longest

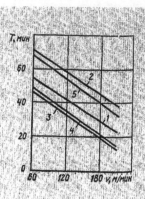

FIG. 2. Tool life with different holder materials 1) St. 104 annealed H_g = 195; 2) St. 1045 quenched in water from 830° H_g = 495; 3) gray cast iron with laminar graphite, annealed H_g = 166; 4) high strength cast iron with spheroidal graphite annealed H_g = 207; 5) high strength cast iron with spheroid graphite, as cast, H_g = 456)

tool life was observed for the toolholder made of hardened steel, 1045, H_B = 495 (line 2); a similar life was observed when high-strength hardened cast iron with spheroidal graphite was used, H_B = 456 (line 5). Line 1 represents annealed steel 1045 (H_B = 196); line 3, annealed cast iron (H_B = 166); and line 4, annealed cast iron with spheroidal graphite (H_B = 207). This near perfect correlation of tool life with the hardness of the holder allows us to assume th the local stiffness in the area of the insert loca-tion determines the lifespan of the insert. The higher (60–70 percent Young's modulus of steel may explain the reversal of this correlation between lines 1 and 4.

On the other hand, tests for milling cutters by Kasiyan et al. and by Simonian, et al. demonstrated that the use of pearlite cast iron for milling cutter housings results in a 50 percent increase in cutting insert lifespan when compared steel housings [11, 12]. This is explained by the reduction of the dynamic effects of the impact of cutting forces from the much higher damping of cast iron as compared to stee

Even better results are reported for toolholders made from special alloys with high damping and/or high Young modulus. The use of toolholders made of a special high damping alloy, Gentalloy, for turning workpieces of 1045 steel with discontinuous surfaces at 140 m/min (465 sfm) resulted in superior performance when compared with both steel and sintered carbide toolholders [13].

Comparative studies of milling cutter tool life as influenced by the design of the clamping system were performed by Shustikov, et al. and by Sakr, et al [14, 15].

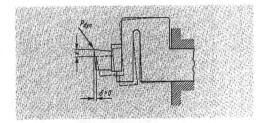

FIG. 3 Parting toolholder deflected x by dynamic force P_{dyn} so that the cutting edge backs away from the workpiece by amount δ

The study was performed on face milling cutters of 125–265.2 mm in diameter from various manufacturers (*T-MAX* from Sandvik, *PENTAX* from USAP, and various Soviet designs) equipped with a single insert of the same tungsten carbide-cobalt material to eliminate the influence of runout, etc. The machined material was cast iron. The results demonstrated a significant influence of cutter design on tool life (up to 2.6 times variation), in addition to a close correlation between tool life and the dynamic characteristics of the cutters.

A study by Minato addressed the influence of end mill stiffness on tool lifespan [16]. In the study, a mill was permanently clamped and the cutting of a narrow titanium workpiece was performed by various segments of the mill along its length. Tool life reduction was correlated with vibratory velocity amplitude, which decreased with increasing stiffness.

Factors Influencing Stiffness and Damping

Realization of the importance of the stiffness and damping characteristics of the attachment system led to studies of factors influencing these parameters. Novoselov, et al. showed that using a steel shim between the insert and the holder reduced the effective stiffness at the edge by half, while a carbide shim produced a much smaller reduction in stiffness [17]. A significant effect on insert life of shims placed under the carbide inserts was also demonstrated by Sergeev, et al. [18]. Studies by Marui, et al. demonstrated that the bending stiffness of a tool is greatly influenced by contact deformations in the joint between the tool and the toolholder, which cause the deviation of clamping conditions from the ideal built-in condition [19, 20]. It was also shown that the damping characteristics of a tool at large vibratory amplitudes occur because of normal contact deformations, and at small ampli-tudes because of tangential contact deformations.

The importance of clamping conditions on cutting insert stiffness and resulting tool life led to several studies using both mechanical means, as in Marui, et al. and state-of-the-

art holographic interferometric methods [19]. Deformations and displacements in turning tools with mechanically clamped inserts as functions of cutting forces were studied by Frankowski, et al. using a *He-Ne* laser [21]. It was shown that the carbide cutting insert was deformed to a lesser extent than the supporting surface and the clamp. As a result, stress concentrations developed near the front edge of the supporting surface of the insert. In addition, the influence of the elasticity moduli of both the insert and the toolholder and of the cutter hardness were found to be quite significant. Similar studies on polymethilacrylate models of tools were described by Isogimi, et al. [22]. A paper by Geniatulin described in detail the methodology of using holographic interferometry for the analysis of cutting tools [23]. In addition, close form and finite element analysis techniques for the study of stress-strain conditions in various cutting tools (e.g., solid, brazed, and with mechanical clamping) were described by Novoselov, et al. [24].

An understanding of the development mechanisms for stress concentrations between cutting inserts and toolholders, together with a perceived need for cutting insert damping enhancement led to the development and proliferation of adhesive attachments using heat-resistant adhesives for cutting inserts on holders, drills, reamers, and milling cutters. Burmistrov, et al. showed after comparative tests between brazed and adhesively assembled end mills, that tool life, surface finish and machining regimes are greatly improved with the latter joining technique [25]. The tests were performed using a hard-to-machine alloyed steel.

Viryuashkin, et al. studied the influence of adhesive line thickness on dynamic characteristics and tool life [26]. Darvish, et al. showed that adhesive bonding also resulted in reduced cracking of carbide parts because the heating procedure (heating to high brazing temperatures) was eliminated and the inevitable impacts were cushioned [27].

In addition, the face mill design described in [28] consists of two ring-shaped parts bonded together to generate a mill housing with a 1 mm adhesive line thickness. Cutting cartridges are mechanically attached to the housing. Tests performed with finishing regimes have demonstrated a significant increase in tool life for this milling cutter as compared with conventional cutters having solid metal housing.

One disadvantage to using adhesive attachments is lower thermoconductivity when compared to a brazed connection. Darvish, et al. demonstrated that the temperature of the cutting insert is higher in the case of adhesive bonding [29]. They proposed reducing the thickness of the adhesive line (which also leads to an enhancement of strength) and using heat-conductive additives in the adhesive.

Machining Systems with Intentionally Reduced Stiffness

Although high stiffness in cutting tools is generally desirable, there are some cases in which significant reductions in tool stiffness were shown to be beneficial. The most common

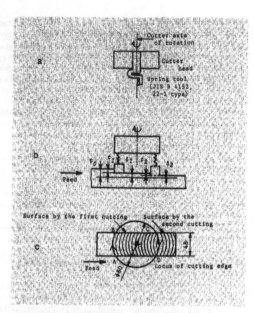

FIG. 4. Face milling with spring-shaped cutters a) model of elastic face milling cutter b) schematics of machining process c) tooth marks and cutting directions in elastic face milling

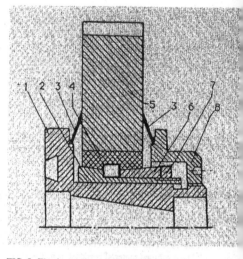

FIG. 5. Elastic attachment of grinding wheel

example is the use of a *goose neck* or *swan neck* tool (see Fig. 3) for turning, planing, and shaping (e.g. [30–32]).

Tobias suggested that a possible explanation for the chatter abatement effect of this type of tool is in the retraction effect of the cutting edge when the dynamic cutting force increases as in Fig. 3 [31]. A successful application of this concept to milling is described by Minato et al. [33]. The face milling cutter in Fig. 4a has spring-shaped cutters. The kinematics of this cutter are presented in Fig. 4b. When the cutter is in front (segment A–B), the nominal depth of cut is large and the spring is deformed by the cutting forces. The surface initially machined is traveled by the cutter again when it is at the back (segment C–D), with a much smaller nominal depth of cut (finishing path). Two passes, with the second in the opposite direction to the first, as shown in Fig. 4c, resulted in a much better surface finish as compared with conventional milling (R_a is reduced to about one third) and in greatly reduced residual stresses.

Taking a different tactic, [34] suggested using the elasticity of the cutting tool to restore the original geometry of the worn insert. Creation of an additional tool microfeed motion and modification of the cutting geometry between the roughing and the finishing cuts were also options examined.

The desirability of reducing radial stiffness in grinding with grinding wheels made of ultrahard (diamond and CBN) abrasive materials was demonstrated by several researchers. Implementation of this concept by Sexton, et al. has shown that a ten-fold reduction in the radial stiffness by using a wheel hub made of a specially designed composite material resulted in a complete elimination of chatter vibrations as well as wheel waviness which can develop as a result of chatter [35]. Because the composite material also exhibits increased damping, it allows us to assume that the enhanced compliance is accompanied by an increasing magnitude of the criterion $k\delta$.

Another approach to this problem is suggested by Burochkin, et al. [36]. As illustrated in Fig. 5, wheel (5) is fit on damping ring (4) sitting on tapered bushings (2) and (6). When tapered bushing (6) is shifted in axial direction by actuating nut (7), damping ring (4) is deformed, thus changing the radial stiffness. Stiffness in axial direction is provided by Belleville springs (3), and can be adjusted by nut (8). Damping in axial direction develops during the slipping of springs (3) against the faces of wheel (5). The design was successfully used for the optimization of electrochemical grinding of hard-to-machine materials.

Application of the concept of intentional reduction of tool stiffness to machining of low stiffness workpieces was proposed and studied by Rivin, et al. [37]. A turning tool was clamped into a fixture assuring a high degree of stiffness in all directions except the radial direction. In this direction a computed degree of compliance was introduced using a high damping elastomeric material. This arrangement resulted in reduced variation of effective stiffness along a slender workpiece and, respectively, a greatly improved cylindricity for a 430 mm long, 12 mm diameter workpiece. Machining without chatter was performed with reasonable regimes without using steady rests, due to damping exchange between the high damping toolholder and the workpiece, facilitated by the appropriate dynamic tuning.

An interesting concept for enhancement of chatter resistance through a reduction in stiffness of the machining system in a direction tangential to the machined surface was proposed by Elyasberg, et al. [38]. This concept was successfully tested for turning and boring applications and demonstrated a significant increase in the chatter-free depth of cut. However, complicated toolholders with precision springs and hydrostatic supports were used. The concept was thoroughly analyzed by Rivin, et al. and a much more compact and versatile design of the so-called torsional compliant head for boring bars was developed and successfully tested [39].

A similar concept for improving chatter resistance of a grinding wheel attached to a robot arm was suggested by Asada, et al. [40]. Here, compliance of the arm in the direction tangential to the grinding surface was artificially increased, while compliance in the normal direction was reduced, for example, by braces.

In other research, Lutsiv, et al. showed that in turning operations, reduction in axial stiffness of the tool-holder might enhance the stability of the cutting process for turning operations by as much as 1.8–2.9 times when compared to a rigid holder [41]. Ryzhov, et al. described a variable stiffness tool clamping system using hydrostatic supports [42]. By changing design parameters and oil pressure, both the stiffness and the damping of the clamping device were affected. This resulted in an altered surface finish and microhardness of the machined surface.

CONCLUSION

The studies surveyed reveal a substantial influence of machining system stiffness on process stability and tool life. How-ever, stiffness is not the only parameter that should be con-sidered and the highest degree of stiffness does not always correlate with the best system performance. Another system parameter that should always be considered is damping. In some cases, the natural frequency of the toolholder is also of importance. In addition, numerous studies have demonstrated that both the degree of stiffness and the damping of toolholders can be varied through a range of values by careful selection of the materials used for toolholders, shims, and joints.

At the same time, several studies have demonstrated that substantial improvements in system performance can be achieved in some cases through an intentional reduction of stiffness in the machining system. In such cases, the directions in which the stiffness is reduced, the degree of damping, etc., are critical for system performance.

ACKNOWLEDGEMENT

Support from the National Science Foundation Grant DDM-9005654 is gratefully acknowledged.

REFERENCES

1. Nilsson K., Adaptive Tool Systems, *Proceedings of the Sixth International Conference on Flexible Manufacturing Systems*, IFS Conferences Ltd, pp. 235–248, 1987.
2. Murata R., Tools and Devices for High Productivity Machining in Unmanned Regimes, *Kika Gidzuzu* (Mechanical Engineering), 32(7): 2–8 (in Japanese), 1989.
3. Anonymous, *Research Priorities for Proposed NSF Strategic Manufacturing Research Initiative*, Report on NSF Workshop conducted by Metcut Research Associates, Inc., pp. 20, 43, 1987.
4. Kocherovskii E. B. and Likhzier G. M., Radial Stiffness of Tool and Machining Accuracy in Feed–Splitting Boring, *Stanki i instrument*, 6: 17–18 (in Russian), 1985.
5. Chryssolouris G., Effects of Machine–Tool–Workpiece Stiffness on the Wear Behaviour of Superhard Cutting Materials, *Annals of the CIRP*, 31(1): 65–69, 1982.
6. Fadeev V. S. and Petridis A. V., Influence of Stiffness of the System Machine–Fixture–Tool–Workpiece on Strength of Carbide Tools, *Stanki i instrument*, 5: 30-31 (in Russian), 1985.
7. Golovin S. A. and Pushkar A., Mikroplastichnost i utalost metallov, (Microplasticity and Fatigue of Metals), *Metallurgia*, Moscow, (in Russian), 1980.
8. Tlusty J., Report of the Machine Tool Task Force 3, *Machine Tool Mechanics*, pp. 8.5–8.32, Lawrence Livermore Laboratory, 1981.
9. Shustikov A. D., and Rastorguev V. V., Parameters of Dynamic Characteristic of Cutting in 1–3kHz Frequency Range, in *Issledovaniya protsessov obrabotiki materialov i dinamiki teknologicheskogo oborudovaniya*, (Studies of Materials Processing and Dynamics of Production Equipment), Peoples Friendship University, Moscow, pp. 9–13 (in Russian), 1982.
10. Solnzev L. A. and Aksenko A. A., Influence of Toolholder Material on Life of Turning Tools, *Stanki i instrument*, 4: 18–19 (in Russian), 1986.
11. Kasiyan M. V., Simonian M. M., and Nadjarian M. T., Fatigue and Damping Capacity of Cast Iron Toolholders, Izvestiya Adademii Nauk Armianskoi SSR, Technicheskie nauki, 38(3): 3–6 (in Russian), 1985.
12. Simonian M. M., Nadjarian M. T., and Posviatenko, E. K., Effectiveness of Using Cutting Tools with Cast Iron Holders, *Sverkhtverdie materiali*, 1: 41–44 (in Russian), 1987.
13. Kitajima K. and Tanaka Y., Damping of Toolholder Vibrations for Interrupted Cutting, Seimitsukikai (Journal of the Japan Society of Precision Engineering, 50(7): 860–865 (in Japanese), 1984.
14. Shustikov A. D., Matveikin V. V., and Khamuda, S. N., Study of Vibration Spectra and Tool Life of Face Milling Cutters with Mechanically Clamped Inserts, in *Issledovaniya dinamiki tekhnologicheskogo oborudovaniya i instrumenta*, (Studies of Dynamics of Production Equipment and Tooling), Moscow, pp. 34–37, 1982.
15. Sakr H., Gromakov K. G., and Shustikov A. D., Experimental Study of Stiffness of Assembled Milling Cutters Using Their Static Characteristics, in *Issledovaniya protsessov obrabotki materialov i dinamiki tekhnologicheskogo oborudovaniya*, (Studies of Materials Processing and Dynamics of Production Equipment), Moscow, pp. 44–49 (in Russian), 1982.
16. Minato J., Ammi S., and Okamoto S., *Bulletin of the Japan Society of Precision Engineering*, 8(1): 7, 1974.
17. Novoselov Y. A. and Mikhailov M. I., Device for Measuring Stiffness of a Tool with Mechanically Clamped Insert, *Mashinostroitel*, 3: 42 (in Russian), 1989.
18. Sergeev L. V., Shishkin P. P., and Potockii G. A., Influence of Interface Stiffness between Carbide Insert and Holder on Tool Performance, *Stanki i Instrument*, 5: 37–38 (in Russian), 1970.
19. Marui E., Ema S., and Kato S., Relative Slip and Damping Characteristic of Turning Tools, *Bulletin of the Japan Society of Precision Engineering*, 18(4): 323–328, 1984.
20. Marui E., Ema S., and Kato S., Contact Rigidity at Tool Shank of Turning Tools, *ASME Journal of Engineering for Industry*, 109(2): 169–172, 1987.

21. Frankowski G., Leopold J., and Zeidler H., Study of Stiffness of Assembled Tools Using Holographic Interferometry, *Wissenschaftliche Zeitschriften Technische Hochschule Karl-Marx-Stadt,* **27**(5): 762–767 (in German), 1985.
22. Isogimi K., Kurita K., and Kondo S., Study of Stresses in Cutting Tools by Method of Optical Interferometry, *Semitsu Kogaku Kaisi* (Journal of the Japan Society of Precision Engineering), **52**(9): 1592–1597 (in Japanese), 1986.
23. Geniatulin A. M., Study of Assembled Cutting Tools by Method of Holographic Interferometry, *Stanki i instrument,* **4**: 24–26 (in Russian), 1987.
24. Novoselov Y. A. and Mikhailov M. I., Analysis of Stress-Strain Conditions of Cutting Tools Considering Their Design Specifics, *Isvestiya VUZov-Mashinostroenie,* **5**: 126–130 (in Russian), 1984.
25. Burmistrov E. V. and Voronov E. N., *Enhancement of Chatter Resistance and Life of End Mills for CNC Machine Tools,* in *Instrumenti dlia Stankov s tchislovim programmim upgravleniem i gibkikh proizvodstvennikh sistem,* (Tooling for CNC Machine Tools and FMS), Leningrad, pp. 51–58 (in Russian), 1985.
26. Viryuaskin A. I., et al., Study of Dynamic Characteristics of Cutting Tools Assembled with Adhesives During Machining of High Strength Steels and Alloys, in *Vysokoeffektivnie metody i instrumenti dlya mekhanicheskoi obrabotki aviatsionnikh materialov,* (High Efficiency Methods and Tools for Machining Aircraft Materials) , Kuybishev, pp. 139–143 (in Russian), 1984.
27. Darvish S. and Davies R., Adhesive Bonding of Metal Cutting Tools, *International Journal of Machine Tools and Manufacture,* **29**(1): 141–152, 1989.
28. Anonymous, Performance of Face Milling Cutter Assembled with Glue, *Seimitsu Kogaku Kaisi,* (Journal of the Japan Society of Precision Engineering), **55**(3): 526–531 (in Japanese), 1989.
29. Darvish S. and Davies R., Investigation of the Heat Flow through Bounded and Brazed Metal Cutting Tools, *International Journal of Machine Tools and Manufacture,* **29**(2): 229–237, 1989.
30. Hölken W., *Untersuchungen von Ratterschwingungen an Drehbänken,* (Study of Chatter During Turning), Diss. TH Aachen and Forschungsbericht 7, Verlag W. Girardet, Essen, Germany 1957.
31. Tobias S. A., *Machine Tool Vibration,* Blackie, London, 1965.
32. Minato J., Ammi S., and Okamoto S., *Bulletin of the Japan Society of Precision Engineering,* **10**(3): 113, 1976.
33. Minato J., Ammi S., and Okamoto S., Application of Spring Tool to Face Milling, *Bulletin of the Japan Society of Precision Engineering,* **18**(1): 35–36, 1984.
34. Anonymous, *Ausgewahlte Fragen der Rauheit von mit flexiblen Werkzeugen bearbeiten Formlachen,* (Shaping of Machined Surface for Machining Using Elastic Tool), Intern. wissenschaftl. Konfer. AUPRO 88, Karl-Marx-Stadt, Bd.2, pp. 458–465 (in German), 1988.
35. Sexton J. S. and Stone B. J., The Development of an Ultrahard Abrasive Grinding Wheel Which Suppresses Chatter, *Annals of the CIRP,* **30**(1): 215–218, 1981.
36. Burochkin Y. P.and Mukhina M. Y., Abrasive Tool, *Mashinostroitel,* **11**: 34 (in Russian), 1989.
37. Rivin E. I. and Kang H.-L., Improvement of Machining Conditions for Slender Parts by Tuned Dynamic Stiffness of Tool, *International Journal of Machine Tools and Manufacture,* **29**(3): 361–376, 1989.
38. Elyasberg M. E., et al., A Method for the Stuctural Improvement of Machine Tool Vibration Stability During Cutting, *Soviet Engineering Research,* **3**(4): 59–63, 1983.
39. Rivin E. I. and Kang H. L., Improving Cutting Performance by Using Boring Bar with Torsionally Compliant Head, *Transactions of NAMRAI-SME,* pp. 230–236, 1990.
40. Asada H. and Goldfine N., *Method of Grinding,* U.S. Patent 4,753,048 1988.
41. Lutsiv I. V. and Nagorniak S. G., Influence of Axial Stiffness of Tool Fixtures on Chatter Resistance at Turning, *Mashinostroenie,* Izvestiya VUSov, No. 3, pp. 146–148 (in Russian), 1990.
42. Ryzhov E. V. and Petrovskii E. A., Tooling Units with Hydrostatic Supports in Holders, *Stanki i instrument,* **1**: 33–34 (in Russian), 1985.

manuscript received 2 April 1991

Eugene Rivin is a professor of Mechanical Engineering and Director of the Machine Tool Research Laboratory at Wayne State University. He performs research and teaches in the areas of machine tools and toolings, robotics, vibration control, and design/machine elements. He is the author of numerous articles, patents, and books, including *Mechanical Design of Robots* (McGraw-Hill) published in 1988. He is an Active Member of the CIRP and a recipient of the 1991 Shingo Prize for Excellence in Manufacturing.

Index